T0329725

Multivalency

Multivalency

Concepts, Research & Applications

Edited by

Jurriaan Huskens
University of Twente
Enschede, the Netherlands

Leonard J. Prins
University of Padova
Italy

Rainer Haag
Freie Universität Berlin
Germany

Bart Jan Ravoo
Westfälische Wilhelms-Universität Münster
Germany

This edition first published 2018
© 2018 John Wiley & Sons Ltd

All rights reserved. No part of this publication may be reproduced, stored in a retrieval system, or transmitted, in any form or by any means, electronic, mechanical, photocopying, recording or otherwise, except as permitted by law. Advice on how to obtain permission to reuse material from this title is available at http://www.wiley.com/go/permissions.

The right of Jurriaan Huskens, Leonard J. Prins, Rainer Haag, and Bart Jan Ravoo to be identified as the author(s) of the editorial material in this work has been asserted in accordance with law.

Registered Office(s)
John Wiley & Sons, Inc., 111 River Street, Hoboken, NJ 07030, USA
John Wiley & Sons Ltd, The Atrium, Southern Gate, Chichester, West Sussex, PO19 8SQ, UK

Editorial Office
9600 Garsington Road, Oxford, OX4 2DQ, UK

For details of our global editorial offices, customer services, and more information about Wiley products visit us at www.wiley.com.

Wiley also publishes its books in a variety of electronic formats and by print-on-demand. Some content that appears in standard print versions of this book may not be available in other formats.

Limit of Liability/Disclaimer of Warranty
In view of ongoing research, equipment modifications, changes in governmental regulations, and the constant flow of information relating to the use of experimental reagents, equipment, and devices, the reader is urged to review and evaluate the information provided in the package insert or instructions for each chemical, piece of equipment, reagent, or device for, among other things, any changes in the instructions or indication of usage and for added warnings and precautions. While the publisher and authors have used their best efforts in preparing this work, they make no representations or warranties with respect to the accuracy or completeness of the contents of this work and specifically disclaim all warranties, including without limitation any implied warranties of merchantability or fitness for a particular purpose. No warranty may be created or extended by sales representatives, written sales materials or promotional statements for this work. The fact that an organization, website, or product is referred to in this work as a citation and/or potential source of further information does not mean that the publisher and authors endorse the information or services the organization, website, or product may provide or recommendations it may make. This work is sold with the understanding that the publisher is not engaged in rendering professional services. The advice and strategies contained herein may not be suitable for your situation. You should consult with a specialist where appropriate. Further, readers should be aware that websites listed in this work may have changed or disappeared between when this work was written and when it is read. Neither the publisher nor authors shall be liable for any loss of profit or any other commercial damages, including but not limited to special, incidental, consequential, or other damages.

Library of Congress Cataloging-in-Publication Data

Names: Huskens, Jurriaan, 1968– editor.
Title: Multivalency : concepts, research & applications / edited by Professor Jurriaan Huskens, University of Twente, Enschede, NL [and three others].
Description: First edition. | Hoboken, NJ : Wiley, 2018. | Includes bibliographical references and index. |
Identifiers: LCCN 2017029790 (print) | LCCN 2017039365 (ebook) | ISBN 9781119143475 (pdf) | ISBN 9781119143499 (epub) | ISBN 9781119143468 (cloth)
Subjects: LCSH: Valence (Theoretical chemistry) | Multivalent molecules.
Classification: LCC QD469 (ebook) | LCC QD469 .M75 2018 (print) | DDC 541/.224–dc23
LC record available at https://lccn.loc.gov/2017029790

Cover design by Wiley
Cover image: Image provided by Rainer Haag

Set in 10/12pt Warnock by SPi Global, Pondicherry, India
Printed and bound in Malaysia by Vivar Printing Sdn Bhd

10 9 8 7 6 5 4 3 2 1

Contents

List of Contributors

Sumati Bhatia
Institute of Chemistry and Biochemistry
Freie Universität Berlin
Germany

Luc Brunsveld
Department of Biomedical
Engineering
Laboratory of Chemical Biology and
Institute of Complex Molecular
Systems
Eindhoven University of Technology
The Netherlands

Maria A. Cardona
Department of Chemical Sciences
University of Padova
Italy

Alessandro Casnati
Department of Chemistry
Life Sciences and
Environmental Sustainability
Università di Parma
Italy

Emanuela Cavatorta
MESA+ Institute for Nanotechnology
& MIRA Institute for Biomedical
Technology and Technical Medicine
University of Twente
Enschede
The Netherlands

Tine Curk
Department of Chemistry
University of Cambridge
UK

Jens Dernedde
Institute of Laboratory Medicine, Clinical
Chemistry, and Pathobiochemistry
Charité–Universitätsmedizin Berlin
Germany

Jure Dobnikar
Institute of Physics & School of
Physical Sciences
Chinese Academy of Sciences
Beijing
China

Daan Frenkel
Department of Chemistry
University of Cambridge
UK

Akash Gupta
Department of Chemistry
University of Massachusetts, Amherst
USA

Rainer Haag
Institute of Chemistry and Biochemistry
Freie Universität Berlin
Germany

Akira Harada
Graduate School of Science
Osaka University
Japan

Akihito Hashidzume
Graduate School of Science
Osaka University
Japan

Zehuan Huang
Department of Chemistry
Tsinghua University
Beijing
China

Jurriaan Huskens
MESA+ Institute for Nanotechnology
University of Twente
Enschede
The Netherlands

Pascal Jonkheijm
MESA+ Institute for Nanotechnology
& MIRA Institute for Biomedical
Technology and Technical Medicine
University of Twente
Enschede
The Netherlands

Ulrike Kauscher
Organic Chemistry Institute
Westfälische Wilhelms-Universität
Münster
Germany

Carlos M. León Prieto
Department of Chemical Sciences
University of Padova
Italy

Bernd Lepenies
Immunology Unit & Research Center for
Emerging Infections and Zoonoses (RIZ)
University of Veterinary
Medicine Hannover
Germany

João T. Monteiro
Immunology Unit & Research Center for
Emerging Infections and Zoonoses (RIZ)
University of Veterinary
Medicine Hannover
Germany

Roland J. Pieters
Department of Chemical Biology & Drug
Discovery
Utrecht Institute for Pharmaceutical
Sciences
Utrecht University
The Netherlands

Leonard J. Prins
Department of Chemical Sciences
University of Padova
Italy

Bart Jan Ravoo
Organic Chemistry Institute
Westfälische Wilhelms-Universität
Münster
Germany

Moumita Ray
Department of Chemistry
University of Massachusetts, Amherst
USA

Vincent M. Rotello
Department of Chemistry
University of Massachusetts, Amherst
USA

Francesco Sansone
Department of Chemistry
Life Sciences and Environmental
Sustainability
Università di Parma
Italy

Hans-Jörg Schneider
FR Organische Chemie
Universität des Saarlandes
Saarbrücken
Germany

Paolo Scrimin
Department of Chemical Sciences
University of Padova
Italy

Marjon Stel
Department of Chemical Biology & Drug
Discovery
Utrecht Institute for Pharmaceutical
Sciences
Utrecht University
The Netherlands

Jens Voskuhl
Institute of Organic Chemistry
University of Duisburg-Essen
Germany

Xi Zhang
Department of Chemistry
Tsinghua University
Beijing
China

Benjamin Ziem
Institute of Chemistry and Biochemistry
Freie Universität Berlin
Germany

Foreword

Scientific challenges come and go; only a few of them remain for a long time. Multivalency is one of those research topics that has been prominent for many years, as this intriguing phenomenon is of profound importance in many biological processes as well as very difficult to understand and mimic. Personally, I became intrigued by the challenge of multivalency when our group entered the field of dendrimers in 1990. The controlled number of end groups – 4, 8, 16, 32, and 64 amines of the polypropylene imines – opened many opportunities for us to explore the controlled use of multiple interactions. However, our ideas were more simple than our experiments in making full use of the potential of multivalency; many of them remained in the realm of dreaming. The broad potential of multivalency as well as its complex mode of action was beautifully illustrated by George Whitesides and coworkers [1] in the seminal *Angewandte Chemie* review paper in 1998. Their review initiated a world-wide search for synthetic mimics of these highly effective natural systems, a search that turned out to be long lasting.

Nature uses both similar interactions (homovalency) and different interactions (heterovalency) to control selectivity and specificity, even leading to ultra-sensitivity. Beautiful examples are found in substrate–cell interactions and immunology. Ever since this elegant mechanism and its importance in biological systems has been recognized, chemists have been intrigued to fully understand the enhancement factors obtained in binding multiple weak interactions through multivalency. Artificial systems are designed, synthesized, and studied, while a number of applications are proposed. Multivalent medication can have lower toxicity while simultaneously having higher medical efficacy.

Although the knowledge on the modus operandi of these systems has increased significantly in time and the systems synthesized have become more active, the full potential of the proposed applications remains. Hence, a number of challenging questions need to be answered before the potential of this intriguing concept can be explored. How to design the ideal structure to arrive at the theoretical maximum avidity and how to obtain scaling with valency are just a few of these intriguing questions. Theoretical and experimental studies of multivalent systems have revealed several design parameters that are critical in obtaining effective multivalent constructs. Next to the binding affinity, linker flexibility plays an important role, as rigid linkers require extremely precise ligand positioning to obtain high binding affinities and selectivity, while flexible linkers offer more freedom in molecular design at the cost of lower affinity and selectivity. Furthermore, additional competing equilibria can be used to enhance binding selectivity or to steer an assembly towards a preferred state. However, the complexity of all these effects and their interference makes the field one of the most challenging areas in the molecular sciences.

Therefore, it is great to see that four outstanding scientists have edited a book on the intriguing topic of multivalent interactions. It is a book full of excellent chapters written by the most active experts in the field, covering all aspects of multivalent interactions with special emphasis on theory, synthesis, surfaces, chemical biology, and supramolecular chemistry. I am convinced that this book will be a great asset for all active in this intriguing field of science.

Eindhoven, May 2017 *E.W. Meijer*

Reference

1 Mammen, M., Choi, S.-K., Whitesides, G. M. Polyvalent interactions in biological systems: Implications for design and use of multivalent ligands and inhibitors. *Angew. Chem. Int. Ed.* **1998**, *37*, 2754–2794.

Preface

Multivalent interactions play a role in molecular and biomolecular systems in which molecules interact by multiple noncovalent bonds. Studying and describing these interactions in a quantitative manner constitute therefore an important way to obtain insight into the functional behavior of the biological and chemical systems in which they are involved. Over the past decades, the research of multivalent interactions has greatly expanded. This growth fits in the overall trends observed in the natural sciences which encompass the merging and overlapping of disciplines, like the biology and chemistry involved here. It also aligns with the emphasis on the study of complex systems, and the development of systems biology and systems chemistry, for example. Therefore, we have observed the need for a book that brings together fundamental aspects of multivalent interactions and relevant current examples of biological as well as chemical multivalent systems.

The disciplines of chemistry and biology are strongly represented in this area of science because they exert a mutual influence on both the understanding of fundamental aspects of multivalency as well as the development of practical research tools and applications. In biology, multivalent interactions play an eminent role in the immune system, but at the same time also describe the interactions between a virus and the host cell which the virus tries to infect. Tools from chemistry and nanotechnology are being developed that assist in studying such complex biological systems, for example, by synthesizing model cell membranes in which the interactions can be studied in a more controllable fashion. Likewise, probe techniques allow quantification of interactions at the single molecule level in individual cells. Conversely, the increase in understanding of the biomolecular interactions in living systems sparks the generation of new types of drugs and inhibitors that can make smart use of the multivalent character to improve both selectivity and activity.

A quantitative understanding of multivalent interactions is essential to promote progress in the field that deals with multivalent systems. Both experimental techniques as well as modeling can be used to stimulate this depth of understanding. Therefore, we decided that chapters with a strong educational character should be an essential part of this book. We present a section (Part I) of four chapters that serve to guide new researchers as well as more experienced researchers in their efforts to contribute to this lively area. These chapters provide a background in thermodynamics, data modeling and the description of multivalent equilibrium systems, numerical modeling of multivalent systems and superselectivity, and an introduction to multivalent biological systems. These chapters build on, and for some aspects briefly review, knowledge that most readers with a background in chemistry or biology will have encountered in their regular academic education, but from there quickly integrate this knowledge into the description of multivalent systems.

Another explicit aim of the book is to expose the active nature of the research on multivalent systems. This is achieved in the two other sections of the book (Parts II and III), dealing with chemical and biological examples of multivalency, respectively. In the chemistry oriented chapters, timely topics such as the host–guest interactions of cyclodextrins and cucurbiturils are covered, as well as soft matter systems, such as vesicles, polymers, and nanoparticles. Not only equilibrium thermodynamics is shown, but also systems in which multivalent interactions control catalysis. In the more biological section, several biological interactions are put forward, such as protein–protein and lectin–glycan interactions. The strong connection between chemistry and biology in this area is emphasized by the examples that describe cell targeting by molecules and nanoparticles, as well as receptor inhibition by multivalent inhibitors.

We hope that this book will serve a need, for new and experienced researchers alike, both for those requiring a deeper understanding as well as those that try to get an overview of existing activities in the field. We thank all contributing authors for their efforts in summarizing and describing their research and that of others, as their joint work makes this book so much more than the individual chapters alone. We also express our gratitude to the Wiley staff for smoothing the pathway for the book that lies before you.

September 2017

Jurriaan Huskens, Leonard J. Prins, Rainer Haag,
and Bart Jan Ravoo

Part I

General Introduction to Multivalent Interactions

1

Additivity of Energy Contributions in Multivalent Complexes

Hans-Jörg Schneider

FR Organische Chemie, Universität des Saarlandes, 66123 Saarbrücken, Germany

1.1 Introduction

Additivity of individual binding contributions is the very basis of multivalency. In classical coordination chemistry such simultaneous actions are described as the chelate effect. They offer almost unlimited ways to enhance the affinity [1,2,3,4,5,6], and therefore within certain limitations also the selectivity [7] of synthetic and natural complexes. Although additivity is often implied in experimental and theoretical approaches it is subject to many limitations which will be also discussed in the present chapter.

1.2 Additivity of Single Interactions – Examples

If only one kind of interaction is present in a complex one can expect a simple linear correlation between the number n of the individual interaction free energies $\Delta\Delta G_i$ and the total ΔG_t (Equation 1.1), as illustrated in Figure 1.1 for salt bridges [8]. Even though the organic ion pair complexes are based on cations and anions of very different size and polarizability one observes essentially additive salt bridges; the slope of the correlation indicates an average of $\Delta\Delta G = (5 \pm 1)$ kJ/mol per salt bridge. The value of (5 ± 1) kJ/mol is observed in usual buffer solution, but varies as expected from the Debye–Hückel equation with the ionic strength of the solution [9]. Scheme 1.1 shows a corresponding value of $K \approx 10\,M^{-1}$ per salt bridge for typical complexes where the affinity depends as expected on the degree of protonation [7].

$$\Delta G_t = n \cdot \Delta\Delta G_i \tag{1.1}$$

The additivity depicted in Figure 1.1 and Scheme 1.1 for salt bridges is in line with the Bjerrum equation, which describes ion pair association as a function of the ion charges z_A and z_B; Figure 1.2 shows for over 200 ion pairs a linear dependence of $\log K$ vs. $z_A z_B$ [3]. For inorganic salts one finds similar $\Delta\Delta G$ values of 5–6 kJ/mol per salt bridge and a similar dependence on charges [10]. At zero ionic strength the stability decreases in the

Multivalency: Concepts, Research & Applications, First Edition. Edited by Jurriaan Huskens,
Leonard J. Prins, Rainer Haag, and Bart Jan Ravoo.
© 2018 John Wiley & Sons Ltd. Published 2018 by John Wiley & Sons Ltd.

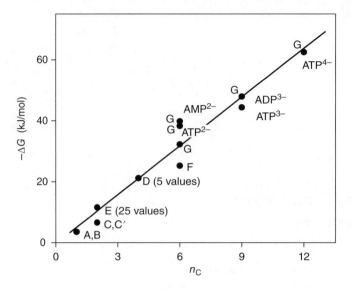

Figure 1.1 Additive ion pair contributions in a variety of complexes with a number n_C of salt bridges. From slope: average (5 ± 1) kJ/mol per salt bridge. A,B and C,C′ – complexes of a tetraphenolate cyclophane (4−) with Me_4N^+ and an azoniacyclophane (4+) with mono- and dianionic naphthalene derivatives; D – anionic (sulfonate or carboxylate) with cationic (ammonio) triphenylmethane derivatives; E – organic dianions with organic dications; F – cationic azamacrocycle (6+ charges) with aliphatic dicarboxylates; G – cationic azacrowns with adenosine mono-, di- and triphosphates. *Source*: Ref. [8]. Reproduced with permission of John Wiley and Sons.

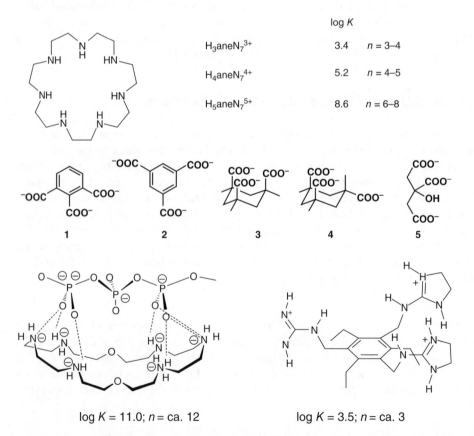

Scheme 1.1 Complexation log K values of anions **1–5** with a macrocyclic amine as function of the degree of protonation of the amine; and ion pairing with some representative complexes; log K values in water; n is the estimated number of salt bridges.

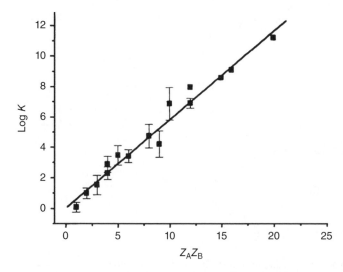

Figure 1.2 Ion pair association constants at zero ionic strength as a function of charge product, calculated for 203 ion pairs. *Source*: Ref. [8]. Reproduced with permission of John Wiley and Sons.

order $Ca^{2+} > Mg^{2+} >> Li^+ > Na^+ > K^+$ and can be described by Equation 1.2 [11]. Additivity is observed although ion pairing in water is determined entirely by entropic contributions[11], unless other contributions dominate [12].

$$\log K = 0.5z + A/z \text{ (where A} = -0.24 \text{ for Li}^+, -0.30 \text{ for Na}^+, -0.43 \text{ for K}^+) \quad (1.2)$$

If there is more than one kind of interaction, Equation 1.3 applies. Often however, only one of the contributions is the same, like salt bridges in complexes of nucleotides with a positively charged host (Scheme 1.2) [13]. Additivity is then observed by the constant stability difference of $2 \times \Delta\Delta G \approx 10\,kJ/mol$ between complexes with charged nucleotides and neutral nucleosides. The $10\,kJ/mol$ reflects the presence of two salt bridges between the phosphate dianion and the host ammonium center, which agrees with structural analyses by NMR spectroscopy.

$$\Delta G_t = n \cdot \Delta\Delta G_A + m \cdot \Delta\Delta G_B \quad (1.3)$$

The complexes shown in Scheme 1.2 exhibit constant single $\Delta\Delta G_A$ values only for the salt bridges, whereas the second contribution $\Delta\Delta G_B$ varies as a function of the different nucleobases. Figure 1.3 illustrates a case where both $\Delta\Delta G_A$ and $\Delta\Delta G_B$ remain constant, the latter reflecting cation–π interactions. In principle one could use Equation 1.3 to derive both $\Delta\Delta G_A$ and $\Delta\Delta G_B$, but more reliable values are obtained if for one interaction a $\Delta\Delta G$ value is used which is known from independent analyses, such as $\Delta\Delta G_A = 5\,kJ/mol$ for each salt bridge (see above). Then one observes a rather linear correlation with the number of phenyl units which shows a contribution of $\Delta\Delta G_B \approx 1.5\,kJ/mol$ for the single $^+N–π$ interaction [14].

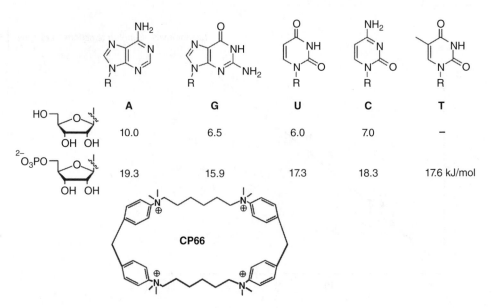

	A	G	U	C	T
HO (nucleoside)	10.0	6.5	6.0	7.0	–
$^{2-}O_3PO$ (nucleotide)	19.3	15.9	17.3	18.3	17.6 kJ/mol

Scheme 1.2 Complexation free energies ΔG of nucleotides and nucleosides with the cyclophane CP66.

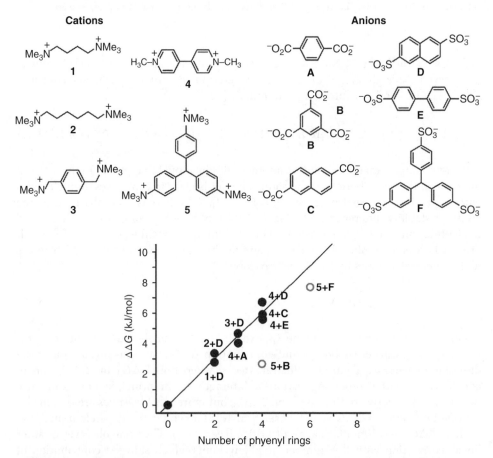

Figure 1.3 Ion pairs exhibiting both salt bridges and cation–π interactions; if $\Delta\Delta G_A = 5$ kJ/mol for each salt bridge are subtracted from ΔG_t of each complex. Outliers (open circles) are due to conformational mismatch. *Source*: Ref. [14]. Reproduced with permission of American Chemical Society.

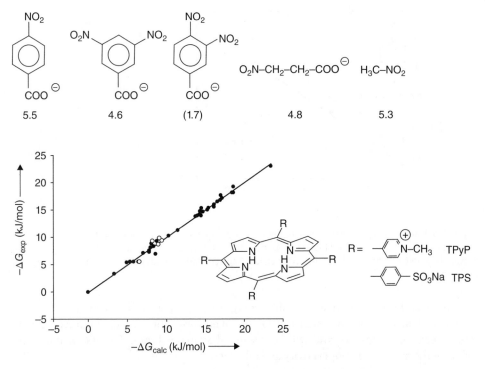

Figure 1.4 Additive $\Delta\Delta G_X$ increments in complexes of porphyrins bearing cationic or anionic substituents R in meso position (TPyP or TPS) in water, after deduction of 5 kJ/mol for ion pair contribution where applicable. $\Delta\Delta G_X$ increments in TPyP complexes for nitro substituents as an example (deviation for *ortho*-dinitro due to steric hindrance); correlation between measured complexation energies ΔG_{exp} and ΔG_{calc} calculated on the basis of experimentally determined averaged single contributions ΔG_S. Filled circles, complexes with TPyP; open circles, complexes with TPS. *Source*: Ref. [15]. Reproduced with permission of John Wiley and Sons.

The effect of nitro substituents on dispersive interactions is another example of additive energy contributions (Figure 1.4) [15,16]. Additivity with respect to substituent effects is observed in Hammett-type linear free energy relationship correlations; Figure 1.5 shows an example for hydrogen bonds with C—H bonds as donor and with hexamethylphosphoramide as acceptor [17].

1.3 Limitations of Additivity

1.3.1 Free Energy Values ΔG Instead of Enthalpic and Entropic Values ΔH, $T\Delta S$

The examples shown above as well as most others in the literature rely on free energy values ΔG, although consideration of the corresponding ΔH and $T\Delta S$ parameters could shed more light on the underlying binding mechanisms. As pointed out earlier by Jencks, the empirical use of ΔG "avoids the difficult or insoluble problem of interpreting observed ΔH and $T\Delta S$ values for aqueous solution" [18]. Furthermore, according to Jencks, there

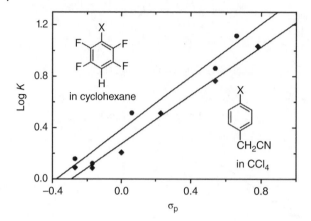

Figure 1.5 Hammett-type correlation of equilibria of hydrogen bonds with hexamethylphosphoramide as acceptor and para-substituted tetrafluorobenzenes or phenylacetonitriles as donor; log K versus Hammett substituent constants. *Source*: Ref. [17]. Reproduced with permission of John Wiley and Sons.

is often an additional "connection Gibbs energy, ΔG_S" (Equation 1.4) which he ascribed largely to changes in translational and rotational entropy. These connection ΔG_S can be either negative or positive and will be discussed as major liming factors for additivity below in the context of cooperativity and allostery.

$$\Delta G_t = \Delta G_A + \Delta G_B + \Delta G_S \tag{1.4}$$

The success of using free energy values instead of enthalpic and entropic values is in an essential part due to entropy–enthalpy compensation which has empirically been found to hold with many complexations, although it is theoretically not well-founded [19,20,21]. Another factor is that in typical supramolecular complexes the loss of translatory freedom is already paid by a single association step. The loss of rotational freedom upon complex formation has been experimentally [9] found to be smaller than theoretically expected (see below).

Entropy contributions pose particular problems, not only for the precise experimental determination, which in the past often relied on the temperature dependence of equilibrium constants (the *Van 'tHoff* method) instead of on more reliable calorimetry techniques. Also their theoretical interpretation is hampered by several factors, for instance because ΔS values depend on the choice of the standard concentration, in contrast to ΔH [8]. Configurational entropy, which refers also to solute motions has been addressed in several papers [22,23,24]. Data for the loss of translatory degrees of freedom in complex formation range from $T\Delta S = 3$ to $9\,kJ/mol$, and depend also on the reaction medium [25]. In multivalent associations this $T\Delta S$ penalty plays, as mentioned above, a minor role as it is paid already by a single interaction. For the loss of rotatory degrees of freedom in complex formation values from $T\Delta S = 1.5$ to $6\,kJ/mol$ were proposed [26], which also should depend on the nature of the bond involved in the rotation [27]. Measurements of complexes involving an increasing number n of single bonds between two binding units furnished values of only $\Delta\Delta G = 0.5$ to $1.3\,kJ/mol$ per single bond (e.g. from the slope in Figure 1.6) [9,28]. Similar small numbers have been found in

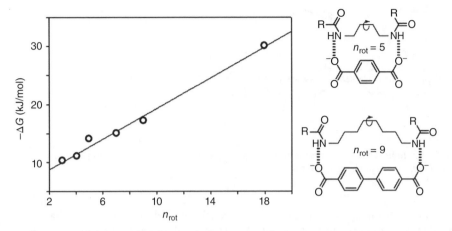

Figure 1.6 Free energies of complex formation between α,ω-diamides and α,ω-dicarboxylates in CHCl$_3$ as a function of the number of rotatable single bonds (n_{rot}) between the terminal amide and carboxylate functions. *Source*: Ref. [28]. Reproduced with permission of VCH/Wiley.

complexes involving peptide- ß-sheets [29], with calcium-EDTA complexes [30], and for example in the coordination of nickel or copper with either *trans*-1,2-diaminocyclohexane or the more flexible ethylene diamine [31]. In line with these rather small numbers it has been found that preorganization of a linker in host molecules has no or a small effect on supramolecular effective molarities [32,33].

1.3.2 Mismatch as Limitation of Additivity

The most obvious limitation for additivity of non-covalent interactions and therefore also for the lock-and-key principle is the necessary geometric fit between host and guest [34]. Insufficient fit between receptor and ligand is a major factor, in particular for a conformationally more rigid polyvalent entity [1]. The steric requirements for an optimal binding between host and guest depend on the nature of the non-covalent bonds. In particular, electrostatic interactions fall off with only with r^{-1} between binding sites whereas dispersive interactions fall off with r^{-6}. In addition, the latter interactions have no or only a small directional dependence, whereas for example the strength of hydrogen or halogen bonds depends on the orientation of donor and acceptor. Exceptions are molecular containers [35] in which the binding of substrates is in most cases controlled by the size of the portals. However, here as in other supramolecular complexes another important restriction is the presence of solvent molecules in a ligand-containing cavity, so that the guest molecule can only use a limited number of interactions which are possible, again depending on the binding mechanism. Thermal motions as well as vibrational and translatory freedom of movement of host and guest are also responsible for the limited fitting; moreover, the surfaces of interacting molecules are characterized by corners and dimples. Recent studies with cryptophanes composed of two bowl-shaped cyclotriveratrylene units showed large solvent molecules such as tetrachloroethane inside the cavity [36]. It has been found earlier [37] that for example some cryptophanes bind, say, chloroform better than methane, although methane fits geometrically as well in the cavity. An occupancy factor or packing coefficient (PC) of 0.886 was calculated for

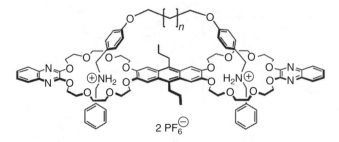

Scheme 1.3 Complex with crown-ammonium pseudorotaxanes [39], with a very large affinity difference between spacer length of either $n=0$ or $n=1$.

the chloroform complex, similar to that in a closely packed crystal. For methane the occupancy factor amounts to a PC of only 0.35. These values are in the range with later systematic evaluations with many container- and capsule-type hosts [38], which were leading to generally observed $55 \pm 9\%$ occupancy of the space available.

Even small geometric changes can have a dramatic impact on the stability of supramolecular complexes, such as in recently described associations with crown-ammonium pseudorotaxanes [39] (Scheme 1.3). Here insertion of just one methylene group in the spacer leads to a drop from $K = 25\,000\,M^{-1}$ for the optimal spacer ($n=0$) to $K = 1100\,M^{-1}$ with the longer spacer ($n=1$), due to differences in both ΔH ($-4.8\,kJ/mol$) and $T\Delta S$ ($2.9\,kJ/mol$).

Frequently one interaction in a supramolecular complex is significantly larger than another one, which then can lead to an induced misfit. Figure 1.7 illustrates schematically the consequences for cyclodextrin complexes as an example [40]. Only in ideal situations like in Case I (Figure 1.7a) one can expect additivity (as for example with the nucleotide complexes in Scheme 1.2). In Case II (Figure 1.7b) the force between D and A is so strong that the second interaction is severely diminished, with an ensuing loss of additivity. Such situations have been seen for example with complexes of nucleotides and cyclodextrins, which bear a different number n of aminoalkyl substituents at the rim [41,42].

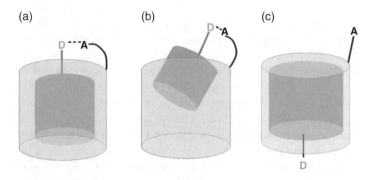

Figure 1.7 Schematic consequences of mismatch: (a) similar interaction in- and outside and sufficient matching (e.g. Case I); (b) stronger interaction outside (Case II); (c) stronger interaction inside cavity (e.g. Case III). *Source*: Ref. [40]. Reproduced with permission of Royal Society of Chemistry. See color section.

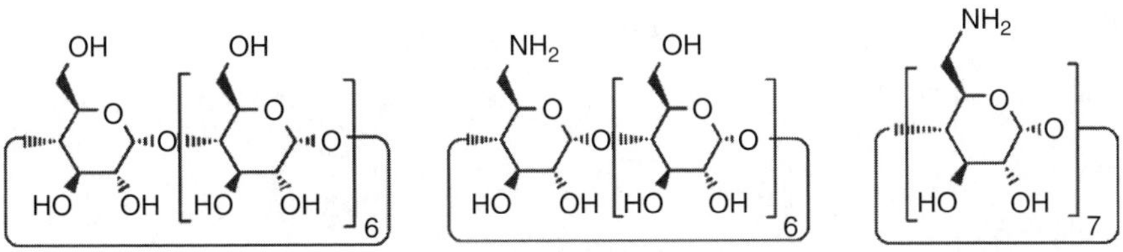

	CD0	CDI	CDII
5-AMP	11	24	29
5-CMP	—	17	24
ATP	—	28	32
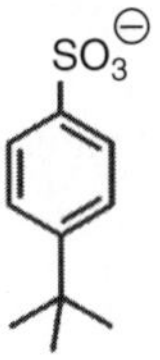 (SO₃⁻ compound)	23	19.5	22.5

Scheme 1.4 Complexation free energies ΔG (kJ/mol) of ß-cyclodextrin derivatives bearing zero, one or seven charges at the rim (CD0, CDI, CDII) with AMP, ATP and *p-tert*-butylphenyl compounds. Data from Ref. [42].

With the monosubstituted cyclodextrin CDI ($n = 1$) the affinity increases from AMP to ATP by only $\Delta\Delta G = 4.7$ kJ/mol (Scheme 1.4), much less than expected by the possible increase of salt bridges between the phosphate residue and the CDI cation, and in contrast to observations with cyclophane complexes (Scheme 1.2). This indicates that the nucleoside residue seeks a sufficient contact with the CDI moiety, resulting in diminished ion pair contacts. Furthermore, there is a moderate selectivity with respect to the nucleobase, but the differences between AMP, GMP, CMP and UMP become smaller with the stronger binder CDII ($n = 7$), for example the $\Delta\Delta G$ between AMP and CMP diminishes from 7 to 4 kJ/mol (Scheme 1.4). This is the result of the then much stronger D···A salt bridge, which allows less contact between the cyclodextrin moiety and the nucleoside residue.

In Case III (Figure 1.7c) one interaction is so strong that the second one can barely materialize. The strong interaction of the butylphenyl residue in the cyclodextrins dominates the binding mode, and prohibits a contact between the anion and cation. This is obvious from the affinity with the positively charged host CDI which strikingly is even smaller in comparison with the neutral CD0, and from the negligible difference between CD0 and CDII complexes [42].

Stereoelectronic effects are also difficult to count as additive contribution, since they strongly depend on orientation, as shown for example for complexes between 1.10-diaza-crown and potassium ions [43]. Here, only after introduction of methyl groups at the nitrogen atoms are the lone pairs enforced towards a diequatorial orientation, and the binding energy increases to much larger affinity (Figure 1.8).

(a)   (b)

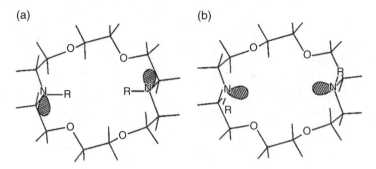

Figure 1.8 Stereoelectronics: the 1.10-diaza-crown with R=H (diaxial lone pair orientation, (a) binds K^+ ions with only $\Delta G = 10$ kJ/mol, with R=Me (diequatorial lone pair orientation, (b) ΔG increases to 26 kJ/mol (in methanol) *Source*: Ref. [43]. Reproduced with permission of John Wiley and Sons.

A similar situation holds for other directional enforcers, in particular for hydrogen bonds, and makes it difficult to simply summarize the number of interactions.

1.3.3 Medium Effects as Limiting Factor

Solvent effects can also significantly limit the possible additivity in multivalent complexes. First, they can decisively change the binding mechanism. Thus, dispersive interactions can be large in water, but are negligible in most organic solvents [16]. The energy for desolvation of host and guest prior to complex formation depends critically on the nature of binding elements, and thus can obscure additivity. In addition, solvophobic contributions can lead to a complete independence of specific non-covalent forces. In particular, water as medium, but also other solvents of low polarizability [44] can lead to dominating solvophobic forces. Especially cucurbituril hosts, which lack binding sites inside their cavity, complex with unsurpassed affinity with many ligands [45,46,47]. It has been shown that these cucurbiturils contain a sizeable number of water molecules which usually can exert only a few inter-water hydrogen bonds. If these are replaced by a suitable guest and freed to the bulk, they enjoy close to four hydrogen bonds. High energy water inside cavities is also present in for example cyclodextrins, cyclophanes, some tweezer or cleft hosts, and so on, and contributes to binding which is difficult to separate from direct non-covalent interactions [48] (Figure 1.9). Crystal structures of cyclodextrin hydrates have indicated the presence of such less coordinated water inside the cavity [49].

1.3.4 Strain and Induced Fit

Many, if not most complex formations occur with some conformational changes for maximizing the pertinent non-covalent interactions. Such an induced fit necessarily costs some strain energy, leading to weaker affinities than they would be if all possible interactions would be simply additive. This poses limits to the evaluation of additive single free energies from the observed total complexation free energies. Such strain effects play a particular role in cooperativity and allostery in multivalent complexes, which are dealt with in the following sections.

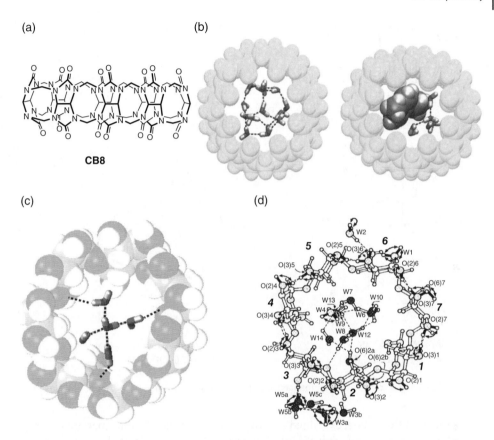

Figure 1.9 Examples of high energy water. (a) Cucurbit[8]uril (CB8) and 14 water molecules. (b) CB8 with viologen as guest and 6 water molecules in the cavity. (c) ß-Cyclodextrin with 5 water molecules, all from molecular dynamics simulations. *Source*: Ref. [48]. Reproduced with permission of John Wiley and Sons. (d) ß-Cyclodextrin dodecahydrate structure derived from neutron diffraction. *Source*: Ref. [49]. Reproduced with permission of American Chemical Society. See color section.

1.4 Cooperativity

Positive cooperativity implies that the binding of one ligand to one of several binding sites in a receptor enhances the affinity at other sites, while negative cooperativity diminishes the affinity [1,4,5,50,51,52]. In classical allosteric systems this is due to conformational coupling between binding sites, as will be discussed in Section 1.5. Cooperativity also occurs if there are direct interactions between the complexed guest molecules. This is typical for ion pair complexation [53,54,55] where the electrostatic forces between anion and cation can lead to significantly enhanced binding constants K (Scheme 1.5). In Case A [56] the presence of Na^+ increases the value of K from 20 to $620\,M^{-1}$, in the crown ether host (Case B) the K increases from 50 to $470\,M^{-1}$ in presence of Na^+ [57].

The cyclopeptide A shown in Scheme 1.6 binds $BuNMe_3X$ salts in chloroform for $X = I$ with $K = 300\,M^{-1}$, while for the tosylate ($X = OTs$) a K increase by 10^4 was observed

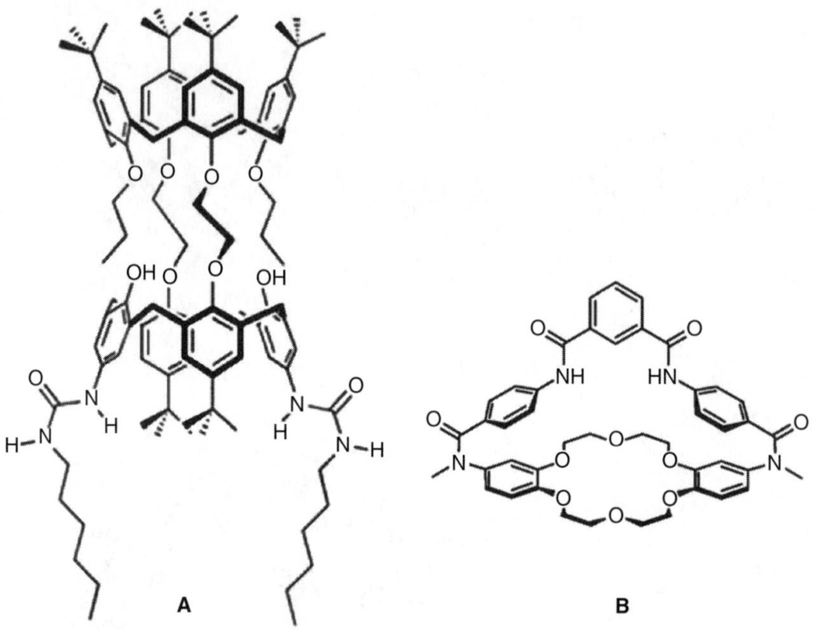

$K_{Na^+}/K_{free} = 31$ for Cl^- in $CDCl_3/CD_3CN$ (2:1)

$K_{Na^+}/K_{free} = 9$ for Cl^- $CDCl_3/CD_3OD$ (9:1)

Scheme 1.5 Positive cooperativity: enhanced anion binding constants in heterotopic complexes.

compared with the iodide, explained also by a tosylate-induced conformational change of the host [58]. A related host B [59] binds very efficiently *N*-methyl-quinuclidinium iodide as ion pair in chloroform with $K = 8.3 \times 10^4\,M^{-1}$. With the host C a 260-fold affinity increase to $K = 1.8 \times 10^4\,M^{-1}$ was observed, with $^{+}H_3NCH(Bn)CO_2Me$ as the cation and nitrate as anion, while tetraalkylammonium salts bind weakly due the steric hindrance of the tetraalkyl residue, with for example $K = 70\,M^{-1}$ with nitrate as the anion [60].

1.5 Allostery

Typical allosteric systems exhibit cooperativity due to conformational coupling between binding sites [61,62]. Case A shown in Scheme 1.6 [59] exemplifies that changes in flexible host structures may often play a large role in limitation of additivity rules. Extreme limitations of observable additivity occur in allosteric systems, which form a binding cavity only in the presence of strongly bound effector, such as metal ions in complexes A and B in Scheme 1.7. In complexes A and B the affinity of the fluorescent dye DNSA (dansylamide) in the absence of the zinc ion is so weak that it cannot be measured, so that the cooperativity ratio amounts to $K_{rel} = K_{Zn/0} > 100$ [63,64].

In complex C (Scheme 1.7) a conformational change induced by Li^+ ions leads to strong binding of [60]fullerene with $K = 2.1 \times 10^3\,M^{-1}$, in comparison with $K = 39\,M^{-1}$ without the metal [65]. A negative cooperativity is seen with Na^+, with $K_{rel} < 10$.

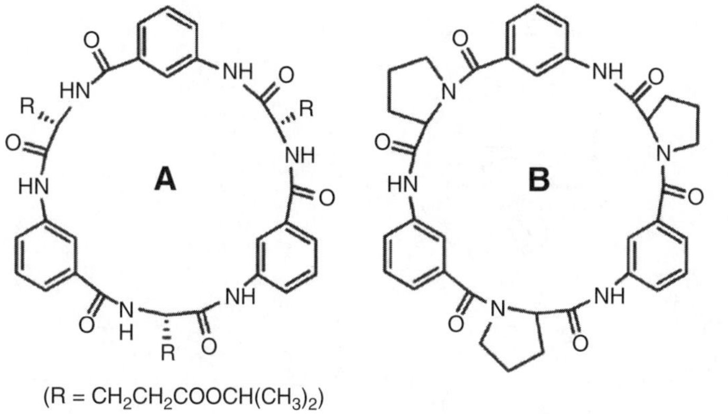

(R = CH₂CH₂COOCH(CH₃)₂)

Scheme 1.6 Cyclopeptide hosts receptors with cooperativity between binding of anions and cations.

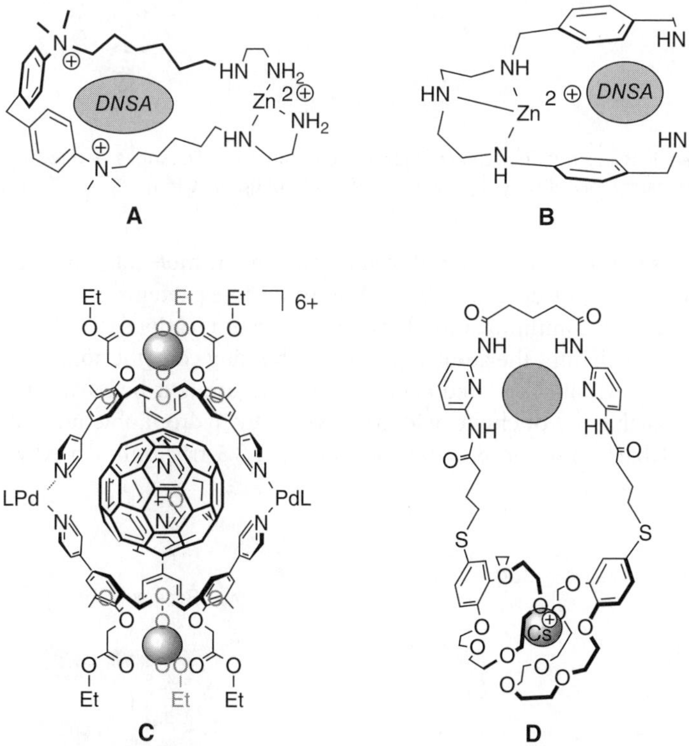

Scheme 1.7 Cooperativity with allosteric ditopic receptor complexes, see text.

The association of anions such as chloride with amide functions in complex D (Scheme 1.7) is significantly enhanced by complexation with Cs⁺ ions, due to interaction with the crown ether units by a conformational rearrangement [66]. In *s*-hydrindacenes conformational changes of binding group orientation and polarity is observed upon association with substituted resorcinols, with a cooperativity ratio K_2/K_1 of up to 30 (Scheme 1.8) [67].

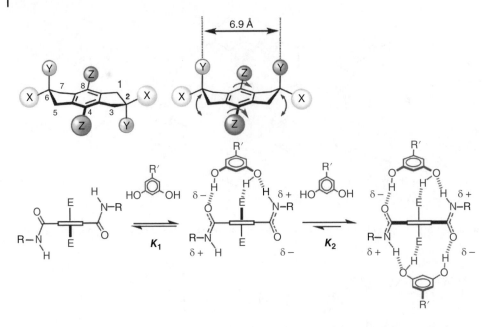

Scheme 1.8 Positive cooperativity in *s*-hydrindacene complexes: conformational changes lead to association with substituted resorcinols with a cooperativity ratio K_2/K_1 of up to 30 [68].

Conformational changes within a receptor, induced by an effector molecule, can lead to reinforced binding at different receptor locations [68]. In flexible proteins correlated rearrangement allows allosteric communication between different locations [69,70,71]; the ensuing entropic factors will limit the additivity of single binding contributions [72]. The host CER (Figure 1.10) exhibits a related allosteric complexation; it bears anionic binding groups attached at the end of cholic acid arms which by hydrophobic interactions between them fold back and complex by ion pairing 1,3,5-tris(amino methyl)

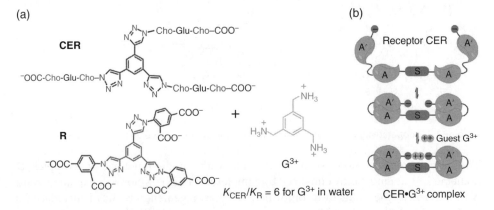

Figure 1.10 (a) Receptor CER with and R without steroidal arms, tricationic guest G^{3+}. (b) Hydrophobic interactions between the steroidal arms of CER preorganizes the CER for binding of guest G^{3+}. *Source*: Ref. [73]. Reproduced with permission of American Chemical Society. See color section.

benzene as guest G with $K = 138\,M^{-1}$, in comparison with only $24\,M^{-1}$ with a parent receptor R lacking the steroidal arms. Both enthalpic and entropic contributions are responsible for the different complexations [73].

Finally, we note that the decisive factor for the efficiency of allosteric systems with positive cooperativity is the conformational energy ΔG_C required for the formation of a suitable cavity for ligand binding in the absence of an effector. ΔG_C is usually dominated by an increase of strain in a folded conformer and/or by high energy solvents within a cavity. A recent analysis of the thermodynamics in synthetic allosteric systems exemplifies how the binding strength of an effector molecule at a second binding site must pay for the energy ΔG_C needed for the binding of the first substrate [74]. Additivity of the binding contributions can only be expected if ΔG_C could be determined independently. Negative cooperativity depends only on the difference $\Delta\Delta G_{A,B}$ between the binding energies at the two sites, which are enhanced or lowered by concomitant changes in ΔG_C. The often small efficiency of synthetic allosteric receptors [61,62], measured by the binding constant ratio $K_A/K_{A(B)}$, in which K_A refers to association of ligand A in the absence of the effector ligand B, and $K_{A(B)}$, is due to small ΔG_C values. Larger efficiency can be expected with increased ΔG_C values, for example by introduction of alkyl substituents in the *ortho*-position of pyridine in the often used [61,62] bipyridyl-based allosteric systems.

1.6 Conclusions

For efficient multivalent complexes it is desirable to preserve as much as possible additivity of all possible binding contributions. Ideally not only the affinity and therefore sensitivity but also the selectivity of such complexes is optimal if additivity of the single binding free energies is materialized. To what degree a geometric fit in the sense of the lock-and-key principle or preorganization is required depends first on the binding mechanism. The distance dependence of the interaction increases distinctly from electrostatic effects or ion pairing to dispersive forces. Mismatch between binding partners leads to a strong decrease of both affinity and selectivity particularly if the binding mechanism is characterized by a steeper distance dependence, and if the components are less flexible. Solvents can strongly influence the binding mechanisms; hydrophobic effects of high energy water inside cavities or clefts can make intermolecular binding contributions unimportant.

If one binding contribution is much stronger than others, the second interaction is often severely weakened due to mismatch; even a complete change of binding modes can occur. High selectivity combined with high affinity, which both require optimal fit, is difficult to attain if binding sites in a receptor are rigidly connected. In principle one can overcome this problem by flexible connections between a primary binding site securing high affinity with another site securing selectivity, provided such sites are available. If a multivalent complex should operate in for example a nanomolar solution the primary interactions should be worth about $50\,kJ/mol$, while at the secondary site values of, say, $\Delta G_X = 15$ and $\Delta G_Y = 5\,kJ/mol$ are enough to achieve a sizeable selectivity for distinction of two compounds X and Y ($\Delta\Delta G \approx 10\,kJ/mol$ or $K_X/K_Y \approx 100$).

In systems with positive cooperativity larger affinity than that predicted by additive single interactions is possible either by attractive forces between nearby bound substrates, or in the classical case by conformational change at one site A induced by occupation at another site B. The efficiency of related allosteric systems can be defined as the ligand A concentration needed for complexation in the absence of the effector binding at B; it depends on the strain energy which would be needed to form an optimal conformation for binding A in the absence of occupation at B. Instead of conformational strain unfavorable solvents in cavities or clefts, such as high energy water, can enhance the efficiency of allosteric systems. It is hoped that the design of synthetic systems for, say, highly sensitive and selective new sensors as well as for drug design can be facilitated by taking into account some of the limitations and possibilities outlined in this chapter.

References

1 Mammen, M.; Choi, S.-K.; Whitesides, G. M. Polyvalent interactions in biological systems: Implications for design and use of multivalent ligands and inhibitors. *Angew. Chem. Int. Ed.* **1998**, *37*, 2754–2794.

2 Persch, E.; Dumele, O.; Diederich, F. Molecular recognition in chemical and biological systems. *Angew. Chem. Int. Ed.* **2015**, *54*, 3290–3327.

3 Schneider, H.-J. Binding mechanisms in supramolecular complexes *Angew. Chem. Int. Ed.* **2009**, *48*, 3924–3977.

4 Hunter, C. A. Quantifying intermolecular interactions: Guidelines for the molecular recognition toolbox. *Angew. Chem. Int. Ed.* **2004**, *43*, 5310–5324.

5 Biedermann, F.; Schneider, H.-J. Experimental binding energies in supramolecular complexes, *Chem. Rev.* **2016**, *116*, 5216–5300.

6 Fasting, C.; Schalley, C. A.; Weber, M.; Seitz, O.; Hecht, S.; Koksch, B.; Dernedde, J.; Graf, C.; Knapp, E. W.; Haag, R. Multivalency as a chemical organization and action principle. *Angew. Chem. Int. Ed.* **2012**, *51*, 10472–10498.

7 Schneider, H.-J.; Yatsimirsky, A. Selectivity in supramolecular host–guest complexes *Chem. Soc. Rev.* **2008**, 263–277.

8 Schneider, H.-J.; Yatsimirski, A. *Principles and Methods in Supramolecular Chemistry*, John Wiley & Sons, Ltd, Chichester, **2000**.

9 Hossain, M. A.; Schneider, H.-J. Flexibility, association constants, and salt effects in organic ion pairs: How single bonds affect molecular recognition. *Chemistry Eur. J.* **1999**, *5*, 1284–1290.

10 De Robertis, A.; De Stefano, C.; Foti, C.; Giuffrè, O.; Sammartano, S. Thermodynamic parameters for the binding of inorganic and organic anions by biogenic polyammonium cations. *Talanta* **2001**, *54*, 1135–1152, and references cited therein.

11 Daniele, P. G.; Foti, C.; Gianguzza, A.; Prenesti, E.; Sammartano, S. Weak alkali and alkaline earth metal complexes of low molecular weight ligands in aqueous solution. *Coord. Chem. Rev.* **2008**, *252*, 1093–1107.

12 Guo, D. S.; Wang, L. H.; Liu, Y. Highly effective binding of methyl viologen dication and its radical cation by p-sulfonatocalix[4,5]arenes. *J. Org. Chem.* **2007**, *72*, 7775–7778.

13 Schneider, H.-J.; Blatter, T.; Palm, B.; Pfingstag, U.; Rüdiger, V.; Theis, I. Complexation of nucleosides, nucleotides, and analogs in an azoniacyclophane *J. Am. Chem. Soc.* **1992**, *114*, 7704–7708.

14 Schneider, H.-J.; Schiestel, T.; Zimmermann, P. The incremental approach to noncovalent interactions: Coulomb and van der Waals effects in organic ion pairs. *J. Am. Chem. Soc.* **1992**, *114*, 7698–7703.

15 Liu, T.J.; Schneider H.-J. Additivity and quantification of dispersive interactions – from cyclopropyl- to nitrogroups. *Angew. Chem. Int. Ed.* **2002**, *41*, 1368–1370.

16 Schneider, H.-J. Dispersive interactions in solution complexes *Acc. Chem. Res.* **2015**, *48*, 1815–1822.

17 Lorand, J. P. Hammett correlations for phenylacetonitriles and tetrafluorobenzenes with HMPA. *J. Phys. Org. Chem.* **2011**, *24*, 267–273.

18 Jencks, W.P. On the attribution and additivity of binding energies. *Proc. Natl. Acad. Sci. USA* **1981**, *78*, 4046–4050.

19 Cornish-Bowden, A. Enthalpy–entropy compensation: A phantom phenomenon. *J. Biosci.* **2002**, *27*, 121–126.

20 Ford, D. M. Enthalpy – entropy compensation is not a general feature of weak association. *J. Am. Chem. Soc.* **2005**, *127*, 16167–16170.

21 Zhou, H.-X.; Gilson, M. K. Theory of free energy and entropy in noncovalent binding. *Chem. Rev.* **2009**, *109*, 4092–4107.

22 Dunitz, J. D. Win some, lose some: Enthalpy–entropy compensation in weak intermolecular interactions. *Chem. Biol.* **1995**, *2*, 709–712.

23 Chodera, J. D.; Mobley, D. L. Entropy–-enthalpy compensation: Role and ramifications in biomolecular ligand recognition and design. *Annu. Rev. Biophys.* **2013**, *42*, 121–142.

24 Sharp, K. Entropy—enthalpy compensation: Fact or artifact? *Protein Sci.* **2001**, *10*, 661–667.

25 Camara-Campos, A.; Musumeci, D.; Hunter, C. A.; Turega, S. Chemical double mutant cycles for the quantification of cooperativity in H-bonded complexes. *J. Am. Chem. Soc.* **2009**, *131*, 18518–18524.

26 Searle, M. S.; Williams, D. H. The cost of conformational order: entropy changes in molecular associations. *J. Am. Chem. Soc.* **1992**, *114*, 10690–10697; Groves, P.; Searle, M. S.; Westwell, M. S.; Williams, D. H. Expression of electrostatic binding cooperativity in the recognition of cell-wall peptide analogues by vancomycin group antibiotics. *J. Chem. Soc., Chem. Commun.* **1994**, 1519–1520.

27 Mammen, M.; Shakhnovich, E. I.; Whitesides, G. M. Using a convenient, quantitative model for torsional entropy to establish qualitative trends for molecular processes that restrict conformational freedom. *J. Org. Chem.* **1998**, *63*, 3168–3175.

28 Eblinger, F.; Schneider, H.-J. Stabilities of hydrogen-bonded supramolecular complexes with various numbers of single bonds. *Angew. Chem. Int. Ed.* **1998**, *37*, 826–829.

29 Nowick, J. S.; Cary, J. M.; Tsai, J. H. A Triply Templated artificial β-sheet. *J. Am. Chem. Soc.* **2001**, *123*, 5176–5180.

30 Christensen, T.; Gooden, D. M.; Kung, J. E.; Toone, E. J. Additivity and the physical basis of multivalency effects: A thermodynamic investigation of the calcium EDTA interaction. *J. Am. Chem. Soc.* **2003**, *125*, 7357–7366.

31 Smith, R.; Martell, A. *Critical Stability Constants*, Vol. 2, Plenum Press, New York, **1975**.

32 Sun, H.; Navarro, C.; Hunter, C. A. Influence of non-covalent preorganization on supramolecular effective molarities. *Org. Biomol. Chem.* **2015**, *13*, 4981–4992.

33 Sun, H.; Hunter, C. A.; Llamas, E. M. The flexibility-complementarity dichotomy in receptor-ligand interactions. *Chem. Sci.* **2015**, *6*, 1444–1453, and references cited therein.

34 Schneider, H.-J. Limitations and extensions of the lock-and-key principle: differences between gas state, solution and solid state structures. *Int. J. Mol. Sci.* **2015**, *16*, 6694.

35 Cram, D. J.; Cram, J. M. *Container Molecules and Their Guests*, Royal Society of Chemistry, Cambridge, **1997**.

36 Haberhauer G.; Woitschetzki, S.; Bandmann, H. Strongly underestimated dispersion energy in cryptophanes and their complexes *Nature Commun.* **2014**, *5*, Article 3542, 1–7.

37 Garel, L.; Dutasta, J.; Collet, A. Complexation of methane and chlorofluorocarbons by cryptophane-A in organic solution *Angew. Chem. Int. Ed.* **1993**, *32*, 1169–1171; Collet, A. in *Comprehensive Supramolecular Chemistry*, Vol. 2 (Eds J. L. Atwood, J. E. Davies, D. D. MacNicol, and F. Vögtle), Elsevier, Dordrecht, **1996**, 325–365.

38 Mecozzi, S.; Rebek Jr, J. The 55% solution: A formula for molecular recognition in the liquid state. *Chem. Eur. J.*, **1998**, *4*, 1016–1022, and references cited therein.

39 Schalley, C. A.; Beizai, K.; Vögtle, F. On the way to rotaxane-based molecular motors: Studies in molecular mobility and topological chirality. *Acc. Chem. Res.* **2001**, *34*, 465–476.

40 Schneider, H.-J. Spatial mismatch, non-additive binding energies and selectivity in supramolecular complexes. *Org. Biomol. Chem.* **2017**, *15*, 2146–2151.

41 Eliseev, A. V.; Schneider, H.-J. Molecular recognition of nucleotides, nucleosides, and sugars by aminocyclodextrins. *J. Am. Chem. Soc.* **1994**, *116*, 6081–6088.

42 Wenz, G.; Strassnig, C.; Thiele, C.; Engelke, A.; Morgenstern, B.; Hegetschweiler, K. Recognition of ionic guests by ionic ß-cyclodextrin derivatives. *Chem. Eur. J.* **2008**, *14*, 7202–7211.

43 Solovev, V. P.; Strakhova, N. N.; Kazachenko, V. P.; Solotnov, A. F.; Baulin, V. E.; Raevsky, O. A.; Rüdiger, V.; Eblinger, F.; Schneider, H.J. Steric and stereoelectronic effects in aza crown ether complexes. *Eur. J. Org. Chem.* **1998**, 1379–1389.

44 Yang, L.; Adam, C.; Cockroft, S. L. Quantifying solvophobic effects in nonpolar cohesive interactions. *J. Am. Chem. Soc.* **2015**, *137*, 10084–10087.

45 Barrow, S. J.; Kasera, S.; Rowland, M. J.; Del Barrio, J.; Scherman, O. A. Cucurbituril-based molecular recognition. *Chem. Rev.* **2015**, *115*, 12320–12406.

46 Shetty, D.; Khedkar, J. K.; Park, K. M; Kim, K. Can we beat the biotin-avidin pair?: Cucurbit[7]uril-based ultrahigh affinity host–guest complexes and their applications *Chem. Soc. Rev.* **2015**, *44*, 8747–8761.

47 Isaacs, L. Stimuli responsive systems constructed using cucurbit[n]uril-type molecular containers. *Acc. Chem Res.* **2014**, *47*, 2052–2062.

48 Biedermann, F.; Nau, W. M.; Schneider, H.-J. The hydrophobic effect revisited *Angew. Chem. Int. Ed.* **2014**, *53*, 11158–11171.

49 Betzel, C.; Saenger, W.; Hingerty, B. E.; Brown, G. M. Circular and flip-flop hydrogen bonding in ß-cyclodextrin undecahydrate: A neutron diffraction study. *J. Am. Chem. Soc.* **1984**, *106*, 7545–7557.

50 Di Cera, E. Site-specific thermodynamics: understanding cooperativity in molecular recognition. *Chem. Rev.* **1998**, *98*, 1563–1592.

51 Hunter, C. A.; Anderson, H. L. What is cooperativity? *Angew. Chem., Int. Ed.* **2009**, *48*, 7488–7499.

52 Mahadevi, A.S.; Sastry, G.N. Cooperativity in noncovalent interactions. *Chem. Rev.* **2016**, *116*, 2775–2825.

53 McConnell, A. J.; Beer, P. D. Heteroditopic receptors for ion-pair recognition. *Angew. Chem. Int. Ed.* **2012**, *51*, 5052–5061

54 Chrisstoffels, L. A. J.; de Jong, F.; Reinhoudt, D. N.; Sivelli, S.; Gazzola, L.; Casnati, A.; Ungaro, R. Facilitated transport of hydrophilic salts by mixtures of anion and cation carriers and by ditopic carriers. *J. Am. Chem. Soc.* **1999**, *121*, 10142–10151.

55 Kim, S. K.; Sessler, J. L. Ion pair receptors. *Chem. Soc. Rev.* **2010**, *39*, 3784–3809.
56 Webber, P. R.; Beer, P. D. Ion-pair recognition by a ditopic calix [4] semitube receptor. *Dalton Trans.* **2003**, 2249–2252.
57 Deetz, M. J.; Shang, M.; Smith, B. D. A macrobicyclic receptor with versatile recognition properties: simultaneous binding of an ion pair and selective complexation of dimethylsulfoxide. *J. Am. Chem. Soc.* **2000**, *122*, 6201–6207.
58 Kubik, S. Large increase in cation binding affinity of artificial cyclopeptide receptors by an allosteric effect. *J. Am. Chem. Soc.* **1999**, *121*, 5846–5855.
59 Kubik, S.; Goddard, R. A new cyclic pseudopeptide composed of (L)-proline and 3-aminobenzoic acid subunits as a ditopic receptor for the simultaneous complexation of cations and anions. *J. Org. Chem.* **1999**, *64*, 9475–9486.
60 Gong, J.; Gibb, B. C. A new macrocycle demonstrates ditopic recognition properties. *Chem. Commun.* **2005**, 1393–1395.
61 Kremer, C.; Lützen, A. Artificial allosteric receptors. *Chem. Eur. J.* **2013**, *19*, 6162–6196.
62 Kovbasyuk, L.; Krämer, R. Allosteric supramolecular receptors and catalysts. *Chem. Rev.* **2004**, *104*, 3161–3188.
63 Schneider, H. J.; Ruf, D. A synthetic allosteric system with high cooperativity between polar and hydrophobic binding sites. *Angew. Chem. Int. Ed.* **1990**, *29*, 1159–1160.
64 Baldes, R.; Schneider, H. J. Complexes from polyazacyclophanes, fluorescence indicators, and metal cations *Angew. Chem. Int. Ed.* **1995**, *34*, 321–323.
65 Ikeda, A.; Udzu, H.; Yoshimura, M.; Shinkai, S. Inclusion of [60]fullerene in a self-assembled homooxacalix[3]arene-based dimeric capsule constructed by a PdII–pyridine interaction. *Tetrahedron* **2000**, *56*, 1825–1832.
66 Nabeshima, T.; Hanami, T.; Akine, S.; Saiki, T. Control of ion binding by cooperative ion-pair recognition using a flexible heterotopic receptor. *Chem. Lett.* **2001**, *30*, 560–561.
67 Kawai, H. Hydrindacenes as versatile supramolecular scaffolds. *Bull. Chem. Soc. Jpn.* **2015**, *88*, 399–409.
68 Otto, S. Reinforced molecular recognition as an alternative to rigid receptors. *Dalton Trans.* **2006**, 2861–2864.
69 DuBay, K. H.; Bowman, G. R.; Geissler, P. L. Fluctuations within folded proteins: Implications for thermodynamic and allosteric regulation. *Acc. Chem. Res.* **2015**, *48*, 1098–1105.
70 Frederick, K. K.; Marlow, M. S.; Valentine, K. G.; Wand, A. J. Conformational entropy in molecular recognition by proteins. *Nature* **2007**, *448*, 325–329.
71 Motlagh, H. N.; Wrabl, J. O.; Li, J.; Hilkser, V. J. The ensemble nature of allostery. *Nature* **2014**, *508*, 331–339.
72 Williams, D. H.; Stephens, E.; O'Brien, D. P.; Zhou, M. Understanding noncovalent interactions: ligand binding energy and catalytic efficiency from ligand-induced reductions in motion within receptors and enzymes. *Angew. Chem. Int. Ed.* **2004**, *43*, 6596–6616.
73 Gunasekara, R. W.; Zhao, Y. Rationally designed cooperatively enhanced receptors to magnify host–guest binding in water. *J. Am. Chem. Soc.* **2015**, *137*, 843–849.
74 Schneider, H.-J. Efficiency parameters in artificial allosteric systems. *Org. Biomol. Chem.*, **2016**, *14*, 7994–8001.

2

Models and Methods in Multivalent Systems

Jurriaan Huskens

Molecular NanoFabrication Group, MESA+ Institute for Nanotechnology, Faculty of Science and Technology, University of Twente, 7500 AE, Enschede, the Netherlands

2.1   Introduction

2.1.1   General Introduction

Self-assembly describes the non-covalent interaction between molecules, biomolecules, nanoparticles, and so on, that leads to larger structures with designed properties and functionalities [1,2,3,4,5]. The development of organic synthesis, which is the methodology development for the creation of new molecules based on covalent bonds, has led to a true revolution in chemistry that started more than a century ago. Literally millions of compounds have been made or can potentially be made based on the methods developed so far, and the field is still developing. One can hardly grasp the idea of taking all of these molecules and to assemble them into larger structures, particles, entities, based on non-covalent interactions: the possible combinations are unimaginable and are orders of magnitude larger in potential. Moreover, the products of self-assembly can in turn be organized in assemblies of a higher order, in a process called hierarchical self-assembly [6].

In chemistry, the notion of the infinite possibilities of self-assembly has grown in particular from the dawn of supramolecular chemistry in the 1960s and 1970s [2,7]. Nowadays, the concepts that have been and still are being developed in this area have pervaded all chemistry and materials science. In biology, self-assembly has been seen for decades as one of the major structuring and compartmentalization forces in nature. The current technological toolbox, with single-molecule techniques, *in-vivo* imaging, and so on, in particular developed within the chemical biology and nanotechnology arenas, now finally allows complete merging of the concepts and methods in the disciplines of chemistry and biology.

These trends put pressure on the current education of undergraduate and graduate students alike. The pervasion of supramolecular and many other concepts throughout materials science, nanotechnology, and chemical biology in its fullest breadth, causes students to need a vastly more multidisciplinary training, sometimes at the expense of monodisciplinary methodological approaches. For young scientists to play a role at the

Multivalency: Concepts, Research & Applications, First Edition. Edited by Jurriaan Huskens,
Leonard J. Prins, Rainer Haag, and Bart Jan Ravoo.
© 2018 John Wiley & Sons Ltd. Published 2018 by John Wiley & Sons Ltd.

forefront of this immensely popular and interesting area of science, they need to be able to grasp broad concepts quickly, oversee the relationship of these concepts with and their importance for other disciplines, yet at the same time be able to go in depth on a subject when needed for a better fundamental understanding.

This chapter aims to provide an understanding of multivalent systems, based on the development of *models* for describing the binding behavior of these systems. In this context, a model is a complete set of species and the equilibria between them, in which the non-covalent bonds are formed in a stepwise manner, and all equilibria are coupled to equilibrium constants (and association and dissociation rate constants, when desired). The primary assumption made here is that a student or scientist will at some point get data on a non-covalently interacting system, and wants to understand what happens, without having to operate the proverbial "black box". Therefore, a direct and intimate connection is made here between real data and models, so that the development of more complex models can be undertaken from easier models and from a basic understanding of how these models can be implemented numerically. Therefore, this chapter starts (Section 2.2) with the development of the numerical treatment of equilibria and the most basic experimental method for assessing equilibria, the titration. This endeavor borrows pieces of knowledge from analytical chemistry and thermodynamics, as well as a bit of numerical mathematics (though hardly escaping the high school level). In practice, many of these methods have been used intensively in other areas of chemistry, such as coordination and analytical chemistry. This section is followed by a section (2.3) on basic models of multivalent systems, introducing the main concepts, such as effective molarity, and the differences compared with non-multivalent systems. Both solution and surface systems are discussed. Thereafter follows a section (2.4) with more specific, and sometimes more complicated, models.

2.1.2 Multivalent versus Cooperative Interactions

Multivalent systems are systems in which molecules interact with each other by more than one non-covalent interaction pair. The hallmark of a multivalent system is the occurrence of one (or more) *intra*molecular[i] interaction. The popularity of the field, and its importance for both chemical and biological systems, has spawned the publication of several reviews on the topic [8,9,10,11].

As will be explained in more detail below, multivalent systems often come with enhanced affinities, and these enhanced affinities have on occasions been taken as a sign of cooperativity. Therefore, a good definition of both terms, multivalency and cooperativity, is needed before the sense or nonsense of their applicability to a particular system can be evaluated.

We here take cooperativity (also called "allosterism" or "allosteric cooperativity") [12] to mean: the change of the affinity (lower or higher) of an interaction pair caused by the presence of a neighboring formed interaction pair. The textbook example of a (positively) cooperative system is hemoglobin, in which the uptake of oxygen by a heme binding site

---

[i] We explicitly take "intramolecular" to also encompass the non-covalent interaction of two ends of a molecular chain that has other non-covalent interactions present in the chain. All interactions are treated in a stepwise manner, and thus the focus on the formation of a particular interaction pair treats already formed non-covalent interactions in the system like normal, covalent bonds.

of the tetravalent protein complex is promoted by the occupation of earlier oxygen molecules at other heme sites within the protein complex. In a cooperative system, at least one of the binding partners needs to be multitopic (have multiple interaction sites), but not necessarily both (the system can also be based on a single, multitopic, self-assembling molecule). In the hemoglobin case, only the protein complex is multitopic, but not the oxygen molecule, therefore, the system is cooperative but not multivalent (between each molecule of oxygen and hemoglobin, there is only one bond formed, and no intramolecular bonds occur).

As already explained above, multivalency (also called "chelate cooperativity") [12,13][ii] requires molecules to form *multiple* interactions between them, and at least one of the interactions needs to be intramolecular. As a result, both molecules (assuming the interaction occurs between two different molecules, possibly with multiple copies within a complex) need to be multitopic. This definition neither includes a statement about the individual affinity between the molecular sites that constitute the interactions, nor about the overall affinity. As such, the definition only says something about numbers of interactions and configuration of the resulting complex. Therefore, multivalent systems can be non-cooperative (i.e., only chelate, but no allosteric cooperativity) or cooperative (both chelate and allosteric cooperativity), but their distinction is far from trivial (see Section 2.3).

2.2 Numerical Data Analysis

When studying equilibria between species, for example in supramolecular chemistry or biochemistry, one desires to understand the population of species involved in these equilibria. Because equilibria are generally established rapidly on common experimental timescales, physical separation of the species is impossible, and analytical techniques need to be used that can assess the equilibria without interfering with them. Thus, data will be gathered, using the analytical technique of choice, that can provide information on the concentrations of the species involved and/or ratios between them. For example, when studying the host–guest binding between a small organic guest molecule and a cyclodextrin (CD) host, ^{1}H NMR signals of the guest or the host can be used to study the inclusion equilibrium. Because the host–guest interaction is fast on the NMR timescale, the signals will shift with changing ratios of the host and guest. Therefore, the more a signal shifts compared with that of the unbound form, the higher the fraction of bound species.

Models to fit data dealing with the thermodynamics of complex equilibria need to be solved numerically. In fact, only a 1:1 model can be solved analytically; the next least complicated model, a 1:2 system, already requires numerical treatment. It is therefore important to understand how more complicated models can be implemented to make simulations of varying species concentrations upon variation of experimental parameters (a "speciation") and how to use these to fit experimental data to a model. The purpose of this analysis can be to assess binding constants, for example of the

---

[ii] I am personally not in favor of attributing two very similar names, "allosteric cooperativity" and "chelate cooperativity", to these very different concepts, therefore the terms "cooperativity" and "multivalency", respectively, will be used in the remainder of this text. See also Ref. [13].

host–guest binding in the previously mentioned CD host–guest complex, but also other aspects can be investigated such as the stoichiometry of the complex, for example using a Job plot.[iii] In such instances, the model that is implemented is presumed to be known and remains unquestioned. In contrast, on other occasions, the analysis can assist in verifying whether an initially assumed model is correct, and changes of or extensions to the model can be implemented to obtain a better description of the experimentally observed behavior.

2.2.1 Model Simulations Using a Spreadsheet Approach

Setting up a simulation of a speciation is the first step of a numerical model analysis. In this simulation, the model (which species are involved, with what stoichiometries, and how are they related) is presumed known, as well as the binding constants that describe the equilibria between the species. A four-step approach is presented here to arrive at a working simulation in a spreadsheet.[iv]

At first, all equilibria involving all species are written out, at best graphically so that all binding sites of all species are visible. Every equilibrium is associated with a forward (association) and a backward (dissociation) step in which one interaction pair is formed or broken. It is preferred to keep all equilibria limited to the formation of one interaction pair only, so that all individual steps are made visible, very much like the explanation of organic chemistry mechanisms in which the stepwise motion of individual electron pairs is explained. All arrows, indicating the association and dissociation steps, are labeled with a rate constant. The equilibrium constants can then be written both as ratios of the corresponding rate constants and as ratios of species concentrations involved in the equilibrium. Generally, all concentrations of all species are regarded as "unknowns" in the numerical process, whereas the equilibrium constants are assumed known and regarded as "model parameters".

Very often in multivalent systems, complexes are formed involving more than one interaction pair of the same type. An important question can then be whether the microscopic affinities of these interactions are identical and independent on the bound or unbound state of neighboring sites or not. As mentioned above, cooperativity occurs when the affinity of a site is influenced by the binding occurring at a neighboring site. When setting up a model, however, one has to start with assuming so-called *independent binding sites*, that is, without mutual influence of affinities between neighboring sites. The model parameters should be adapted to make the model cooperative only when needed. This is explained in more detail below. When assuming independent binding sites, many of the equilibrium and rate constants that interrelate the species involved in the model, become correlated. It can then be convenient to relate all rate constants to that of the so-called *intrinsic interaction*, which describes the equilibrium between a receptor and a ligand (or between a host and a guest), each with a single

---

[iii] Note that a Job plot, or an inflection point in a titration, provides (at best) the overall ratio of building blocks represented in the complex. The real stoichiometry needs to be inferred from this and other data, often in combination with structural information on the building blocks and complex.

[iv] The "tips and tricks" provided here apply to the use of Microsoft's Excel, but other spreadsheet programs can be used as well. In the author's opinion, a spreadsheet is preferred over existing packages when teaching model building (or even when making new models in scientific research), because most packages are "black box" and therefore do not easily reveal occasional mistakes made in the formulas that describe the model.

(a)

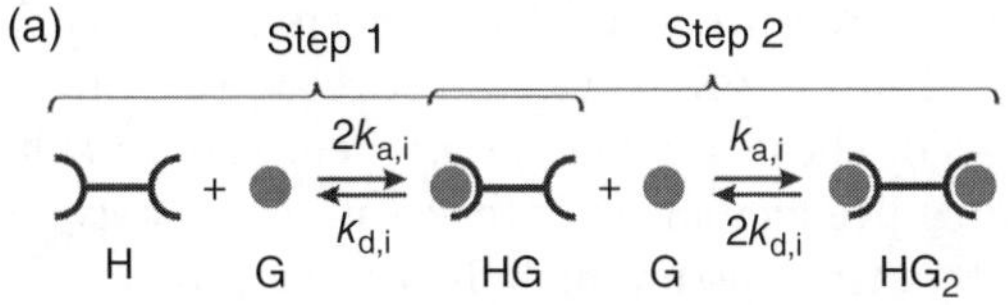

Step 1   Step 2

$$H + G \underset{k_{d,i}}{\overset{2k_{a,i}}{\rightleftharpoons}} HG + G \underset{2k_{d,i}}{\overset{k_{a,i}}{\rightleftharpoons}} HG_2$$

$$K_1 = \frac{k_{a,1}}{k_{d,1}} = \frac{2k_{a,i}}{k_{d,i}} = 2K_i \qquad K_2 = \frac{k_{a,2}}{k_{d,2}} = \frac{k_{a,i}}{2k_{d,i}} = \tfrac{1}{2}K_i$$

(b)
$$[H]_{tot} = [H] + [HG] + [HG_2]$$
$$[G]_{tot} = [G] + [HG] + 2[HG_2]$$
$$K_1 = \frac{[HG]}{[H][G]} \qquad K_2 = \frac{[HG_2]}{[HG][G]}$$

(c)
$$[H]_{tot} = [H](1+K_1[G](1+K_2[G]))$$
$$[G]_{tot} = [G](1+K_1[H](1+2K_2[G]))$$
$$\Rightarrow [H] = \frac{[H]_{tot}}{(1+K_1[G](1+K_2[G]))}$$
$$[G] = \frac{[G]_{tot}}{(1+K_1[H](1+2K_2[G]))}$$

Figure 2.1 Model description of a 1:2 host–guest complex: (a) equilibria; (b) mass balances and equilibrium constant equations; and (c) substituted and rewritten mass balances used in a spreadsheet simulation.

interaction site. The rate constants of equilibria of more complex systems can then be related to the intrinsic rate constants using statistical pre-factors. These pre-factors indicate the numbers of possibilities by which the one species can be converted into the other when following the arrow. When applying this strictly to all arrows of all equilibria described by the model, this "kinetic analysis" gives insight into the relationships between binding constants (again assuming independent binding sites).[v]

As an example, Figure 2.1a shows a 1:2 system, involving the stepwise binding of two monovalent guest molecules G, each with one binding site g, to a divalent host H which has two binding sites h. In this system, two equilibria can be discerned: (i) the binding of the first guest to the host, leading to HG; and (ii) the binding of the second guest molecule to HG, leading to HG_2. The intrinsic interaction pair h–g is governed by an association ($k_{a,i}$) and a dissociation rate constant ($k_{d,i}$). Therefore, the kinetic analysis described above indicates that the first binding of G to H has two possibilities (there are two sites h free for binding, i.e., a statistical pre-factor of 2), providing the rate constant for the first association step to be $k_{a,1} = 2k_{a,i}$. In contrast, the corresponding dissociation step has a statistical pre-factor of only 1, because the dissociation can only occur for the one h–g interaction present in HG, leading to $k_{d,1} = k_{d,i}$. Likewise, the association step

---

[v] The same insight can be obtained using symmetry arguments. The use of pre-factors is not done in the same way by all authors. Look carefully how they are implemented when studying a model in the literature, and make sure the mass balances take the pre-factors into account in the proper manner.

from HG to HG_2 has a pre-factor of 1, while the dissociation gets a pre-factor 2. Overall, therefore, the equilibrium constants K_1 and K_2 can be related to K_i ($K_1 = 2K_i$ and $K_2 = \frac{1}{2}K_i$) and to each other ($K_1 = 4K_2$). The factor 4 in the latter equation, resulting from multiplication of all corresponding pre-factors, gives the relationship between the equilibrium constants in this 1:2 system when assuming independent binding sites. This approach can be easily extended to larger systems, for example, for a host with n sites h binding to n monovalent guests G (each with one g site), see Section 2.4.1.

Important to note is the number of unknowns, that is, the number of species involved in the model. In the 1:2 system shown in Figure 2.1a, we can discriminate four species, the unbound guest G and host H, as well as the 1:1 complex HG and the 1:2 complex HG_2. In order to calculate all concentrations of these four species at a given set of conditions (e.g., at the particular total concentrations of host and guest, as these are commonly controlled by the experimenter when setting up a titration), one needs (at least) the same number of equations that relate the concentrations of these species.

The equations needed to make the model numerically solvable, come from two sources. First, for each free species, in the current example H and G, a mass balance can be written down. Secondly, for each equilibrium, the equilibrium constant can be written as a ratio of concentrations. In the 1:2 model, we can therefore derive two mass balances and two equilibrium constant equations, as shown in Figure 2.1b. This provides the necessary total of four equations, equal to the number of unknowns, which makes the system numerically solvable.[vi]

The set of equations needs to be implemented in a mathematical program to be able to solve them numerically. As explained above, systems with multiple equilibria become complicated so quickly that analytical treatments become impossible. In the spreadsheet approach propagated here, the equations need to be rewritten in order to derive expressions that provide one of the unknowns. An approach is described here that allows easy extension to more complicated systems. First, all equilibrium constant equations are rewritten in a form that gives access to the concentrations of the species formed in the corresponding equilibrium. For example, in the first step between H and G in the 1:2 model, the equation for that equilibrium is rewritten as Equation 2.1.

$$[HG] = K_1[H][G] \tag{2.1}$$

By substituting equations of earlier steps into the equations of later steps (or using equations directly with so-called overall binding constants), all equations have only unbound species as unknowns in the right part of the equation. For example, the second step is initially described by Equation 2.2.

$$[HG_2] = K_2[HG][G] \tag{2.2}$$

Equation 2.2 can be rewritten, when substituting Equation 2.1 into it, as Equation 2.3.

$$[HG_2] = K_1K_2[H][G]^2 \tag{2.3}$$

[vi] In more complicated interrelated equilibria (examples will be shown later), equilibria between species can sometimes provide a "cyclic" system. In that case, one of the equilibrium constants can be written as a function of the others involved in this cycle, which makes the equation of that equilibrium constant "redundant", that is, it contains no new information, and will not play a role in the solving of the total set of equations. Such redundancies can best be identified and removed before running a speciation simulation.

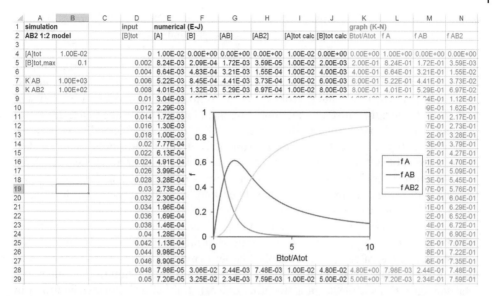

Figure 2.2 Screenshot of a spreadsheet simulation for a 1:2 model, with general input parameters in columns A and B, a varying mass balance in column D, the numerical species concentrations in columns E–J, and the output values for graphical presentation in columns K–N, resulting in the graph presented in the center. The equations for [A] and [B] are derived as explained in Figure 2.1 and the accompanying text.

The resulting equations are then substituted into the mass balances, which results in a set of equations (here: two; see Figure 2.1c) which have as unknowns only the concentrations of the free species involved in the system, here: [G] and [H]. This set of equations, corresponding to the number of mass balances in the system, is the minimal set needed to run the simulation. To allow their implementation in a spreadsheet, these are rewritten as "[G] = [G]$_{tot}$/..." from the mass balance of G and likewise as "[H] = [H]$_{tot}$/..." from the mass balance of H, see Figure 2.1c. Since every term of a mass balance has at least one molecule of that species included, rewriting the mass balance in this way can always be performed, also for more complicated systems. More complex models only give lengthier denominators in these expressions.

For implementing the set equations in a spreadsheet, a layout needs to be set. In a typical titration experiment, the mass balances are varied in a systematic manner. Therefore, the set of equations (Figure 2.1c) can best be implemented in neighboring columns in a spreadsheet, directly adjacent to columns of the experimental mass balances (Figure 2.2). As can be noted from the set of equations in Figure 2.1c, the equation for [H] refers to the one for [G], and vice versa. This creates so-called "circular references" which can be solved by a spreadsheet program in an iterative fashion.[vii] In subsequent columns, the concentrations of the complexes, here [HG] and [HG$_2$], are calculated using the equilibrium constant equations described above. As an internal control of the validity of the equations and the proper convergence of the iteration process, calculated mass

---

[vii] In Excel, "circular references" provide an explicit error of the same name. This can be circumvented by (i) changing the calculation to "manual" and (ii) checking the iteration box. Commonly, iteration numbers of 100–1000 are sufficient, and non-linear scaling should be applied.

balances can be added as subsequent columns according to the original mass balance equations while referring in the spreadsheet to the calculated concentrations of the free and complexed species.

Alternatively, algorithms which provide an iteration routine, for example the Simplex algorithm, can be implemented explicitly [14]. In this case, initial values for [G] and [H] are chosen (e.g., the total values), from which all complex concentrations are calculated using the equilibrium constant equations. The mass balances of G and H are used subsequently to create a control criterion by which [G] and [H] are changed in an iterative fashion to match the calculated and experimental (or simulated) mass balances. The latter, more robust, strategy has been chosen in cases when convergence problems occurred with the former method, usually when more complicated models were implemented with larger numbers of complexes or high stoichiometries [15], leading to higher-order concentration dependencies.

The final step is to run the simulation. As the model requires the stability constants to be known, their values need to be introduced in the spreadsheet. Commanding the spreadsheet to (re)calculate all unknowns is performed,[viii] and graphical representations of the speciation can be prepared at will, for example as a function of one of the total concentrations while keeping the other one constant (as is performed routinely in titrations) or as a function of the fraction of total host or guest while keeping the sum of total host and guest together constant (as is used in Job plots). Moreover, related parameters or variables can be calculated, such as concentration ratios (e.g., fractions of free and bound guest) and the degree of occupancy (fraction of all sites that occur in the bound form, a parameter, Y, used extensively in cooperative systems and their treatment, for example in the Hill equation) [16,17]. An additional advantage of the spreadsheet approach is that a parameter can be changed at will, and a simple recalculation makes the result of that change visible in all aspects of the spreadsheet (species concentrations, figures) in one go.

2.2.2 Setting Up and Assessing Titrations

Titrations are the most common experimental methodology to assess binding constants and to study complex equilibria in general [18]. A titration refers to a process by which one component is added to a solution containing another component (which can be dissolved in the solution or immobilized onto a surface), so that their interaction can be studied while gradually varying their ratio. A large range of analytical techniques can be directly applied or adapted to study the outcome of a titration. Examples of the most commonly applied techniques are: calorimetry (or its direct derivative suited for titrations, isothermal titration calorimetry, ITC), potentiometry, and UV/Vis absorbance, fluorescence, and NMR spectroscopy for solution systems, and surface plasmon resonance (SPR) and quartz crystal microbalance (QCM) for surface systems.

As explained above, the main purpose of the use of the analytical technique is to assess species concentrations and/or ratios thereof, therefore by definition in a quantitative manner. Some techniques, in particular the spectroscopic ones such as UV/Vis, fluorescence and NMR, make use of molecular signatures of molecules directly employed in the equilibria of the system of choice. Other techniques study derived properties resulting from the equilibration process, for example heat effects using

---

[viii] In Excel, this is achieved by the command Calculate Now or F9.

calorimetry, pH effects using potentiometry, and variations in the refractive index near a surface using SPR. A full description of techniques used for titrations as well as a detailed account of how to properly conduct titrations experimentally, fall outside the scope of this chapter.

One aspect that deserves attention is the choice of the correct concentrations of the components that are added to each other in a titration. When reliable binding constants are to be obtained, the workable concentration range is rather narrow. Two criteria are in that respect important: (i) concentrations of free species should traverse the range of 20–80% of the total species concentration during the titration; and (ii) there needs to be a range of data points for which the concentrations of all three species involved in an equilibrium, for example, HG formed from H and G, are of similar magnitude (less than an order of magnitude different). Both aspects can be visualized using the simulation tool described in Section 2.2.1. The first puts restrictions on the lower concentration limit, whereas the second puts restrictions on both the upper and lower limits, and basically can be interpreted as that the concentration needs to be "just right". What is "just right"? A rough and easy estimate can be obtained as follows: the equilibrium equation for a 1:1 complex can be written as Equation 2.1. Noting that all three species concentration need to be of the same order of magnitude can be approximated by assuming all to be equal to x. Then the equation is reduced to $K_1 = x/x \cdot x = 1/x$, which leads to $x = 1/K_1$.[ix] Common sense then prescribes that one can best work at roughly 10 times higher starting concentrations, so that, when complexation starts to reduce the concentration of free species while the bound fraction goes up, all go through the workable concentration range simultaneously, leading to the largest number of "good" data points (i.e., for which the second criterion holds; see the example given in Figure 2.3).[x]

A caveat lies in the estimate for K_1. One normally sets out to do a titration to determine K (here K_1), so its value is by definition unknown. Yet, one must have a good estimate of the value in order to do a proper titration in the right concentration range. In practice, this is usually solved by looking for a literature example that is structurally related as close as possible and using the equilibrium constant for that case as an estimate for the one that needs to be determined. When no comparison exists, one has to start at a practically achievable concentration, preferably on the high side, and dilute the stock solutions if the titration turns out to be performed at a too high concentration. The latter situation is visualized by a linear increase of the complex concentration during the titration which abruptly turns into a plateau when the complex stoichiometry is met (Figure 2.3a). It can easily be reasoned that such a titration violates the second criterion for a good titration: Assume that only G is present in the cell at the start, and that H is added to the cell leading to the formation of HG.[xi] Then as long as $[H]_{tot}/[G]_{tot} < 1$, so before reaching the equivalence point, only [G] and [HG] have comparable values,

---

[ix] Biochemists commonly use (apparent) dissociation constants, which already have the units of concentration, and provide the concentration at which a complex is formed for 50%. These constants are (often) equal to the inverse of the here used association constants.

[x] Simulations can make visible that fits obtained at different values for the targeted affinity constant lead to large deviations only in these "good" data points, but hardly lead to any changes in the other data points. Therefore, all data points are equal, but some are more equal than others.

[xi] There is no general preference for which component should be placed in the cell or in the buret. Practical issues usually provide the choice, for example, molecules that have a tendency to aggregate can usually best be placed in the cell, as the starting concentration in the cell is lower than in the buret (the latter is usually 10× higher).

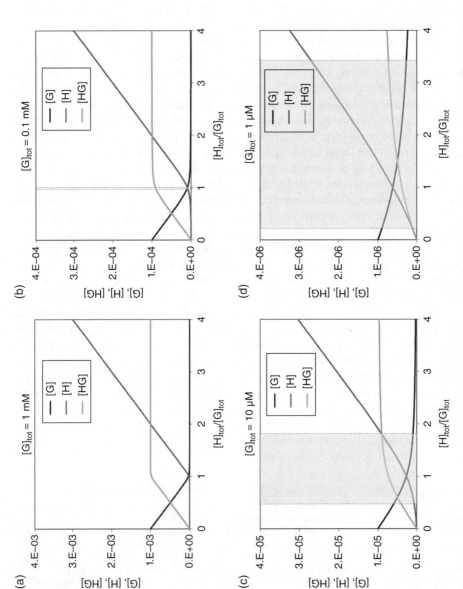

Figure 2.3 Simulations of speciations for a 1:1 complex ($K_1 = 10^6 \, M^{-1}$) formed in titrations performed at varying total guest concentrations. Gray boxes indicate the concentration range for which the concentrations of all species involved in the equilibrium are within one order of magnitude.

but the concentration of free H is much lower (can be regarded as "absent") because all new H that is added gets complexed to form HG. At the point of equivalence there is only HG with negligible concentrations of G and H, and thereafter the cell contains basically a mixture of H and HG but then free G is absent because it has all been complexed by H. The only way to get all three concentrations of H, G, and HG in the same range is to lower the starting concentrations, in particular that of $[G]_{tot}$ in the cell. This is visualized graphically in Figure 2.3. In all panels, $K_1 = 10^6 M^{-1}$. In Figure 2.3a, $[G]_{tot} = 1\,mM$, which is $1000/K_1$. By choosing to start the first titration at $1\,mM$, one implicitly assumes a binding constant of approximately $10^3 M^{-1}$, which is clearly an underestimation. In Figure 2.3b–d, the total concentration of G is lowered in steps of a factor 10, reaching $[G]_{tot} = 1\,\mu M = 1/K_1$ in Figure 2.3d. The gray boxes in Figure 2.3b–d indicate the concentration range in which the second criterion (concentrations of all species involved in the equilibrium within one order of magnitude) is fulfilled. It starts from a single data point (at equivalence) in Figure 2.3b when $[G]_{tot} = 100/K_1$, while achieving a nice broad working range in Figure 2.3c and d. When taking into account also the first criterion (concentrations of free species should traverse the range of 20–80% of the total species concentration), Figure 2.3d does not fulfill this for H. This becomes only worse when even lower concentrations (e.g. of $0.1\,\mu M$) are chosen. When combining the two criteria, it follows: $4/K_1 < [G]_{tot} < 100/K_1$, in which $[G]_{tot} = 10/K_1$ (Figure 2.3c) is the most ideal.

The data generated by an analytical technique used in a titration experiment needs an equation that describes the relationship between the measured quantity and one or more of the unknowns, that is, the species concentrations, of the model. For spectroscopic data, there is normally an easily accessible equation that provides this relationship, such as the Lambert–Beer law for UV/Vis absorbance A, which treats the measured absorbance as a sum of absorbances of a total number n of individual species k, each of which is related to the concentration of that species (c_k), so that the total absorbance A_{tot} is given by Equation 2.4.

$$A_{tot} = \sum_{k=1}^{n} A_k = \sum_{k=1}^{n} \varepsilon_k c_k l \tag{2.4}$$

Here, ε_k and l are the specific molar absorptivity and the optical path length, respectively. Similar, also often linear, relationships can be written for fluorescence intensities and NMR chemical shifts (or integrals in case of slow exchange). Although not measuring species individually by molecular signatures, the adsorption monitoring techniques SPR and QCM can often be easily (and linearly) related to (adsorbed) amounts of a species as well.

A huge exception to the techniques described above and the implementation of their measured quantities, is ITC. ITC works so differently, yet its importance is so paramount for the field of thermodynamics of complex equilibria, that some specific attention is warranted here. The importance of ITC becomes obvious in a glance: it is the *only* technique that provides direct access to complexation enthalpies. A properly conducted ITC titration can therefore yield the complete thermodynamic picture in a single titration experiment: not only the equilibrium constant K is obtained by the systematic variation of species concentrations, but the enthalpies are obtained simultaneously from fitting the experimental heat data. This is visualized in Figure 2.4, which shows

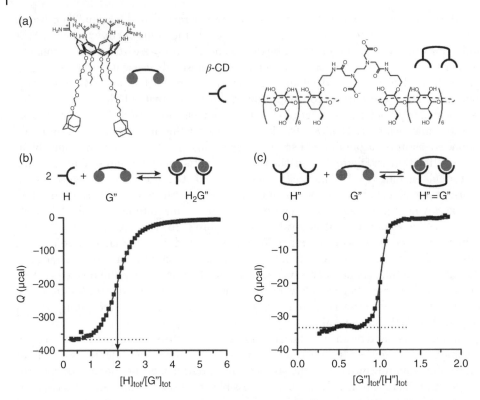

Figure 2.4 Examples of calorimetric titrations for complexes between mono- and divalent β-CD host compounds and a ditopic adamantyl guest [19]: (a) compounds used in this study; (b) formation of a 2:1 complex and its ITC titration data; and (c) formation of a divalent 1:1 complex (H″=G″ where the "=" indicates the divalent nature of the interaction) and its ITC titration data. The arrows at the inflection points indicate the binding stoichiometry, the dashed lines at the y-axes indicate the heat plateau, from which the enthalpy of binding is readily achieved, and the curvature and steepness around the inflection point qualitatively indicates the affinity, which can be calculated by a data fitting procedure.

two enthalpograms (Figure 2.4b and c) for the 2:1 complex and a (divalent) 1:1 complex of a ditopic guest with a mono- and a divalent β-CD derivative, respectively (Figure 2.4a) [19]. In these graphs, the binding stoichiometry can be read directly from the position of the inflection point, when the host–guest stoichiometry is plotted on the x-axis. The binding enthalpy is read from the abscissa of the integrated heat flows at the y-axis: in the graphs shown here, the plateau heat values only need to be divided by the number of moles of titrant (here, H or G″, respectively) administered during each addition. The binding affinity is obtained (after fitting) from the shape (or steepness/slope) of the curve around the inflection point. This makes ITC graphs very useful, and instructive at the same time.

The main differences between ITC and other techniques arise from the nature of the data (evolved heat), the way titrations are normally set up in commercial titration calorimeters (in particular, working under overflow conditions), and their consequences for the numerical treatment of the data. An important aspect that sometimes remains underappreciated is that the evolved heat is a *transient* parameter: the heat starts evolving (or is being consumed) right at the moment that the titrant is added to the measurement cell, and must therefore be measured instantly and integrated over the time over which the heat effect lasts. In practice the actual measurement is performed by a heater coil which keeps the temperature in the cell constant by providing a specific heat flux to the cell. When an aliquot of titrant is added from the buret, the evolution (or consumption) of heat induced by the exothermic (or endothermic) complexation leads to a temporary decrease (or increase) of the heat flux delivered by the heater coil. The change in heat flux delivered by the heater coil provides the raw signal, that is, the measured heat flow versus time. Integration of this difference in heat flow over the time while the effect lasts compared with the baseline gives the heat Q produced (or consumed) during the addition of an aliquot of titrant (as plotted on the y-axis in Figure 2.4b and c).

In contrast, all other techniques mentioned above provide a view at the equilibrium state after an aliquot of titrant has been added and equilibrium has been re-established. Therefore, the timing of the measurement is unimportant (as long as it can be safely assumed that the equilibration process has been completed). As a consequence of the differences discussed above, heat effects measured by calorimetry are related to con-centration *changes* (Δc_k) occurring between before and after addition of an aliquot and thus constitute *increments* of the total heat evolved between the start of the titration and the respective aliquot. In contrast, other techniques give access to actual species concentrations corresponding to the *total* amount of titrant added so far. Visually, this difference can be seen in the fitted graphs. The integrated heat values plotted versus the molar ratio (as seen in the graphs in Figure 2.4b and c) resembles the *derivative* of a fitted graph obtained by other techniques (as seen by the curves for [HG] in Figure 2.3). When one would create a graph from the *cumulative* heat effects reported in Figure 2.4b or c, a graph like Figure 2.3b (for [HG]) would be obtained.

Additionally, commercial titration calorimeters are run under overflow to provide similar accuracies in heat determinations over all data points. That means that upon addition of an aliquot of titrant an equal volume of cell contents is pushed out of the measurement cell and is not contributing anymore to the equilibrium nor to the meas-ured heat effect. This causes *exponential* changes of the concentrations of the individual and complex species in the cell (at constant volume), instead of the dilution dependencies at increasing aliquots of titration and likewise increasing total volume observed using other techniques. Both aspects, the observation of incremental heat changes and over-flow conditions, have a strong influence on the numerical implementation of the equa-tions in the spreadsheet, but the exact way this is done is outside the scope of this chapter.

The equation that provides the relationship between the concentration changes of the species upon addition of an aliquot and the complexation enthalpies is simply given by Equation 2.5.

$$Q = \sum_{k=1}^{n} \Delta H^{\circ}_k \, \Delta c_k V \tag{2.5}$$

Here, Q is the heat evolved after addition of an aliquot of titrant, ΔH°_k is the compl-exation enthalpy of complex k, Δc_k is the concentration change of species k, and V is the cell volume of the titration equipment. Data fitting of calorimetric data can therefore be achieved by converting simulated species concentrations into calculated Q values, after which K is optimized in a least-squares minimization routine, as described in more detail below, by variation of K and ΔH°_k (often, also a dilution heat, observed as the residual heat effects at high molar ratios in Figure 2.4b and c, is fitted simultaneously, although this can also be determined independently in a separate titration). When the fitted K and enthalpy have been obtained, the free energy ΔG^{o}_{k} and the entropy ΔS^{o}_{k} of the equilibria can be deduced as well.

Other methods can also provide complexation enthalpies, in particular by using the Van 't Hoff method [20]. However, the practical performance of this method is cumbersome (multiple titrations need to be performed at varying temperatures), the interpretation of the data is often difficult (e.g., in case of temperature-dependent enthalpies, i.e., non-zero heat capacities), and relative errors of one or more of the thermodynamic parameters remain large (because the width of the applicable temperature range at the Kelvin scale is often limited by practical constraints, e.g., the liquid phase range of the solvent and solubilities of the components). Therefore, ITC measurements are highly preferred when the molecular system under investigation allows this.

2.2.3 Using Spreadsheet Simulations to Fit Experimental Data to a Model

When a model needs to be fitted to an experimental data series, the model itself is assumed known, and only the model parameters (typically the binding constants) are varied until an optimal fit is obtained. Technically, this means that (i) the simulated species concentrations need to be fed into an equation that calculates ("predicts") the experimental quantity using the equation that corresponds with the technique of choice (see previous paragraph); (ii) the difference ("error") between measured and calculated quantity is determined; and (iii) the sum of the squared errors is minimized, using the so-called least-squares minimization, by variation of the model parameters. In a spreadsheet approach, steps (i) and (ii) are simple calculations performed in separate columns in the spreadsheet, whereas step (iii) requires an iteration process.[xii] As the simulation itself needs the model parameters (the equilibrium constants governing all equilibria) to be "known", one needs to start with an estimate of all binding constants including those that are practically (but not numerically!) unknown, usually the one(s) that initiated the performance of the titration in the first place. As explained above, this is a paradox which is solved by choosing the (here assumed 1:1) equilibrium constant to be the inverse of the starting concentration. If the titration has yielded a "good curve" (Figure 2.3), this implies that the concentration x has been chosen correctly and that $K_1 = 1/x$ is a good starting assumption.

An NMR titration example is shown in Figure 2.5. In this titration, a solution of the host γ-CD (H) was added to a solution of a calix[4]arene guest G with two appending phenylpropyl chains both of which interact simultaneously with the γ-CD cavity, yielding a simple 1:1 complex HG [21]. A proton signal of the guest was used as the spectroscopic marker, and its shift δ was monitored at increasing concentrations of the host. In this example, the equation that provides calculated values of the experimental quantity δ is Equation 2.6.

$$\delta_{calc} = f_G \cdot \delta_G + f_{HG} \cdot \delta_{HG} \tag{2.6}$$

[xii] In Excel, the Solver add-in allows this functionality, and model parameters that are optimized in the least-squares minimization routine can be chosen almost at will.

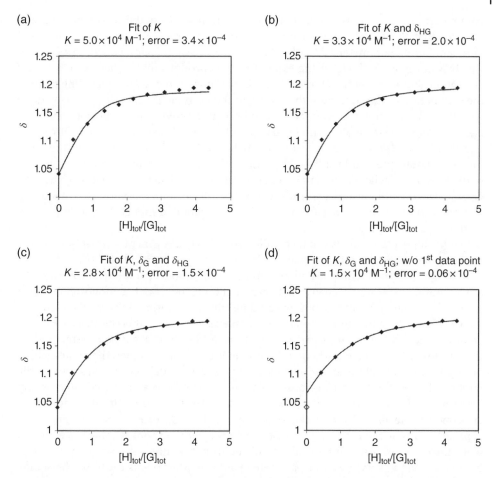

Figure 2.5 Fits (solid lines) of a 1:1 model to ^{1}H NMR titration data (markers) of the binding of a calix[4]arene guest by γ-CD as a host [21], when fixing (a) δ_G (1.041 ppm) and δ_{HG} (1.194 ppm); (b) δ_G (1.041 ppm); (c) no parameters; (d) no parameters, while ignoring the 1st data point in the summed error.

Here, f_G (= $[G]/[G]_{tot}$) and f_{HG} (= $[HG]/[G]_{tot}$) are the fractions of guest in the free and bound state, respectively, while δ_G and δ_{HG} are the chemical shifts of the free and bound guest, respectively. Obviously, the boundary chemical shifts can be assessed experimentally, and in particular δ_G (here: 1.041 ppm) is usually included in the titration (obtained at zero H). However, exact measurement of δ_{HG} is often difficult since HG tends to dissociate to some extent when dissolved in the solvent, and the value is commonly approximated by measuring the chemical shift at a large excess of H (here, the last data point provides a value of 1.194 ppm, which was initially fixed for δ_{HG}). In the least-squares minimization routine, one can select more parameters to be optimized than just the binding constant K. Also δ_{HG} can be added as a parameter, and even δ_G, so that the data point at zero host gets an equal statistical weighing as the other data points.

Figure 2.5 shows the consecutive fits when using only K (Figure 2.5a), when using K and δ_{HG} (Figure 2.5b), and when using K, δ_G and δ_{HG} (Figure 2.5c) as the optimized parameters. It can be appreciated by visual inspection and from the decreasing summed fit error that the fit quality improves, but this is in large part due to the increased number of fitted parameters. Also the changes in K are significant, but within a factor 2, which is normally well within the experimental reproducibility. When (valid) doubts arise about some data points, these can be simply "unselected" by not including them in the summation of the squares of the errors. In the fit in Figure 2.5d this has been done with the first data point (at zero H). An additional improvement of the apparent fit is clear, although the mismatch with the data point at zero H is evident as well. Obviously, one should have good reasons to exclude data points from a fit (and even then, the markers should preferably still be shown); in the current example, the guest G was partly forming micelles at the starting concentration, which led to a deviating value [21]. Overall, the open nature of a spreadsheet offers full flexibility in parameter handling, and one can cross the border from data *fitting* to data and model *interpretation* in a gradual and natural manner.

More complicated models can be handled equally well. A general description of how to set up such models and the equations involved has been provided above. Generally (but not always), more complicated models have more model parameters. For example, a 1:2 complex, formed stepwise between a ditopic host and a monovalent guest (Figure 2.1a), employs not one but two binding constants, for both the 1:1 (HG) and 1:2 (HG$_2$) complexes, and, quite often, these complexes have also different physico-chemical properties (different enthalpy of formation, NMR shift, absorbance, etc.). When there is no evidence at hand whether the system is cooperative or not, the model should be approached initially as an *independent* binding sites model. As explained above, the physico-chemical independence of the binding sites leads to a numerical relationship between the K values of the complexes ($K_1 = 4K_2$). In that case, relevant when using calorimetry, the enthalpies of the sites are equal ($\Delta H^\circ_1 = \Delta H^\circ_2$). So, the model is approached in the following way: the data are fitted with only one K value (K_i, K_1 or K_2) while the other ones are numerically calculated from the fitted one. Similarly, only one binding enthalpy is used as a fit parameter (and the other one is set to be the same). Only when the fit quality is deemed insufficient, one can consider to "relax" the model by allowing the two binding constants and the two binding enthalpies to be fitted (numerically) independently. It only makes sense to fit the model with two or four free parameters, but not with three. Although it is mathematically possible to fit the data with three parameters (e.g., K_1, K_2, and ΔH°_1, while setting the second enthalpy to be the same as the first), it does not make sense physically: a system is either cooperative (with all four parameters freely varied in the fit routine) or non-cooperative (i.e., independent, only two free parameters, and the other two restricted by $K_1 = 4K_2$ and $\Delta H^\circ_1 = \Delta H^\circ_2$).

An example is shown in Figure 2.6. It gives the calorimetric enthalpograms and their fits for the binding of two 4-*t*-butylbenzoate ligands (G) to a β-CD dimer (L), in which the two cavities have been linked by an EDTA spacer (Figure 2.6a) [22]. In the first case, the EDTA is uncomplexed (H = L), while in the second a lanthanide metal ion (Eu^{3+}) has been coordinated to the EDTA unit (H = Eu^{3+}L). It was an open question whether the carboxylate group of the guest G would somehow interact with the EDTA moiety (then probably repulsive as both are negatively charged) or, only in the second case, with the lanthanide ion (as the lanthanide still has free coordination sites after binding to EDTA). The binding data for the guest interacting with the CD dimer with the uncoordinated EDTA unit (H = L) could be perfectly fitted to an independent

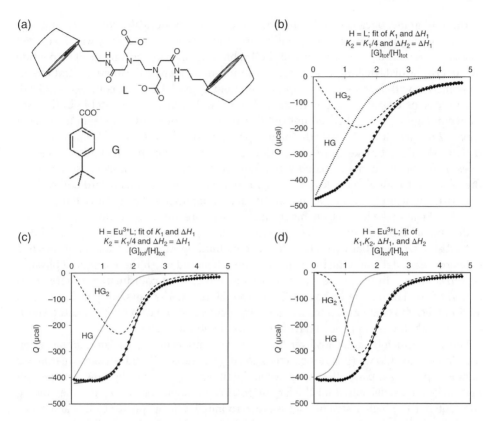

Figure 2.6 Cooperativity effects for the binding of a monovalent guest to a divalent host [22]: (a) host and guest compounds; (b–d) fits (solid lines) to a 1:2 model of ITC data (markers) for the binding of G to L (b: independent binding sites model) and for the binding of G to $Eu^{3+}L$ (c: independent binding sites model; d: cooperative model). Dotted and dashed lines indicate contributions to the overall heat effect from HG and HG_2, respectively, while the fitted overall heat effect (solid lines) is the sum of these effects plus the dilution heat.

binding sites model (Figure 2.6b) and the binding constant and enthalpy compared well to those of the binding of one guest molecule to native β-CD, indicating that interactions between the carboxylate group of the guest and the EDTA unit were not detected. In contrast, the fit of the data for the case with the lanthanide ion bound ($H = Eu^{3+}L$) to the independent binding sites model was poorer, in particular in the first part ($[G]_{tot}/[H]_{tot}$ between 0 and 1) of the graph (Figure 2.6c). When the model was relaxed to the cooperative situation (both binding constants and both binding enthalpies varied freely), a significantly better fit was obtained (Figure 2.6d). Interestingly, a small negative cooperativity was found ($K_1 > 4K_2$), also witnessed by a clearer separation of the heat effects of formation of HG and HG_2. This affinity difference was paired to a lower (less exothermic) enthalpy of binding for the first guest ($\Delta H°_1 > \Delta H°_2$). This was attributed to additive effects of an exothermic binding of the aromatic part of the guest to the CD cavity and a (smaller) endothermic binding of the carboxylate moiety of the guest to the lanthanide ion. Apparently, this hetero-ditopic mode of binding occurred only for the first guest, but not for the second; the latter only bound to the second CD cavity and not to the lanthanide, because the lanthanide ion was already sterically congested after the binding of the first guest. More recently, a more extensive methodology was developed to assess the possible cooperativity in the 1:2 binding of (two equivalents of) a peptide by cucurbit[8]uril, making use of concentration-dependent calorimetric titrations in combination with NMR spectroscopic assignment of the 1:1 and 1:2 complexes [23,24].

The reasons to start fitting with the independent binding sites model in both cases shown in Figure 2.6b and c, and not right away with a cooperative model in the case of Figure 2.6c, are twofold. First, the philosophical principle of Ockham's razor dictates that one should not make more assumptions or introduce more parameters than needed to explain a phenomenon. Secondly, the more free parameters a model has, the better the fit becomes but also the less meaningful. The best quote in this regard comes from the mathematician John von Neumann: "With four parameters I can fit an elephant, and with five I can make him wiggle his trunk".[xiii] Therefore, one should always start fitting with a model that has the smallest number of free parameters. In the 1:2 model discussed above, one initially tries the independent binding sites model, and only goes to a cooperative model when the fit quality is insufficient and when there are good physical reasons to do so. In the particular example, incomplete first coordination sphere occupation of the lanthanide by the EDTA spacer hinted at the possibility of binding an additional carboxylate group from a guest molecule, and the data confirmed the endothermic enthalpy effect expected from such a contribution [22].

The same principles hold, however, for many other model questions. For example, when the binding stoichiometry is not exactly known, one starts with the smallest reasonable number, and only expands the model when the fit is poor. Technically, however, one can best make the model suited for the largest possible number, and simply put the binding constants for the higher stoichiometries to zero initially. That way no technical amendments need to be made to the model when a higher stoichiometry is needed, but simply the K values can be co-optimized in the fit routine. As an example, the files used to fit the data presented in Figure 2.6, can also be used to fit data to a 1:1 model, simply by fixing $K_2 = 0$.

Also in models that are well known to be cooperative, placing restrictions on the number of free parameters can be helpful. For example, in the case of hemoglobin (Hb), four oxygen molecules (O_2) are known to bind in a positively cooperative fashion to the protein complex. If one attempts to fit oxygen saturation data to a 1:4 model, one can keep all four binding constant parameters (K_n for $Hb(O_2)_n$ with $n = 1$–4) free, but one will find out soon that the data can be fitted to many combinations of K_1 to K_4 with similar correlation coefficients. Instead, it can be very instructive to compare the fits of models in which relationships between K_1 and K_4 are fixed in accordance with a physical model. For example, if one assumes that the binding of the first oxygen molecule leads to the payment of an energy penalty so that the whole protein complex becomes pre-organized for the uptake of subsequent oxygen molecules, one can assume the latter oxygen molecules to bind independently. This fixes relationships between K_3 and K_2 and between K_4 and K_2 according to statistical pre-factors, so that the model has only two free fitting parameters (K_1 and K_2). Alternatively, one could assume that the binding energy increases gradually, for example, linearly. In that case, now ignoring statistical

---

xiii Attributed to von Neumann by Enrico Fermi, as quoted by Freeman Dyson in "A meeting with Enrico Fermi", *Nature* **2004**, *427*, 297.

pre-factors for clarity, $K_{k+1} = aK_k$, in other words, all stability constants increase with a factor a compared with the constant of the previous step. In that case, the model also has two free fit parameters, K_1 and a. When converting a speciation into an occupancy graph ($Y = \Sigma n[HG_n]/4[H]_{tot}$), one can start to appreciate the dependence between a and the Hill coefficient in this model, or between the K_2/K_1 ratio and the Hill coefficient in the former case.

2.3 Models for Multivalent Systems

Multivalent systems pose additional challenges to the setup of models, their numerical implementation, and their interpretation. The main characteristic of multivalent systems compared with other systems is the formation of one or more species with one or more intramolecular interaction pairs. The equilibrium constants describing the intramolecular steps have different units than those describing intermolecular interactions, which prohibits their direct comparison. The often observed higher apparent overall binding constants associated with multivalent systems have therefore sometimes been regarded as signs of positive cooperativity. Here, we strictly adhere to a definition of cooperativity that is limited to changes in affinity induced by the occupation of neighboring sites, while we use multivalency to describe systems with combinations of inter- and intramolecular bonds.

Different formalisms exist to deal with these differences [8,12,25]. Here, we choose the effective molarity/effective concentration formalism [26,27,28], described in more detail below, because it provides a molecular picture for understanding and estimating effective concentrations in a quantitative manner and because it allows the use of a rigorous procedure to evaluate whether a multivalent system is also cooperative or not. The effective molarity (EM) is a parameter well known from polymer chemistry to evaluate probabilities for ring-closing reactions (e.g., ring-closing metathesis) [29,30,31]. In that context, EM is defined as the ratio between the rate constants of the intra- versus intermolecular reactions. Analogously, EM has been used in supramolecular chemistry to describe the ratio between intra- and intermolecular equilibria [32,33]. In both systems, EM is primarily an empirical parameter which allows easy comparison between various systems. Being a concentration term with units of molarity (M), physically unrealistic values (of up 10^6 M) have been reported for covalent systems, while much lower and physically attainable values have been found for equilibria. Here we choose to treat EM and the very similar effective concentration, C_{eff} [34], separately [19] in order to provide a methodology for the assessment of cooperativity in multivalent systems.

2.3.1 The Simplest Multivalent System: A 1:1 Complex with Two Interaction Sites

Here, we provide a stepwise development of a model for the simplest and basic multivalent complex: the divalent interaction between a single divalent host and a single divalent guest. Figure 2.7 shows an example in which a β-CD ditopic host (H") interacts with a ditopic bis-adamantyl (Ad) guest (G") (Figure 2.7a) [19]. To appreciate the differences between regular and multivalent complexes, the binding behavior of the divalent 1:1 complex (H" = G", Figure 2.7d) is compared with the parent intrinsic

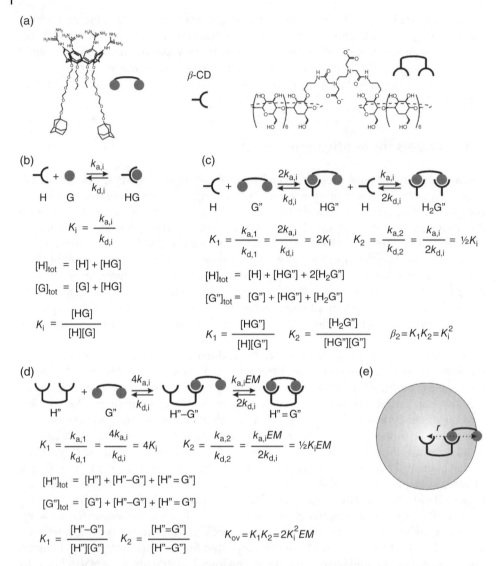

Figure 2.7 The archetypical multivalent complex: formation of a divalent 1:1 complex between a ditopic host and a ditopic guest: (a) host and guest molecules used in this study [19]; (b–d) equilibria, equilibrium constants and equations (for mass balances and equilibrium constants) for the parent monovalent intrinsic interaction (b), the non-cooperative 2:1 system (c), and the divalent 1:1 system (d). A cartoon (e) shows a conceptual picture of the effective concentration (C_{eff}): the two unbound moieties in the intermediate H"-G" complex have a probability of finding each other to form an intramolecular bond determined by the probing volume defined as a sphere with radius r.

(monovalent) 1:1 complex (HG, Figure 2.7b) and a 2:1 complex between the divalent guest and monovalent host (HG" and H$_2$G", Figure 2.7c). ITC data for the intrinsic system and the 2:1 complex were straightforward: the intrinsic binding affinities for the interaction between an Ad guest moiety and a β-CD cavity were highly comparable (approx. $5 \times 10^4 \, M^{-1}$), as well as their enthalpies of binding (approx. −7 kcal/mol) [19]. This confirmed that the 2:1 system (Figure 2.7c) behaves fully as an independent (non-cooperative) binding sites system. Obviously all stepwise binding constants have the same units (M^{-1}) and correspond to intermolecular interactions, while the product of K_1 and K_2 in the 2:1 system provides the overall binding constant β_2 (with units M^{-2}).

In contrast, the divalent 1:1 H"=G" complex (Figure 2.7d) provided different ITC data, an affinity of $1.2 \times 10^7 M^{-1}$ and an enthalpy of binding of $-14.8 \, kcal/mol$, while the stoichiometry was clearly 1:1 [19]. Obviously these data correspond to the *overall* formation of the divalent complex, that is, the formation of H"=G" from H" and G". The enthalpy was, within experimental error, the double of the enthalpy of the intrinsic interaction, which clearly indicated the divalent nature of the interaction within the complex. Although no solid proof, this nice quantitative match can be regarded as a sign of non-cooperativity in this system as well. The main question, therefore, that remained was: how does the observed overall affinity K_{ov} (in M^{-1}) depend on the intrinsic affinity constant K_i?

The key to understand this relationship is to dissect the formation process of H"=G" into two steps (Figure 2.7d). The first step, the formation of H"-G" from H" and G", is a normal intermolecular step, associated in this case with a pre-factor 4, resulting from the product of two ditopic interaction opportunities for both H" and G". The second step, that is, the formation of H"=G" from H"-G", however, is an *intra*molecular binding event. The equilibrium constant equation for this step $(K_2 = [H"=G"]/[H"-G"])$ must be unitless, and can therefore not be compared directly with the intrinsic affinity constant. The common way to solve this is to multiply the intrinsic affinity constant K_i (in M^{-1}) with a concentration term that, by default, must have the units of concentration (M). As explained above, this term is called the effective molarity (*EM*). Since the dissociation step (from H"=G"to H"-G") is normal and comparable with the dissociation steps of all other equilibria presented in Figure 2.7b–d, *EM* ends up in the association step. As a result, the overall affinity constant, K_{ov}, can now explicitly be related to K_i by Equation 2.7.

$$K_{ov} = K_1 K_2 = 2 K_i^2 EM \tag{2.7}$$

Yet, at the same time, one must note that the problem has only been shifted: the mathematical formalism is now correct, but a physico-chemical interpretation, in particular of *EM*, is not evident. In short, when presented this way, *EM* is an "empirical quantity", or simply a fudge factor.

How can we grasp the physical meaning of *EM*? In order to do this, another, but very similar, quantity is introduced, the effective concentration (C_{eff}) [19,34]. Formally, the effective concentration represents the probability for one unbound moiety in H"-G" to find the other. A good approximation of this probability is obtained when viewing the cartoon in Figure 2.7e: in the intermediate H"-G", the unbound host site, which is looking for the free guest site, is connected to same guest site by a spacer of length r, which describes a sphere around the host site. This sphere, called the probing volume, contains one complementary site, hence, the concentration of that site is one (site/molecule) divided by the probing volume, which is a concentration term.

In the specific example shown in Figure 2.7, the value of *EM* was calculated from Equation 2.7, which led to a value of approximately 3 mM [19]. Using the approximation for C_{eff} explained above, a (minimal) value of 2 mM was estimated for C_{eff}. Because of the excellent match between *EM* and C_{eff}, it was concluded that the divalent complex was formed from two, independent host–guest interactions, each fully comparable with the intrinsic interaction in the intermolecular parent case.

Thus, the effective concentration, C_{eff}, does not only provide a rationale for the "fudge factor" (EM) introduced before, it also provides a way to estimate whether the consecutive binding steps in the formation of a multivalent complex behave in an independent or cooperative manner. The assessment of cooperativity in multivalent systems is otherwise notoriously difficult. For multivalent systems, experimental data usually provide access only to the final product, in other words, the concentration(s) of the intermediate(s) is/are too low to be observed directly. In the current example, this can be estimated from the binding constant of the second step, $K_2 = \frac{1}{2}K_i EM$ (Figure 2.7d), which is, using the values given above, approximately 50. This means that in the system at hand 98% of the complexes exists in the form of the divalent end-product H"=G", while only 2% occurs as the intermediate H"-G", even though the binding has been shown to be non-cooperative. This is in sharp contrast to normal intermolecular systems, in which case the absence of appreciable amounts of intermediate(s) is taken as an explicit sign of (positive) cooperativity. More importantly, since the concentration ratio of a product and an intermediate that are coupled by an intramolecular interaction is unitless (as is the equation for K_2 as described above), the ratio of their concentrations cannot be varied by a titration. Moreover, analytical methods to assess cooperativity in normal intermolecular systems, such as the Hill equation and the Scatchard plot, have been shown to fail in multivalent systems [35]. These reasons combined have led to several erroneous conclusions in the literature about the cooperativity of the systems under study.

The EM–C_{eff} formalism introduced here provides a more general method to assess cooperativity in multivalent systems. As explained above, a system should initially always be fitted to an independent binding sites model. The set of equations then provides a relationship between the overall affinity constant and the intrinsic affinity constant, also involving EM. Independent experimental assessment of K_i thus allows determination of EM. On the other hand, simple molecular (e.g., ball-and-stick) models usually suffice to provide a reasonable estimate of the probing length r, and thus of the order of magnitude of C_{eff}. Two criteria can thus have a say about the applicability of the assumption of independent binding sites: (i) the overall binding enthalpy of the multivalent complex is the product of the valency and the enthalpy of the monovalent intrinsic interaction; and (ii) the effective molarity, EM, and the effective concentration, C_{eff}, are of the same order of magnitude. In case both criteria are met, and keeping in mind the principles of Ockham's razor and von Neumann's elephant, there is no reason to assume the system to be cooperative, and thus the system must be treated as a non-cooperative system. Only when one criterion (or both) is not met, and one has good physico-chemical reasons to suspect so, cooperativity may be involved in the analysis and interpretation of the system at hand.

When returning to the divalent 1:1 complex of Figure 2.7 and the equilibria provided in Figure 2.7d, one can ask: Why is the intermediate H"-G" not polymerizing into long linear complexes? The answer lies again in the intermediate H"-G": this complex has the choice either to bind to a free guest (or host) moiety within its own complex (the remaining unbound site), so in an *intra*molecular fashion, or to bind a new molecule of G" (or H") in an *inter*molecular fashion. By the former route, the cycle H"=G" is formed, while the latter (and subsequent additions) lead to supramolecular polymerization. The cyclization is concentration independent, because of its intramolecular nature, as also witnessed by the constant concentration ratio of cycle and intermediate. On the other hand, the polymerization is highly concentration dependent. Therefore, qualitatively, one can expect that cyclization is preferred at low concentrations while polymerization occurs preferentially at high concentrations. At the same time, one can appreciate that the comparison between *EM* and the free, uncomplexed monomer concentration plays a role in the probabilities whether a binding site prefers to bind intra- or intermolecularly.

The dependence of the speciation on the effective molarity *EM* is viewed in a model simulation (Figure 2.8), assuming the presence of equal, but increasing, total concentrations of H" and G", which can form the cycle H" = G" and linear polymers (Figure 2.8a). All binding events are presumed to be independent and governed by the same intrinsic affinity constant ($5 \times 10^4 \, \text{M}^{-1}$), and higher order cycles are ignored here. As explained above for Figure 2.7, the unitless parameter $\frac{1}{2}K_iEM$ determines the probability for cycle formation. At a low *EM* (0.02 mM, at which $\frac{1}{2}K_iEM = 0.5$), Figure 2.8b shows the occurrence of only a small fraction of cycle, while the polymer is formed at

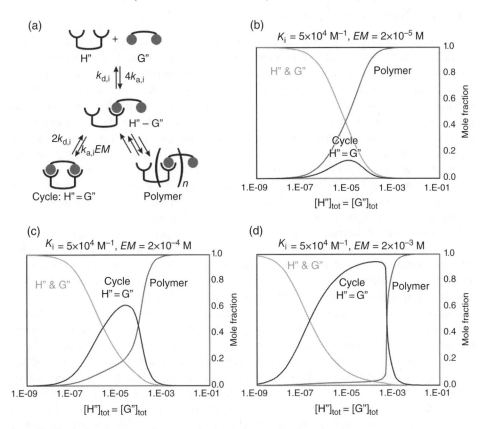

Figure 2.8 Equilibria (a) and simulations of speciations (b–d) for the interaction between a ditopic host and a ditopic guest at increasing concentrations, but equimolar ratio, of the two components, assuming independent interactions ($K_i = 5 \times 10^4 \, \text{M}^{-1}$) and only once cyclic product (H"=G") at different values of EM: 0.02 mM ($\frac{1}{2}K_iEM = 0.5$; $K_{ov} = 10^5 \, \text{M}^{-1}$) (b); 0.2 mM ($\frac{1}{2}K_iEM = 5$; $K_{ov} = 10^6 \, \text{M}^{-1}$) (c); and 2 mM ($\frac{1}{2}K_iEM = 50$; $K_{ov} = 10^7 \, \text{M}^{-1}$) (d).

concentrations which are normal for a 1:1 complex with the given affinity constant [36].[xiv] The polymer formation is thus hardly affected by the cycle formation under these conditions. For higher *EM* (Figure 2.7c and d), fractions of cycle increase, while the rising of the polymer fraction is delayed to higher monomer concentrations. Consequently, the higher *EM*, the wider the concentration range in which the cycle is the predominant species, both at lower and at higher concentrations.

Figure 2.8d corresponds to the values discussed above for the system shown in Figure 2.7 (for which $\frac{1}{2}K_iEM = 50$). It can now be appreciated that the observed overall binding affinity of the cycle was on the order of $10^7 \, M^{-1}$ (which corresponds to a concentration of $10^{-7} \, M$ at which the cycle starts to dominate over the free monomers), and that polymers do not play any significant role until the mM range (which is close to *EM*). This qualitatively indicates that, when the concentration of the free components approaches a value close to *EM*, the probability of intermolecular binding (polymerization) becomes equally likely as intramolecular binding (cyclization).

2.3.2 Multivalent Binding at Surfaces

Many biomolecular interactions occur at interfaces instead of freely in solution [8,9]. Therefore, expanding the model formation toolbox to the simulation and understanding of interactions at interfaces is of high importance. For monovalent interactions, the Langmuir isotherm [37] is often used, although other models, which include lateral interactions, varying adsorption energies, and mass-transport issues such as rebinding [38], and so on, have been developed as well. Here we focus exclusively on multivalent interactions, and the comparison with the normal, intrinsic 1:1 parent complex (which is described by the Langmuir isotherm). The description given below largely follows the methodology for the treatment of multivalent interactions at interfaces developed earlier [39].

Technically, models with surface species can at first look a bit more complicated. The primary point to note is that surface species are different from solution species, and are usually expressed in units of moles per surface area (instead of per volume) or in relative coverages. Experimental surface techniques can provide quantitative information on the (relative or absolute) amounts that are adsorbed during a titration, in which different concentrations of a solution species are put in contact with a surface functionalized with a molecular component that is complementary to the solution species. Mostly widely used techniques are SPR [40] and QCM [41,42].

A large difference in model treatment occurs depending on whether the analytical technique is performed on a cuvette or a flow cell. The cuvette case is the most complicated: Assuming the case in which a guest is supplied from solution onto a host-covered substrate, working with a fixed volume of guest solution can lead to depletion of guest from the solution upon adsorption, in particular at low guest concentrations and strong host–guest affinity. Thus, the amounts of guest in solution and on the surface are coupled, and one cannot assume that the concentration of the guest in the stock solution is the concentration that is actually in equilibrium with the substrate after

---

[xiv] One must note that supramolecular polymerization based on independent binding sites can be modeled with only a 1:1 complex with the intrinsic affinity constant, when one recognizes that the concentrations of host and guest sites for this 1:1 complex are equal to twice the concentrations of the ditopic host and guest molecules. Therefore, the position of this equilibrium provides direct access to the average degree of polymerization for this system.

adsorption has occurred. That means that the whole system, substrate and solution, must be modeled together, and that the mass balance of the guest must encompass both surface and solution species. Consequently, the surface concentrations must be made compatible with solution concentrations, and this is done by dividing the moles of adsorbed host by the cuvette volume so that volumetric concentrations can be obtained for all surface species. In the past [19,39], we have performed and modeled many SPR titrations for a cuvette system and, although doable, the mass balance equations that follow from the methodology described above (Figure 2.1 and accompanying text) become lengthy and complicated quickly.

Besides these technical numerical issues, also practical issues exist: the absolute amount of host adsorbed on the surface needs to be known in order to obtain a reliable solution concentration of this species and thus provide the correct coupling between the solution and surface guest species, in particular when solution depletion is expected to play a serious role. That means that both the total sample area (not only the macroscopic area, but also the sample roughness, for example) and the host density on the substrate need to be well known, both of which are difficult to assess and come with relatively large experimental errors.

Most modern analytical equipment for surface adsorption studies, however, nowadays make use of flow cells. When a flow of guest is supplied to a host surface, such an amount of guest will be adsorbed until the surface reaches equilibrium with the solution above it. Because the guest solution is continuously refreshed, it can be assumed to be of infinite volume, and thus depletion of the guest from the solution does not occur. As a result, the guest species in the solution when in contact with the surface are identical as directly after the making of the stock solution. This simplifies model building considerably, and relaxes the necessary experimental knowledge of the absolute amount of host on the surface. When focusing on the numerical aspects, the mass balances of the solution and the surface, and thus of the host and the guest, become decoupled. The mass balance for the guest applies only to the solution species, while the mass balance of the host, which can now be fully expressed in (relative or absolute) surface coverages, covers all surface species.

As an example, we here describe the direct follow-up of the system described in Figure 2.7: the adsorption of the ditopic bis-Ad calix[4]arene guest onto a β-CD host-covered substrate (Figure 2.9a) [19]. The host-covered surface was based on a β-CD derivative with seven thioether chains, which adsorbs to Au surfaces to yield well-packed β-CD self-assembled monolayers (SAMs) [43,44], also called molecular print-boards [45,46], with a period of approximately 2 nm. (Monolayers on glass with similar densities have been made as well [47].) Important to note is that such an immobilization of an otherwise simple monovalent host onto a surface makes the resulting host surface a multitopic entity, capable of forming multivalent interactions with multitopic guests. Because initial titrations showed a high apparent affinity (resembling the situation sketched in Figure 2.3a, which does not allow the proper determination of the affinity constant) and because dilution of the receptor density at the surface was not an option, competition with β-CD in solution was used to tune down the apparent affinity of the guest for adsorption to the surface [19].

In the case of using a flow system, the modeling of this system works as follows. The solution only contains the ditopic guest and β-CD. As noted above, working under flow allows assuming the species concentrations of the guest, that can be obtained from species calculations of the solution phase only, to be in equilibrium with the surface.

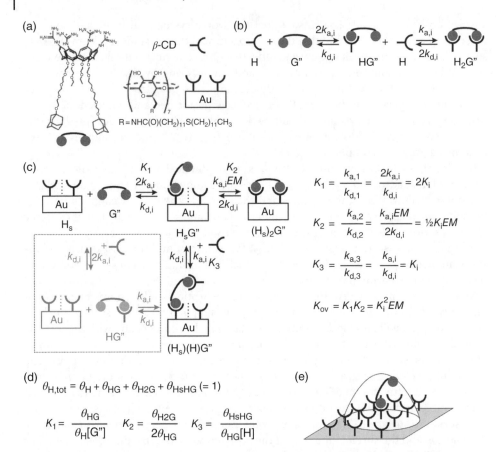

Figure 2.9 Multivalent binding at a surface: formation of a divalent complex between a ditopic guest and a host-covered surface: (a) host and guest molecules used in this study [19]; (b) equilibria for the non-cooperative 2:1 system in solution; (c) equilibria, equilibrium constants and (d) equations (for mass balance and equilibrium constants) for the divalent system at the surface, including interaction with monovalent host from solution as a competitor. In the equilibria of (c), the gray section, leading to/from the monovalently capped HG" in solution, indicates redundant equilibria not incorporated in the model description. A cartoon (e) shows a conceptual picture of the effective concentration (C_{eff}): the unbound guest moiety in the intermediate H_sG'' complex has a probability of finding another host site to form an intramolecular bond determined by the density of free hosts and the probing volume.

Therefore, the species and equilibria occurring in solution (Figure 2.9b), and thus also the equations resulting from these, are identical to the 2:1 system given in Figure 2.7c. In a titration of the guest onto the host surface, one can then vary the guest concentration in the solution, as well as the β-CD concentration.

On the surface, four species exist, logically all containing the surface-immobilized host (H_s). These are: unbound H_s (coverage ϑ_H), H_s monovalently bound to G" (H_sG'', coverage ϑ_{HG}), H_s monovalently bound to G" of which the other guest site is bound to β-CD from solution (($H_s)(H)G''$, coverage ϑ_{HsHG}), and two H_s divalently bound to one G" (($H_s)_2G''$, coverage ϑ_{H2G}). Note that the mixed species ($H_s)(H)G''$ can also be obtained from adsorption of HG" from solution onto the H_s surface, but these equilibria in this cycle of equilibria are redundant (see above). All coverages can be either expressed in absolute coverages (e.g., in mol/cm^2) or in relative coverages (in %). Obviously, these are connected by the total coverage. Note that we use in all species calculations the coverages based on host, hence the factor 2 in the denominator of the equation for K_2 ($K_2 = \vartheta_{H2G}/2\vartheta_{HG}$); the coverage of guest for the divalently bound $(H_s)_2G''$ is half of that of the host.

As is evident from the equilibria shown in Figure 2.9c, the monovalently bound H_sG'' is again the pivotal species. It is formed from the adsorption of free guest, G'', from solution onto a surface host site, and it leads to $(H_s)(H)G''$ or $(H_s)_2G''$ by subsequent intermolecular binding of a β-CD from solution or by intramolecular binding of an additional surface host site, respectively. All intermolecular steps can be expressed in the affinity of the intrinsic interaction, K_i, when assuming independent binding sites. In analogy to the divalent binding observed in solution as described in Figure 2.7d, the intramolecular formation of $(H_s)_2G''$ (Figure 2.9c) is associated with an *EM* term.

When viewing a molecular picture of the intramolecular step, a complicating factor regarding the effective concentration, C_{eff}, is that there is usually more than one cavity in the vicinity of the monovalently bound H_sG'' that can be reached by the uncomplexed guest site of H_sG''. This is, like in the solution case, determined by the length of the linker between the two guest sites in relation with the host–host spacing (or simply the host density or total coverage). Although the host coverage may, on some occasions, be seen as a regular crystal lattice, the flexibility of the linker allows to use a number of hosts determined by the *average* host coverage multiplied by the projected area that can be reached by the unbound guest [39]. This is schematically seen in Figure 2.9e: only the host sites that are within the half-dome described by the linker-bound guest site can be reached by that guest site. Moreover, not all of those surface host sites have to be free, but can already have been occupied by other guest molecules. Therefore, the effective concentration (C_{eff}) can be regarded as the number of free, accessible host cavities in the probing volume determined by the linker length.

In the numerical treatment of such intramolecular steps, we therefore use a fixed, maximal effective molarity, EM_{max}, to fit experimental data. In the equation for K_2 given in Figure 2.9c, EM is related to this parameter by multiplying the latter with the fraction of free surface sites, that is, the relative coverage of free hosts: $EM = EM_{max} \times (\vartheta_H/\vartheta_{H,tot})$, which reduces to $EM = EM_{max}\vartheta_H$ when fractional (relative) surface coverages are used. When fitting titration data, either from SPR [40] or QCM [41,42], one needs an equation that relates the measured parameter to the adsorbed amount of guest. Since both techniques usually provide signal changes that depend linearly on the adsorbed mass, defining such an equation is straightforward. For example, the relative SPR angle change, which is the observed angle change, $\Delta\alpha$, divided by the maximal angle change, $\Delta\alpha_{max}$, can be related to the relative coverage of guest by Equation 2.8.

$$\frac{\Delta\alpha}{\Delta\alpha_{max}} = \frac{\theta_{HG} + \theta_{HsHG} + \frac{1}{2}\theta_{H2G}}{\theta_{H,tot}} \tag{2.8}$$

Equation 2.8 reduces to Equation 2.9 when fractional (relative) surface coverages are used (i.e., when $\vartheta_{H,tot} = 1$).

$$\frac{\Delta\alpha}{\Delta\alpha_{max}} = \theta_{HG} + \theta_{HsHG} + \frac{1}{2}\theta_{H2G} \tag{2.9}$$

Two issues need to be considered, though. First, the host coverages used in the numerical simulation need to be converted to guest coverages, and, as already addressed above, the latter are equal to the ratio of the host coverages and the valencies of the respective complexes (hence the factor ½ in the ½ϑ_{H2G} term in Equation 2.8 and Equation 2.9). Secondly, one needs to recognize that the maximum coverage of guest on the surface can theoretically be achieved when all surface sites are covered by monovalently bound guest, whereas one will normally measure surface adsorption in the range where the multivalent complex is prevailing. For the data fitting itself, this is no issue, as the parameter $\Delta\alpha_{max}$ is optimized together with the surface affinity in the least-squares minimization routine. However, when comparing fits of the here described multivalent model with those of a Langmuir model, the fitted limiting values of the experimental parameter (here, $\Delta\alpha_{max}$) of the two models are off by a factor equal to the valency of the main complex.

As an example, the SPR titration data of the ditopic Ad guest (Figure 2.9a) binding to the β-CD surface under competition with β-CD in solution is presented in Figure 2.10a. These measurements (and fits) were performed in a cuvette system [19], but the trends for a flow system are very similar. The fits shown in Figure 2.10a were obtained by fitting the data to the multivalency model shown in Figure 2.9, while fixing EM_{max} at 0.2 M and varying $\Delta\alpha_{max}$ and the interaction for the individual, intrinsic Ad–β-CD interaction at the surface ($K_{i,s}$) as fit parameters. The fits for all curves shown in Figure 2.10a gave a value of $K_{i,s}$ on the order of $10^5 \, M^{-1}$, which is close to the solution value of $5 \times 10^4 \, M^{-1}$.

Data obtained at low (or zero) β-CD concentration in solution (Figure 2.10a, top curve) showed practically complete saturation of surface sites, hinting at very strong binding [19]. Actually, fitting the top curve in Figure 2.10a to a Langmuir isotherm yielded a binding constant of approximately $10^9 - 10^{10} \, M^{-1}$. This was much larger than the affinity of $10^7 \, M^{-1}$ observed for the divalent complex (H" = G") of the same bis-Ad guest and the β-CD dimer in solution discussed above (Figure 2.7). The rationale for this large difference in affinity is found in the effective molarity: when comparing the cartoons for the effective concentration provided in Figure 2.9e and Figure 2.7e, one can quickly appreciate that the effective concentration in the surface case is much higher than in the solution case. Not only are there multiple available cavities within the probing volume, the probing volume is also smaller, partly because half of the volume is occupied by the substrate, and partly because the linker between the complementary sites is smaller. Overall, we estimated the (maximal) effective molarity to be about 2 orders of magnitude larger (0.1–0.4 M) than the EM observed for the solution system [19,39].

The simulations in Figure 2.10b–d, performed for a flow system, show trends for changes of EM_{max} and of the competitor concentration, $[H]_{tot}$, in solution. Figure 2.10b shows that, when using the parameters found by the experimental data (host–guest affinities in solution and at the surface equal at $5 \times 10^4 \, M^{-1}$, $EM_{max} = 0.2 \, M$) and in the absence of competitor in solution, the divalent complex $(H_s)_2G"$ is the prevalent surface species over a very large concentration range of the guest. In line with the Langmuir fit discussed above, the species already starts to dominate in the nM regime and only when approaching 0.1 M, not coincidentally close to EM_{max}, the monovalent $H_sG"$ takes over, which further increases the coverage of guest on the surface. In an SPR titration, therefore, one would typically only observe the adsorption of the divalent complex in the nM–μM range, upon which one typically stops measuring because a signal plateau

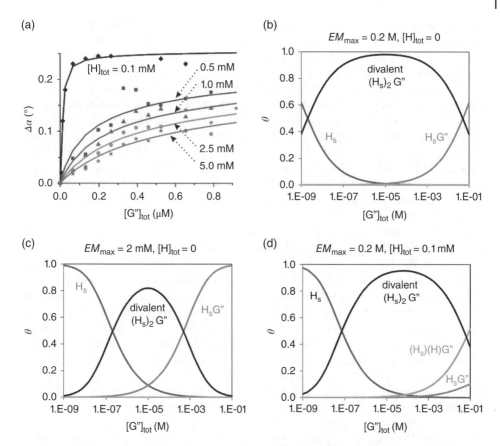

Figure 2.10 SPR data (a) [19] and simulations of speciations (b–d) for the interaction between a ditopic guest and a host-covered surface, assuming independent interactions ($K_i = 5 \times 10^4 \, M^{-1}$) and in the presence of monovalent competing host in solution (H): (a) SPR angle change as a function of guest concentration in solution at different $[H]_{tot}$, and (b–d) simulations of relative coverages of surface host species at varying EM and $[H]_{tot}$. *Source*: Adapted from Ref. [19]. Reproduced with permission of American Chemical Society.

is reached. To our knowledge, such a well-separated transition from multivalent to monovalent binding for the same host–guest system has never been observed experimentally. In a rare example [48], monovalent and divalent binding have been observed to occur for different steroid guests binding to a β-CD monolayer (albeit with a different architecture than the one shown in Figure 2.9a), by comparing differences in maximal SPR angle changes for these steroids.

In the case EM_{max} would have been lower (2 mM instead of 0.2 M; thus becoming equal to the EM observed in solution for the interaction between the ditopic guest and the ditopic β-CD derivative described in Figure 2.7), the range in which the divalent $(H_s)_2G''$ prevails becomes narrower, both at the left and right sides of the graph. Basically, the lowering of EM_{max} causes a decrease of the driving force for intramolecular ring closure, which causes both a later adsorption of the divalent

$(H_s)_2G''$ and a more efficient competition by monovalent H_sG''. When EM_{max} would decrease even further (not shown), the maximum amount of divalent complex would drop even further, its stability range would narrow further, and the adsorption would primarily take place by the monovalent H_sG'', at the concentration range (µM) corresponding to the intrinsic affinity constant. Practically, lowering of EM_{max} can be achieved by increasing the linker length in the multitopic guest and/or increasing the spacing (i.e., lowering the surface density) of the surface-bound host. When focusing on variation of the linker length, it can be reasoned that the linker length dependence of EM_{max} is less steep at an interface than in solution [39], and even in solution the dependence appears to be less than theoretically expected [49,50]. To our knowledge, a clear linker length dependence of EM_{max} has not been shown for densely packed host surfaces, although a convincing example has been shown for a tetratopic protein receptor in a membrane [27].

The effect of competition by β-CD in solution is shown in Figure 2.10d. The simulation performed at $[H]_{tot} = 0.1$ mM shows that the apparent binding affinity goes down in comparison with Figure 2.10b (the adsorption of the divalent complex starts at around 100 nM in the former compared with at a few nM in the latter case). However, the onset of the monovalent complex remains at 0.1 M, even though it now primarily takes the form of the mixed complex $(H_s)(H)G''$. At even higher $[H]_{tot}$ (1 mM, not shown here) this trend continues: the range in which the divalent $(H_s)_2G''$ prevails becomes narrower, but primarily at the expense of the adsorption (left) side of the graph. This behavior is explained by the fact that the complexation by β-CD from solution, occurring with equal probability at any free Ad group, has a larger effect on the decrease of $[G'']$ (by complexation into HG'' and H_2G''), than in the promotion of the mixed complex $(H_s)(H)G''$, because in the latter case only one β-CD can be bound compared with two in case of H_2G'' in solution.

Generally speaking, these studies provide the following lessons. First, the driving force for intramolecular, and thus multivalent, binding can be orders of magnitude larger at surfaces than in solution. Obviously, the primary governing factor is the EM_{max}, and that is in turn dependent on the surface density of the host and on steric molecular factors of the multivalent guest that is adsorbing onto the surface. Secondly, competition by monovalent host in solution works primarily in preventing adsorption of the guest, by lowering of the free guest concentration in solution. However, kinetic studies have shown that blockage of the free guest site of the monovalent H_sG'' by host from solution, although hardly visible in a change of the species concentration of the divalent complex, has a large effect on the surface diffusion and desorption rates of the guest at/from the surface [51]. Thirdly, as shown by this system, but even more so by the higher valent dendrimers discussed below, multivalent molecules prefer to adsorb to surfaces with their highest valency that is sterically possible. This is primarily caused by the multivalent driving force, best expressed by the unitless factor K_iEM, which dictates that the overall affinity of a multivalent complex increases by that factor compared with the complex with one bound site less. In other words, ignoring statistical pre-factors, the overall affinity can be written as Equation 2.10, where n is the valency.

$$K_{ov} = K_i \left(K_i EM \right)^{n-1} \tag{2.10}$$

2.4   Special Multivalent Systems

This section discusses different examples of multivalent systems, with a strong focus on interface systems. The choice on interfacial systems is governed by their high relevance for both biology and materials science. Already early on, we noted that multivalency provides a way to anchor molecular entities onto surfaces in a stable manner [45,52,53,54], although the individual interactions remain labile and dynamic. This has caused large efforts in materials assembly, including on our molecular printboards [9,46,55].

Because, however, most of these systems only need a rudimentary understanding of multivalency, basically reducing all the knowledge provided above to "many = strong", we chose the examples presented below because these describe systems for which detailed modeling has provided more insight. For space issues, systems will be presented in a coarse overview only, together with the main conclusions; more details are provided in the references supplied with the examples.

2.4.1   Increasing the Valency of Interfacial Assemblies: Dendrimers, Oligomers, and Polymers

The valency of a multivalent complex is one of the main determinants of the complex's overall binding affinity and dynamics. At the same time, the synthesis of organic molecules allows an almost infinite variation of the number of binding sites that can be placed on a scaffold, as well as their relative positions and distances. Therefore, the synthesis and interaction of a variety of guest molecules on a variety of scaffolds, with different numbers of interaction sites and different spatial configurations has led to a series of studies that have allowed the elucidation of their effects on overall valency and binding behavior.

In particular, dendrimers offer a versatile toolbox to vary the valency of binding in a systematic manner by functionalizing a series of different generations of dendrimers with interaction sites at their exterior. At the same time, the spherical nature of the dendrimers can lead to steric constraints in how many of the interaction sites can actually be in contact with an interface, in particular for the higher generations. This is caused in part by the fact that the number of end groups grows exponentially with generation, while the outer area, assuming a sphere with a diameter determined by the maximal spacing between end groups, grows only quadratically. As a result, larger dendrimers can reach more binding sites on a surface, yet at the same time the density of their own binding moieties is increasing.

For the 1,4-butanediamino poly(propylene imine) (PPI) dendrimers, we observed already in solution an interesting steric effect of the binding of the monotopic β-CD [56]. For generations 1–4, having 4–16 end groups all functionalized with an Ad guest moiety, all end groups could be bound with β-CD. In contrast, the generation-5 dendrimer, with 64 Ad end groups, showed only partial binding, because steric hindrance caused by the high Ad density at the periphery prevented full occupancy of all sites.

At β-CD-functionalized monolayer surfaces, guest-functionalized dendrimers proved to be a powerful way to introduce functionality at these interfaces. In early work [45,57], using Ad-functionalized dendrimers, qualitative aspects of the multivalent binding of these dendrimers could already be appreciated: lower generations showed reversible binding, requiring increasing amounts of competing native β-CD in solution to accomplish desorption, while the highest generations 4 and 5 provided basically irreversible adsorption. A basic question that remained at that time, however, was with how many guest sites these dendrimers interacted with the β-CD surface. Steric and density issues like those observed in solution as well as conformational restrictions imposed by the branching nature of the dendrimer scaffold were expected to provide binding valencies of the dendrimer to the surface that would be well below the total number of guest sites available in the dendrimers.

A breakthrough in quantitative understanding was achieved by using ferrocenyl (Fc)-functionalized dendrimers (Figure 2.11) as the multivalent guest family [58]. Fc does not only bind weaker to β-CD than Ad, leading to a wider range of generations that could be bound reversibly, but, more importantly, the surface binding of the redox-active Fc can be assessed in a quantitative manner using electrochemical techniques. Electrochemistry provided a direct comparison between the coverage of adsorbed Fc and the coverage of β-CD in the monolayer for all dendrimer generations. This showed in perfect correlation integer numbers of the binding valency of the dendrimers at the surface: the generation-1 dendrimer used 2 of 4 Fc sites in binding to the surface, generation-2 used 3 out of 8, and so on. These numbers were in full agreement with the match between the dimensions of the dendrimers and the number of β-CD sites that could be reached for the respective dendrimer sizes.

For the lower generations 1–3, the electrochemistry data could be correlated to SPR titrations [58].The SPR titration data were fitted to the multivalent surface model outlined above, in which the main assumption was that all intramolecular binding steps were associated with the same *EM* value. This might seem plausible because the end groups are all at the same distance from the center of the dendrimer, but it ignores, amongst others, effects of increasing steric hindrance upon the binding of subsequent guest sites to the surface and differences in probability of binding for sites that reside at the same or different branches of the dendrimer structure compared with already bound sites. Yet, optimal fits for titrations across the dendrimer series could be achieved by assuming one *EM* and one intrinsic affinity parameter only. The SPR titration data were thus fitted to models in which *EM* was kept constant while the surface-binding valency n_s was varied and the intrinsic binding affinity at the surface, $K_{i,s}$, was optimized in the fit procedure (as well as the maximal SPR angle change, as described above). Thus sets of fits were obtained, all with equally good correlation coefficients, in which higher valencies yielded lower fitted $K_{i,s}$ values, basically following Equation 2.10 discussed above. The values of n_s for which $K_{i,s}$ was close to the measured intrinsic affinity in solution $(1200\,M^{-1})$, matched perfectly well with the valencies found for the corresponding dendrimers by electrochemistry [58].

This three-fold match, between electrochemistry, SPR and simple model-based steric considerations, thus gave a quantitative understanding of valency, intrinsic affinity and effective molarity, and allowed extension of the electrochemical method to dendrimers that bound irreversibly and that could therefore not be assessed by SPR anymore [58]. In subsequent studies, the relationship between steric match and binding stoichiometry was confirmed both for dendrimers with longer linkers and different scaffolds [also for the commonly used poly(amido amine), PAMAM, dendrimer type] [59], as well as for a different redox-active guest moiety, the biferrocenyl group [60].At the same time, the

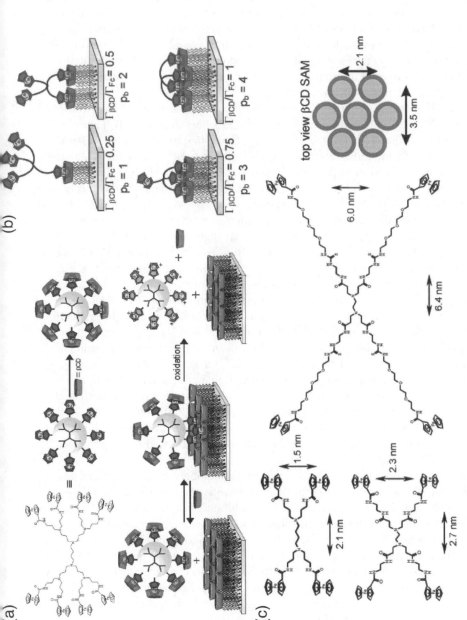

Figure 2.11 The binding of the redox-active ferrocenyl (Fc) dendrimers to β-CD monolayers [58,59]: (a) adsorption scheme of Fc dendrimers by initial binding in solution by β-CD followed by adsorption onto the β-CD surface, with subsequent oxidation-induced desorption; (b) schematic representation of the four possible binding modes of the generation-1 dendrimer with the numbers, p_b of bound sites and the predicted coverage ratios of β-CD and Fc depicted below; (c) comparison between dimensions of the tetra-topic PPI (left top), PAMAM (left bottom), and oligo(ethylene glycol)-extended PAMAM (center) on the one hand and the β-CD lattice period (right) on the other hand, predicting the (confirmed) binding valencies of 2, 3, and 4, respectively. *Source:* Adapted from Refs [58,59]. Reproduced with permission of American Chemical Society. See color section.

redox-active properties of Fc and its oxidation-induced loss of binding affinity for β-CD has provided inspiration for stimulus-induced control over the binding affinity of surface assemblies on β-CD surfaces, ranging from molecules to nanoparticles and even cells [61,62,63,64,65].

Linear polymers can provide also a very powerful, and arguably even more versatile scaffold for the engineering of multivalent ligands compared with dendrimers. At the same time, however, they also introduce more "fuzziness" as the polymer length, the exact number of binding moieties on a polymer chain, and the inter-moiety spacing cannot be controlled to the same extent as in the molecularly controlled dendrimers. Commonly, the degree of functionalization needs to be tuned because functionalization of every monomer unit would lead to steric congestion upon binding. Therefore, most common synthetic procedures proceed by the random copolymerization of functionalized and unfunctionalized monomers or by the random partial functionalization of a pre-synthesized polymer with reactive groups.

In an early study [66], we used the linear poly(isobutene-*alt*-maleic acid) (PiBMA) modified with hydrophobic guest moieties (Ad and *p-tert*-butylphenyl) (Figure 2.12). The two moieties differ in their intrinsic binding affinity ($5 \times 10^4\,M^{-1}$ versus $1 \times 10^4\,M^{-1}$), and the degree of functionalization was varied between 10% and 40%. The highest degree of functionalization used here corresponds approximately to an inter-moiety spacing commensurate with the β-CD surface lattice period. All guest polymers bound to the β-CD surface in an irreversible fashion, and gave very thin adsorbed layers. This indicates a very effective use of most or all guest units in the polymers, even though the polymers need to adapt their conformation strongly to do so. At the same time, these results point at very high overall binding affinities with concomitantly low dissociation rate constants. However, no further quantification was possible. Yet, these guest polymers and similarly prepared β-CD-functionalized polymers appeared to be powerful building blocks in size-controlled supramolecular nanoparticles [67,68,69,70].

A marriage between the molecular definition of dendrimers and the linear scaffold properties of polymers was recently found in the multivalent binding of linear peptides [71]. Peptide chemistry allows the design and synthesis of linear chains of amino acids with practically unlimited structural control. In the context of multivalent binding, we designed multivalent linear peptides with hydrophobic amino acids that function as guest moieties for β-CD spaced with hydrophilic amino acids that provide both control over the distance between the guest moieties and water solubility. These linear peptides can be regarded as molecularly defined oligomers with a specific number of binding sites. Mixed linear-dendritic systems were obtained by sticking these peptides to the ends of PAMAM dendrimers.

The family of amino acids contains three aromatic members, tyrosine (Tyr, Y), phenylalanine (Phe, F), and tryptophan (Trp, W), that have a small (monovalent) affinity for β-CD (130, 68, and $17\,M^{-1}$, respectively) [72]. The binding valency of the linear peptides was tuned from 3 to 6, by the use of peptides that contained these numbers of aromatic amino acid residues in their chain, spaced by other amino acids [71]. In this manner peptides were created that had a measurable affinity for β-CD surfaces, but all of these peptides showed reversible binding. By combination of various numbers of Tyr and Phe residues the overall binding affinity could be programmed precisely. For the larger peptide-dendrimer constructs, the range over which the overall affinity and the surface-binding valency could be tuned was increased.

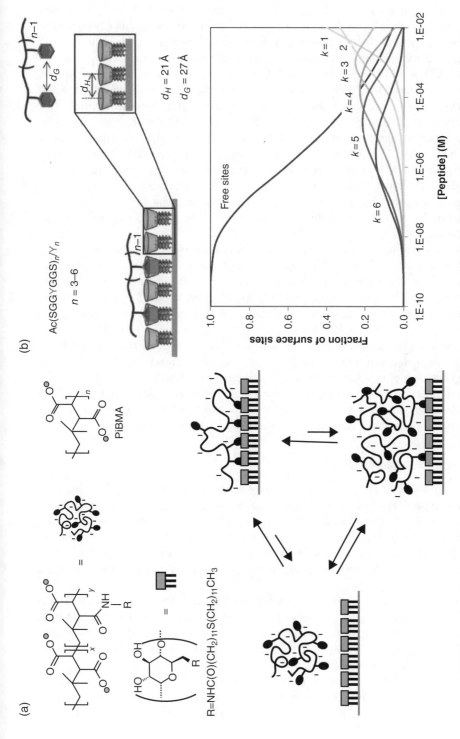

Figure 2.12 The binding of linear multivalent guests to β-CD-modified surfaces: (a) poly(isobutene-*alt*-maleic acid) (PiBMA) modified with guest groups (R = Ad or *p*-tert-butylphenyl) (top) and the possible modes of adsorption to the β-CD surface (bottom) [66]; (b) peptides with *n* (=3–6) Tyr units incorporated at regular distances from each other (top) and simulated speciation diagram of all surface host sites as a function of the concentration of the Y_6 peptide, in which *k* indicates the summed concentrations of all species with the given number of bonds to the surface (bottom). *Source:* Adapted from Ref. [66]. Reproduced with permission of John Wiley and Sons.

SPR titrations allowed the determination of the affinities of all linear (and dendrimeric) peptides [71]. In contrast with the Ad and Fc dendrimers described above, the linear nature of the peptides was taken into account in the multivalent binding model by assuming that the effective molarity scaled with the inverse of the distance between the already bound site and the one that was at the point of binding intramolecularly. As a result, each aromatic amino acid in the sequence can exist in three states: unbound; bound to β-CD in solution; and bound to β-CD at the surface. For a hexa-topic peptide, this creates a family of $3^6 = 729$ species, with 64 solution species and 665 surface species. Nevertheless, the fit procedure for this model is very strict, because the only free parameters are the intrinsic affinity constants for Tyr ($K_{i,Y}$) and Phe ($K_{i,F}$) at the surface, and *EM*. Simultaneously fitting of the SPR titration data for eight different peptides with the numbers of Tyr and Phe moieties ranging from 2 to 6 and 0 to 4, respectively, whereby the total valency ranged from 3 to 6, gave a single set of optimized parameters. The fitted affinity constants $K_{i,Y}$ ($120\,M^{-1}$) and $K_{i,F}$ ($60\,M^{-1}$) were very close to the values found before in solution. The value for *EM* (43 mM) was somewhat lower than observed for the Ad derivatives discussed above, but appeared to be applicable to both peptide designs tested in this study.

Another important factor of these peptides was their weakly multivalent binding nature. We defined weak multivalency as the multivalent effect occurring when the multivalent enhancement factor $K_i EM$ is between 1 and 10. For the Tyr and Phe units used in these peptides it was shown [71] that the $K_i EM$ factor was 5.1 and 2.6, respectively, clearly in this range. As a result, the binding valency is never equal to the maximum valency, as observed before for the strong binding Ad guest ($K_i EM = 5000$), but instead takes intermediate values. Moreover, the valency is gradually reduced upon increase of the surface coverage, which is a direct result of the lowering of the *EM* by the decreased fraction of free surface sites, again in contrast with the binding behavior of strong binding guests, such as Ad, which have a large concentration range over which the maximum valency is observed. The implications of this weak multivalent binding on the dynamics of surface systems are currently under investigation.

2.4.2 Heterotropic Interactions

So far mostly homotropic systems have been discussed, that is, systems in which the interaction type is the same for all bonds formed in the equilibrium process. Only the binding of the first guest to the Eu^{3+}-bound β-CD dimer shown in Figure 2.6a can be viewed as a ditopic interaction in which the two bonds, the β-CD host–guest and the Eu^{3+}–carboxylate interactions, are of a different chemical nature. However, this system has not been analyzed this way, but it has served as an inspiration for a surface antenna system described below.

Heterotropic systems provide more versatility and complexity compared with homotropic ones. We discriminate here between systems in which the different interaction motifs do not influence each other because they take place in sequential, intermolecular events, and those in which the motifs do have a mutual influence because they are linked by intramolecular bonds.

As a direct extension of the molecular printboard concept, a calixarene with four Ad legs, very similar to the ditopic one shown in Figure 2.7a, was used as one half of a capsule, to which the second half was bound to the guanine moieties on the top of the calixarene, thus yielding the full capsule [73]. In a sequential process, the tetra-Ad guanine-modified calixarene was bound to the β-CD surface, followed by adsorption of the tetra-sulfonato calixarene on top of the first calixarene. The tetravalent host–guest interaction made the first calixarene bind practically irreversibly, so that the surface could be extensively rinsed before placing the second capsule half on top of the first. The binding affinity of the capsule formation at the surface ($3.5 \times 10^6 \, M^{-1}$) was close to the value obtained in solution, which confirmed the expected independent and orthogonal binding behavior of the two motifs.

This hierarchical build-up process provided a powerful paradigm for protein assembly onto the molecular printboards. Bifunctional linkers were designed with Ad (1 or 2 moieties) and biotin or nitrilo-triacetate (NTA) on the other end [74,75,76,77,78,79,80]. This way, streptavidin (SAv) and His-tagged proteins, respectively, could be assembled in a multistep process, in which the streptavidin surface could in turn be used as a platform for the binding of further biotinylated proteins, antibodies, and even cells [74,79]. In both cases, intramolecular bonds are formed. In the case for SAv, however, the effect was only investigated in a qualitative manner. Here [74] it was shown that having two Ad units per biotin linker was preferred over one, as: (i) the divalent linker allowed preadsorption onto a surface followed by adsorption of SAv allowing a stepwise buildup of the assemblies; and (ii) the assembly of two linker molecules and SAv became tetravalently bound to the β-CD surface providing a high kinetic stability to the assembly.

The inherently reversible nature of the multivalent self-assembly of these protein constructs allowed the development of a new, supramolecular way to suppress the nonspecific adsorption of proteins onto the β-CD surface [75]. A monovalent hexa(ethylene glycol) molecule with an Ad moiety (Ad-HEG) was used as an agent that temporarily blocks the β-CD cavities and at the same time equips the surface with the protein-repelling oligo(ethylene glycol) group. Because of its monovalent nature and the intrinsic Ad-β-CD affinity ($5 \times 10^4 \, M^{-1}$), supplying Ad-HEG at a concentration of 0.1–1 mM was sufficient to provide a dynamic poly(ethylene glycol)-like layer to the β-CD surface. When a protein was administered together with its linker, in the presence of Ad-HEG, the multivalent interactions of the protein-linker construct were sufficiently strong to replace the monovalent Ad-HEG, leading to the desired protein construct in the absence of non-specific adsorption. This competition procedure was also successfully used in the adsorption of a bio-engineered light-harvesting protein complex equipped with, on average, three Ad moieties to achieve a dense, well-packed and oriented mimic of the light-harvesting complex arrays found in purple bacteria [81,82].

The trivalent nature of the binding of His_6-tagged proteins onto β-CD surfaces by the NiNTA-Ad linker was investigated in more detail (Figure 2.13) [77]. A full description of all species and equilibria is given in Figure 2.13c, and includes a statistical treatment of the binding of all possible neighboring His_2 units to the Ni^{2+} center. A good fit of the model to SPR titrations was observed (Figure 2.13d), whereby the micromolar apparent affinity was explained by the combination of intrinsic affinity constants of the Ad-β-CD and the His_2-NiNTA interactions. The model confirmed that at the surface the complex employed three NiNTA-Ad linkers between the protein and the surface was the most stable, whereas in solution it was only a minor species (Figure 2.13e).

More recently, we used His-tagged growth factors to tune the cell differentiation of adsorbed cells [83]. A crucial performance parameter appeared to be the very slow spontaneous desorption rate of the His-tagged growth factor from the surface,

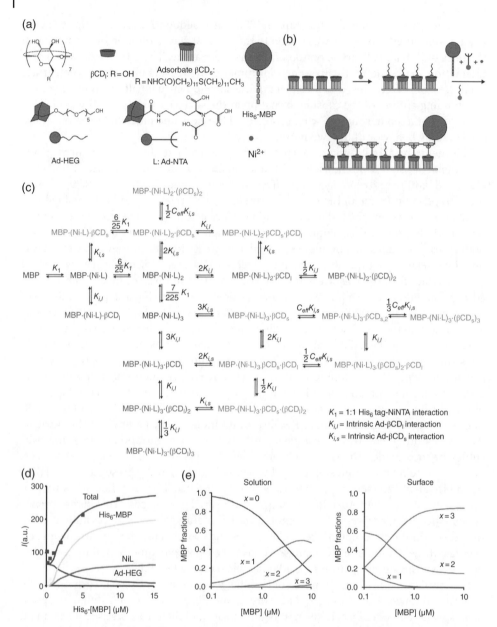

Figure 2.13 Multivalent assembly of His-tagged proteins on a receptor surface [77]. (a) Compounds used in this study (MBP: mono-His$_6$-tagged maltose binding protein). (b) Adsorption scheme for the assembly of MBP·(Ni·L)$_3$ at a β-CD SAMs in the presence of Ad-HEG. (c) Equilibria for all species (solution and surface) for the attachment of His$_6$-MBP at the molecular printboard (charges are omitted for clarity). Subsequent complexation steps of Ni·L to MBP in solution are shown in red, other solution species in blue, and all surface species are given in green. (d) Equilibrium values of the SPR intensities (markers) of a titration of MBP onto the β-CD SAM and corresponding fit to the model and contributions by different components to the signal (solid lines). (e) Speciation modeling showing fractions of His$_6$-MBP·(Ni·L)$_x$ (i.e., complexed to different numbers, x, of Ni·L) as a function of [His$_6$-MBP] in solution (left) and at the surface (right). *Source*: Adapted from Ref. [77]. Reproduced with permission of John Wiley and Sons. See color section.

providing a prolonged delivery of the growth factor to the adsorbed cells over the course of several days, right in the vicinity needed for uptake by the cells. This allowed a reduction of the required amount of growth factor of orders of magnitude, and the supramolecular construct showed superior cell differentiation behavior in comparison with a variety of controls, including delivery via the culture medium and by covalently anchored growth factor. In principle, the desorption kinetics of such constructs can be controlled to a large extent because of its multivalent nature, including, for example, by the pH of the solution, the valency, and competing agents.

Also proteins with multiple His-tags have been employed, and they showed an enhanced affinity in line with the occurrence of multiple (2–3) His-tags in contact with the surface [77]. This system of a protein with multiple His-tags bound to the surface with a hetero-ditopic linker provides a type of "nested multivalency" that has so far not been analyzed in full detail. A similar system occurs for the peptide-dendrimer constructs adsorbed onto the β-CD surface, as discussed above [71]. In both cases, a single multivalent side-arm, either a His-tag or a linear peptide, interacts with multiple linker molecules which interact with the surface. For a His$_6$-tag or a trivalent peptide, three linker molecules are used in this interaction, which provide two intramolecular bonds, associated with an effective molarity. Another side-arm, positioned further away in the protein or dendrimer, interacts similarly, but when the first arm has already been bound, all interactions of the second arm are intramolecular, whereby the first interaction probably has a lower *EM* because it is governed by the probability by which the second side-arm reaches the surface once the first one is in contact, and thus by the distance between the arms in the protein or dendrimer. The consequences of this nested multivalency for the thermodynamics, the speciation of the linker and surface coverage, and the dynamics have not been studied in detail.

In the His-tag protein assembly discussed above, it was observed that the surface promotes the adsorption of species that are hardly formed at all in solution. The driving force of that is the formation of multiple intramolecular bonds between the species and the surface, governed by the high *EM* at the surface, which enhances the overall stability of that species. A similar observation was made when a hetero-ditopic Ad-ethylene diamine (Ad-en) ligand was used together with a transition metal ion M (Cu^{2+} or Ni^{2+}) [84]. In solution, a mixture of various possible species was obtained because, at the given concentrations and pH, also the M-en complexes were weak. However, both systems led to the preferential formation of $M(Ad-en)_2$ upon adsorption onto the surface. This phenomenon was coined "supramolecular expression" [84]. When this mixed host–guest/metal–ligand system was applied to β-CD vesicles, strong intra-vesicle interactions leading to vesicle aggregation were observed for Ni^{2+}, but not for Cu^{2+} [85]. This difference in behavior was attributed to the weaker metal ligand interactions for Ni^{2+}, which gave vacant coordination sites at the metal ion as well as unbound en units at the vesicle interface, which allowed a velcro-like attachment of the interfaces upon the approach of two vesicle surfaces.

The EuEDTA coordination system, as discussed above for a solution system (Figure 2.6), was also applied on a β-CD surface, and gave very similar supramolecular expression as observed for the M(Ad-en) system [86]. A bis-Ad-functionalized EDTA, complexed to Eu^{3+} was used as a fluorogenic agent, while a bis-Ad-functionalized naphthalene carboxylate was used as an antenna (Figure 2.14a). Upon coordination of the carboxylate group of the antenna to the Eu^{3+} center, the complex became bright luminescent. The luminescence behavior was used to signal the supramolecular

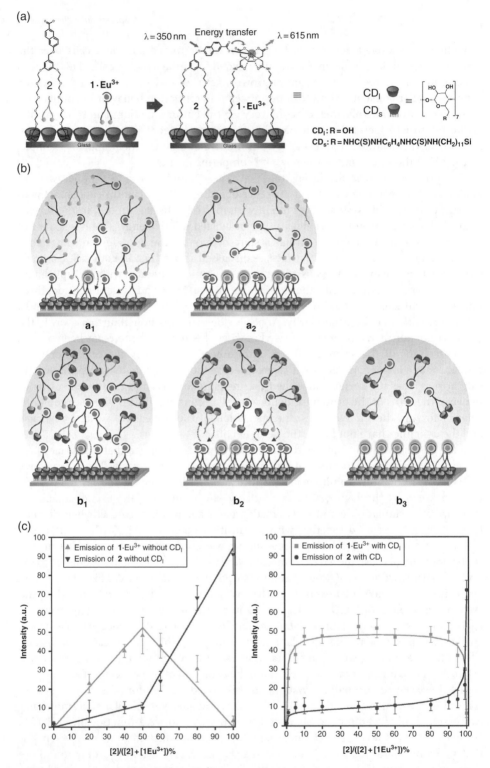

Figure 2.14 Non-linear supramolecular expression of a luminescent signaling complex at a receptor surface [87]. (a) Complexation of antenna **2** and fluorogenic **1**·Eu^{3+} on a CD monolayer and the occurrence of sensitized luminescence upon coordination of **2** to the Eu^{3+} center.

expression of the complex at the surface, and also in this case the complex was the major species at the surface even though the Eu^{3+}–carboxylate interaction was weak. By variation of the ratio of the antenna and fluorogenic molecules, which adsorbed with equal probabilities from a solution onto the surface because of their identical bis-Ad structure, a Job plot on the surface was obtained and visualized by the luminescence of the product. As expected, the Job plot confirmed the 1:1 stoichiometry of the antenna-EuEDTA complex at the surface. In a follow-up study [87], the adsorption of different ratios of the antenna and fluorogenic molecules was followed by a prolonged equilibration step in the presence of β-CD in solution to promote exchange of the divalent molecules (Figure 2.14b). Thus, the weak Eu–antenna interaction drove the system to almost exclusive formation of the 1:1 complex at the surface, while the solution composition could be varied from 95:5 to 5:95 (Figure 2.14c). This constitutes a clear case of non-linear amplification, here caused by the intramolecular nature of the Eu–antenna interaction at the surface. Competition between the antenna and dipicolinic acid, which is a biomarker for anthrax, was employed to turn the system into a sensing platform with ratiometric fluorescent detection by the simultaneous monitoring of Eu^{3+} emission and naphthalene fluorescence [88].

2.4.3 Kinetics and Dynamics

Dynamics is an indispensable property of self-assembly. In biology, thermodynamics is not the main ruling factor, but primarily structure (including compartmentalization) and dynamics, and many biological systems function far out of equilibrium. Correspondingly, dynamic aspects of self-assembly and supramolecular systems have received increased attention. Yet, an overview of these activities is outside the scope of this chapter. Instead we introduce here a few examples in which the kinetics and dynamics of multivalent systems have been addressed in a quantitative manner.

Probe methods have become a powerful tool to study the dynamic aspects of single molecule behavior and molecular interactions. In particular force spectroscopy, in which a probe (typically an AFM tip) is brought in close contact with a surface followed by a retraction of the same tip during which the bending behavior of the tip is monitored, has become the method of choice for many supramolecular systems.

In the past, we have used AFM force spectroscopy to study β-CD host–guest interactions at interfaces at the single-molecule level [89,90]. Various guests for β-CD were immobilized on the AFM tip and brought into contact with the β-CD surface. As discussed above, surfaces functionalized with a monovalent interaction moiety form a

---

Figure 2.14 (Cont'd) (b) Schematic representation of two assembly procedures of $1 \cdot Eu^{3+}$ and **2** onto CD monolayers: without (kinetically controlled assembly, a_1 and a_2) or with (thermodynamically controlled assembly, b_1–b_3) CD_I present in the solution mixtures: (top) by rapid filling of the empty CD_s surface (a_1) in the absence of CD_I, the two different adamantyl-functionalized ligands are assembled on the surface, with the ratio linearly corresponding to the solution ratio (a_2); (bottom) after the rapid filling of the empty CD surface (b_1) leading to a filled surface (b_2), the self-organization by dynamic exchange is promoted by the presence of CD_I, thus promoting the formation of the stable tetravalent complex ($1 \cdot Eu^{3+} \cdot 2$) under thermodynamic equilibrium (b_3). (c) Fluorescence intensities of $1 \cdot Eu^{3+}$ and **2**, obtained from solution assembly, are shown as a function of the solution composition. Solid curves indicate linear fits (left) and the corresponding fits to the multivalent, sequential binding model (right). *Source*: Adapted from Ref. [87]. Reproduced with permission of American Chemical Society.

multivalent entity upon immobilization. This holds in the AFM system for both the host and the guest. By definition, that makes the interaction multivalent once the interaction area and receptor–ligand densities are sufficiently large so that multiple bonds can be formed. Indeed, force–displacement curves were recorded that showed unbinding events of both single and multiple interactions. Interestingly, the unbinding forces for the multiples appeared to scale linearly with the number of bonds. Importantly, the unbinding force of single events appeared to be independent of loading rate, indicating that the complexes were probed under thermodynamic equilibrium, and thus that the dynamics of binding and unbinding was fast on the AFM timescale. By measuring a series of different guests with different binding affinities for β-CD, we were able to define a relationship between the measured dissociation force and the binding energy [90]. Here it is important to note that β-CD complexes generally have a diffusion-controlled, that is barrier-less, association, which is therefore practically identical for all guests studied here. Thus the unbinding force, which is related to the dissociation rate constant, provides access to the binding affinity.

A real attempt to assess multivalent interactions by force spectroscopy was made by the design and synthesis of a series of Ad-functionalized adsorbates, with one, two, or three Ad moieties connected to an aromatic ring and functionalized with an adsorbing group for attachment to an AFM tip [91]. A polymeric spacer was used to avoid the convolution of any non-specific interaction between tip and sample with the unbinding forces of the complexes. The detachment behavior of the functionalized tips showed a much richer force behavior. As observed for the monovalent immobilized guests, the monovalent Ad derivative showed a force corresponding well with earlier data, and its unbinding force was independent of loading rate (Figure 2.15a). However, the di- and trivalent derivatives showed a loading rate dependence, indicative of an orders of magnitude slower overall dissociation rate, in agreement with the higher affinity expected for multivalent complexes. Yet, also force peaks that corresponded to lower valent complexes were observed in the force histograms. The loading rate dependence of these di- and trivalent systems was fitted to the Bell–Evans model [92,93], which basically looks at unbinding only, thus ignoring any contribution from re-association. This fit procedure gave dissociation rate constants that were in excellent agreement with the overall affinities predicted by the multivalent model described above when assuming equal association rate constants (Figure 2.15b). These data confirmed therefore the validity of the multivalency model, and showed that reliable dissociation rate data can also be obtained for multivalent systems.

Surface diffusion is important for various biological systems, for example in cell membranes. Surface diffusion is a two-dimensional or sometimes even one-dimensional process (through pores) and can therefore be seen as a system with reduced dimensionality. So far, the surface assemblies discussed above were only studied regarding their adsorption from solution and possible desorption from the surface.

To modify the β-CD surface system described above for studying surface diffusion, mono-, di- and trivalent Ad-functionalized guest molecules were equipped with a fluorophore to allow fluorescence microscopic visualization (Figure 2.16a) [51]. These molecules were printed onto the β-CD surface into micro-sized lines by microcontact printing using polydimethylsiloxane stamps. The spreading of the molecules from the densely packed, printed lines into the empty β-CD areas was monitored by fluorescence imaging. These spreading experiments were performed with a layer of water on top, in the absence or presence of increasing amounts of competing β-CD in solution. Because

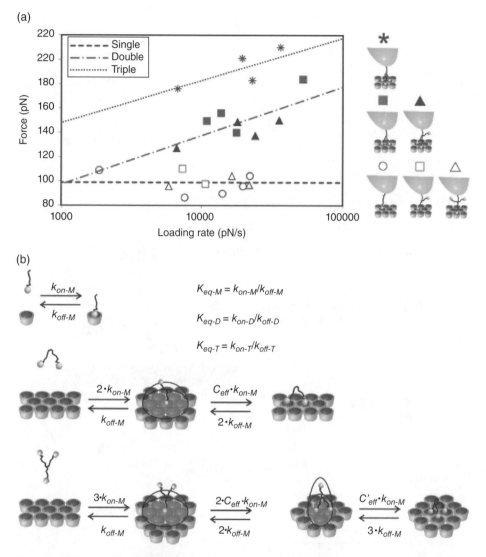

Figure 2.15 Single molecule interactions of mono- and multivalent host–guest complexes by AFM [91]. (a) Measured snap-off forces for mono-, di- and trivalent Ad guests immobilized on an AFM tip as a function of loading rate and grouped per surface species depending on their binding valency with the β-CD surface. (b) Equilibria and rate constants for the binding of the different guests to the surface, used as a model to provide predicted values for the observed k_{off} values obtained by fitting the force data to the Bell–Evans model. *Source*: Adapted from Ref. [91]. Reproduced with permission of American Chemical Society.

competition is expected to have a rate-accelerating effect on desorption [39], increased spreading was also expected. However, the divalent guest, and to a lesser extent the trivalent one, showed a more complex behavior (Figure 2.16e) [51]. The divalent guest started spreading at a small but finite spreading rate, which was first enhanced with increasing β-CD in solution, then dropped, and subsequently increased again, but now with concomitant desorption. This counter-intuitive behavior was explained by a set of

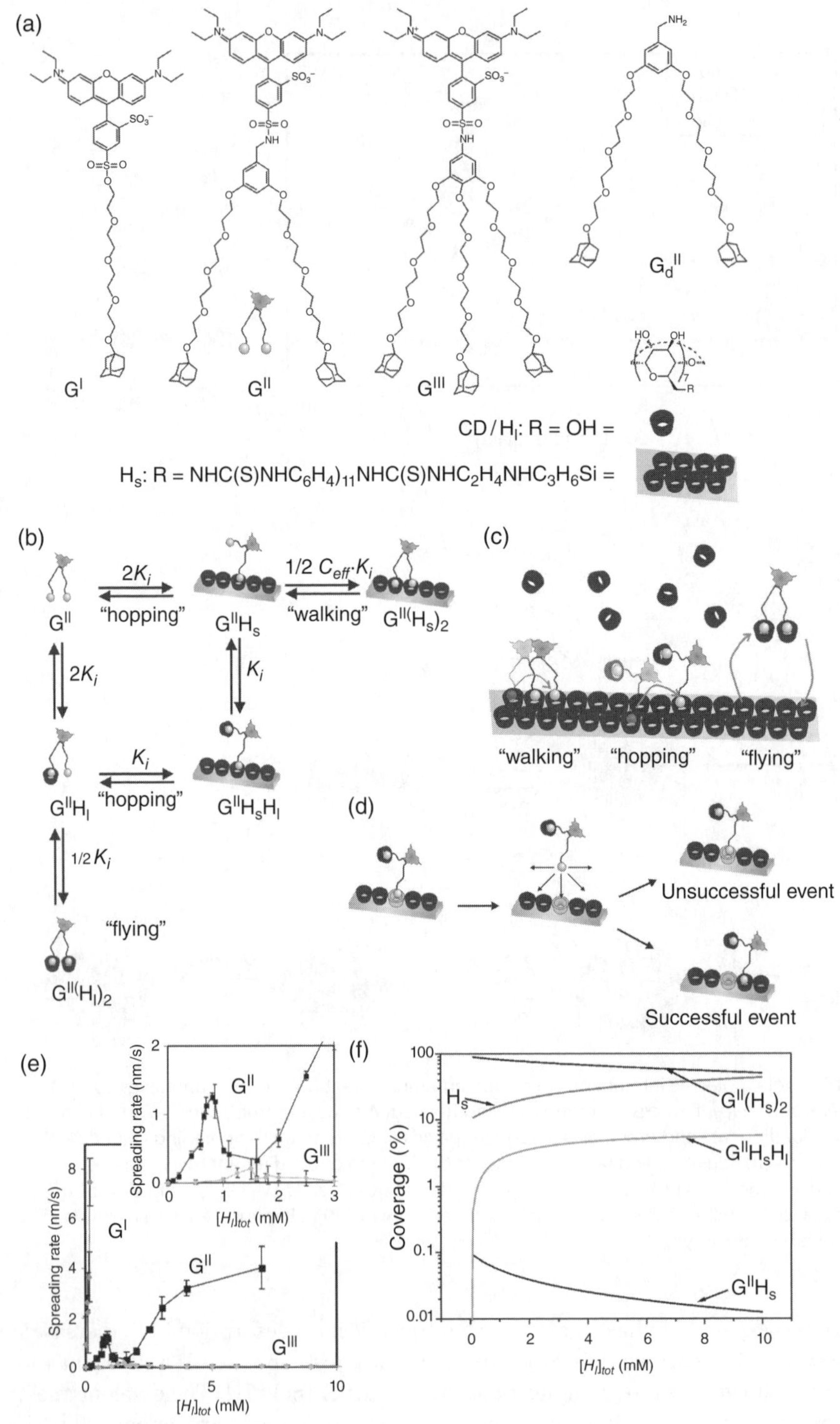
(a)
G^I
G^II
G^III
G_d^II
CD / H_l: R = OH =
H_s: R = NHC(S)NHC6H4)11NHC(S)NHC2H4NHC3H6Si =
(b)
2K_i
"hopping"
1/2 C_eff·K_i
"walking"
G^II
G^IIH_s
G^II(H_s)2
K_i
G^IIH_l
G^IIH_sH_l
1/2 K_i
"flying"
G^II(H_l)2
(c)
"walking"
"hopping"
"flying"
(d)
Unsuccessful event
Successful event
(e)
Spreading rate (nm/s)
[H_l]tot (mM)
G^I
G^II
G^III
(f)
Coverage (%)
[H_l]tot (mM)
H_s
G^II(H_s)2
G^IIH_sH_l
G^IIH_s

surface-spreading mechanisms, coined "walking", "hopping", and "flying" (Figure 2.16c). The primary diffusive mechanism appeared to be "hopping", by which a monovalently bound species, with the second Ad site capped by a β-CD from solution, moves from cavity to cavity. The transition from "hopping" to "flying" was given as the cause for the decrease of the spreading rate, and it was primarily attributed to a decrease of the probability of hopping to a neighboring cavity at the expense of rebinding to the same site, as a result of a reduced lifetime of the species (Figure 2.16d).

Most of the reasoning was established using thermodynamic modeling of the species distributions, but single-molecule Monte Carlo simulations and molecular dynamics simulations provided valuable information as well. Thermodynamic equilibrium considerations were found valid because it was recognized that the diffusion-controlled association caused the solution species to be in local equilibrium with the surface. For the divalent guest, the model for the speciation simulations (Figure 2.16b) was very similar to the one shown in Figure 2.9. Trends in the species concentrations as a function of competing β-CD in solution (Figure 2.16f) were discussed to explain the trends in the relative contributions of the "walking", "hopping", and "flying" mechanisms to the overall spreading rate.

Experiments on a sample backfilled using a guest molecule without a dye, showed a lower spreading rate under the same conditions, which led to the conclusion that the main driving force for the surface diffusion was a gradient of a concentration of free surface β-CD cavities orthogonal to the line features, which caused an increase in EM and affinity for the guest molecules that moved toward the higher fraction of free cavities, reminiscent of chemotaxis. Similar spreading experiments with peptides are currently underway [71].

More recently, we used competition with a ferrocene (Fc)-based guest, instead of β-CD in solution, to create non-equilibrium supramolecular surface gradients of bis-Ad guest molecules on the β-CD surface [94]. Because the Fc guest loses affinity for β-CD upon oxidation, we used a gradient of Fc, achieved by spatially separated oxidation and reduction at interdigitated microelectrode arrays, to drive the local desorption of the bis-Ad guest molecules near the sites with the highest Fc concentrations. Also backfilling and re-equilibration were shown in this system. Full thermodynamic speciation modeling confirmed the gradient formation in a quantitative manner.

Reduction of dimensionality leading to enhanced diffusion was observed for the interaction and diffusion of nuclear transport receptors along surfaces functionalized with nucleoporins, the proteins that coat the inner wall of nucleopores [95]. Interestingly, AFM force spectroscopy showed that the interaction affinity of the nuclear transport

Figure 2.16 Surface diffusion of mono- and multivalent guests across a molecular printboard [51]. (a) Guest and host molecules used in this study: fluorescently labeled mono-, di- and trivalent Ad guests (G^{I}, G^{II}, and G^{III}, respectively), a divalent guest without dye, and β-CD in solution (H_I) and at the surface (H_s). (b) Equilibria of the solution and surface species for the divalent guest. (c) Mechanisms of surface spreading observed in this study. (d) Schematic view of hopping leading to a successful (hopping to neighboring cavity) or unsuccessful (rebinding to original cavity) event. (e) Spreading rates observed by fluorescence microscopy for microcontact printed line features of the multivalent guests onto the molecular printboard as a function of competing host H_I in solution. (f) Speciation of the surface species for the divalent guest as a function of competing host H_I in solution. *Source*: Adapted from Ref. [51]. Reproduced with permission of Nature Publishing Group.

receptors became less when the nucleoporin layer was more saturated with the receptors, at the same time allowing enhanced diffusion. It is expected that this enhanced and regulated diffusion also plays a role in the biological transport through nucleopores, and a further reduction of the dimensionality to the one-dimensional channel is expected to provide an additional transport rate enhancement [96].

2.5 Conclusions

Multivalent interactions provide a rich and rewarding subject of research in which knowledge of thermodynamics, molecular and supramolecular structure, and physico-chemical properties and functions go hand-in-hand. Some properties of multivalent systems appear to "emerge" from the interconnection of many species, leading to complex behavior. In particular, in this area of complex systems, the self-assembly principles of chemistry and biology merge and provide mutually enhancing understanding.

Large steps have been made in the past decades to understand the behavior of multivalent systems. Progression of this understanding has been from solution to surface systems, and from thermodynamic equilibrium to out-of-equilibrium. Steps towards a more detailed understanding of complex systems are being undertaken, but these will require increasing interdisciplinary experimental efforts coupled to quantitative interpretation. Models for such systems will therefore only gain in importance, and will have to encompass more aspects such as kinetics, mass transport, coupling to covalent reactions, and so on. Therefore, the design and study of multivalent systems will remain attractive, interdisciplinary, and imaginative.

Acknowledgments

Funding from the European Union Horizon 2020 research and innovation programme under the Marie Sklodowska-Curie grant agreement No. 642793 (Multi-App) and from the Netherlands Science Foundation NWO (NWO-CW, TOP project 715.015.001) is gratefully acknowledged.

References

1 R. F. Service, How far can we push chemical self-assembly? *Science* **2005**, *309*, 95.
2 D. N. Reinhoudt, M. Crego-Calama, Synthesis beyond the molecule. *Science* **2002**, *295*, 2403–2407.
3 G. M. Whitesides, M. Boncheva, Beyond molecules: Self-assembly of mesoscopic and macroscopic components. *Proc. Natl. Acad. Sci. USA* **2002**, *99*, 4769–4774.
4 G. M. Whitesides, B. Grzybowski, Self-assembly at all scales. *Science* **2002**, *295*, 2418–2421.
5 Y. Xia, T. D. Nguyen, M. Yang, B. Lee, A. Santos, P. Podsiadlo, Z. Tang, S. C. Glotzer, N. A. Kotov, Self-assembly of self-limiting monodisperse supraparticles from polydisperse nanoparticles. *Nature Nanotech.* **2011**, *6*, 580–587.

6 C. Rest, R. Kandanellia, G. Fernández, Strategies to create hierarchical self-assembled structures via cooperative non-covalent interactions. *Chem. Soc. Rev.* **2015**, *44*, 2543–2572.

7 J.-M. Lehn, Supramolecular chemistry –Scope and perspectives, Nobel lecture, December 8, **1987**; http://www.nobelprize.org/nobel_prizes/chemistry/laureates/1987/lehn-lecture.html. Accessed June 3, 2017.

8 M. Mammen, S.-K. Choi, G. M. Whitesides, Polyvalent interactions in biological systems: Implications for design and use of multivalent ligands and inhibitors. *Angew. Chem. Int. Ed.* **1998**, *37*, 2754–2794.

9 A. Mulder, J. Huskens, D. N. Reinhoudt, Multivalency in supramolecular chemistry and nanofabrication. *Org. Biomol. Chem.* **2004**, *2*, 3409–3424.

10 J. D. Badjic, A. Nelson, S. J. Cantrill, W. B. Turnbull, J. F. Stoddart, Multivalency and cooperativity in supramolecular chemistry. *Acc. Chem. Res.* **2005**, *38*, 723–732.

11 C. Fasting, C. A. Schalley, M. Weber, O. Seitz, S. Hecht, B. Koksch, J. Dernedde, C. Graf, E.-W. Knapp, R. Haag, Multivalency as a chemical organization and action principle. *Angew. Chem. Int. Ed.* **2012**, *51*, 10472–10498.

12 C. A. Hunter, H. L. Anderson, What is cooperativity? *Angew. Chem. Int. Ed.* **2009**, *48*, 7488–7499.

13 G. Ercolani, L. Schiaffino, Allosteric, chelate, and interannular cooperativity: A mise au point. *Angew. Chem. Int. Ed.* **2011**, *50*, 1762–1768.

14 J. Huskens, H. van Bekkum, J. A. Peters, A convenient spreadsheet approach to the calculation of stability constants and the simulation of kinetics. *Comp. Chem.* **1995**, *19*, 409–416.

15 M. G. J. ten Cate, J. Huskens, M. Crego-Calama, D. N. Reinhoudt, Thermodynamic stability of hydrogen-bonded nanostructures: a calorimetric study. *Chem. Eur. J.* **2004**, *10*, 3632–3639.

16 M. L. Coval, Analysis of Hill interaction coefficients and the invalidity of the Kwon and Brown equation. *J. Biol. Chem.* **1970**, *245*, 6335–6336.

17 H. d'A. Heck, Statistical theory of cooperative binding to proteins. Hill equation and the binding potential. *J. Am. Chem. Soc.* **1971**, *93*, 23–29.

18 H.-J. Schneider, A. Yatsimirsky, *Principles and Methods in Supramolecular Chemistry*, John Wiley & Sons, Ltd, **1999**.

19 A. Mulder, T. Auletta, A. Sartori, S. Del Ciotto, A. Casnati, R. Ungaro, J. Huskens, D. N. Reinhoudt, Divalent binding of a bis(adamantyl)-functionalized calix[4]arene to β-cyclodextrin-based hosts: an experimental and theoretical study on multivalent binding in solution and at self-assembled monolayers. *J. Am. Chem. Soc.* **2004**, *126*, 6627–6636.

20 P. Atkins, J. De Paula, *Physical Chemistry* (8th edn). W.H. Freeman and Co., **2006**, p. 212.

21 J. J. Michels, J. Huskens, J. F. J. Engbersen, D. N. Reinhoudt, Probing the interactions of calix[4]arene-based amphiphiles and cyclodextrins in water. *Langmuir* **2000**, *16*, 4864–4870.

22 J. J. Michels, J. Huskens, D. N. Reinhoudt, Non-covalent binding of sensitizers for lanthanide(III) luminescence in an EDTA-bis(β-cyclodextrin) ligand. *J. Am. Chem. Soc.* **2002**, *124*, 2056–2064.

23 E. Cavatorta, P. Jonkheijm, J. Huskens, Assessment of cooperativity in ternary peptide-cucurbit[8]uril complexes. *Chem. Eur. J.* **2017**, *23*, 4046–4050.

24 E. Cavatorta, Cucurbit[8]uril ternary complexes for biomolecular assemblies in solution and at interfaces. *PhD Thesis*, University of Twente, **2016**.

25 J. Rao, J. Lahiri, R. M. Weis, G. M. Whitesides, Design, synthesis, and characterization of a high-affinity trivalent system derived from vancomycin and L-Lys-D-Ala-D-Ala. *J. Am. Chem. Soc.* **2000**, *122*, 2698–2710.

26 J. M. Gargano, T. Ngo, Y. Kim, D. W. K. Acheson, W. J. Lees, Multivalent inhibition of AB$_5$ toxins. *J. Am. Chem. Soc.* **2001**, *123*, 12909–12910.

27 R. H. Kramer, J. W. Karpen, Spanning binding sites on allosteric proteins with polymer-linked ligand dimers. *Nature* **1998**, *395*, 710–713.

28 P. I. Kitov, H. Shimizu, S. W. Homans, D. R. Bundle, Optimization of tether length in nonglycosidically linked bivalent ligands that target sites 2 and 1 of a Shiga-like toxin. *J. Am. Chem. Soc.* **2003**, *125*, 3284–3294.

29 J. A. Kirby, Effective molarities for intramolecular reactions. *Adv. Phys. Org. Chem.* **1980**, *17*, 183–278.

30 C. Galli, L. Mandolini, The role of ring strain on the ease of ring closure of bifunctional chain molecules. *Eur. J. Org. Chem.* **2000**, 3117–3125.

31 L. Mandolini, Intramolecular reactions of chain molecules. *Adv. Phys. Org. Chem.* **1986**, *22*, 1–111.

32 G. Ercolani, A model for self-assembly in solution. *J. Phys. Chem. B.* **2003**, *107*, 5052–5057.

33 G. Ercolani, Physical basis of self-assembly macrocyclizations. *J. Phys. Chem. B* **1998**, *102*, 5699–5703.

34 M. A. Winnik, Cyclization and the conformation of hydrocarbon chains. *Chem. Rev.* **1981**, *81*, 491–524.

35 G. Ercolani, Assessment of cooperativity in self-assembly. *J. Am. Chem. Soc.* **2003**, *125*, 16097–16103.

36 S. A. Schmid, R. Abbel, A. P. H. J. Schenning, E. W. Meijer, R. P. Sijbesma, L. M. Herz, Analyzing the molecular weight distribution in supramolecular polymers. *J. Am. Chem. Soc.* **2009**, *131*, 17696–17704.

37 Wikipedia, Langmuir adsorption model; https://en.wikipedia.org/wiki/Langmuir_adsorption_model. Accessed June 3, **2017**.

38 B. C. Lagerholm, N. L. Thompson, Theory for ligand rebinding at cell membrane surfaces. *Biophys. J.* **1998**, *74*, 1215–1228.

39 J. Huskens, A. Mulder, T. Auletta, C. A. Nijhuis, M. J. W. Ludden, D. N. Reinhoudt, A model for describing the thermodynamics of multivalent host-guest interactions at interfaces. *J. Am. Chem. Soc.* **2004**, *126*, 6784–6797.

40 R. B. M. Schasfoort, A. J. Tudos (Eds), *Handbook of Surface Plasmon Resonance*, RSC Publishing, **2008**.

41 D. A. Buttry, M. Ward, Measurement of interfacial processes at electrode surfaces with the electrochemical quartz crystal microbalance. *Chem. Rev.* **1992**, *92*, 1355–1379.

42 K. A. Marx, Quartz crystal microbalance: a useful tool for studying thin polymer films and complex biomolecular systems at the solution-surface interface. *Biomacromolecules* **2003**, *4*, 1099–1120.

43 M. W. J. Beulen, J. Bügler, B. Lammerink, F. A. J. Geurts, E. M. E. F. Biemond, K. G. C. van Leerdam, F. C. J. M. van Veggel, J. F. J. Engbersen, D. N. Reinhoudt, Self-assembled monolayers of heptapodant β-cyclodextrins on gold. *Langmuir* **1998**, *14*, 6424–6429.

44 M. W. J. Beulen, J. Bügler, M. R. de Jong, B. Lammerink, J. Huskens, H. Schönherr, G. J. Vancso, B. A. Boukamp, H. Wieder, A. Offenhäuser, W. Knoll, F. C. J. M. van Veggel, D. N. Reinhoudt, Host-guest interactions at self-assembled monolayers of cyclodextrins on gold. *Chem. Eur. J.* **2000**, *6*, 1176–1183.

45 J. Huskens, M. A. Deij, D. N. Reinhoudt, Attachment of molecules at a molecular printboard by multiple host-guest interactions. *Angew. Chem. Int. Ed.* **2002**, *41*, 4467–4471.

46 M. J. W. Ludden, D. N. Reinhoudt, J. Huskens, Molecular printboards: versatile platforms for the creation and positioning of supramolecular assemblies and materials. *Chem. Soc. Rev.* **2006**, *35*, 1122–1134.

47 S. Onclin, A. Mulder, J. Huskens, B. J. Ravoo, D. N. Reinhoudt, Molecular printboards: monolayers of β-cyclodextrin on silicon oxide surfaces. *Langmuir* **2004**, *20*, 5460–5466.

48 M. R. de Jong, J. Huskens, D. N. Reinhoudt, Influencing the binding selectivity of self-assembled cyclodextrin monolayers on gold through their architecture. *Chem. Eur. J.* **2001**, *7*, 4164–4170.

49 V. M. Krishnamurthy, V. Semetey, P. J. Bracher, N. Shen, G. M. Whitesides, Dependence of effective molarity on linker length for an intramolecular protein-ligand system. *J. Am. Chem. Soc.* **2007**, *129*, 1312–1320.

50 V. M. Krishnamurthy, G. K. Kaufman, A. R. Urbach, I. Gitlin, K. L. Gudiksen, D. B. Weibel, G. M. Whitesides, Carbonic anhydrase as a model for biophysical and physical-organic studies of proteins and protein–ligand binding. *Chem. Rev.* **2008**, *108*, 946–1051.

51 A. Perl, A. Gomez-Casado, D. Thompson, H. H. Dam, P. Jonkheijm, D. N. Reinhoudt, J. Huskens, Gradient-driven motion of multivalent ligand molecules along a surface functionalized with multiple receptors. *Nature Chem.* **2011**, *3*, 317–322.

52 O. Crespo-Biel, B. Dordi, D. N. Reinhoudt, J. Huskens, Supramolecular layer-by-layer assembly: alternating adsorptions of guest- and host-functionalized molecules and particles using multivalent supramolecular interactions. *J. Am. Chem. Soc.* **2005**, *127*, 7594–7600.

53 C. M. Bruinink, C. A. Nijhuis, M. Péter, B. Dordi, O. Crespo-Biel, T. Auletta, A. Mulder, H. Schönherr, G. J. Vancso, J. Huskens, D. N. Reinhoudt, Supramolecular microcontact printing and dip-pen nanolithography on molecular printboards. *Chem. Eur. J.* **2005**, *11*, 3988–3996.

54 O. Crespo-Biel, B. J. Ravoo, D. N. Reinhoudt, J. Huskens, Noncovalent nanoarchitectures on surfaces: from 2D to 3D nanostructures. *J. Mater. Chem.* **2006**, *16*, 3997–4021.

55 X. Y. Ling, D. N. Reinhoudt, J. Huskens, From supramolecular chemistry to nanotechnology: Assembly of 3D nanostructures. *Pure Appl. Chem.* **2009**, *81*, 2225–2233.

56 J. J. Michels, M. W. P. L. Baars, E. W. Meijer, J. Huskens, D. N. Reinhoudt, Well-defined assemblies of adamantyl-terminated poly(propylene imine) dendrimers and β-cyclodextrin in water. *J. Chem. Soc., Perkin Trans. 2* **2000**, 1914–1918.

57 T. Auletta, B. Dordi, A. Mulder, A. Sartori, S. Onclin, C. M. Bruinink, C. A. Nijhuis, H. Beijleveld, M. Péter, H. Schönherr, G. J. Vancso, A. Casnati, R. Ungaro, B. J. Ravoo, J. Huskens, D. N. Reinhoudt, Writing patterns of molecules on molecular printboards. *Angew. Chem. Int. Ed.* **2004**, *43*, 369–373.

58 C. A. Nijhuis, J. Huskens, D. N. Reinhoudt, Binding control and stoichiometry of ferrocenyl dendrimers at a molecular printboard. *J. Am. Chem. Soc.* **2004**, *126*, 12266–12267.

59 C. A. Nijhuis, F. Yu, W. Knoll, J. Huskens, D. N. Reinhoudt, Multivalent dendrimers at molecular printboards: influence of dendrimer structure on binding strength and stoichiometry, and their electrochemically induced desorption. *Langmuir* **2005**, *21*, 7866–7876.

60 C. A. Nijhuis, K. A. Dolatowska, B. J. Ravoo, J. Huskens, D. N. Reinhoudt, Redox-controlled interaction of biferrocenyl-terminated dendrimers with β-cyclodextrin molecular printboards. *Chem. Eur. J.* **2007**, *13*, 69–80.

61 C. A. Nijhuis, B. J. Ravoo, J. Huskens, D. N. Reinhoudt, Electrochemically controlled supramolecular systems. *Coord. Chem. Rev.* **2007**, *251*, 1761–1780.

62 C. A. Nijhuis, B. Boukamp, B. J. Ravoo, J. Huskens, D. N. Reinhoudt, Electrochemistry of ferrocenyl dendrimer – β-cyclodextrin assemblies at the interface of an aqueous solution and a molecular printboard. *J. Phys. Chem. C* **2007**, *111*, 9799–9810.

63 C. A. Nijhuis, J. K. Sinha, G. Wittstock, J. Huskens, B. J. Ravoo, D. N. Reinhoudt, Controlling the supramolecular assembly of redox active dendrimers at molecular printboards by scanning electrochemical microscopy. *Langmuir* **2006**, *22*, 9770–9775.

64 X. Y. Ling, D. N. Reinhoudt, J. Huskens, Reversible attachment of nanostructures at molecular printboards through supramolecular glue. *Chem. Mater.* **2008**, *20*, 3574–3578.

65 A. Qi, J. Brinkmann, J. Huskens, S. Krabbenborg, J. de Boer, P. Jonkheijm, A supramolecular system for the electrochemically controlled release of cells. *Angew. Chem. Int. Ed.* **2012**, *51*, 12233–12237.

66 O. Crespo-Biel, M. Péter, C. M. Bruinink, B. J. Ravoo, D. N. Reinhoudt, J. Huskens, Multivalent host-guest interactions between β-cyclodextrin self-assembled monolayers and poly(isobutene-*alt*-maleic acid)s modified with hydrophobic guest moieties. *Chem. Eur. J.* **2005**, *11*, 2426–2432.

67 L. Graña-Suárez, W. Verboom, J. Huskens, Cyclodextrin-based supramolecular nanoparticles stabilized by balancing attractive host-guest and repulsive electrostatic interactions. *Chem. Commun.* **2014**, *50*, 7280–7282.

68 R. Mejia-Ariza, Gavin A. Kronig, J. Huskens, Size-controlled and redox-responsive supramolecular nanoparticles. *Beilstein J. Org. Chem.* **2015**, *11*, 2388–2399.

69 L. Graña-Suárez, W. Verboom, J. Huskens, Fluorescent supramolecular nanoparticles signal the loading of electrostatically charged cargo*Chem. Commun.* **2016**, *52*, 2597–2600.

70 L. Graña-Suárez, W. Verboom, T. Buckle, M. Rood, F. W. B. van Leeuwen, J. Huskens, Loading and release of fluorescent oligoarginine peptides into pH-responsive anionic supramolecular nanoparticles. *J. Mater. Chem. B* **2016**, *4*, 4025–4032.

71 T. Satav, he self-assembly and dynamics of weakly multivalent, peptide-based, host-guest systems. *PhD Thesis*, University of Twente, **2015**.

72 M. Rekharsky, Y. Inoue, Chiral recognition thermodynamics of β-cyclodextrin: the thermodynamic origin of enantioselectivity and the enthalpy – entropy compensation effect. *J. Am. Chem. Soc.* **2000**, *122*, 4418–4435.

73 F. Corbellini, A. Mulder, A. Sartori, M. J. W. Ludden, A. Casnati, R. Ungaro, J. Huskens, M. Crego-Calama, D. N. Reinhoudt, Assembly of a supramolecular capsule on a molecular printboard. *J. Am. Chem. Soc.* **2004**, *126*, 17050–17058.

74 M. J. W. Ludden, M. Péter, D. N. Reinhoudt, J. Huskens, Attachment of streptavidin to β-cyclodextrin molecular printboards via orthogonal host-guest and protein-ligand interactions. *Small* **2006**, *2*, 1192–1202.

75 M. J. W. Ludden, A. Mulder, R. Tampé, D. N. Reinhoudt, J. Huskens, Molecular printboards as a general platform for protein immobilization: a supramolecular solution to nonspecific adsorption. *Angew. Chem. Int. Ed.* **2007**, *46*, 4104–4107.

76 M. J. W. Ludden, X. Y. Ling, T. Gang, W. P. Bula, H. J. G. E. Gardeniers, D. N. Reinhoudt, J. Huskens, Multivalent binding of small guest molecules and proteins to molecular printboards inside microchannels. *Chem. Eur. J.* **2008**, *14*, 136–142.

77 M. J. W. Ludden, A. Mulder, K. Schulze, V. Subramaniam, R. Tampé, J. Huskens, Anchoring of histidine-tagged proteins to molecular printboards: self-assembly, thermodynamic modeling and patterning. *Chem. Eur. J.* **2008**, *14*, 2044–2051.

78 M. J. W. Ludden, J. K. Sinha, G. Wittstock, D. N. Reinhoudt, J. Huskens, Control over binding stoichiometry and specificity in the supramolecular immobilization of cytochrome c on a molecular printboard. *Org. Biomol. Chem.* **2008**, *6*, 1553–1557.

79 M. J. W. Ludden, X. Li, J. Greve, A. van Amerongen, M. Escalante, V. Subramaniam, D. N. Reinhoudt, J. Huskens, Assembly of bionanostructures onto β-cyclodextrin molecular printboards for antibody recognition and lymphocyte cell counting. *J. Am. Chem. Soc.* **2008**, *130*, 6964–6973.

80 A. González-Campo, B. Eker, H. J. G. E. Gardeniers, J. Huskens, P. Jonkheijm, A supramolecular approach to homogeneous enzyme assays in micro-channels. *Small* **2012**, *8*, 3531–3537.

81 M. Escalante, Y. Zhao, M. J. W. Ludden, R. Vermeij, J. D. Olsen, E. Berenschot, C. N. Hunter, J. Huskens, V. Subramaniam, C. Otto, Nanometer arrays of functional light harvesting antenna complexes by nanoimprint lithography and host-guest interactions. *J. Am. Chem. Soc.* **2008**, *130*, 8892–8893.

82 M. Escalante, A. Lenferink, Y. Zhao, N. Tas, J. Huskens, C. N. Hunter, V. Subramaniam, C. Otto, Long-range energy propagation in nanometer arrays of light harvesting antenna complexes. *Nano Lett.* **2010**, *10*, 1450–1457.

83 J. Cabanas-Danés, E. D. Rodrigues, E. Landman, J. van Weerd, C. van Blitterswijk, T. Verrips, J. Huskens, M. Karperien, P. Jonkheijm, A supramolecular host-guest carrier system for growth factors employing VHH fragments. *J. Am. Chem. Soc.* **2014**, *136*, 12675–12681.

84 O. Crespo-Biel, C. W. Lim, B. J. Ravoo, D. N. Reinhoudt, J. Huskens, Expression of a supramolecular complex at a multivalent interface. *J. Am. Chem. Soc.* **2006**, *128*, 17024–17032.

85 C. W. Lim, O. Crespo-Biel, M. C. A. Stuart, D. N. Reinhoudt, J. Huskens, B. J. Ravoo, Intravesicular and intervesicular interaction by orthogonal multivalent host-guest and metal-ligand complexation. *Proc. Natl. Acad. Sci. USA* **2007**, *104*, 6986–6991.

86 S.-H. Hsu, M. D. Yilmaz, C. Blum, V. Subramaniam, D. N. Reinhoudt, A. H. Velders, J. Huskens, Expression of sensitized Eu^{3+} luminescence at a multivalent interface. *J. Am. Chem. Soc.* **2009**, *131*, 12567–12569.

87 S.-H. Hsu, M. D. Yilmaz, D. N. Reinhoudt, A. H. Velders, J. Huskens, Nonlinear amplification of a supramolecular complex at a multivalent surface. *Angew. Chem. Int. Ed.* **2013**, *52*, 714–719.

88 M. D. Yilmaz, S.-H. Hsu, D. N. Reinhoudt, A. H. Velders, J. Huskens, Ratiometric fluorescent detection of an anthrax biomarker at molecular printboards. *Angew. Chem. Int. Ed.* **2010**, *49*, 5938–5941.

89 H. Schönherr, M. W. J. Beulen, J. Bügler, J. Huskens, F. C. J. M. van Veggel, D. N. Reinhoudt, G. J. Vancso, Individual supramolecular host-guest interactions studied by dynamic single molecule force spectroscopy. *J. Am. Chem. Soc.* **2000**, *122*, 4963–4967.

90 T. Auletta, M. R. de Jong, A. Mulder, F. C. J. M. van Veggel, J. Huskens, D. N. Reinhoudt, S. Zou, S. Zapotoczny, H. Schönherr, G. J. Vancso, L. Kuipers, Cyclodextrin host-guest complexes probed under thermodynamic equilibrium: thermodynamics and AFM force spectroscopy. *J. Am. Chem. Soc.* **2004**, *126*, 1577–1584.

91 A. Gomez-Casado, H. H. Dam, M. D. Yilmaz, D. Florea, P. Jonkheijm, J. Huskens, Probing multivalent interactions in a synthetic host-guest complex by dynamic force spectroscopy. *J. Am. Chem. Soc.* **2011**, *133*, 10849–10857.

92 E. Evans, K. Ritchie, Dynamic strength of molecular adhesion bonds. *Biophys. J.* **1997**, *72*, 1541–1555.

93 G. I. Bell, Models for the specific adhesion of cells to cells. *Science* **1978**, *200*, 618–627.

94 S. O. Krabbenborg, J. Veerbeek, J. Huskens, patially controlled out-of-equilibrium host-guest system under electrochemical control. *Chem. Eur. J.* **2015**, *21*, 9638–9644.

95 K. D. Schleicher, S. L. Dettmer, L. E. Kapinos, S. Pagliara, U. F. Keyser, S. Jeney, R. Y. H. Lim, Selective transport control on molecular velcro made from intrinsically disordered proteins. *Nature Nanotech.* **2014**, *9*, 525–530.

96 J. Huskens, Diffusion: Molecular velcro in Flatland. *Nature Nanotech.* **2014**, *9*, 500–502.

3

Design Principles for Super Selectivity using Multivalent Interactions

Tine Curk[1], Jure Dobnikar[2], and Daan Frenkel[1]

[1] *Department of Chemistry, University of Cambridge, Lensfield Road, Cambridge CB2 1EW, UK*
[2] *Institute of Physics & School of Physical Sciences, Chinese Academy of Sciences, Beijing 100190, China*

3.1 Introduction

Multivalent particles have the ability to form multiple bonds to a substrate. Hence, a multivalent interaction can be strong, even if the individual bonds are weak. However, much more interestingly, multivalency greatly increases the sensitivity of the particle–substrate interaction to external conditions, resulting in an ultra-sensitive and highly non-linear dependence of the binding strength on parameters such as temperature, pH or receptor concentration.

In this chapter we focus on super selectivity: the high sensitivity of the strength of multivalent binding to the number of accessible binding sites on the target surface (see the schematic drawing in Figure 3.1). For example, the docking of a multivalent particle on a cell surface can be very sensitive (super selective) to the concentration of the receptors to which the multiple ligands can bind.

We present a theoretical analysis of systems of multivalent particles and describe the mechanism by which multivalency leads to super selectivity. We introduce a simple analytical model that allows us to predict the overall strength of interactions based on physicochemical characteristics of multivalent binders. Finally, we formulate a set of simple design rules for multivalent interactions that yield optimal selectivity.

3.1.1 Background: Ultra-sensitive Response

Many processes in biology depend ultra-sensitively on variations in one or more of the parameters that control the process. Such ultra-sensitivity manifests itself as an almost switch-like, sigmoidal change in the 'output' when the control parameter crosses a threshold value. Understanding such switch-like behaviour is obviously important to understand many regulatory processes in living systems, but such understanding will also help us design synthetic systems that combine weak supramolecular interactions with high selectivity.

Multivalency: Concepts, Research & Applications, First Edition. Edited by Jurriaan Huskens,
Leonard J. Prins, Rainer Haag, and Bart Jan Ravoo.
© 2018 John Wiley & Sons Ltd. Published 2018 by John Wiley & Sons Ltd.

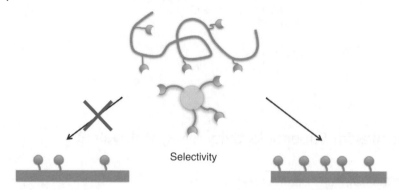

Figure 3.1 Selectivity denotes the ability of multivalent entities to distinguish between substrates depending on the surface density of binding sites.

The best known example of ultra-sensitivity dates back to Hill who, in the beginning of the twentieth century, studied the binding of oxygen to haemoglobin. He found the that the relation between bound oxygen and partial pressure was sigmoidal [1]. Today this phenomenon is explained in terms of allosteric cooperativity whereby the four binding sites on haemoglobin do not act independently but are 'cooperative', that is binding of the first oxygen molecule increases the probability that the second oxygen molecule will bind. Hence, haemoglobin is likely to be either fully loaded with oxygen or empty, which makes haemoglobin an efficient transporter of oxygen between lungs and peripheral tissues. Other examples of ultra-sensitivity include the switch-like response of bacterial motors [2], or the switch-like behaviour in gene regulation due to positive feedback loops in nucleosome modification [3]. For more information on this broad topic, the reader is referred to a review by Ferrell [4, 5, 6] and references therein.

Ultra-sensitive response is usually characterized by a so-called Hill curve:

$$\text{Output} = \frac{\text{Input}^n}{K^n + \text{Input}^n}, \tag{3.1}$$

where the Hill coefficient n quantifies the degree of cooperativity of the process: the higher the Hill coefficient, the more sensitive the response.[1]

Due to cooperativity, blocks that, individually, have limited selectivity can form units that interact selectively. For example, DNA base pairing is highly specific, even though underlying interactions (hydrogen bonding and base-stacking) are not. Multivalent (or polyvalent) interactions can also lead to an ultra-sensitive response, for example, the aggregation of multivalent DNA-coated colloids depends sensitively on temperature [7]. Moreover, ligand–receptor or antibody–antigen interactions, are very sensitive to temperature, but also to ion concentration and pH. Internal protein interactions are also multivalent: protein folding and unfolding depend critically on temperature and

---

1 In order to make this chapter accessible to a broad audience, we keep mathematical expressions in the main text to an absolute minimum: well known relations, such as the Langmuir isotherm, and our final design principles are included because they are needed to understand super selectivity. However, all other equations and mathematical derivations are enclosed in boxes for the aficionado: readers less interested in the mathematical background can skip these without risk.

other external conditions. The functioning of the biochemical machinery in cells relies (mostly) on multivalent supra-molecular interactions. These interactions are very sensitive to external conditions which helps explain why the properties of living matter (cells, tissues) are very sensitive to temperature, while those of 'formerly living' matter (say, a piece of wood) are not.

Multivalent Interactions: Why So Sensitive?

Imagine two multivalent entities at a fixed distance that are connected by a number of bonds (say k). The two entities can dissociate only when all k bonds are broken. We denote the probability that an individual bond is broken by p_1^{unbound} and the probability that all k bonds are broken by p_k^{unbound}. If different bonds do not influence each other, the probability of unbinding is

$$p_k^{\text{unbound}} \sim \left(p_1^{\text{unbound}} \right)^k . \tag{3.2}$$

Note that for large 'valencies' k, the relation between p_1^{unbound} and p_k^{unbound} is highly non-linear. In fact, the expression for the ratio between probabilities $p_k^{\text{unbound}}/p_k^{\text{bound}}$ can be written in a form reminiscent of the Hill equation:

$$\frac{p_k^{\text{unbound}}}{p_k^{\text{bound}}} \sim \frac{\left(p_1^{\text{unbound}} \right)^k}{1 - \left(p_1^{\text{unbound}} \right)^k} , \tag{3.3}$$

where the exponent k plays a role similar to that of the Hill coefficient [Eq. (3.1)]. The probability of a single bond spontaneously breaking p_1^{unbound} will depend not only on control parameters such as bond strength, temperature, pH of the solution and so on, but also on the number of possible bonding arrangements. Clearly, the unbinding probability, Eq. (3.2), tends to be very sensitive to any parameter that influences p_1^{unbound}. This example illustrates the physical origins of ultra-sensitive response in multivalent interactions. We shall see below that competition between different bonds modifies the response but retains ultra-sensitivity.

In what follows, we focus on the ultra-sensitivity of multivalent interactions to the density of 'receptors' on the substrate surface. In particular, we will derive expressions that show how the binding strength of a multivalent entity (say a ligand-decorated nanoparticle or a multivalent polymer) to a substrate changes with the concentration of receptors[2] on the substrate surface (Figure 3.2). It will turn out that multivalent interactions can be designed such that they result in an almost step-like switch from unbound to bound as the receptor concentration exceeds a well-defined threshold value. In the remainder of this chapter, we will use the term 'super selectivity' to denote this kind of sharp response.

---

2 A brief comment on the use of terminology: we make liberal use of the terms 'ligand' and 'receptor' with which we shall denote individual binding partners. 'Receptors' will be found on the substrate surface whilst individual 'ligands' are attached to the multivalent entity (say, a nano-particle) that binds to the substrate, shown in Figures 3.2 and 3.3. We use the term 'multivalent entity' to denote any moiety that is able to form multiple bonds. The term 'binding site' always denotes an individual monovalent interaction site, equivalent to a single 'ligand' or 'receptor'.

(a)

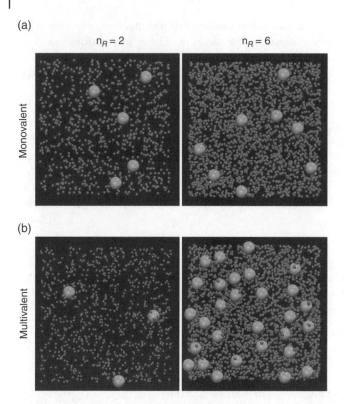

(b)

Figure 3.2 Simulation snapshots comparing the targeting selectivity of monovalent and multivalent guest nanoparticles. We compare the adsorption onto two host surfaces with receptor concentrations (n_R) that differ by a factor of three. (a) The monovalent guests provide little selectivity: increasing by three times the receptor coverage just increases the average number of bound guests by 1.8 (i.e. from 5.4 to 9.7 bound particles on average). (b) The multivalent nanoparticles behave super selectively: an increase of the receptor coverage by a factor three causes a 10-fold increase in the average number of adsorbed particles. The multivalent guests have ten ligands per particle. The individual bonds of the multivalent case (b) are weaker than those in the monovalent case (a). *Source*: Ref. [8]. Reproduced with permission of National Academy of Sciences of the United states of America. See colour section.

The remainder of this chapter is structured as follows: First, we show how the description of simple chemical equilibria and Langmuir adsorption can be extended to multivalent interactions. We then discuss the conditions under which super selectivity appears and formulate simple design principles to achieve super selectivity. We include an appendix where we discuss how, in simple cases, our approach reduces to the widely used 'effective molarity' picture.

3.2   Super Selectivity: An Emergent Property of Multivalency

We first focus on a prototypical system of multivalent particles in solution that can adsorb to a receptor-decorated surface (Figure 3.2). For simplicity, we assume that the surface is flat and much larger than the multivalent particles. Furthermore, we assume that these particles are larger than the surface receptors such that each particle can attach to many receptor sites simultaneously. Adsorption of particles is governed by the well-known Langmuir isotherm which states that the fraction of the surface occupied by particles is

$$\theta = \frac{\rho K_A^{av}}{1 + \rho K_A^{av}},$$
(3.4)

with ρ the molar concentration of particles in solution[3], K_A^{av} is the equilibrium avidity association constant of particles adsorbing to a surface. Note that K_A^{av} is different from the affinity equilibrium constant K_A which specifies chemical equilibria of individual ligand–receptor binding. Avidity (functional affinity) is the accumulated strength of multiple affinities [9].

We aim to understand how the overall avidity constant K_A^{av} depends on the properties of the system, that is individual bond affinities K_A[4], the ligand valency k and number of receptors n_R. The avidity constant includes all possible bound states, and is written as a sum over bonds

$$K_A^{av} = \Omega_1 K_A + \Omega_2 K_A K_{intra} + \Omega_3 K_A K_{intra}^2 + \dots$$
(3.5)

The first term on the right-hand side takes into account all states with a single formed bond, the second term represents all doubly bound states, the third term triply bound states and so on K_{intra} is a constant specifying the internal equilibrium between singly and doubly bonded states. We have assumed that individual bonds form independently and K_{intra} is a constant, that is we ignore (allosteric) cooperative effects. We do this to clearly distinguish multivalent effects (the subject of this chapter) from cooperative effects [10].[5]

Ω_i is the degeneracy pre-factor, it measures the number of ways in which i bonds can be formed between two multivalent entities, see Figure 3.3 for representative cartoons. Degeneracy Ω is often labelled as a 'statistical pre-factor' which denotes something that should be included for rigour but is otherwise not essential. However, as we will show, it is precisely this degeneracy that gives rise to super selectivity. The focus of the majority of theoretical papers [9, 11, 12, 13, 14] is on the calculation of the internal equilibrium constant K_{intra}. Here, instead, we focus on the degeneracy Ω. We will simply assume that K_{intra} is (or can be) known.

The degeneracy Ω depends on the spatial arrangement of both ligands and receptors. However, it is instructive to consider first the binding of flexible ligands, where all k ligands on a particle can bind to n_R receptors (Figure 3.3b). In this case the degeneracy given by Eq. (3.6) becomes a very steep and non-linear function of k and n_R. This form was first considered by Kitov and Bundle [15] and has been applied, among others, to super-selective targeting [8] and modelling the adhesion of influenza virus [16].

---

3 For non-ideal solutions the density ρ in the Langmuir isotherm [Eq. (3.4)] should be replaced by the fugacity.
4 K_A is the association equilibrium constant between a monovalent particle (a single ligand attached to a particle) and a single receptor, we assume it can be determined experimentally.
5 Some authors [11] use the term 'chelate cooperativity' to denote multivalent effects.

Degeneracy Ω

In the 'flexible' binding case where each of the k ligands can bind to every one of the n_R receptors that are available, the number of ways (degeneracy) to form i bonds is

$$\Omega_i = \binom{n_R}{i}\binom{k}{i} i! = \frac{n_R!\ k!}{(n_R - i)!\ (k - i)!\ i!}. \tag{3.6}$$

We need to choose i ligands out of k and choose i receptors out of n_R, then there are $i!$ (that is i factorial) ways of binding the chosen ligands/receptors together.

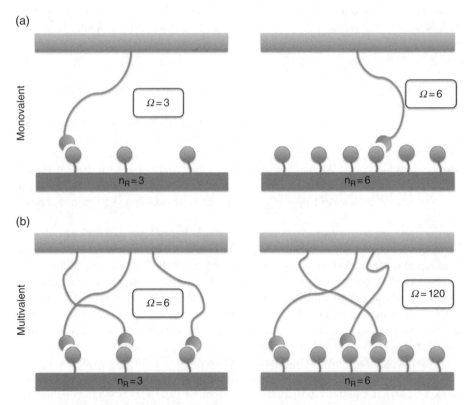

Figure 3.3 Entropic origin of super selectivity. The cartoons give a schematic representation of the simulation snapshots in Figure 3.2. The pictures show the binding of monovalent (a) and multivalent (b) entities (represented as a bar with attached flexible ligands). Receptors are shown as spheres tethered to the bottom surface. The left panels show a low receptor density ($n_R = 3$) and the panels on the right show a receptor density that is twice as high. In the monovalent case the number of distinct ways (Ω) to link ligands and receptors grows linearly with the number of receptors n_R, while the multivalent case show a highly non-linear response: changing n_R from 3 to 6 increases Ω by a factor of 20. In general, the number of binding combinations (degeneracy) Ω is calculated using Eq. (3.6).

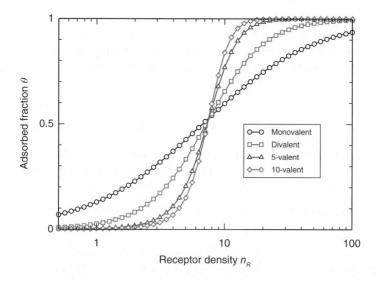

Figure 3.4 Adsorption profile of multivalent particles computed using Eqs (3.4) and (3.7). Monovalent adsorption (circles) $k = 1$ yields the familiar Langmuir isotherm. In contrast, multivalent particles display a steep, sigmoidal response. In the case shown, we have chosen the dimensionless activity in solution to be $z \equiv \rho \dfrac{K_A}{K_{intra}} = 0.001$, the binding affinity of individual bonds decreases as the valency increases from mono-valent to 10-valent: $\log(K_{intra}) = 5, 1.5, -1, -2$, such that the overall avidity K_A^{av} at 50% bound fraction ($\theta = 0.5$) is kept constant for all valencies.

A low fraction of bound receptors in the system can arise either because the number of receptors is greater than the number of available ligands: $n_R \gg k$ or when individual bonds are weak: $K_{intra} \ll 1/k^6$. In this case the avidity constant [Eq. (3.5)], using degeneracy [Eq. (3.6)], can be rewritten[7] to yield a simple form:

$$K_A^{av} \approx \frac{K_A}{K_{intra}}\left[\left(1 + n_R K_{intra}\right)^k - 1\right],\tag{3.7}$$

where, as before, K_A is the monomeric single-bond affinity constant, K_{intra} the internal association constant, and n_R and k are the number of receptors and ligands respectively. For our purpose it is important to note that for multivalent binding ($k > 1$), K_A^{av} is a steep, non-linear function of n_R (Figure 3.4).

Equation (3.7) could have also been obtained directly by reasoning that for non-saturated receptors (fraction of bound receptors is low), competition for the same receptor can be ignored. Each of the k ligands can then bind independently to any of the n_R receptors with an equilibrium constant K_{intra} (weight $n_R K_{intra}$). Alternatively, the ligand is unbound (weight 1). Hence, for systems with a low fraction of bound receptors, the factor $(1 + n_R K_{intra})^k$ accounts (approximately) for all possible states. Furthermore,

---

6 The largest term in Eq. (3.5) is obtained by $\Omega(i)K_{intra}{}^i \approx \Omega(i+1)K_{intra}{}^{i+1}$, which results in $K_{intra} \approx \dfrac{i}{(k-i)(n_R-i)}$. If the bonds are sufficiently weak: $K_{intra} < 1/k$, the largest term will always arise when the fraction of occupied receptors is low: $\dfrac{i}{n_R} < 0.5$.

7 Equation (3.5) becomes a binomial expansion series that we can sum [8].

we subtract 1 because we use the convention that at least a single bond needs to be formed for the multivalent particle to be considered bound. The avidity constant has units of inverse molar concentration. To obtain the correct limiting behaviour in the limit $k = 1$, where $K_A^{av} = n_R K_A$, we must multiply the expression in square brackets by $\dfrac{K_A}{K_{intra}}$.

We note that the ratio $\dfrac{K_A}{K_{intra}} = v_{eff}$ has the dimension of an effective volume v_{eff}. The form of Eq. (3.7) suggests that we can view the multivalent particle adsorption as a two-step process. First, the particle adsorbs from the solution to the surface and comes into a position to start forming bonds, the equilibrium constant of this process is given by the ratio $\dfrac{K_A}{K_{intra}}$. Once the particle is in this position, all of the k ligands can independently form bonds with surface receptors.

In the monovalent case ($k = 1$) the avidity constant reduces to $K_A^{av} = n_R K_A$ and the standard Langmuir isotherm is obtained. Furthermore, expanding Eq. (3.7) in a binomial series and using a maximum term approximation we can insert the maximum term in Eq. (3.4) and obtain the phenomenological Hill equation [Eq. (3.1)]. In the case of very strong individual bonds ($n_R K_{intra} \gg 1$) virtually all k bonds are formed and the avidity becomes[8]: $K_A^{av} \approx n_R^k K_A K_{intra}^{k-1}$.

Notation

In this chapter we choose to work with equilibrium constants and densities as our quantities of choice. However, in earlier work we used a notation based on statistical mechanics. In that case, the central quantities are binding free energies and partition functions. This box provides a translation cheat-sheet between the chemical and statistical mechanical language:

- Gibbs free energy of forming the first bond: $e^{-\beta \Delta G} = K_A \rho_0$
- Binding free energy of subsequent bonds: $e^{-\beta f} = K_{intra}$
- Bound state partition function: $q_b = K_A^{av} \dfrac{K_{intra}}{K_A}$
- Dimensionless activity of multivalent ligands in solution: $z = \rho \dfrac{K_A}{K_{intra}}$

where $\rho_0 = 1M$ is the standard concentration, $\beta = 1/k_B T$ is the inverse of temperature T and k_B the Boltzmann constant. Using these identifications, we can rewrite the surface coverage Eq. (3.4) as

$$\theta = \frac{z q_b}{1 + z q_b}, \tag{3.8}$$

---

8 This holds for $n_R \gg k$ when Eq. (3.7) is applicable even for strong bonds, in general [using Eq. (3.5)] the expression would be $K_A^{av} = \dfrac{n_R!}{(n_R - k)!} K_A K_{intra}^{k-1}$.

where the bound partition function is given by

$$q_b = \left(1 + n_R e^{-\beta f}\right)^k - 1. \tag{3.9}$$

This dimensionless notation was used in Refs [8, 17, 18, 19]. We will use it below when formulating general design principles.

We have shown how combinatorial entropy (also called 'avidity entropy' [9]) gives rise to sharp switching behaviour upon a change in receptor concentration n_R (Figure 3.4). Next, we introduce a measure of the sensitivity of the binding of multivalent particles to the surface concentration of receptors:

$$\alpha = \frac{d\log\theta}{d\log n_R}. \tag{3.10}$$

α is the slope of the adsorption profile in a log–log plot (Figure 3.5). For mono-valent binding the selectivity α is never larger than one, while in the multivalent case the selectivity can reach values greater than one, indicating a supra-linear response. Note that for low surface coverage ($\rho K_A^{av} \ll 1$) the selectivity α is equivalent to the effective Hill coefficient n from Eq. (3.1). However, because we consider all terms (all possible number of bonds) in calculating avidity [Eq. (3.7)], α is not a constant. At very low receptor concentrations the avidity shows a linear dependence on n_R, and $\alpha \approx 1$[9]. Selectivity then grows with increasing receptor concentration n_R until reaching a peak just before the saturation of the surface ($\rho K_A^{av} \approx 1$). We refer to the region with $\alpha > 1$ as the 'super-selective' region. In this region, a small change in the receptor density n_R causes a faster-than-linear change in adsorption θ.

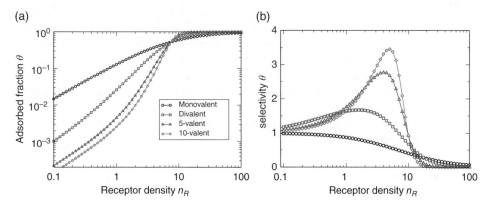

Figure 3.5 Selectivity. (a) shows the log–log plot of Figure 3.4, and (b) shows its slope, that is the selectivity $\alpha = \dfrac{d\log\theta}{d\log n_R}$. We observe that selectivity is typically less than one for monovalent particles indicating at most linear response. Multivalent particles, on the other hand, exhibit a region with values of α significantly greater than one, thus demonstrating that the number of adsorbed ligands increases faster than linearly with the receptor concentration: in this regime, the system is super selective.

---

9 At low receptor concentration for $n_R k K_{intra} \ll 1$, expanding Eq. (3.7) to first order we obtain $K_A^{av} \approx n_R k K_A$.

3.3   Multivalent Polymer Adsorption

To validate the model for super-selective adsorption described above, we now compare its predictions with experimental data on polymer adsorption. Multivalent glycopolymers have been used as selective probes for protein–carbohydrate interactions in a biochemical setting [20, 21, 22]. More recently, super-selective targeting was demonstrated in a synthetic system based on host–guest chemistry [17, 18]. We briefly describe multivalency effects in the case of polymers functionalized with many ligands.

We consider a flexible polymer with a contour length much larger than the persistence length. Ligands are randomly attached along the polymer chain (Figure 3.6a). Similar to the nanoparticles case above, a reasonable first assumption is that, due to polymer-chain flexibility, all k ligands on a polymer can bind to any of the n_R receptors within a domain on the surface with lateral dimensions comparable with those of the polymer. For simplicity, we describe the surface as a square lattice. The cells of the lattice have linear dimensions comparable with the radius of gyration R_g of the polymer. As in the case of soft multivalent particles, any ligand on the polymer can bind to any receptor in one (and only one) lattice cell, see Figure 3.6. The model is expected to offer a faithful description of the real system if the mean distance between ligands is larger than the Kuhn segment length such that even consecutive ligands along the polymer chain can be treated as uncorrelated.

Multivalent Polymer: A Cloud of Ideal Ligands

The calculation of the avidity constant, via Eq. (3.7), is the same for multivalent polymers or particles. In the case of flexible polymers we can also estimate the intra association constant as $K_{intra} = K_A c_{eff}$, with the effective concentration $c_{eff} \approx 1/N_A a^3$, the lattice size $a = R_g(4\pi/3)^{1/3}$ and R_g the polymer radius of gyration, N_A is Avogadro's number. This model (and the choice of effective concentration) effectively describes a multivalent polymer as a cloud of ideal gas ligands. Ligands are uncorrelated (can bind independently) but must stay within a lattice site with volume a^3, see Figure 3.6. The number of receptors that a polymer can see is then $n_R = \Gamma N_A a^2$, where Γ denotes the molar surface density of receptors.

Using the above definitions, we find the following expression for the avidity constant of a multivalent polymer:

$$K_A^{av} = a^3 N_A \left[\left(1 + \frac{\Gamma K_A}{a} e^{-\beta U_{poly}} \right)^k - 1 \right],$$ (3.11)

which is the equation used to obtain adsorption profiles in Figure 3.7. We have added a correction term U_{poly} which takes into account the deviation of the real system to our 'cloud of ideal ligands' approximation. This approximation neglects the polymeric degrees of freedom and, consequently, any spatial correlations between ligands. Moreover, we ignore the fact that the binding free energy of ligands to receptors is changed by the coupling of the ligands to the polymer backbone. These approximations will result in an error of order $k_B T$ and we expect U_{poly} to be $\mathcal{O}(k_B T)$.

(a)

(b)

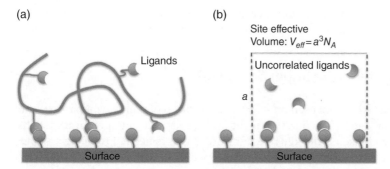

Figure 3.6 Cartoon of the multivalent polymer model. (a) Flexible multivalent polymer close to the receptor decorated surface is modelled as (b) uncorrelated ligands within a lattice site with volume $V_{eff} = \dfrac{K_A}{K_{intra}} = a^3 N_A$ and a the linear lattice size. The ligands can move and bind to receptors independently within the lattice site, but cannot escape the site individually.

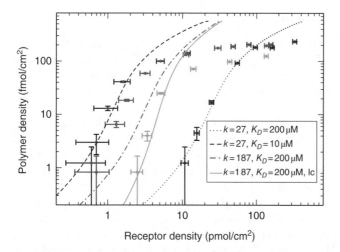

Figure 3.7 Multivalent polymer adsorption. Experimental adsorption profiles (points with error bars) for hyaluronic acid polymers functionalized with β-cyclodextrin hosts (HA-β-CD) binding to surface attached adamantene (affinity $K_D = 1/K_A = 10\,M$) or ferrocene ($K_D = 200M$) guests. As can be seen, the theoretical adsorption profiles (dashed, dotted or solid lines) match the experimental data well for all valencies (k), affinities (K_D) and polymer concentration studies. In the key, 'lc' denotes lower concentration of polymers in solution. One parameter [U_{poly} in Eq. (3.11)] was fitted, the value $U_{poly} = 4.6k_BT$ provides a good fit to all data points. *Source:* Adapted from Refs [17, 18].

The analytical expression given by Eq. (3.11) captures the essentials of multivalent polymer adsorption[10]. Importantly the model allows us to predict adsorption profiles and selectivities depending on the physicochemical properties of multivalent polymers, shown in Figure 3.7. Hence, use of the simple theoretical expression given by Eq. (3.11) allows us

---

10 Reference [18] also considered the effect of interpenetration of adsorbed polymers. This effect yielded a slightly more complicated theoretical expression. However, the important results, scaling relations and design guidelines, are fully captured by Eq. (3.11).

to design a multivalent polymer such that it will selectively target a desired receptor density. In other words, Eq. (3.11) offers a tool for the rational design of selective targeting.

3.4 Which Systems are Super Selective?

The discussion thus far has focused on selective adsorption of multivalent particles and polymers. We now generalize our treatment and discuss various practical systems. In particular, we will discuss the key role of disorder that is needed to observe super-selective behaviour in multivalent interactions. Specifically, what is needed is that a multivalent entity can bind in many different ways to a receptor-decorated substrate. This kind of disorder is usually not possible for multivalent interactions on the angstrom or nanometre scale, as the interacting units tend to be effectively rigid on that scale. In contrast, larger supramolecular systems (e.g. the binding of a multivalent polymer to a receptor decorated membrane) can sustain the 'disordered' interactions.

3.4.1 Rigid Geometry Interactions

A prototypical example of multivalent interactions is the fixed (rigid) geometry multi-valency shown in Figure 3.8. Two rigid, multivalent entities bind via multiple bonds: as the geometry is rigid, individual bonds either fit together, or they do not. Examples of this kind of interaction include the base pairing between nucleotides in complementary sequences of single-stranded DNA.

Another well-known example of a rigid multivalent interaction is the binding between an enzyme and a substrate. The interaction between a pair of proteins is multivalent, as it involves a number of local interactions of various types (hydrogen bonding, hydrophobic, Van der Waals, electrostatic etc.). To a first approximation the enzyme and substrate can be described as rigid objects. This is a simplification as proteins, even in their native state, are not entirely rigid. In any given relative orientation of the ligand to a substrate we find a two-dimensional equivalent of the Figure 3.8. We name this class of multivalent interactions 'rigid geometry multivalency'.

Due to the lack of flexibility of individual bonds, rigid multivalency will generally not show super-selective behaviour. To understand this, consider a simple one-dimensional example of a sequence of rigidly positioned ligands that bind to a commensurate

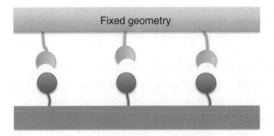

Figure 3.8 Rigid geometry multivalency. The cartoon presents a prototypical fixed geometry interaction where bonds are commensurate (e.g. DNA base pairing or enzyme–substrate interactions). Such systems generally do not exhibit super-selective behaviour as we cannot increase the (binding site) density on one multivalent entity without breaking the commensurability of the bonds. See also Eq. (3.12) and the discussion in the corresponding box.

sequence of receptors. One cannot increase the binding site density on the substrate without breaking the commensurability of the binding. Hence, increasing the receptor density will normally decrease the binding strength. In other words: commensurate lock-and-key interactions are not super selective. Interestingly, it seems that the ability of rigid multivalent particles to detect commensurate structures is exploited in nature, for instance in the activation of certain Toll-like receptors [19].

Rigid Geometry (Commensurate) Multivalency Can Be Super Selective, But Usually Is Not

The simplest mean-field model for the commensurable binding case (Figure 3.8) is that every bond pair is equivalent and can be either formed (weight e^{-f/k_BT}) or not (weight 1), and all I bond pairs are independent. The avidity constant $K_A^{av\,fix}$ of the multivalent interaction is proportional to the bound partition function q_b^{fix} taking into account all possible states

$$K_A^{av\,fix} \propto q_b^{fix} \approx \left(1 + e^{-f/k_BT}\right)^I. \tag{3.12}$$

Evidently the avidity constant is very sensitive to the number of possible bond pairs I, the temperature T and the individual bond strength f. The number of possible pairs I depends on the geometry of the interaction. In the simplest model the number of pairs is given by $I = \min[n_R, k]$, limited by whichever substrate or the multivalent entity has a smaller number of sites [8]. Hence, rigid geometry multivalent interactions can show super-selective behaviour, but only when the multivalent construct initially had an excess number of binding sites compared with the substrate. Furthermore, when increasing the number of binding sites on the substrate, geometric constraints (commensurability) must be obeyed.

3.4.2 Disordered Multivalency

Super-selective behaviour can be exhibited by multivalent systems that can increase the number of possible bonds as the density of receptors increases. As we saw above, fully ordered multivalent systems only bind optimally to commensurate receptor arrangements. To achieve super selectivity, we typically need some kind of disorder or randomness in the geometry of binding. The ability to increase the number of bonds with increasing receptor density can be due to: (i) long, flexible binders; (ii) mobile receptors; or (iii) random binder positions. Figure 3.9 shows schematic examples of these three cases. Different types of bond disorder will result in different expressions for the bound partition functions (and therefore, for the avidity constants), see Eqs (3.13–3.15). However, they all show similar super-selective behaviour (Figure 3.10).

Disordered Systems: Different, Yet Similar

Different forms of disorder may cause super-selective behaviour in multivalent systems. The theoretical expressions for the partition function (and hence the avidity constant) of the bound state will depend on the nature of the disorder. Below, we list a few examples:

- Long flexible ligands, Figure 3.9a); the number of ligands and receptors is fixed and all k ligands can reach any of the n_R receptors

$$q_b\left(n_R, k\right) = \sum_{i=1}^{\min(n_R, k)} \binom{n_R}{i}\binom{k}{i} i! \, e^{-\beta i f},$$ (3.13)

which is the expression that we have already used above [see Eqs (3.5,3.6)].
- Mobile receptors (Figure 3.9b); the number n_R of accessible receptors fluctuates and is Poisson distributed with mean $\tilde{n}_R$. Poisson averaging of Eq. (3.13) over n_R, we find

$$q_b\left(\tilde{n}_R, k\right) = \left(1 + \tilde{n}_R e^{-\beta f}\right)^k - 1,$$ (3.14)

which is the same as expressions (3.7, 3.9) already considered above.
- Large colloids (or cells) with disordered or mobile ligand positions (Figure 3.9c); both the number of ligands and the number of receptors are Poisson distributed with mean $\tilde{k}$ and $\tilde{n}_R$, respectively. Poisson averaging Eq. (3.14) over k, we find

$$q_b\left(\tilde{n}_R, \tilde{k}\right) = e^{\tilde{n}_R \tilde{k} e^{-\beta f}} - 1.$$ (3.15)

A comparison between the predicted behaviour of these different systems is shown in Figure 3.10. In the limit of high valency ($k \gg 1$, $n_R \gg 1$) and weak bonds ($n_R e^{-\beta f} < 1$, $k e^{-\beta f} < 1$) the behaviour of all systems converges to the same form.

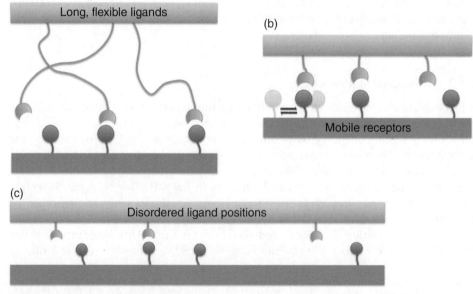

(a)
Long, flexible ligands

(b)

Mobile receptors

(c)
Disordered ligand positions

Figure 3.9 Disordered multivalent systems. Three characteristic types of multivalent interactions are shown: (a) long, flexible binders, (b) mobile receptors, (c) disordered, random positions of individual binders. Different types can behave slightly differently, see the adsorption profiles in Figure 3.10. However, they all exhibit super selectivity, and consequently, any practical system that is similar to at least one of them, will be super selective.

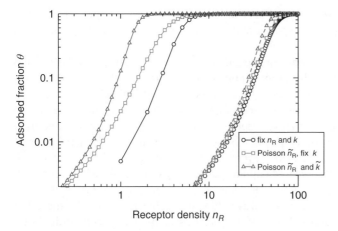

Figure 3.10 Adsorption profile for different disordered systems, depicted in Figure 3.9. Different systems show qualitatively similar super-selective behaviour. For large number of bonds the adsorption profiles converge. The plots shown were generated using the Langmuir expression for the adsorption isotherm Eq. (3.8) with activity $z = 0.001$. We used expressions (3.13, 3.14, 3.15) with $k = 5$ and $\beta f = 0$ to compute the adsorption isotherm in the case of a few strong bonds (solid lines). To represent the case of many weak bonds (dashed lines), we used the same equations but assumed $k = 25$ and $\beta f = 5$.

At first sight, it would seem that the case of mobile receptors shown in Figure 3.9b should be rather different from the immobile case. However, since the receptors are mobile, each ligand can, in principle, bind to any receptor. In this light the two problems become very similar. Another way of looking at the system with mobile receptors is to consider the receptors as a (two-dimensional) 'ideal gas' of particles that can bind to the ligands with an interaction strength f. Up to a concentration-independent term μ_R^0, the chemical potential of these receptors is given by $\mu_R \approx k_B T \log(n_R)$. A small change in the receptor concentration n_R leads to a small change in the chemical potential μ_R, which alters the probability of each and every individual ligand binding. For multivalent particles a small change per ligand adds up to a large change per particle.[11] Clearly, the binding probability depends on n_R, see Refs [19, 23] for practical examples of super selectivity with mobile receptors. We note that for dilute receptors the chemical potential is dominated by the translational entropy. Hence, the origin of super selectivity is entropic, also for mobile receptors.

Finally, for immobile randomly distributed binders shown in Figure 3.9c the intuitive reasoning for super selectivity follows from our initial discussion in the introduction. Let us consider two ligand/receptor-decorated multivalent nanoparticles, A and B that can attach through ligand–receptor binding. The binding moieties are randomly distributed on both nanoparticles. From a point of view of a particular ligand on particle A, the probability of it binding, p_{1A}, is to a first approximation linear in the density n_R of complementary receptors on particle B. The number of possible bonds in the

---

11 If we assume that there are many more receptors than ligands, we can then write the bound-state partition function for k ligands as $q_b \approx (1 + c n_R e^{-\beta f})^k - 1$, where the constant c depends only on the concentration-independent part of the chemical potential μ_R as $c = -k_B T \ln \mu_R^0$. In the case of many weak binders: $c n_R e^{-\beta f} \ll 1$ and we can approximate $q_b \approx e^{k c n_R e^{-\beta f}} - 1$.

contact area is proportional to the number of ligands k in that area. The net result is that the binding probability depends exponentially on the product of k and n_R, as would follow from Eq. (3.15).

We note that in the cases of fixed short ligands we have only illustrated and discussed the two limiting cases: (i) perfectly complementary rigid interaction (Figure 3.8); and (ii) disordered interaction case (Figure 3.9c). Practical systems will fall between these two extremes. As a rule of thumb, small molecules and macromolecules, such as DNA or proteins, or virus capsids have a rather well defined geometry and we expect their interactions to be closer to the rigid geometry case. On the other hand, the spatial distribution of binders (ligands) on entities larger than a few nanometres is, in general, more disordered; be they man-made such as DNA coated colloids [24, 25, 26], or natural such as cells.

We have presented simple analytical models that can be used to rationalize and understand super selectivity in various multivalent systems. In the case of polymers, the simple model works very well (3.7). However, certain systems have been studied in a greater detail. For these cases, more sophisticated (and more complex) models have been developed. For example, cell endocytosis of a virus is mediated by a multivalent interaction between membrane proteins (receptors) and virus capsid proteins (ligands). But to model the process, one should account for membrane elasticity and, in some cases, also for active processes [27]. More detailed models of multivalent polymer adsorption have recently been developed [28, 29]. A theory of valence-limited interactions explicitly taking into account specific positions and different types of tethered binders requires the self-consistent solution of a system of equations [30, 31], the framework was also extended to mobile ligands [32]. A complementary approach is based on a saddle-point approximation for the binding free energy [33]. We note that the results presented in these papers support the conclusions about super-selective behaviour that we have obtained here using much simpler models.

3.5 Design Principles for Super-Selective Targeting

Clearly, super-selective targeting has important practical applications (as even viruses seem to 'know'). It is therefore important to formulate design principles for achieving optimal super selectivity. To formulate design rules, we start once again from the simple model described above: multivalent particle docking to a receptor-decorated surface (e.g. a cell). The density of receptors on the surface is again measured by n_R, the mean number of receptors in the contact area (i.e. the area accessible to a docked particle). In many cases of practical interest, we aim to target only those surfaces (e.g. a cell surface) that have a receptor concentration above a certain threshold. How should we design the particle to target this surface optimally? Our control parameters are the valency k, the ligand–receptor binding strength f, and the activity of particles in solution z.

Optimizing the Selectivity

In terms of the theoretical expressions, Eqs (3.8) and (3.9), we aim to maximize the selectivity

$$\alpha\left(n_R\right) = \frac{\partial \log \theta}{\partial \log n_R} \tag{3.16}$$

at a given desired receptor density n_R. We note that partition function q_b [Eq. (3.9)] and its derivative are increasing functions of n_R, k and $-f$. Hence, we expect the selectivity (slope) to be the highest just before denominator in Eq. (3.8) becomes important and the maximal selectivity will be found when $zq_b \approx 1$. Using Eq. (3.9) we can solve this equation, which yields a relation between k and f

$$k = \frac{-\log(z)}{\log\left(1 + n_R e^{-\beta f}\right)}. \tag{3.17}$$

When $zq \approx 1$ we also have approximately $\theta \approx \frac{1}{2} zq$ and the selectivity becomes

$$\alpha \approx k \frac{n_R e^{-\beta f}}{1 + n_R e^{-\beta f}} = -\log(z) \frac{n_R e^{-\beta f}}{\left(1 + n_R e^{-\beta f}\right)\log\left(1 + n_R e^{-\beta f}\right)}, \tag{3.18}$$

where, in the last step, we used Eq. (3.17).

Expanding the above function to first order for weak/strong binding we find the characteristic behaviour: In the case of strong binding the selectivity is

$$\alpha\left(n_R e^{-\beta f} > 1\right) \approx \frac{-\log(z)}{-\beta f + \log(n_R)}, \tag{3.19}$$

and in the weak binding limit

$$\alpha\left(n_R e^{-\beta f} < 1\right) \approx -\log(z). \tag{3.20}$$

Clearly, the selectivity is maximal in the weak-binding limit and is determined by the logarithm of the activity, see landscape plots in Figure 3.11. In the strong-binding limit, the selectivity decreases with increasing strength of the individual bonds. We remember that $z = \rho \dfrac{K_A}{K_{intra}}$ and $e^{-\beta f} = K_{intra}$.

The landscape plots of selectivity as a function of the valency k and bond strength f are shown in Figure 3.11. We immediately notice three features: (i) High selectivity appears only in a small region of the parameter space, along the curve predicted by Eq. (3.17). (ii) The selectivity reaches a plateau value at large valencies k and weak individual bonds. (iii) Maximum selectivity is limited by the activity z; lowering the activity (or density) of multivalent particles yields a higher selectivity.

The dimensionless activity $z = \rho \dfrac{K_A}{K_{intra}}$ depends on the density ρ, but also on the ratio of the equilibrium constants for the formation of the first bond, and for the formation of subsequent ligand–receptor bonds in a particle–substrate complex [see Eq. (3.5)].

Therefore, even at large densities, selectivity can be substantial if the ratio $\dfrac{K_A}{K_{intra}}$ is small.

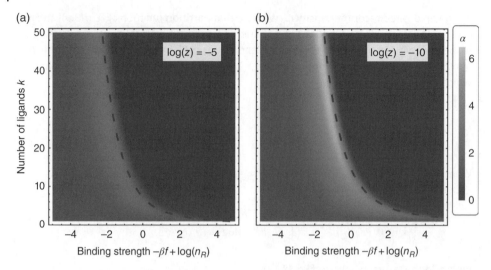

Figure 3.11 The selectivity landscape as function of the valency k and the rescaled binding strength $-\beta f + \log(n_R)$. The landscape was obtained by calculating the selectivity α using Eq. (3.16). The activity of multivalent particles was chosen as: (a) $z = \exp(-5)$; and (b) $z = \exp(-10)$. Both plots use the same colour scale. The dashed curves represent the approximate optimal selectivity relation given by Eq. (3.17), which rather accurately fits the maximum selectivity region. See colour section.

This can be achieved by adding a non-specific repulsion between the multivalent entities (for instance, by coating the particle with inert polymer that provides steric repulsion [34]). Such a repulsion would present a barrier to particle association but would not prevent additional bonds from forming once the barrier is overcome: the result would be a reduction in K_A due to repulsion, but as K_{intra} would be less affected, this steric repulsion would decrease the ratio $\dfrac{K_A}{K_{intra}}$.

Our calculations show that selectivity is suboptimal when using high affinity bonds. However, strong affinity multivalent constructs can still behave super selectively ($\alpha > 1$) if their activity (concentration) in the solution is low enough, see Eq. (3.19). This suggests that, although in principle it is possible to design a super-selective system based on very strong affinity interactions, such as the biotin–streptavidin pair, such a system would only be super selective at extremely low concentrations where the kinetics would be too slow for practical applications.

Multivalency leads to super selectivity, but it also leads to high sensitivity of binding to the variation in other relevant quantities. Therefore, in practical applications, it is important to control (or, at least know) parameters such as temperature, pH, ionic binding strength when using multivalent particles for selective targeting. The parameter range that yields high selectivity is rather small, see Figure 3.11b. A brute-force 'random' search in design-parameter space is, therefore, unlikely to find the optimal selectivity region. We hope that the theoretical guidelines and design principles set forth in this chapter will enable a more rational design of particles for super-selective targeting.

We condense the results shown in Figure 3.11 and our theoretical considerations, Eq. (3.18), in a set of simple design rules for multivalent binding that yield maximum selectivity.

We use our dimensionless statistical mechanics notation, which can be straightforwardly converted to chemical equilibrium units using $z = \rho \dfrac{K_A}{K_{intra}}$ and $e^{-\beta f} = K_{intra}$, as discussed in the Notation box.

1) The maximal possible selectivity α is limited by the activity of multivalent particles in solution: $\alpha_{max} - \log(z)$ so the activity z of multivalent binders should be small.
2) Many weak bonds are better than a few strong ones. The selectivity is also limited by the valency k, until a point of saturation given by $k \sim -\log(z)$. The first two design rules together state that the maximal selectivity is limited by either the valency k or the $-\log(z)$, whichever is smaller.
3) The relationship between the ligand number k and binding strength f should be obeyed: $k = \dfrac{-\log(z)}{\log\left(1 + n_R e^{-\beta f}\right)}$. Together with the above rule, this one states that to achieve maximal selectivity individual bonds should be very weak $K_{intra} = e^{-\beta f} < 1/n_R$. In other words, the fraction of bound receptors/ligands should always remain small.

The main assumptions used to arrive at these design rules are: (i) ligands are identical and bind independently; (ii) all ligands of a (surface bound) multivalent construct can reach all surface attached receptors within a lattice site, but cannot bind to any receptor outside of the site (Figure 3.6); and (iii) receptors, ligands or particles have no interactions except for the steric repulsion and ligand–receptor affinity.

3.6 Summary: It is interesting, but is it useful?

We have shown that weak, multivalent interactions can result in a super-selective behaviour where the overall interaction strength becomes very sensitive to the concentration of individual binders (receptors). We presented a simple yet powerful analytical model with good predictive power for designing multivalent interactions. We expect that, even in cases where the simple model fails quantitatively, the above design rules will still provide a good starting point for designing super selectivity in practical multivalent systems. Figure 3.12 summarizes advantages of weak multivalent interactions in selective targeting.

We can imagine effective purification devices where nano objects of different valencies are passed through super-selective sieves. In the field of material self-assembly, multivalent supramolecular entities could be designed to hierarchically assemble depending on the valency, thus enhancing the precision of self-assembled constructs [25].

The ability to target diseased cells pathogens based on the surface concentration of certain (over)expressed receptors would be of huge practical importance. At present, the delivery of pharmaceutical compounds to specific cells is usually based on the existence of a specific marker (e.g. a sugar or a peptide fragment) that is unique to the targeted cell type. The current wisdom seems to be to functionalize drugs or drug carriers such that they bind strongly to the specific marker. This strategy is fine if the target cells (e.g. bacteria) are very different from the cells of the host, and carry very different markers.

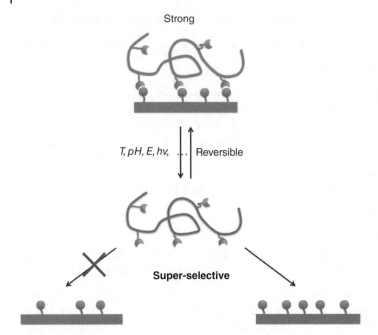

Figure 3.12 Advantages of using weak bonds. Contrary to strong monovalent antibody–antigen interactions and covalent bonds, multiple weak complexes can be disassembled (one by one) using different environmental stimuli (temperature, interaction strength, pH, light), which provides flexibility and reversibility. Examples of systems that exploit multivalency are dendrimers [35], stimulus-responsive coatings [36], renewable sensors for biomolecules [37], reversible gels [38] and gel-particle glue [39]. Importantly, external stimuli can be used to tune the super-selectivity region to the desired surface density of receptors. For example, one could exploit the acidic extracellular environment of tumour tissues to improve the efficiency of drug targeting using multivalent particles.

However, the strong-binding strategy becomes problematic if one wishes to target, say, cancerous cells, which are usually very similar to our healthy cells. Cancerous cells typically over-express markers that are also present, be it in smaller quantities, on healthy cell surfaces. Examples are the CD44 ('don't eat me' receptor) or the folic receptor. In such cases, a compound that binds strongly to the over-expressed marker will also bind to (and kill) healthy cells. The insensitivity of strong binders to the surface concentration of markers is one of the main reasons why antibiotics can be efficient with few side effects (in most patients), while chemotherapy is directly harmful to our body.

As outlined in this chapter, carefully designed multivalent drugs could be targeted super selectively only to cells with cognate receptor concentration above a certain threshold value [8, 40, 41]. Furthermore, in a living cell, receptor interactions and signalling play a major role which can further enhance the non-linear response of the system [42, 43, 44, 45, 46].

Multivalency extends the sensitivity of interactions into the receptor density domain. Moreover, it enables the design of specific, highly selective interactions based on the concentration of ligands or binders, as well as on their chemical nature, thus opening up the possibility for selective targeting with minimal side effects.

Appendix 3.A: What Is Effective Molarity?

Effective molarity (*EM*) is an empirical concept that is commonly used to relate the kinetics and equilibria of intramolecular and intermolecular reactions [9, 10, 11]. It is defined as

$$EM = \frac{K_A^{intra}}{K_A^{inter}}, \tag{3.A.1}$$

where K_A^{intra} and K_A^{inter} are the equilibrium association constants. *EM* has units of molar concentration and is a useful measure of multivalent interactions efficacy, see Figure 3.A.1. For example, when the concentration ρ of multivalent ligands in solution is high $\rho \gg EM$ multivalent effects are suppressed and ligands will bind monovalently. On the other hand when $\rho \ll EM$ multivalent interactions dominate over monovalent binding. Additionally, *EM* allows us to de-convolute the intra equilibrium constant into

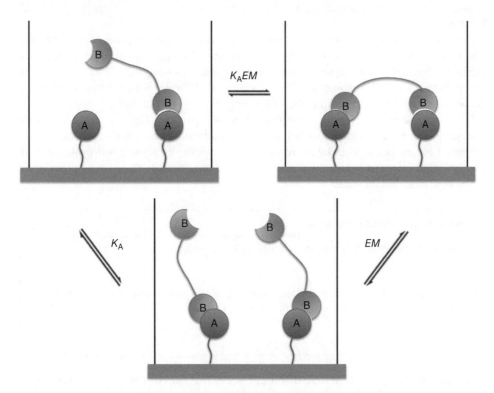

Figure 3.A.1 The concept of effective molarity. The above cycle shows the three different states that divalent ligands (*BB*) can bind to two receptors (*AA*) (unbound state is omitted). We have three distinct states and, therefore, need two equilibrium constants to characterize the equilibrium properties of the system, K_A and *EM*. A product of the two is often called an intra association constant $K_A^{intra} = K_A EM$. A useful reference point is that for a divalent ligand/receptor system and saturated receptors, *EM* determines the concentration of divalent ligands [*BB*] in solution at which we expect equal number of singly and doubly bonded ligands.

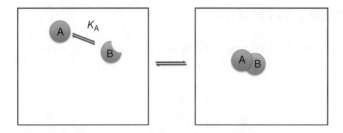

Figure 3.A.2 Dimerization reaction in a small box. We have two particles (a single A-type and a single B-type) in a box with volume V. We assume that, although the particles can bind, they do otherwise behave as an ideal gas. We wish to calculate the relation between the probability of dimerization and equilibrium association constant K_A. Simply calculating effective concentrations of $[A]$, $[B]$ and $[AB]$, and using standard chemical equilibrium equation $\frac{[AB]}{[A][B]} = K_A$ gives a wrong answer, see boxes on dimer reactions.

a simple part (K_A) due to bond formation, and a complicated part (EM) related to the change of conformational entropy and free energy upon binding, see Refs [9, 10, 12, 13] for more discussion.

However, it is important not to over-interpret the meaning of 'effective' concentrations. The name suggests that we can calculate the internal chemical equilibria of multivalent interactions simply by using some effective concentrations of ligands. That, however, is not quite the case, as the expressions for association equilibrium between two compounds do not carry over to the situation when the numbers involved are small.

Let us consider a prototypical system: Only two particles (ligands) in a box with volume V. The particles can associate with an equilibrium constant K_A that was predetermined for us, see Figure 3.A.2. We wish to calculate the association probability of these two particles. To obtain the correct result we can calculate the partition functions of the bound/unbound state.

Dimerization: Correct Calculation

The unbound partition function of two molecules in the box is

$$q_u = V^2,\tag{3.A.2}$$

since we assume both particles are non-interacting and can independently explore the entire box volume V. The bound partition function is

$$q_b = Vv_0 e^{-\beta\Delta G},\tag{3.A.3}$$

with $e^{-\beta\Delta G} = K_A\rho_0$ the dimerization free energy and $v_0 = \dfrac{1}{\rho_0 N_A}$ the microscopic volume of the bond and $\rho_0 = 1M$ the standard concentration. The ratio of the partition functions determines the probability that a dimer is formed.

$$\frac{1-p_u}{p_u} = \frac{q_b}{q_u} = \frac{K_A}{VN_A},\tag{3.A.4}$$

with p_u denoting the unbound probability and the probability that two particles are bound is simply $p_b = 1-p_u$.

On the other hand, if we naively make use of the expression for chemical equilibrium in a bulk mixture binary chemical equilibrium, we do not reproduce the correct result.

Dimerization: Wrong Calculation

We could simply rationalize that the effective (time averaged) concentration of unbound chemicals is

$$[A] = [B] = \frac{p_u}{VN_A},$$

(3.A.5)

where p_u is the probability that A and B are unbound, V is the box volume and we have added the Avogadro's number N_A to make $[A]$ and $[B]$ a molar concentration. Similarly for

the dimerized state $[AB] = \frac{1 - p_u}{VN_A}$. Hence, in line with standard chemical dimerization

reaction, we could reason that

$$K_A = \frac{[AB]}{[A][B]} = \frac{1 - p_u}{p_u^2} VN_A,$$

(3.A.6)

which is clearly different from Eq. (3.A.4).

Treating the system as a bulk binary reaction is not valid for only two dimerizing particles. The approach is valid in the thermodynamic limit where the chemical potential of a molecular species can be related to the logarithm of its concentration. What it boils down to is that Stirling's approximation is valid only for large number of particles $\log N! \approx N \log N - N$, it is clearly wrong when N equals 1 or 2. The same problem occurs when trying to calculate equilibrium constant from molecular dynamics simulations using small system sizes [47].

The above example might seem rather abstract. However, it exposes a potential pitfall of misusing 'effective' concentrations. The same pitfall is encountered when calculating binding probabilities of multivalent ligands, because the reactions shown in Figures 3.A.1 and 3.A.2 are very similar. For example, one could naively argue that both the unbound ligand (A) and receptor (B) in Figure 3.A.1 are flexible and can explore some effective volume V and have some effective concentration within this volume. One then applies a 'Local chemical equilibrium' (LCE) assumption [24, 48] which, in our simple system is given by Eqs (3.A.5, 3.A.6). But this procedure does not generally give a correct result. It becomes a good approximation only in the limit of weak binding[12] or a very large valency where the Stirling's approximation becomes applicable.

It should be clear that effective molarity is not really a concentration.[13] Rather, it is a quantity with the dimensions of concentration, defined by Eq. (3.A.1). We can view the effective molarity as a measure for the probability that an unbound ligand and receptor

---

12 For weak binding $p_u \approx 1$ and Eq. (3.A.6) becomes a very good approximation to Eq. (3.A.4)
13 The effective molarity can be calculated via relative concentrations of singly and doubly bound states in solution, see Figure 3.A.1, but *EM* as such is determined only by the interaction between the two multivalent binders.

would overlap in space (and hence come into position to bind). In an idealized system, neglecting the effects of the linker and orientational correlations in the unbound state, this probability is related to an effective concentration of, say, a ligand (*B*) as experienced by its complementary receptor (*A*) [9, 12, 14]. This is exactly the 'cloud of ideal ligands' approximation we have used as a starting point for our theory of multivalent polymer adsorption, Eq. (3.11). In the case of our simplified system of two dimerizing particles (Figure 3.A.2) the effective concentration c_{eff} of type-*A*, as experienced by type-*B* (or vice versa) is

$$c_{eff} = 1/(VN_A), \tag{3.A.7}$$

where we recall that *V* is the box volume. We can think of particle *A* adsorbing to particle *B* and the ratio of probabilities of being bound to unbound becomes

$$p_b = K_A c_{eff} p_u, \tag{3.A.8}$$

which is consistent with the correct result, Eq. (3.A.4). We could view $c_{eff} p_u$ as the concentration of unbound *A*.

Applying this concept to dimer adsorption (Figure 3.A.1) we would find that the empirically calculated effective molarity [Eq. (3.A.1)] is similar to the theoretical effective concentration $EM \sim c_{eff}$ (in our idealized system they are equal). Therefore, effective concentration, when applied properly, is a useful concept when attempting to theoretically predict equilibria of multivalent binding.

Acknowledgements

We thank Galina V. Dubacheva and Stefano Angioletti-Uberti for useful suggestions for the manuscript and help with designing figures. This work on multivalency was supported by the ERC Advanced Grant 227758 (COLSTRUCTION), ITN grant 234810 (COMPPLOIDS) and by EPSRC Programme Grant EP/I001352/1. T.C. acknowledges support form the Herchel Smith fund.

References

1 Hill, A.V. (1910) Proceedings of the Physiological Society: January 22, 1910. *The Journal of Physiology*, **40** (suppl.), i–vii.
2 Tu, Y. (2008) The nonequilibrium mechanism for ultrasensitivity in a biological switch: Sensing by Maxwell's demons. *Proceedings of the National Academy of Sciences*, **105** (33), 11 737–11 741.
3 Sneppen, K., Micheelsen, M.A. and Dodd, I.B. (2008) Ultrasensitive gene regulation by positive feedback loops in nucleosome modification. *Molecular Systems Biology*, **4** (1), 182.
4 Ferrell, J.E. and Ha, S.H. (2014) Ultrasensitivity part I: Michaelian responses and zero-order ultrasensitivity. *Trends in Biochemical Sciences*, **39** (10), 496–503.
5 Ferrell, J.E. and Ha, S.H. (2014) Ultrasensitivity part II: multisite phosphorylation, stoichiometric inhibitors, and positive feedback. *Trends in Biochemical Sciences*, **39** (11), 556–569.

6 Ferrell, J.E. and Ha, S.H. (2014) Ultrasensitivity part III: cascades, bistable switches, and oscillators. *Trends in Biochemical Sciences*, **39** (12), 612–618.

7 Dreyfus, R., Leunissen, M.E., Sha, R., Tkachenko, A.V., Seeman, N.C., Pine, D.J. and Chaikin, P.M. (2009) Simple quantitative model for the reversible association of DNA coated colloids. *Physical Review Letters*, **102**, 048 301.

8 Martinez-Veracoechea, F.J. and Frenkel, D. (2011) Designing super selectivity in multivalent nano-particle binding. *Proceedings of the National Academy of Sciences of the United States of America*, **108** (27), 10 963–10 968.

9 Krishnamurthy, V.M., Estroff, L.A. and Whitesides, G.M. (2006) *Multivalency in Ligand Design*, Wiley-VCH Verlag GmbH and Co. KGaA, pp. 11–53.

10 Ercolani, G. and Schiaffino, L. (2011) Allosteric, chelate, and interannular cooperativity: A mise au point. *Angewandte Chemie International Edition*, **50** (8), 1762–1768.

11 Hunter, C. and Anderson, H. (2009) What is cooperativity? *Angewandte Chemie International Edition*, **48** (41), 7488–7499.

12 Huskens, J., Mulder, A., Auletta, T., Nijhuis, C.A., Ludden, M.J.W., and Reinhoudt, D.N. (2004) A model for describing the thermodynamics of multivalent host-guest interactions at interfaces. *Journal of the American Chemical Society*, **126** (21), 6784–6797.

13 Mulder, A., Huskens, J. and Reinhoudt, D.N. (2004) Multivalency in supramolecular chemistry and nanofabrication. *Organic & Biomolecular Chemistry*, **2**, 3409–3424.

14 Krishnamurthy, V.M., Semetey, V., Bracher, P.J., Shen, N. and Whitesides, G.M. (2007) Dependence of effective molarity on linker length for an intramolecular protein-ligand system. *Journal of the American Chemical Society*, **129** (5), 1312–1320.

15 Kitov, P.I. and Bundle, D.R. (2003) On the nature of the multivalency effect: A thermodynamic model. *Journal of the American Chemical Society*, **125** (52), 16 271–16 284.

16 Xu, H. and Shaw, D.E. (2016) A simple model of multivalent adhesion and its application to influenza infection. *Biophysical Journal*, **110** (1), 218–233.

17 Dubacheva, G.V., Curk, T., Mognetti, B.M., Auzély-Velty, R., Frenkel, D. and Richter, R.P. (2014) Superselective targeting using multivalent polymers. *Journal of the American Chemical Society*, **136** (5), 1722–1725.

18 Dubacheva, G.V., Curk, T., Auzély-Velty, R., Frenkel, D. and Richter, R.P. (2015) Designing multivalent probes for tunable superselective targeting. *Proceedings of the National Academy of Sciences*, **112** (18), 5579–5584.

19 Schmidt, N.W., Jin, F., Lande, R., Curk, T., Xian, W., Lee, C., Frasca, L., Frenkel, D., Dobnikar, J., Gilliet, M. and Wong, G.C.L. (2015) Liquid-crystalline ordering of antimicrobial peptide-DNA complexes controls TLR9 activation. *Nature Materials*, **14** (7), 696–700.

20 Mortell, K.H., Weatherman, R.V., and Kiessling, L.L. (1996) Recognition specificity of neoglycopolymers prepared by ring-opening metathesis polymerization. *Journal of the American Chemical Society*, **118** (9), 2297–2298.

21 Carlson, C.B., Mowery, P., Owen, R.M., Dykhuizen, E.C. and Kiessling, L.L. (2007) Selective tumor cell targeting using low-affinity, multivalent interactions. *ACS Chemical Biology*, **2** (2), 119–127.

22 Kiessling, L.L. and Grim, J.C. (2013) Glycopolymer probes of signal transduction. *Chemical Society Reviews*, **42**, 4476–4491.

23 Albertazzi, L., Martinez-Veracoechea, F.J., Leenders, C.M.A., Voets, I.K., Frenkel, D., and Meijer, E.W. (2013) Spatiotemporal control and superselectivity in supramolecular polymers using multivalency. *Proceedings of the National Academy of Sciences*, **110** (30), 12 203–12 208.

24 Rogers, W.B. and Crocker, J.C. (2011) Direct measurements of DNA-mediated colloidal interactions and their quantitative modeling. *Proceedings of the National Academy of Sciences*, **108** (38), 15 687–15 692.

25 van der Meulen, S.A.J. and Leunissen, M.E. (2013) Solid colloids with surface-mobile DNA linkers. *Journal of the American Chemical Society*, **135** (40), 15 129–15 134.

26 Wang, Y., Wang, Y., Zheng, X., Ducrot, É., Yodh, J.S., Weck, M. and Pine, D.J. (2015) Crystallization of DNA-coated colloids. *Nature Communications*, **6**, 7253 EP –.

27 Gao, H., Shi, W. and Freund, L.B. (2005) Mechanics of receptor-mediated endocytosis. *Proceedings of the National Academy of Sciences of the United States of America*, **102** (27), 9469–9474.

28 Gernier, R.D., Curk, T., Dubacheva, G.V., Richter, R.P. and Mognetti, B.M. (2015) A new configurational bias scheme for sampling supramolecular structures. *Journal of Chemical Physics*, **141**, 244909.

29 Tito, N.B. and Frenkel, D. (2014) Optimizing the selectivity of surface-adsorbing multivalent polymers. *Macromolecules*, **47** (21), 7496–7509.

30 Varilly, P., Angioletti-Uberti, S., Mognetti, B.M. and Frenkel, D. (2012) A general theory of DNA-mediated and other valence-limited colloidal interactions. *The Journal of Chemical Physics*, **137** (9), 094108.

31 Angioletti-Uberti, S., Varilly, P., Mognetti, B.M., Tkachenko, A.V. and Frenkel, D. (2013) Communication: A simple analytical formula for the free energy of ligand–receptor-mediated interactions. *The Journal of Chemical Physics*, **138** (2), 021102.

32 Angioletti-Uberti, S., Varilly, P., Mognetti, B.M. and Frenkel, D. (2014) Mobile linkers on DNA-coated colloids: Valency without patches. *Physical Review Letters*, **113**, 128 303.

33 Tito, N.B., Angioletti-Uberti, S., and Frenkel, D. (2016) Communication: Simple approach for calculating the binding free energy of a multivalent particle. *The Journal of Chemical Physics*, **144** (16), 161101.

34 Wang, S. and Dormidontova, E.E. (2012) Selectivity of ligand-receptor interactions between nanoparticle and cell surfaces. *Physical Review Letters*, **109**, 238 102.

35 Ling, X.Y., Reinhoudt, D.N., and Huskens, J. (2008) Reversible attachment of nanostructures at molecular printboards through supramolecular glue. *Chemistry of Materials*, **20** (11), 3574–3578.

36 Dubacheva, G.V., Heyden, A.V.D., Dumy, P., Kaftan, O., Auzély-Velty, R., Coche-Guerente, L. and Labbé, P. (2010) Electrochemically controlled adsorption of Fc-functionalized polymers on beta-CD-modified self-assembled monolayers. *Langmuir*, **26** (17), 13 976–13 986.

37 Dubacheva, G.V., Galibert, M., Coche-Guerente, L., Dumy, P., Boturyn, D. and Labbe, P. (2011) Redox strategy for reversible attachment of biomolecules using bifunctional linkers. *Chemical Communications*, **47**, 3565–3567.

38 Yamaguchi, H., Kobayashi, Y., Kobayashi, R., Takashima, Y., Hashidzume, A. and Harada, A. (2012) Photoswitchable gel assembly based on molecular recognition. *Nature Communications*, **3**, 603.

39 Rose, S., Prevoteau, A., Elziere, P., Hourdet, D., Marcellan, A. and Leibler, L. (2014) Nanoparticle solutions as adhesives for gels and biological tissues. *Nature*, **505** (7483), 382–385.

40 Duncan, G.A. and Bevan, M.A. (2015) Computational design of nanoparticle drug delivery systems for selective targeting. *Nanoscale*, **7**, 15 332–15 340.

41 Satav, T., Huskens, J. and Jonkheijm, P. (2015) Effects of variations in ligand density on cell signaling. *Small*, **11** (39), 5184–5199.

42 Grove, J. and Marsh, M. (2011) The cell biology of receptor-mediated virus entry. *The Journal of Cell Biology*, **195** (7), 1071–1082.

43 Duke, T.A.J. and Bray, D. (1999) Heightened sensitivity of a lattice of membrane receptors. *Proceedings of the National Academy of Sciences*, **96** (18), 10 104–10 108.

44 Monine, M.I., Posner, R.G., Savage, P.B., Faeder, J.R. and Hlavacek, W.S. (2010) Modeling multivalent ligand-receptor interactions with steric constraints on configurations of cell-surface receptor aggregates. *Biophysical Journal*, **98** (1), 48–56.

45 Wu, H. (2013) Higher-order assemblies in a new paradigm of signal transduction. *Cell*, **153** (2), 287–292.

46 Collins, B.E. and Paulson, J.C. (2004) Cell surface biology mediated by low affinity multivalent protein-glycan interactions. *Current Opinion in Chemical Biology*, **8** (6), 617–625.

47 De Jong, D.H., Schäfer, L.V., De Vries, A.H., Marrink, S.J., Berendsen, H.J.C. and Grubmüller, H. (2011) Determining equilibrium constants for dimerization reactions from molecular dynamics simulations. *Journal of Computational Chemistry*, **32** (9), 1919–1928.

48 Angioletti-Uberti, S., Mognetti, B.M. and Frenkel, D. (2016) Theory and simulation of DNA-coated colloids: a guide for rational design. *Physical Chemistry Chemical Physics*, **18**, 6373–6393.

4

Multivalency in Biosystems

Jens Dernedde

Institute of Laboratory Medicine, Clinical Chemistry, and Pathobiochemistry, Charité–Universitätsmedizin Berlin, 13353 Berlin, Germany

4.1 Introduction

Multivalency is the action of multiple recognition of the same type occurring simultaneously between two entities. This concept is a common principle to increase the affinity and specificity of ligand–receptor interactions in nature and results in a cooperative, over-additive enhancement of binding affinity [1,2]. In biosystems mostly low affinity binders are clustered to achieve the required binding strength. From the evolutionary point of view, it seems to be convenient to merge multiple single low affinity interactions to generate a collectively stronger output. This strategy avoids the tedious *de novo* protein design and resorts to already existing interaction networks. Furthermore, a tunable multivalent receptor–ligand interaction allows for a graded response to biological signals.

Based on pioneering work from the disciplines of chemistry and biochemistry reviewed by Mammen *et al*. [1] and Fasting *et al*. [3] great progress has been achieved to describe multivalent effects in recent years, mainly from the perspective of synthetic chemistry. Nevertheless, our current understanding is still fragmentary. Novel concepts include highly interdisciplinary research to unravel the impact of multivalent molecular recognition in biological systems and to comprehend diverse complex tasks such as molecular assembly, cell–cell adhesion, and signal transduction. Although a reductionistic approach is first necessary to prove multivalent effects at the molecular level *in vitro*, analyses of more complex biosystems remain so far elusive. Experimental set-ups that meet the natural environment are therefore required.

A fascinating feature of living cells is the precise control over biological activity in space and time. The cell as a functional unit relies on such high-fidelity spatiotemporal control of molecular interactions, wherein thousands of different components, namely lipids and proteins, self-assemble [4]. A driving force that contributes to the recruitment of molecules and therefore to cellular organization is multivalency. As an example, the cell boundary layer, the cytoplasmic membrane, is assembled into a two-dimensional fluid architecture and lipids and proteins seem to be heterogeneously distributed.

Multivalency: Concepts, Research & Applications, First Edition. Edited by Jurriaan Huskens, Leonard J. Prins, Rainer Haag, and Bart Jan Ravoo.
© 2018 John Wiley & Sons Ltd. Published 2018 by John Wiley & Sons Ltd.

In response to internal or external stimuli that lead to membrane protein cross-linking, proteins can be segregated into nanometer-sized domains referred to as lipid rafts. This clustering of biomolecules is highly dynamic and can also rapidly disassemble [5]. Inside the cell preorganization of functional units is realized by compartmentation. Membrane surrounded organelles like the nucleus, endoplasmic reticulum, and Golgi complex define functional units to perform special tasks; however, vesicles also involved in transport, storage, and degradation constitute individual subcellular reaction centers with limited diffusion and thereby foster protein–protein interaction. Recently, it was shown that also in the crowded cytoplasm areas of higher organization can develop due to phase separation, a phenomenon induced by multivalent interactions of proteins [6]. The principles that underlie control over the molecular composition of the cellular microenvironment in space and time are therefore a subject of great scientific interest as they are of essential importance for proper cell function.

4.2 Cell–Cell Adhesion

Intercellular adhesion is probably the best-studied example of multivalency in biosystems. Cell–cell adhesion can be classified by different scenarios (Figure 4.1) to realize multivalent binding. The connection of cells from the same type in tissues is mentioned as a homotypic interaction of proteins; in contrast, a heterotypic interaction connects cells from different origin (Figure 4.1a). A homophilic interaction is caused by attraction forces between identical molecules, whereas different molecules are joined in a heterophilic manner (Figure 4.1b). Trans-binding connects facing cell surfaces,

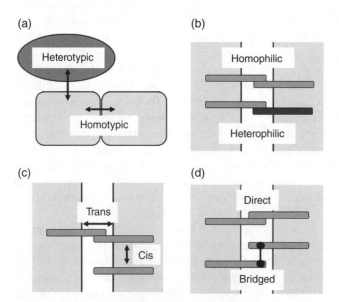

Figure 4.1 Types of cell–cell adhesion and interaction of cell adhesion molecules. (a) Hetero- and homotypic interaction. (b–d) show the extracellular adhesion of transmembrane proteins of two adjacent cells. (b) Homophilic and heterophilic binding. (c) Trans- and cis-binding. (d) Direct and bridged interaction.

cis-binding connects proteins that reside on the same cell surface (Figure 4.1c), whether they cluster in a homophilic or heterophilic manner. Further, molecular interaction can be direct or bridged by additional adaptor proteins inside or outside the cell (Figure 4.1d).

4.2.1 Homotypic Interactions, Cadherins Keep Cells Together

Cell adhesion molecules that mediate lateral cell–cell contact in epithelia are organized within adhesion complexes. Individual protein clusters are segregated to defined areas and constitute tight junctions, adherens junctions, desmosomes, and gap junctions to bridge the intercellular space. This overall spatial organization leads to a polarization of epithelial cells into an apical and basal side.

Multivalent interactions at adherens junctions have been intensively studied [7–10]. The single pass transmembrane cadherin connects at adherens junctions neighboring cells by homophilic trans-binding (Figure 4.2). Essential for adhesive function is the stabilization of the five ectodomains (ECs) of classical cadherins by Ca^{2+} ions. Calcium binding sites are located between the EC domains [11,12]. Calcium binding rigidifies the cadherin ectodomain overall structure and the protein retains a crescent-shaped architecture. Homophilic binding of membrane-distal EC1 domains from opposing cell surfaces is realized by a special mechanism. Adhesive binding arises via an intermediate X-dimer structure through the exchange of β strands between EC1 domains of cadherins from adjacent cells [13–15]. Key to this mechanism is the vice versa docking of a hydrophobic tryptophan anchor residue into a conserved hydrophobic pocket of the partnering EC1 domain. This strand-swap binding mode is common to classical and desmosomal cadherins and known for the oligomerization of several monomeric proteins [16] (Figure 4.2b).

Binding constants of homophilic cadherin interactions are of low affinity, a dissociation constant (K_D) of 0.7 mM was detected for endothelial E-cadherin [17]. For the vascular

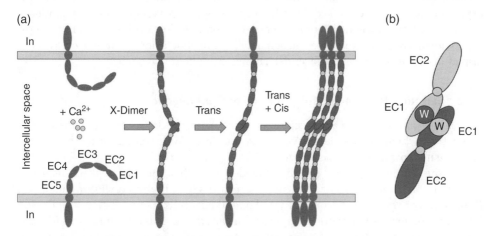

Figure 4.2 Formation of homophilic cadherin interactions at adherens junctions. (a) Calcium dependent trans- and cis-binding is established via the intermediate X-dimer structure. EC1–5 are the extracellular domains of a classical cadherin. (b) Strand swap of trans-interacting cadherins, adjacent EC1 domains are connected via the exchange of a hydrophobic loop containing a conserved tryptophan (W).

endothelial VE-cadherin fused to the fragment crystallizable (Fc) region of an antibody, the dimeric Fc-cadherin resulted in K_D values ranging from 10^{-3} to 10^{-5} M, representing still a weak interaction [18]. Therefore, to maintain cadherin function in cell–cell adhesion, a multivalent presentation that strengthens cell-to-cell contact at adherens junctions is obvious. The physiological relevance of cadherin trans-binding has been confirmed in several studies by single site mutagenesis, electron microscopy, structural and cell-based analyses [14,19–21]. Further, cis-interactions between neighboring cadherins seem to be important for packing the protein in adherens junctions [22–24]. Although cis-interactions were so far not proven by determination of binding constants, they are estimated to be in the mM range. Structural data and mutational analyses point to their existence. The intracellular linkage to the actin skeleton is not essential to enable molecular interactions of EC1 domains, but solely determines density of cadherin distribution at adherens junctions. Cell culture experiments show that directed actin polymerization is the driving force for epithelial cell adhesion. A zipper-like architecture initiated by membrane protrusions tipped with cadherin of opposing cells mediates initial attachment [25].

Cadherins constitute a diverse family of different adhesion molecules that exhibit spatial and temporal expression patterns during development and therefore contribute to tissue morphogenesis. The conclusion that cadherins are homophilic adhesion molecules came from aggregation experiments, in which cells expressing different types of cadherins were observed to sort out and form distinct aggregates [26]. Segregation of cells into specific tissue layers and the formation of tissue boundaries rely on preferred homophilic interactions of individual cadherins [27]. Furthermore, differential levels of expression of a single cadherin type also contribute to cell segregation [28–30]. In this context it is of note that a varying density of cadherin adhesive clusters in the intercellular space is a critical determinant of endothelial macromolecular sieving [31]. This task of controlling passive paracellular mass transport was earlier primarily allocated to the protein network at the apical tight junction barrier.

Given the complexity of their functions, it is not surprising that impaired cadherin expression has also been linked directly to a variety of diseases including its dominant role in metastatic cancer development. Partial or complete loss of endothelial E-cadherin expression correlates with cancer malignancy [32]. In conclusion, a variable multivalent display of cadherins on cell surfaces determines cell function and fate.

4.2.2 Selectins, Heterotypic Cell Adhesion to Fight Infections

A classic example of glycan–receptor interaction is the leukocyte–endothelial cell interaction, which initiates leukocyte extravasation from the blood during homing to the bone marrow and in disease (Figure 4.3).

Here the family of calcium coordinating C-type lectins, the selectins, are displayed in clusters on either specialized membrane protrusions, the microvilli of leukocytes (L-selectin), or the flat surface of activated endothelial cells (E- and P-selectin). During inflammation pro-inflammatory stimuli prime endothelial cells and leukocytes for subsequent recognition [33,34]. The selectins interact with the common tetrasaccharide sialyl Lewis X (sLeX) binding epitope presented on proteoglycans of both cell surfaces and thereby enable leukocyte–endothelial cell interaction. Individual selectin–sLeX interactions are of low affinity in the high μM to mM range [35], but multiple interactions allow for sufficient strong binding to capture leukocytes from the blood stream to

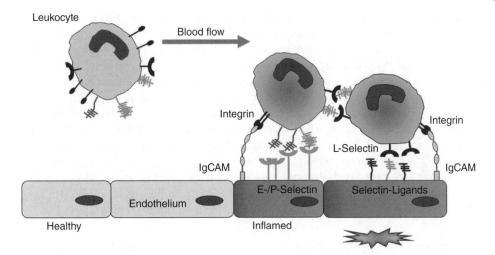

Figure 4.3 Leukocyte extravasation during inflammation. Clustered selectins and carbohydrate ligands on activated cell surfaces mediate initial contacts and slow down the velocity of leukocytes. Firm adhesion is based on integrin binding to cell adhesion molecules of the immunoglobulin family (IgCAM).

sites of inflammation and initiate the adhesion cascade. The rapid binding and release events slow down leukocyte velocity from the blood flow and mediate leukocyte rolling on the vascular endothelial cell surface [36,37]. A dominant ligand of all selectins is the proteoglycan P-selectin glycoprotein-ligand-1 (PSGL-1). The transmembrane protein is a cysteine bridged dimer and displays in addition to the carbohydrate sLeX three sulfotyrosines that contact basic residues in proximity to the glycan binding site of P-selectin [38] and L-selectin [39]. In the case of P-selectin binding the bipartite protein–glycan ligand PSGL-1 shifts the binding constant ~1000-fold to 800 nM in comparison with sLeX and identifies PSGL-1 as a high affinity ligand [38]. In this context it is interesting to note that during deceleration of leukocytes to a certain threshold of shear force, cell rolling appears and a primary low affinity interaction turns to high affinity binding [40]. At least P- and L-selectin show this special "catch" bond behavior that relies on intramolecular structural rearrangements of the receptor. Upon ligand recognition the N-terminal lectin domain is separated from the adjacent epidermal growth factor (EGF)-like domain by shear force which improves ligand binding and results in longer bond lifetime [41,42]. In addition, homotypic leukocyte–leukocyte interactions, which are realized by the heterotypic interaction of P-selectin and PSGL-1 generate cell clusters at the site of adhesion and thereby further facilitate endothelial targeting of activated leukocytes. Furthermore, the initial carbohydrate–ligand recognition triggers intracellular signaling that activates leukocyte integrins, cell adhesion molecules essential to mediate firm adhesion [43–45] and a prerequisite for subsequent cell extravasation.

Over recent years multiple selectin targeting compounds have been developed to counteract leukocyte adhesion [1,3]. The rationale behind an interventional strategy is that during severe injury and acute or chronic inflammatory diseases the unbalanced extravasation of leukocytes from the blood vessel contributes to further tissue destruction. Here multivalent compounds that present the pan selectin targeting natural sLeX epitope or a synthetic sLeX mimetic successfully interfere with the selectin–ligand binding. In addition, polysulfates like the dendritic polyglycerol sulfate (dPGS) that target L- and P-selectin show in competitive binding assays *in vitro* and cell culture half-maximal inhibitor concentration (IC_{50}) values down to the picomolar range [46,47]. Although dPGS lacks target specificity, *in vivo* experiments demonstrate their overall anti-inflammatory characteristic and indicate effective leukocyte shielding [48,49].

So far the translation of multivalent compounds targeting selectins into the clinic has not been achieved. In contrast, a successful rational design inspired by the sLeX and PSGL-1 architectures yielded a synthetic bifunctional pan-selectin antagonist. The molecule targets the carbohydrate binding site of all selectins and in addition basic amino acid residues present at the binding sites of L- and P-selectin [50,51]. This glycomimetic GMI-1070 (Rivipansel) is currently being tested in clinical trials.

4.2.3 Bacterial Adhesion by FimH

A second example for a multivalent lectin based cell–cell interaction is the binding of uropathogenic *Escherichia coli* (UPEC) to the urinary tract epithelium. Upon binding, the bacteria get internalized in an active process that is similar to phagocytosis [52] and cause tissue colonization. Urinary tract infections (UTIs) are among the most prevalent inflammatory diseases that are caused by pathogens [53,54] and are therefore of primary interest for a medical intervention.

The transmembrane protein uroplakin Ia (UPIa) is the primary cellular target of UPEC adhesion. UPIa is abundantly expressed on the superficial epithelial cells of the urinary tract, especially in the bladder. Highly mannosylated *N*-glycan structures with six to nine mannose residues linked to UPIa provide an ideal surface for mannose-specific lectin targeting [55]. Bacterial attachment relies on the highly conserved FimH lectins, which are located at the tip of the type 1 fimbriae [56]. A structure–function analysis showed that the residues of the FimH mannose binding pocket are invariant across 200 UPEC strains [57], which highlights that UPEC cause more than 80% of all infections.

A polymannose surface on one site and hundreds of FimH lectin domains on the other allow for successful adhesion. To escape clearance by urine excretion, FimH binding to mannose is a further example for adaptation to shear force. FimH binding to mannose is enhanced at higher shear stress. The molecular basis to shear adaptation relies on the interaction of the two subdomains of FimH: the N-terminal mannose binding lectin domain $FimH_L$ and the C-terminal pilin domain ($FimH_P$). In the absence of shear $FimH_P$ tightly connects to $FimH_L$ and the lectin shows only low affinity to mannose [58–60]. Shear force stretches the pilus, allosterically separates the subdomains and increases thereby $FimH_L$ binding affinity to mannose. Recently, from crystallographic data, binding kinetics analyses, and molecular simulations it was found that $FimH_L$ in the separated state shows a 3300-fold higher affinity for the model ligand heptylmannoside when compared with the associated state of full-length FimH [61].

A second intermolecular interaction that is required for successful firm UPEC adhesion and paves the way for bacterial entry is the association with α3β1 integrins [61]. FimH is able to bind matrix proteins such as fibronectin, laminin, and collagen, which in turn recognize integrins [62,63]. Besides this mediated contact recent results also indicate a direct interaction of FimH to high mannose-type glycans linked to α3 and β1 integrin chains [61].

In addition to UTIs, FimH-mediated adhesion plays an important role in ileal lesions of patients suffering from Crohn´s disease (CD), a chronic and disabling inflammatory disorder of the intestine. In that case adherent-invasive *E. coli* (AIEC) strains promote inflammation and lead to colitis. AIEC bacteria strongly adhere to and invade intestinal epithelial cells. As a consequence, secretion of inflammatory cytokines is induced [64]. The molecular target of FimH in the intestine are mannosylated glycans linked to the carcinoembryonic antigen-related cell adhesion molecule 6 (CECAM6), overexpressed on intestine epithelial cells from CD patients [65].

The development of competitive FimH antagonists as anti-adhesive compounds is a promising strategy for a potential medical application. The vast spreading of antibiotic resistances among UPEC and AIEC strains requires alternative treatment options. Different multivalent presentations of mannose derivatives have been explored and nanomolar affinities reported [66–68].

The treatment of UTIs by UPEC requires orally active FimH antagonists. Compounds should get absorbed in the intestine and renally excreted. This implies well-defined pharmacokinetic properties to achieve an optimal balance between target affinity and retention in the bladder. Monovalent aryl mannosides seem to be well suited to target the FimH binding site by their mannose component and high affinity to a secondary binding site, the tyrosine gate, by an extended lipophilic aglycone structure that contributes to target specificity [69–73]. Oral treatment of UTIs with aryl mannosides in mouse models has already revealed promising results [74–76]. Additionally, in the context of urinary catheterization of patients, intervening concepts are proposed to reduce the bacterial load by wash out strategies with FimH antagonists [77,78].

For the treatment of Crohn´s disease an anti-adhesive strategy that reduces the level of adherent-invasive *E. coli* is also applicable [79]. A main advantage of this complementary approach to current antibiotic treatments is the maintenance of gut microbiota. In this respect it is interesting to note that a therapy with a probiotic yeast strain in a model for Crohn´s disease prevented colitis induced by AIEC [80]. Obviously the high mannose content in the yeast cell wall outcompetes FimH binding of AIEC bacteria to CEACAM6 on the epithelial cell surface.

4.3 Phase Transition, Multivalent Intracellular Assemblies

Electron microscopy and advanced methods in light microscopy have demonstrated a subdivision and local organization of membrane-less biomolecular organelles in the cytoplasm and nucleus at the nanometer to micrometer scale. These macromolecular dynamic aggregates include for instance P granules, Cajal bodies, promyelocytic leukemia bodies, paraspeckles, and the nucleolus [81]. Recently, it has been proposed, that these structures set up via liquid–liquid demixing phase transitions of their constituent molecules, such as proteins, DNA, and RNA [82]. Nevertheless, the molecular mechanism of how these liquid colloid particles assemble and execute specific functions is largely unknown. The interaction between multivalent molecules in a dilute solution is governed by their concentrations, and depends on the association and dissociation constants. Forces that lead to an attraction of multivalent molecules by electrostatic, hydrophobic, van der Waals interactions, or hydrogen bonding increase the local effective molarity of the resulting receptor–ligand complex and result in a higher probability of molecular recognition. Similarly, lipophilic proteins that partition into membranes, or hydrophilic proteins enclosed in an organelle or vesicle, will interact more often if not dispersed evenly throughout the system. Thus, multivalent interactions of distinct proteins might generate highly ordered assemblies in aqueous solution that phase separate at a critical concentration and lead to liquid droplet formation [83].

Characteristics of intracellular multivalent interacting proteins are their high valency and modest affinity of binding elements, which are connected via long, flexible linker elements [84]. Li *et al.* [83] recently followed phase separation initially *in vitro* on engineered proteins that contain two low affinity binding modules, the SRC homology 3 (SH3) domain and its proline-rich motif (PRM) ligand. Both domains are widely distributed among signaling proteins and often displayed in tandem arrays [84,85]. Binding of the monomeric SH3 to PRM is of low affinity ($K_D = 350\,\mu M$). Titration series of oligovalent binding partners resulted in a concentration dependent sharp sol–gel transition, in which two protein solutions after reaching a critical concentration separate from the environment to produce diffractive droplets of $1\,\mu m$ to $>50\,\mu m$ in diameter. The concentration of droplet constituent thereby increased up to ~100-fold.

In a second *in vitro* experiment the authors demonstrated how phase transition can organize signaling and initiate a biochemical response. A three-component system consisting of the proteins nephrin, NCK (non-catalytic region of tyrosine kinase), and N-WASP (neuronal Wiskott–Aldrich syndrome protein) is involved in actin polymerization. The proteins assemble into complexes (Figure 4.4) prone to phase transition. Nephrin is a transmembrane adhesion receptor mainly expressed in the kidney and participates in the assembly of the cortical actin skeleton necessary for proper formation of the filtration barrier in the kidney [86]. The cytoplasmic tail of nephrin contains three tyrosine residues, which can be phosphorylated upon receptor activation and bind to the SH2 domain of the adaptor protein NCK [87]. The three SH3 domains of

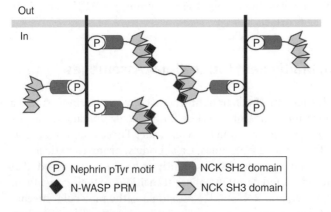

Figure 4.4 Multivalent interactions of nephrin, NCK, and N-WASP lead to intracellular complex formation and phase separation. *Source:* Adapted from Ref. [83]. Reproduced with permission of Nature Publishing Group.

NCK, in turn, can interact with the six PRMs of N-WASP [88]. As a physiological consequence of protein assembly and droplet formation, actin polymerization is initiated in cooperation with the actin related protein (Arp2/3) complex [89].

Phase transition was also observed in cell culture experiments. Coexpression of pentavalent fluorescent fusion proteins [mCherry-$(SH3)_5$ and eGFP-$(PRM)_5$] in eukaryotic cells resulted in the formation of micrometer-sized puncta in the cytoplasm where both fluorophores colocalize. High recovery rates after photobleaching indicate the dynamic behavior of these complexes. At the molecular level, this correlates to a transformation of small complexes into large polymers that were detected via dynamic light scattering, small angle X-ray scattering, light- and cryo-electron microscopy. Interestingly, further experiments indicate that not only the interacting complementary protein domains and their valency are essential, but weak interactions of disordered interdomain linkers synergistically contribute to phase separation due to homotypic self-association. Reorganization of the linker molecules seems to condense the multivalent complexes and increase the local protein concentrations. Moreover, Pak *et al.* [90] have shown that non-specific charge-mediated interactions of the nephrin intracellular domain (NICD) and their intrinsic disordered sequences can promote phase separation in the absence of NCK and N-WASP, when expressed in cells. The same results were obtained *in vitro* when they titrated a solution of the acidic NICD with highly positively charged proteins [90]. The authors explain this unexpected phenomenon with coacervation, an electrostatically driven liquid–liquid phase separation process, resulting from association of oppositely charged molecules. Thus, non-specific charge-mediated interactions, together with specific modular domain interactions, trigger the formation of liquid droplets by the ternary nephrin/ Nck/N-WASP complex. Since a number of transmembrane signaling proteins contain disordered cytoplasmic regions and are linked via phosphorylation to basic motifs of adaptor proteins, it is conceivable that coacervation represents a more general concept of cytoplasmic protein clustering and phase separation.

4.4 Multivalency in the Fluid Phase, Pathogen Opsonization

Invading pathogens that reach the blood have to be immediately recognized and eliminated by the immune system in order to avoid them spreading fast throughout the body. As the first line of defense, clearance of pathogen is achieved by phagocytosis from cells of the innate immune system. Mainly macrophages, dendritic cells and neutrophils are capable of recognizing conserved structures on pathogens, termed pathogen-associated molecular patterns [91] by means of their complementary pattern recognition receptors [92]. Furthermore, the pathogen surface can be labeled by soluble plasma derived proteins and in this way prepared for uptake and degradation (opsonization). Multivalent proteins, such as antibodies and lectins accomplish this task.

Lectins bind to carbohydrates present on pathogen surfaces. Abundant in human serum are the mannose binding lectin (MBL) [93,94] and members of the ficolin family [95,96] that bind to acetylated sugars (*N*-acetylglucosamine). They act as pathogen sensors. Both lectins have a comparable shape, they assemble into trimers via their collagen domain and further oligomerize. Their oligomeric structure compensates for the general low binding affinity to carbohydrates that is for the trimer in the millimolar range, but can reach nanomolar affinity in the oligomeric state [97]. Pathogen binding induces intrinsic conformational changes in linked MBL-associated serine proteases (MASPs) [98], which in turn activate the lectin pathway of the complement cascade. The final product is the membrane attack complex (MAC) [99]. A pore-forming complex inserts into the membrane and lyses the pathogen. Comparable with MBL and ficolin in structure and function is the C1 complex that activates the classical antibody-dependent complement pathway [100] (Figure 4.5 and see below).

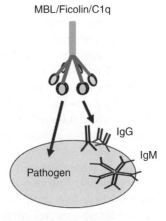

Figure 4.5 Simplified common structure of multivalent pathogen sensors. MBL and ficolin contact the pathogen directly. Targeting of C1q is mediated by IgM or IgG aggregates.

Antigen binding is at least a monovalent recognition but all immunoglobulins contain multiple receptor binding sites. The monomers IgA, IgD, IgE, and IgG have two valencies; the subclass IgA can also appear as dimer (most common), trimer, and tetramer aggregates with resulting valencies of 4, 6, and 8, respectivley. The pentamer IgM displays 10 binding sites. Multivalent epitopes will obviously strengthen binding if addressable epitopes are available. Karulin and Dzantiev [101] have shown that targeting of a bacterial surface with a bivalent antibody is more effective compared with a monovalent antibody fragment. While in this case the antibody association rates did not differ dramatically, the lower affinity of the monovalent protein originates from its enhanced dissociation rate which was about 40 times higher. At least for some immunoglobulins of the IgA subclass, multimerization even seems to be essential for affine binding [102]. The authors showed that a monomeric IgA was not able to target lipopolysaccharides (LPS) efficiently; in contrast, the dimeric IgA binds LPS with an affinity in the low nanomolar range. At least for further signaling antibody multimerization is required.

Ligand recognition by the fragment antigen binding (Fab) site of the antibody might induce conformational changes in the distant Fc part of the molecule. This can be observed by enhanced Fc reactivity, namely Fc receptor binding and complement activation. Here it is remarkable that single recognition events barely contribute to signal transduction but multivalent interactions are essential. For instance, multiple immunoglobulins attached to a pathogen surface create an immune complex that binds to several Fc receptors on a macrophage and trigger pathogen uptake. Mast cell activation only occurs when IgE cross-linked antigens bind to the cell surface receptors and then induce degranulation of histamine and heparin containing vesicles. Moreover, classical complement activation is only mediated by binding of the C1 complex consisting of the hexavalent pattern recognition molecule C1q and a heterotetramer of proteases C1r and C1s [103]. Here the spatial arrangements of binding partners seem to play a pivotal role. C1q binds a single IgG Fc only with very low affinity ($K_D \sim 10^{-4}$ M) [104,105].

The C1 to IgG antibody stoichiometry so far remains poorly understood [106]. Potent complement activation by monoclonal antibodies is restricted to certain antigens, presumably because antigen size, epitope density and geometry affect activation [107–109]. Polyclonal antibodies in contrast appear to be less sensitive to such constraints [110,111] because binding of antibodies to a variety of epitopes facilitates clustered Fc presentation.

Recently it was shown, that some IgG species are able to self-assemble upon ligand binding to hexamers and thereby initiate a strong complement response [112]. Certain amino acids were identified that contribute to the Fc–Fc contact formation and are obviously necessary to organize the defined well-ordered structure. Although the hexamer arrangement was earlier detected in crystal structures [113,114] the physiological implication was not realized. The hexavalent antibody structure perfectly matches the C1q architecture and enables subsequent high affinity multivalent binding of the six C1q headpieces to the targeted antibody Fc regions. Highest binding affinities of C1q to antibody-opsonized cells are in the low nanomolar range and binding strength positively correlates with complement-dependent cytotoxicity (CDC) activity [112]. Due to the structural similarity, it is not surprising that the pentavalent IgM generally elicits strong complement activation and promotes equivalent signaling through conformational change.

Antibody coating of targets has been shown to mediate different potent killing mechanisms. CDC occurs via MAC formation, as described before. In the case of antibody-dependent cellular cytotoxicity (ADCC) the antibody labeled pathogen is recognized by Natural Killer cells which consequently secrete membrane penetrating peptides and proteases to destroy the pathogen. Finally, the antibody-dependent cellular phagocytosis (ADCP) is accomplished by macrophages and monocytes. All of these effector functions are mediated by the antibody Fc region. From the medical point of view, engineering of the Fc region to enhance the cytotoxic activity of therapeutic antibodies is currently a subject of intense investigation [115–119].

4.5 Conclusion

Multivalent interactions are widespread in biological systems. The best characterized are receptor–ligand interactions that mediate receptor clustering at cell surfaces. However, research in recent years has shown that this apparently simple study area is challenging. The dynamic character of interactions that leads to assembly and disassembly of molecular complexes in space and time is a difficult task and often remains obscure. Nevertheless, there is an increasing interest from different disciplines to unravel how multivalent interactions of biomolecules confer biological functions. Recent advances that demonstrated how membrane-less organelles are generated are a good example to comprehend how a graded biological response relies on the multivalent intermolecular assembly of biomolecules (see Section 4.3) and there is much more to explore. Experiments aimed at interfering with biological recognition have to be validated in natural environments. Future research should address questions of how multivalency can be applied to control cell behavior or mediate specific targeting. Potential applications in the medical field are broad and include but are not limited to cell differentiation, imaging, pathogen shielding, and tumor therapy.

Acknowledgment

I gratefully acknowledge the German Research Foundation (DFG) for financial support within the Collaborative Research Centre 765.

References

1 Mammen M, Choi SK, Whitesides GM. Polyvalent interactions in biological systems: Implications for design and use of multivalent ligands and inhibitors. Angew. Chem. Int. Ed. 1998; 37(20):2755–94.

2 Hunter CA, Anderson HL. What is cooperativity? Angew. Chem. Int. Ed. 2009; 48(41):7488–99.

3 Fasting C, Schalley CA, Weber M, *et al.* Multivalency as a chemical organization and action principle. Angew Chem Int Ed Engl. 2012; 51(42):10472–98.

4 Singer SJ, Nicolson GL. The fluid mosaic model of the structure of cell membranes. Science 1972; 175(4023):720–31.

5 Simons K, Sampaio JL. Membrane organization and lipid rafts. Cold Spring Harb. Perspect. Biol. 2011; 3(10):a004697.

6 Banjade S, Wu Q, Mittal A, *et al.* Conserved interdomain linker promotes phase separation of the multivalent adaptor protein Nck. Proc. Natl Acad. Sci. USA 2015; 112(47):E6426–35.

7 Tomschy A, Fauser C, Landwehr R, Engel J. Homophilic adhesion of E-cadherin occurs by a co-operative two-step interaction of N-terminal domains. EMBO J. 1996; 15(14):3507–14.

8 Brieher WM, Yap AS, Gumbiner BM. Lateral dimerization is required for the homophilic binding activity of C-cadherin. J. Cell Biol. 1996; 135(2):487–96.

9 Ahrens T, Lambert M, Pertz O, *et al.* Homoassociation of VE-cadherin follows a mechanism common to "classical" cadherins. J. Mol. Biol. 2003; 325(4):733–42.

10 Leckband D, Sivasankar S. Mechanism of homophilic cadherin adhesion. Curr. Opin. Cell Biol. 2000; 12(5):587–92.

11 Nagar B, Overduin M, Ikura M, Rini JM. Structural basis of calcium-induced E-cadherin rigidification and dimerization. Nature 1996; 380(6572):360–4.

12 Harrison OJ, Jin X, Hong S, *et al.* The extracellular architecture of adherens junctions revealed by crystal structures of type I cadherins. Structure 2011; 19(2):244–56.

13 Shapiro L, Fannon AM, Kwong PD, *et al.* Structural basis of cell–cell adhesion by cadherins. Nature 1995; 374(6520):327–37.

14 Harrison OJ, Corps EM, Berge T, Kilshaw PJ. The mechanism of cell adhesion by classical cadherins: the role of domain 1. J. Cell Sci. 2005; 118(Pt 4):711–21.

15 Posy S, Shapiro L, Honig B. Sequence and structural determinants of strand swapping in cadherin domains: do all cadherins bind through the same adhesive interface? J. Mol. Biol. 2008; 378(4):954–68.

16 Bennett MJ, Schlunegger MP, Eisenberg D. 3D domain swapping: a mechanism for oligomer assembly. Protein Sci. 1995; 4(12):2455–68.

17 Haussinger D, Ahrens T, Aberle T, *et al.* Proteolytic E-cadherin activation followed by solution NMR and X-ray crystallography. EMBO J. 2004; 23(8):1699–708.

18 Baumgartner W, Hinterdorfer P, Ness W, *et al.* Cadherin interaction probed by atomic force microscopy. Proc. Natl Acad. Sci. USA 2000; 97(8):4005–10.

19 Shapiro L, Weis WI. Structure and biochemistry of cadherins and catenins. Cold Spring Harb. Perspect. Biol. 2009; 1(3):a003053.

20 Meng W, Takeichi M. Adherens junction: molecular architecture and regulation. Cold Spring Harb. Perspect. Biol. 2009; 1(6):a002899.

21 Patel SD, Chen CP, Bahna F, *et al.* Cadherin-mediated cell-cell adhesion: sticking together as a family. Curr. Opin. Struct. Biol. 2003; 13(6):690–8.

22 Troyanovsky RB, Sokolov E, Troyanovsky SM. Adhesive and lateral E-cadherin dimers are mediated by the same interface. Mol. Cell Biol. 2003; 23(22):7965–72.

23 Harris TJ, Tepass U. Adherens junctions: from molecules to morphogenesis. Nat. Rev. Mol. Cell Biol. 2010; 11(7):502–14.

24 Hong S, Troyanovsky RB, Troyanovsky SM. Spontaneous assembly and active disassembly balance adherens junction homeostasis. Proc. Natl Acad. Sci. USA 2010; 107(8):3528–33.

25 Vasioukhin V, Bauer C, Yin M, Fuchs E. Directed actin polymerization is the driving force for epithelial cell–cell adhesion. Cell. 2000; 100(2):209–19.

26 Nose A, Nagafuchi A, Takeichi M. Expressed recombinant cadherins mediate cell sorting in model systems. Cell 1988; 54(7):993–1001.

27 Takeichi M. Morphogenetic roles of classic cadherins. Curr. Opin. Cell Biol. 1995; 7(5):619–27.

28 Steinberg MS. Does differential adhesion govern self-assembly processes in histogenesis? Equilibrium configurations and the emergence of a hierarchy among populations of embryonic cells. J. Exp. Zool. 1970; 173(4):395–433.

29 Friedlander DR, Mege RM, Cunningham BA, Edelman GM. Cell sorting-out is modulated by both the specificity and amount of different cell adhesion molecules (CAMs) expressed on cell surfaces. Proc. Natl Acad. Sci. USA 1989; 86(18):7043–7.

30 Steinberg MS, Takeichi M. Experimental specification of cell sorting, tissue spreading, and specific spatial patterning by quantitative differences in cadherin expression. Proc. Natl Acad. Sci. USA 1994; 91(1):206–9.

31 Quadri SK, Sun L, Islam MN, *et al.* Cadherin selectivity filter regulates endothelial sieving properties. Nat. Commun. 2012; 3:1099.

32 Berx G, Cleton-Jansen AM, Nollet F, *et al.* E-cadherin is a tumour/invasion suppressor gene mutated in human lobular breast cancers. EMBO J. 1995; 14(24):6107–15.

33 Ley K, Laudanna C, Cybulsky MI, Nourshargh S. Getting to the site of inflammation: the leukocyte adhesion cascade updated. Nat. Rev. Immunol. 2007; 7(9):678–89.

34 Kolaczkowska E, Kubes P. Neutrophil recruitment and function in health and inflammation. Nat. Rev. Immunol. 2013; 13(3):159–75.

35 Poppe L, Brown GS, Philo JS, *et al.* Conformation of sLe(x) tetrasaccharide, free in solution and bound to E-, P-, and L-selectin. J. Am. Chem. Soc. 1997; 119(7):1727–36.

36 Varki A. Selectin ligands. Proc. Natl Acad. Sci. USA 1994; 91(16):7390–7.

37 Vestweber D, Blanks JE. Mechanisms that regulate the function of the selectins and their ligands. Physiol. Rev. 1999; 79(1):181–213.

38 Somers WS, Tang J, Shaw GD, Camphausen RT. Insights into the molecular basis of leukocyte tethering and rolling revealed by structures of P- and E-selectin bound to SLe(X) and PSGL-1. Cell 2000; 103(3):467–79.

39 Woelke AL, Kuehne C, Meyer T, *et al.* Understanding selectin counter-receptor binding from electrostatic energy computations and experimental binding studies. J. Phys. Chem. B 2013; 117(51):16443–54.

40 McEver RP. Selectins: lectins that initiate cell adhesion under flow. Curr. Opin. Cell Biol. 2002; 14(5):581–6.

41 Marshall BT, Long M, Piper JW, *et al*. Direct observation of catch bonds involving cell-adhesion molecules. Nature 2003; 423(6936):190–3.

42 Sarangapani KK, Yago T, Klopocki AG, *et al*. Low force decelerates L-selectin dissociation from P-selectin glycoprotein ligand-1 and endoglycan. J. Biol. Chem. 2004; 279(3):2291–8.

43 Zarbock A, Lowell CA, Ley K. Spleen tyrosine kinase Syk is necessary for E-selectin-induced alpha(L)beta(2) integrin-mediated rolling on intercellular adhesion molecule-1. Immunity 2007; 26(6):773–83.

44 Zarbock A, Ley K, McEver RP, Hidalgo A. Leukocyte ligands for endothelial selectins: specialized glycoconjugates that mediate rolling and signaling under flow. Blood 2011;118(26):6743–51.

45 Pruenster M, Kurz AR, Chung KJ, *et al*. Extracellular MRP8/14 is a regulator of beta2 integrin-dependent neutrophil slow rolling and adhesion. Nat. Commun. 2015; 6:6915.

46 Weinhart M, Groger D, Enders S, *et al*. The role of dimension in multivalent binding events: structure–activity relationship of dendritic polyglycerol sulfate binding to L-selectin in correlation with size and surface charge density. Macromol. Biosci. 2011; 11(8):1088–98.

47 Weinhart M, Groger D, Enders S, *et al*. Synthesis of dendritic polyglycerol anions and their efficiency toward L-selectin inhibition. Biomacromolecules 2011; 12(7):2502–11.

48 Dernedde J, Rausch A, Weinhart M, *et al*. Dendritic polyglycerol sulfates as multivalent inhibitors of inflammation. Proc. Natl Acad. Sci. USA 2010; 107(46):19679–84.

49 Oishi K, Hamaguchi Y, Matsushita T, *et al*. A crucial role of L-selectin in C protein-induced experimental polymyositis in mice. Arthritis Rheumatol. 2014; 66(7):1864–71.

50 Magnani JL, Ernst B. Glycomimetic drugs – a new source of therapeutic opportunities. Discov. Med. 2009; 8(43):247–52.

51 Chang J, Patton JT, Sarkar A, *et al*. GMI-1070, a novel pan-selectin antagonist, reverses acute vascular occlusions in sickle cell mice. Blood 2010; 116(10):1779–86.

52 Palmer LM, Reilly TJ, Utsalo SJ, Donnenberg MS. Internalization of *Escherichia coli* by human renal epithelial cells is associated with tyrosine phosphorylation of specific host cell proteins. Infect. Immun. 1997; 65(7):2570–5.

53 Fihn SD. Clinical practice. Acute uncomplicated urinary tract infection in women. N. Engl. J. Med. 2003; 349(3):259–66.

54 Mak RH, Kuo HJ. Pathogenesis of urinary tract infection: an update. Curr. Opin. Pediatr. 2006; 18(2):148–52.

55 Zhou G, Mo WJ, Sebbel P, *et al*. Uroplakin Ia is the urothelial receptor for uropathogenic *Escherichia coli*: evidence from in vitro FimH binding. J. Cell Sci. 2001; 114(Pt 22):4095–103.

56 Krogfelt KA, Bergmans H, Klemm P. Direct evidence that the FimH protein is the mannose-specific adhesin of *Escherichia coli* type 1 fimbriae. Infect. Immun. 1990; 58(6):1995–8.

57 Hung CS, Bouckaert J, Hung D, *et al*. Structural basis of tropism of *Escherichia coli* to the bladder during urinary tract infection. Mol. Microbiol. 2002; 44(4):903–15.

58 Thomas WE, Trintchina E, Forero M, *et al*. Bacterial adhesion to target cells enhanced by shear force. Cell 2002; 109(7):913–23.

59 Sokurenko EV, Vogel V, Thomas WE. Catch-bond mechanism of force-enhanced adhesion: counterintuitive, elusive, but … widespread? Cell Host Microbe 2008; 4(4):314–23.

60 Le Trong I, Aprikian P, Kidd BA, *et al.* Structural basis for mechanical force regulation of the adhesin FimH via finger trap-like beta sheet twisting. Cell 2010; 141(4):645–55.

61 Eto DS, Jones TA, Sundsbak JL, Mulvey MA. Integrin-mediated host cell invasion by type 1-piliated uropathogenic *Escherichia coli*. PLoS Pathog. 2007; 3(7):e100.

62 Pouttu R, Puustinen T, Virkola R, *et al.* Amino acid residue Ala-62 in the FimH fimbrial adhesin is critical for the adhesiveness of meningitis-associated *Escherichia coli* to collagens. Mol. Microbiol. 1999; 31(6):1747–57.

63 Kukkonen M, Raunio T, Virkola R, *et al.* Basement membrane carbohydrate as a target for bacterial adhesion: binding of type I fimbriae of *Salmonella enterica* and *Escherichia coli* to laminin. Mol. Microbiol. 1993; 7(2):229–37.

64 Eaves-Pyles T, Allen CA, Taormina J, *et al. Escherichia coli* isolated from a Crohn's disease patient adheres, invades, and induces inflammatory responses in polarized intestinal epithelial cells. Int. J. Med. Microbiol. 2008; 298(5–6):397–409.

65 Barnich N, Carvalho FA, Glasser AL, *et al.* CEACAM6 acts as a receptor for adherent-invasive *E. coli*, supporting ileal mucosa colonization in Crohn disease. J. Clin. Invest. 2007; 117(6):1566–74.

66 Touaibia M, Wellens A, Shiao TC, *et al.* Mannosylated G(0) dendrimers with nanomolar affinities to *Escherichia coli* FimH. ChemMedChem 2007; 2(8):1190–201.

67 Almant M, Moreau V, Kovensky J, *et al.* Clustering of *Escherichia coli* type-1 fimbrial adhesins by using multimeric heptyl alpha-D-mannoside probes with a carbohydrate core. Chemistry 2011; 17(36):10029–38.

68 Gouin SG, Wellens A, Bouckaert J, Kovensky J. Synthetic multimeric heptyl mannosides as potent antiadhesives of uropathogenic *Escherichia coli*. ChemMedChem 2009; 4(5):749–55.

69 Bouckaert J, Berglund J, Schembri M, *et al.* Receptor binding studies disclose a novel class of high-affinity inhibitors of the Escherichia coli FimH adhesin. Mol. Microbiol. 2005; 55(2):441–55.

70 Sperling O, Fuchs A, Lindhorst TK. Evaluation of the carbohydrate recognition domain of the bacterial adhesin FimH: Design, synthesis and binding properties of mannoside ligands. Org. Biomol. Chem. 2006; 4(21):3913–22.

71 Han Z, Pinkner JS, Ford B, *et al.* Structure-based drug design and optimization of mannoside bacterial FimH antagonists. J. Med. Chem. 2010; 53(12):4779–92.

72 Jiang X, Abgottspon D, Kleeb S, *et al.* Antiadhesion therapy for urinary tract infections – a balanced PK/PD profile proved to be key for success. J. Med. Chem. 2012; 55(10):4700–13.

73 Scharenberg M, Schwardt O, Rabbani S, Ernst B. Target selectivity of FimH antagonists. J. Med. Chem. 2012; 55(22):9810–6.

74 Klein T, Abgottspon D, Wittwer M, *et al.* FimH antagonists for the oral treatment of urinary tract infections: from design and synthesis to in vitro and in vivo evaluation. J. Med. Chem. 2010; 53(24):8627–41.

75 Cusumano CK, Pinkner JS, Han Z, *et al.* Treatment and prevention of urinary tract infection with orally active FimH inhibitors. Sci. Transl. Med. 2011; 3(109):109ra15.

76 Kleeb S, Pang L, Mayer K, *et al.* FimH antagonists: bioisosteres to improve the in vitro and in vivo PK/PD profile. J. Med. Chem. 2015; 58(5):2221–39.

77 Wellens A, Garofalo C, Nguyen H, *et al.* Intervening with urinary tract infections using anti-adhesives based on the crystal structure of the FimH-oligomannose-3 complex. PLoS One 2008; 3(4):e2040.

78 Guiton PS, Cusumano CK, Kline KA, *et al.* Combinatorial small-molecule therapy prevents uropathogenic *Escherichia coli* catheter-associated urinary tract infections in mice. Antimicrob. Agents Chemother. 2012; 56(9):4738–45.

79 Alvarez Dorta D, Sivignon A, Chalopin T, *et al.* The antiadhesive strategy in Crohn's disease: Orally active mannosides to decolonize pathogenic *Escherichia coli* from the gut. ChemBioChem 2016; 17(10):936–52.

80 Sivignon A, de Vallee A, Barnich N, *et al. Saccharomyces cerevisiae* CNCM I-3856 prevents colitis induced by AIEC bacteria in the transgenic mouse model mimicking Crohn's disease. Inflamm. Bowel Dis. 2015; 21(2):276–86.

81 Spector DL. SnapShot: Cellular bodies. Cell 2006;127(5):1071.

82 Hyman AA, Simons K. Cell biology. Beyond oil and water – phase transitions in cells. Science 2012; 337(6098):1047–9.

83 Li P, Banjade S, Cheng HC, *et al.* Phase transitions in the assembly of multivalent signalling proteins. Nature 2012; 483(7389):336–40.

84 Jin J, Xie X, Chen C, Park JG, *et al.* Eukaryotic protein domains as functional units of cellular evolution. Sci. Signal 2009; 2(98):ra76.

85 Pawson T, Nash P. Assembly of cell regulatory systems through protein interaction domains. Science 2003; 300(5618):445–52.

86 Jones N, Blasutig IM, Eremina V, *et al.* Nck adaptor proteins link nephrin to the actin cytoskeleton of kidney podocytes. Nature 2006; 440(7085):818–23.

87 Blasutig IM, New LA, Thanabalasuriar A, *et al.* Phosphorylated YDXV motifs and Nck SH2/SH3 adaptors act cooperatively to induce actin reorganization. Mol. Cell Biol. 2008; 28(6):2035–46.

88 Rohatgi R, Nollau P, Ho HY, *et al.* Nck and phosphatidylinositol 4,5-bisphosphate synergistically activate actin polymerization through the N-WASP-Arp2/3 pathway. J. Biol. Chem. 2001; 276(28):26448–52.

89 Banjade S, Wu Q, Mittal A, *et al.* Conserved interdomain linker promotes phase separation of the multivalent adaptor protein Nck. Proc. Natl Acad. Sci. USA 2015; 112(47):E6426–35.

90 Pak CW, Kosno M, Holehouse AS, *et al.* Sequence determinants of intracellular phase separation by complex coacervation of a disordered protein. Mol. Cell 2016; 63(1):72–85.

91 Mogensen TH. Pathogen recognition and inflammatory signaling in innate immune defenses. Clin. Microbiol. Rev. 2009; 22(2):240–73.

92 Kumar H, Kawai T, Akira S. Pathogen recognition in the innate immune response. Biochem. J. 2009; 420(1):1–16.

93 Kawasaki T, Etoh R, Yamashina I. Isolation and characterization of a mannan-binding protein from rabbit liver. Biochem. Biophys. Res. Commun. 1978; 81(3):1018–24.

94 Drickamer K, Dordal MS, Reynolds L. Mannose-binding proteins isolated from rat liver contain carbohydrate-recognition domains linked to collagenous tails. Complete primary structures and homology with pulmonary surfactant apoprotein. J. Biol. Chem. 1986; 261(15):6878–87.

95 Ichijo H, Hellman U, Wernstedt C, *et al.* Molecular cloning and characterization of ficolin, a multimeric protein with fibrinogen- and collagen-like domains. J. Biol. Chem. 1993; 268(19):14505–13.

96 Lu J, Teh C, Kishore U, Reid KB. Collectins and ficolins: sugar pattern recognition molecules of the mammalian innate immune system. Biochim. Biophys. Acta 2002; 1572(2–3):387–400.

97 Kawasaki N, Kawasaki T, Yamashina I. Isolation and characterization of a mannan-binding protein from human serum. J. Biochem. 1983; 94(3):937–47.

98 Takahashi M, Mori S, Shigeta S, Fujita T. Role of MBL-associated serine protease (MASP) on activation of the lectin complement pathway. Adv. Exp. Med. Biol. 2007; 598:93–104.

99 Kolb WP, Haxby JA, Arroyave CM, Muller-Eberhard HJ. The membrane attack mechanism of complement. Reversible interactions among the five native components in free solution. J. Exp. Med. 1973; 138(2):428–37.

100 Gaboriaud C, Teillet F, Gregory LA, *et al.* Assembly of C1 and the MBL- and ficolin-MASP complexes: structural insights. Immunobiology 2007; 212(4–5):279–88.

101 Karulin A, Dzantiev BB. Polyvalent interaction of antibodies with bacterial cells. Mol. Immunol. 1990; 27(10):965–71.

102 Lullau E, Heyse S, Vogel H, *et al.* Antigen binding properties of purified immunoglobulin A and reconstituted secretory immunoglobulin A antibodies. J. Biol. Chem. 1996; 271(27):16300–9.

103 Kishore U, Reid KB. C1q: structure, function, and receptors. Immunopharmacology 2000; 49(1-2):159–70.

104 Hughes-Jones NC, Gardner B. Reaction between the isolated globular sub-units of the complement component C1q and IgG-complexes. Mol. Immunol. 1979; 16(9):697–701.

105 Feinstein A, Richardson N, Taussig MI. Immunoglobulin flexibility in complement activation. Immunol. Today 1986; 7(6):169–74.

106 Gaboriaud C, Thielens NM, Gregory LA, *et al.* Structure and activation of the C1 complex of complement: unraveling the puzzle. Trends Immunol. 2004; 25(7):368–73.

107 Bindon CI, Hale G, Waldmann H. Importance of antigen specificity for complement-mediated lysis by monoclonal antibodies. Eur. J. Immunol. 1988; 18(10):1507–14.

108 Cragg MS, Morgan SM, Chan HT, *et al.* Complement-mediated lysis by anti-CD20 mAb correlates with segregation into lipid rafts. Blood 2003; 101(3):1045–52.

109 de Weers M, Tai YT, van der Veer MS, *et al.* Daratumumab, a novel therapeutic human CD38 monoclonal antibody, induces killing of multiple myeloma and other hematological tumors. J. Immunol. 2011; 186(3):1840–8.

110 Hughes-Jones NC, Gorick BD, Howard JC, Feinstein A. Antibody density on rat red cells determines the rate of activation of the complement component C1. Eur. J. Immunol. 1985; 15(10):976–80.

111 Dechant M, Weisner W, Berger S, *et al.* Complement-dependent tumor cell lysis triggered by combinations of epidermal growth factor receptor antibodies. Cancer Res. 2008; 68(13):4998–5003.

112 Diebolder CA, Beurskens FJ, de Jong RN, *et al.* Complement is activated by IgG hexamers assembled at the cell surface. Science 2014; 343(6176):1260–3.

113 Saphire EO, Parren PW, Pantophlet R, *et al.* Crystal structure of a neutralizing human IGG against HIV-1: a template for vaccine design. Science 2001; 293(5532):1155–9.

114 Wu Y, West AP, Jr, Kim HJ, Thornton ME, Ward AB, Bjorkman PJ. Structural basis for enhanced HIV-1 neutralization by a dimeric immunoglobulin G form of the glycan-recognizing antibody 2G12. Cell Rep. 2013; 5(5):1443–55.

115 Ying T, Gong R, Ju TW, Prabakaran P, Dimitrov DS. Engineered Fc based antibody domains and fragments as novel scaffolds. Biochim. Biophys. Acta 2014; 1844(11):1977–82.

116 Caaveiro JM, Kiyoshi M, Tsumoto K. Structural analysis of Fc/FcgammaR complexes: a blueprint for antibody design. Immunol. Rev. 2015; 268(1):201–21.
117 Park HI, Yoon HW, Jung ST. The highly evolvable antibody Fc domain. Trends Biotechnol. 2016; 34(11):895–908.
118 Lobner E, Traxlmayr MW, Obinger C, Hasenhindl C. Engineered IgG1-Fc – one fragment to bind them all. Immunol. Rev. 2016; 270(1):113–31.
119 Brezski RJ, Georgiou G. Immunoglobulin isotype knowledge and application to Fc engineering. Curr. Opin. Immunol. 2016; 40:62–9.

Part II

Multivalent Systems in Chemistry

5

Multivalency in Cyclodextrin/Polymer Systems

Akihito Hashidzume and Akira Harada

Graduate School of Science, Osaka University, 1-1 Machikaneyama-cho, Toyonaka, Osaka 560-0043, Japan

5.1   Introduction

Cyclodextrins (CDs) are cyclic oligosaccharides composed of D-(+)-glucopyranose units linked through α-1,4-glycoside bonding. CDs have a narrower rim possessing primary hydroxy groups and a wider rim possessing secondary hydroxy groups. The most popular CDs are α-cyclodextrin (αCD), β-cyclodextrin (βCD), and γ-cyclodextrin (γCD), which are composed of six, seven, and eight D-(+)-glucopyranose units, respectively. The molecular dimensions of these CDs are summarized in Table 5.1 [1]. CDs are soluble in water, but they possess a rather hydrophobic cavity. On the basis of the unique hydrophile–lipophile balance, CDs recognize hydrophobic guest compounds, of which the shape and size match their cavities, to form inclusion complexes in water [2–6]. Thus CDs have been attracting considerable interest from researchers as simple biomimetic compounds for decades. CDs have been of increasing importance along with increasing interest in environmentally benign systems not only because CDs are produced enzymatically from starch and are nontoxic but also because CDs are usually used preferentially in aqueous media.

In the earlier stages, CDs were studied in detail focusing on their formation of inclusion complexes with low molecular weight compounds. However, systems containing CDs and polymers have been investigated intensively for a few decades, since polymers carrying CD moieties were synthesized and inclusion complexes were formed from CDs with polymer side chains, as well as the main-chain type poly-*pseudo*-rotaxanes were synthesized from CDs and linear polymers, for example, αCD and poly(ethylene glycol) [7–16].

The system composed of CDs and polymers can be divided into five categories: (1) CD/guest-polymer systems; (2) CD-polymer/guest systems; (3) CD-polymer/ guest-polymer systems; (4) CD-guest-polymer systems; and (5) supramolecular polymer systems (Figure 5.1). In these systems, Categories 3 and 4 can exhibit the effect of multivalency. In these systems, inclusion complexes act as noncovalent crosslinkers between polymer chains to form polymer aggregates.

Multivalency: Concepts, Research & Applications, First Edition. Edited by Jurriaan Huskens, Leonard J. Prins, Rainer Haag, and Bart Jan Ravoo.
© 2018 John Wiley & Sons Ltd. Published 2018 by John Wiley & Sons Ltd.

Table 5.1 Molecular dimensions of cyclodextrins.

Cyclodextrin	Number of D-(+)-glucopyranose units	Molecular weight	Cavity diameter (Å)	Height (Å)
αCD	6	972	4.7 – 5.3	7.9 ± 1
βCD	7	1135	6.0 – 6.5	7.9 ± 1
γCD	8	1297	7.5 – 8.3	7.9 ± 1

Source: Ref. [1]. Reproduced with permission of American Chemical Society.

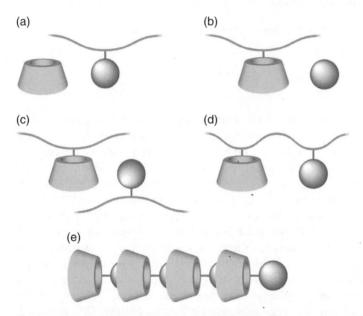

Figure 5.1 Conceptual illustration of cyclodextrin/polymer systems: (a) cyclodextrin/guest-polymer systems; (b) cyclodextrin-polymer/guest systems; (c) cyclodextrin-polymer/guest-polymer systems; (d) cyclodextrin-guest-polymer systems; and (e) supramolecular polymer systems.

We have investigated CD/polymer systems as simple models for biological molecular recognition (i.e., "macromolecular recognition") [17, 18]. We have revealed the role of polymer chains; the selectivity of interaction of CDs with side chains of water soluble polymers can be controlled by the steric effect of the main chain [19–22], the conformational effect of the main chain [23–26], and the effect of multivalency [27]. Among these effects, it is especially worth describing the effect of multivalency in detail. This is thus the aim of this book.

This chapter reviews multivalency in CD/polymer systems. In the next section (Section 5.2), general perspectives of multivalency in CD/polymer systems are described. And then, only typical examples of multivalency in CD/polymer systems are reviewed briefly in Section 5.3. Other multivalent systems involving cyclodextrins, such as "Molecular printboards" [28] and vesicles, are described in Chapters 2 and 8.

5.2 General Perspectives of Multivalency in Cyclodextrin/Polymer Systems

The most important effect of multivalency is the local concentration of moieties. In solutions of polymers carrying a number of CD or guest moieties, these moieties are localized on the polymer chains. When a CD moiety on a CD-polymer chain forms an inclusion complex with a guest moiety on a guest-polymer chain, the residual free CD and guest moieties on the polymer chains are localized in the limited space of the polymer aggregate. At an increased local concentration, the probability of encounter of the free CD and guest moieties are also increased, resulting in more favored formation of inclusion complexes (Figure 5.2). It should be noted here that the intrinsic binding constant of CD and guest moieties is practically unchanged.

Here we consider polymer aggregates formed from a CD-polymer and two guest-polymers carrying guest moieties of different binding constants, respectively. When a CD moiety forms an inclusion complex with a guest moiety, free CD and guest moieties in the polymer aggregate undergo more favored formation of inclusion complexes. The difference in binding constants for the guest moieties leads to different degrees of inclusion, resulting in different local concentrations of CD and guest moieties. Thus the accumulative difference causes a marked difference in the size of the polymer aggregates (Figure 5.3). This can be considered as a selectivity enhanced by multivalency.

Once a number of inclusion complexes are formed in a polymer aggregate, a polymer chain can be dissociated only upon dissociation of all the inclusion complexes of moieties on the chain. An inclusion complex in the polymer aggregate is dissociated quickly with a shorter lifetime of inclusion complex. However, the lifetime of a polymer aggregate is much longer than that of each inclusion

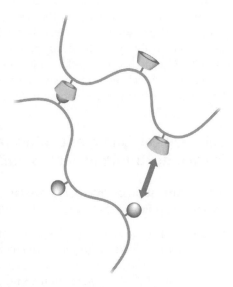

Figure 5.2 Conceptual illustration for entropically favored inclusion complex formation.

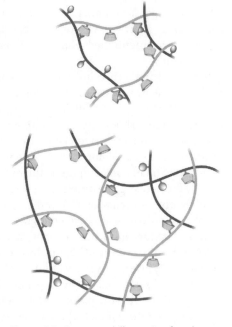

Figure 5.3 Conceptual illustration for a large difference in the size of polymer aggregates.

complex, depending strongly on the number of inclusion complexes in the aggregate (Figure 5.4). When the polymer aggregates formed from a number of inclusion complexes act as a crosslinker for polymer gels, the rearrangement of crosslinking points should be much slower than those of typical physical gels formed from hydrophobically modified associative thickeners [29, 30]. Such gels should exhibit properties intermediate between chemical gels and physical gels. This effect of multivalency can be considered as cooperativity in a broad sense.

5.3 Typical Examples of Multivalency in Cyclodextrin/Polymer Systems

5.3.1 Formation of Polymer Aggregates from Cyclodextrin-Polymers and Guest-Polymers

CD-polymer/guest-polymer systems or CD-guest-polymer systems form polymer aggregates, which often cause a significant change in solution properties. This subsection describes typical examples of the formation of polymer aggregates from CD-polymers and guest-polymers.

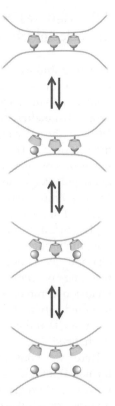

Figure 5.4 Conceptual illustration for an elongated lifetime of a polymer aggregate.

To the best of our knowledge, the formation of polymer aggregates from CD-polymers and guest-polymers through inclusion complexation was first reported by Amiel *et al.* in 1995 [31]. After this report, a number of research groups have studied this topic using a variety of polymer backbones [32–53]. These reports have mentioned almost nothing about the effect of multivalency.

When polymers carrying a number of CD moieties and guest moieties are mixed in aqueous solution, these polymers form polymer aggregates through the formation of inclusion complexes. These polymer aggregates often take an extended network structure, resulting in an increase in solution viscosity. Thus the formation of polymer aggregates from CD-polymers and guest-polymers is usually characterized by viscometry or by rheometry. When the CD and guest moieties form one-to-one inclusion complexes, the most expanded or largest polymer aggregates are usually formed at a one-to-one molar ratio of CD and guest moieties [43, 54].

When the contents of CD and guest moieties are optimized and the molecular weights of polymers are high enough, the mixtures of CD-polymer and guest-polymer form a self-standing hydrogel at higher concentrations. Recently, poly(acrylic acid)s (pAAs) modified with βCD and ferrocene (Fc) moieties (pAA/βCD and pAA/Fc in Scheme 5.1), respectively, were mixed at 2 wt% to form a self-standing hydrogel (Figure 5.5) [55]. Since the hydrogel obtained is a kind of physical gel, it shows a self-healing ability. After a piece of the hydrogel was cut in half by a knife and the two halves came into contact, the cut pieces adhered to each other to form a gel piece accompanied by disappearance of the contact interface. After 24 h, the stress at rupture of the healed gel recovered to 85 % of that of the virgin sample. It is known that a pair

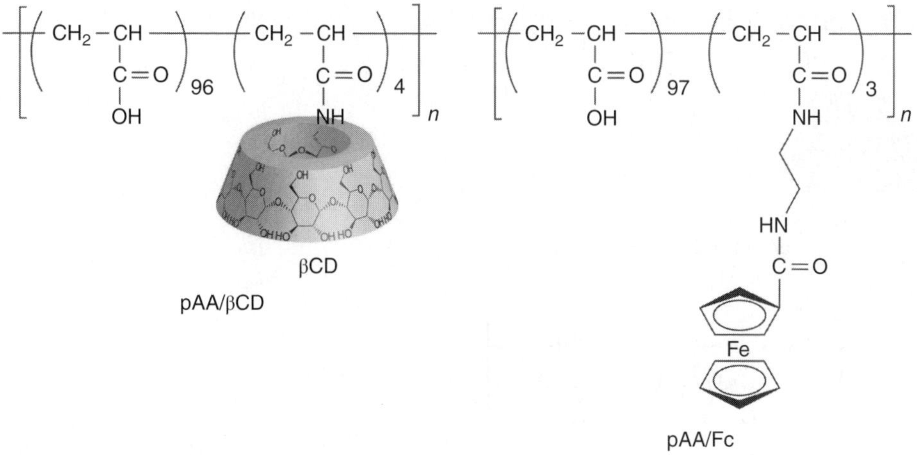

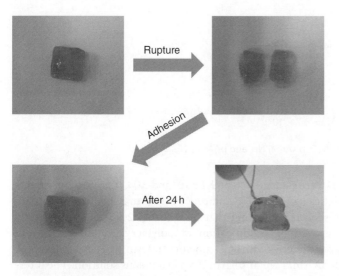

Scheme 5.1 Structures of pAA/βCD and pAA/Fc.

Figure 5.5 Self-healable hydrogel formed from pAA/βCD and pAA/Fc. *Souce*: Ref. [55]. Reproduced with permission of John Wiley and Sons.

of βCD and Fc exhibits redox-responsiveness; βCD includes Fc favorably but not ferrocenium (Fc$^+$), which is obtained by oxidation of Fc. Thus, when the Fc moieties on the cut surface were oxidized with sodium hypochlorite, the cut gel piece did not adhere. However, when the Fc$^+$ moieties were reduced to Fc moieties with glutathione, the gel regained the self-healing ability.

5.3.2 Selectivity of Interaction Enhanced by Multivalency

It is known that βCD forms inclusion complexes with both 1-naphthylmethyl (1 Np) and 2-naphthylmethyl (2 Np) moieties with moderate binding constants (ca. $10^2 \, \text{M}^{-1}$).

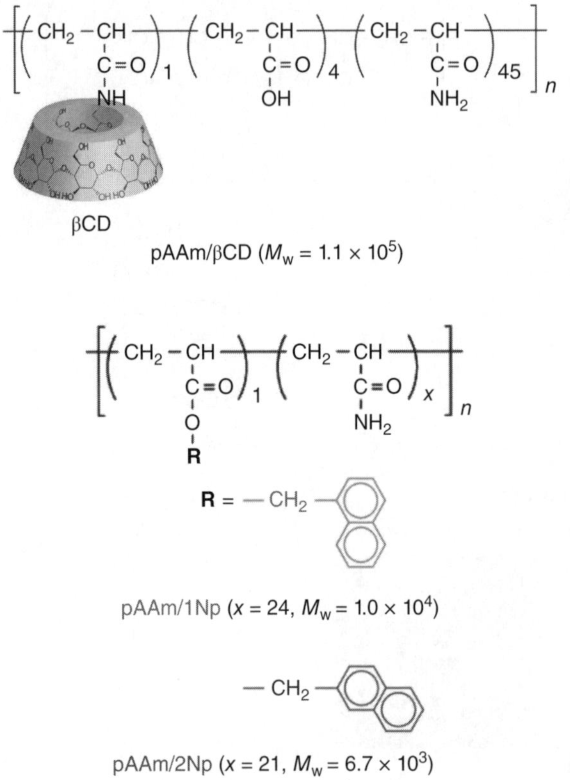

βCD

pAAm/βCD ($M_w = 1.1 \times 10^5$)

$R = - CH_2$

pAAm/1Np ($x = 24$, $M_w = 1.0 \times 10^4$)

$- CH_2$

pAAm/2Np ($x = 21$, $M_w = 6.7 \times 10^3$)

Scheme 5.2 Structures of pAAm/βCD, pAAm/1 Np, and pAAm/2 Np.

The binding constants of βCD were evaluated to be 1.1×10^2 and $2.0 \times 10^2 \, M^{-1}$ for polymer-carrying 1 Np and 2 Np moieties (pAAm/1 Np and pAAm/2 Np in Scheme 5.2), respectively, by steady-state fluorescence, indicating that βCD favors 2 Np moieties slightly [56]. βCD-polymer was prepared by modification of poly(acrylamide) (pAAm) of $M_w = 1.1 \times 10^5$ carrying carboxyl moieties with βCD moieties (pAAm/βCD in Scheme 5.2). pAAm/βCD was mixed with pAAm/1 Np or pAAm/2 Np in aqueous solution. The interaction was investigated by steady-shear viscosity measurements [27]. All the mixtures were Newtonian, indicating that polymer aggregates did not take an extended network structure. This is because the binding constants and the M_w values of pAAm/1 Np and pAAm/2 Np are rather low. At a constant concentration ($100 \, gl^{-1}$) of pAAm/βCD and varying concentrations of pAAm/1 Np and pAAm/2 Np, zero-shear viscosities (η_0) for the mixture were determined by viscometry and plotted in Figure 5.6 against the concentration of guest-polymer (C_{gp}). As reference, the η_0 values for the mixture of unmodified pAAm were also plotted. It is noteworthy that only the mixture of pAAm/2 Np exhibits a significant η_0 increase, whereas that of pAAm/1 Np shows η_0 values which are almost the same as those of the mixture of pAAm. These data indicate that pAAm/βCD and pAAm/2 Np form larger polymer aggregates through the formation of inclusion complexes, whereas pAAm/βCD and pAAm/1 Np do not form significant polymer aggregates, even though the binding constants of βCD with 1 Np and with 2 Np are on the same order (Figure 5.7). The multivalency is responsible for the large difference in size of the polymer aggregates. It is thus considered that multivalency enhances the selectivity.

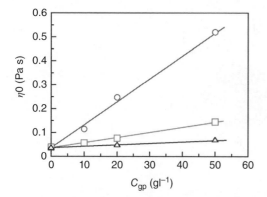

Figure 5.6 Zero-shear viscosities (η_0) as a function of the guest-polymer concentration (C_{gp}) for mixtures of $100\,g\,l^{-1}$ pAAm/βCD and guest-polymers [pAAm/1 Np (square) and pAAm/2 Np (circle), as well as pAAm (triangle)]. *Source*: Ref. [27]. Reproduced with permission of John Wiley and Sons.

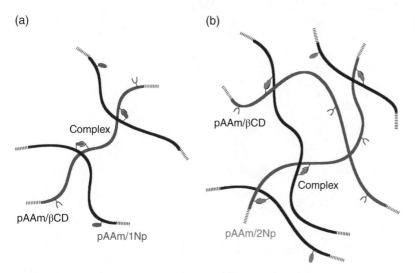

Figure 5.7 Conceptual illustration for polymer aggregates of pAAm/βCD–pAAm/1 Np (a) and pAAm/βCD–pAAm/2 Np (b) showing a considerable difference in the size. *Source*: Ref. [27]. Reproduced with permission of John Wiley and Sons.

Recently, our group has first visualized molecular recognition using pAAm-based gel possessing CD and guest moieties [57, 58]. Pieces of gel possessing βCD moieties (βCD-gel) adhere to pieces of gel possessing adamantane (Ad) moieties (Ad-gel) to form a gel assembly. Since gel pieces strongly adhere to each other, the gel assembly can be picked up by tweezers. When pieces of gels possessing αCD, βCD, *n*-butyl (*n*Bu), and *t*-butyl (*t*Bu) moieties (αCD-gel, βCD-gel, *n*Bu-gel, and *t*Bu-gel, respectively) are mixed in water, αCD-gel pieces adhere only to *n*Bu-gel pieces whereas βCD-gel pieces adhere only to *t*Bu-gel pieces. In these macroscopic self-assemblies based on molecular recognition, a number of inclusion complexes are formed between CD and guest moieties on the gel interface, in which the effect of multivalency works.

Similar to the polymer/polymer systems, βCD, 1 Np, and 2 Np moieties have been chosen as recognition groups for investigation of the enhanced selectivity of gel

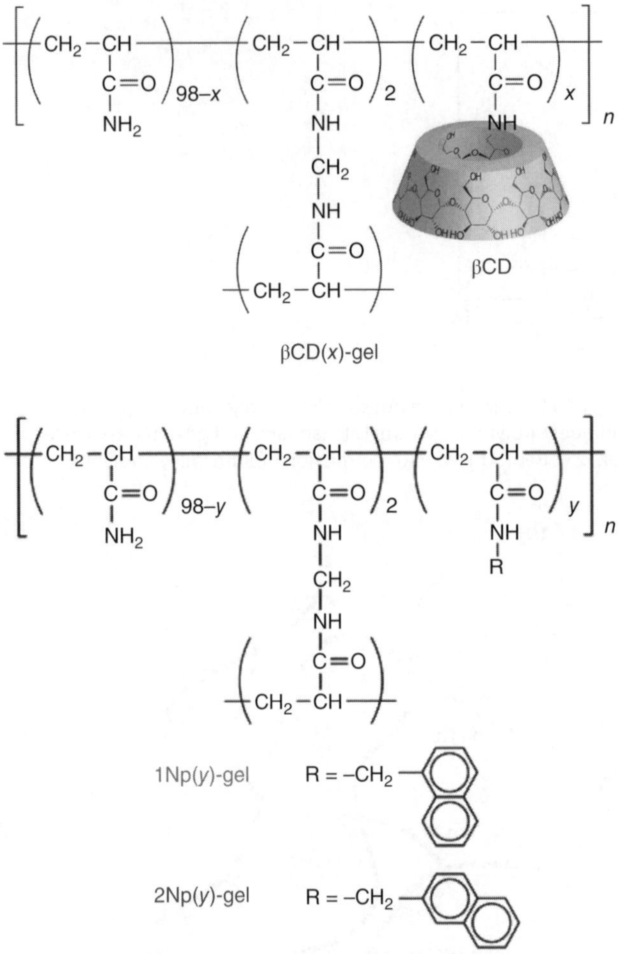

βCD(x)-gel

1Np(y)-gel   R = –CH$_2$–

2Np(y)-gel   R = –CH$_2$–

Scheme 5.3 Structures of βCD(x)-gel, 1Np(y)-gel, and 2Np(y)-gel.

interaction [59]. Scheme 5.3 shows the chemical structures of gels prepared [i.e., βCD(x)-gel, 1Np(y)-gel, and 2Np(y)-gel]. Here x and y denote the mol% contents of βCD and 1Np (or 2Np) moieties in gel, respectively. When pieces of βCD(5)-gel and 1Np(5)-gel [or 2Np(5)-gel] were mixed in water with a mixer, a gel assembly was formed (Figure 5.8). On the other hand, when pieces of βCD(5)-gel, 1Np(2)-gel, and 2Np(2)-gel were mixed in water, βCD(5)-gel pieces discriminated in favor of 2Np(2)-gel pieces to form a gel assembly (Figure 5.9a). Similarly, when pieces of βCD(1)-gel, 1Np(5)-gel, and 2Np(5)-gel were mixed in water, βCD(1)-gel pieces discriminated in favor of 2Np(5)-gel pieces (Figure 5.9b). These observations are suggestive of the selectivity being enhanced by multivalency. The strength of interaction of gel pieces was estimated by a tensile tester. The stresses at rupture were evaluated and plotted in Figure 5.10 against x and y at constant y and x (=5 mol%), respectively. The stress at rupture increases with x and y, showing a weak tendency of saturation at higher x and y. The data indicate that βCD(x)-gel discriminates the substitution position on the Np moiety at appropriate pairs of x and y, for example, $x = 5$ mol% and $y = 2$ mol%, and

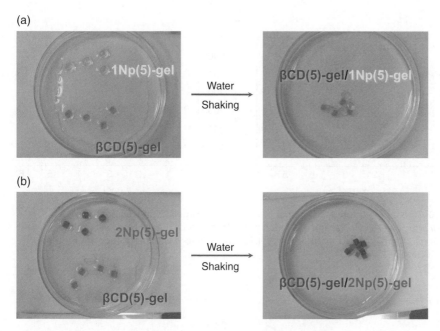

Figure 5.8 Photographs for gel assemblies formed from βCD(5)-gel/1 Np(5)-gel (a) and βCD(5)-gel/2 Np(5)-gel (b). *Source*: Ref. [59]. Reproduced with permission of American Chemical Society. See color section.

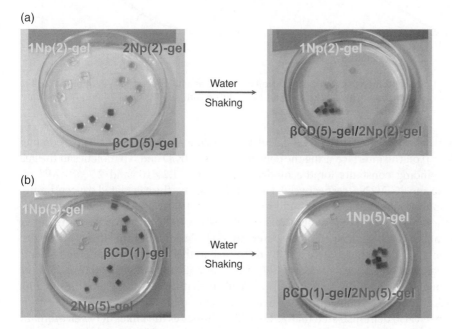

Figure 5.9 Photographs for gel assemblies formed from βCD(5)-gel/2 Np(2)-gel in the presence of 1 Np(2)-gel (a) and βCD(1)-gel/2 Np(5)-gel in the presence of 1 Np(5)-gel (b). *Source*: Ref. [59]. Reproduced with permission of American Chemical Society. See color section.

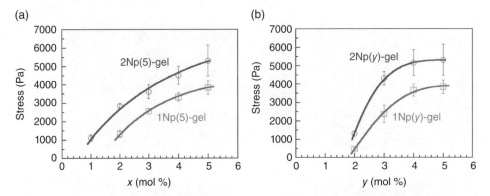

Figure 5.10 Stresses at rupture as a function of *x* for the βCD(*x*)-gel/1 Np(5) gel (square) and βCD(*x*)-gel/2 Np(5)-gel (circle) (a) and as a function of *y* for the βCD(5)-gel/1 Np(*y*) gel (square) and βCD(5)-gel/2 Np(*y*)-gel (circle) (b). *Source*: Ref. [59]. Reproduced with permission of American Chemical Society.

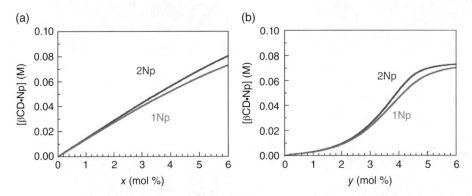

Figure 5.11 Concentration of complex ([βCD•Np]) calculated as a function of *x* at *y* = 5 mol% (a) and as a function of *y* at *x* = 5 mol% (b). *Source*: Ref. [59]. Reproduced with permission of American Chemical Society.

$x = 1$ mol% and $y = 5$ mol%. Here we calculated the concentration of inclusion complex ([βCD•Np]) on the interface using the concentrations of βCD and Np moieties in the gels and the binding constants for the model systems (i.e., 1.2×10^2 and $2.7 \times 10^2 \, \mathrm{M^{-1}}$ for βCD/1 Np and βCD/2 Np, respectively). Figure 5.11 displays the calculated concentrations as functions of *x* and *y* at constant *y* and *x* (=5 mol%), respectively. The *x* and *y* dependencies of the stress at rupture resemble those of [βCD•Np], indicating that the formation of gel assemblies can be explained partly based on a simple equilibrium. However, two distinct points should be noted here: (1) the difference in the stress at rupture for 1 Np and 2 Np is significantly larger than that in [βCD•Np]; and (2) the stress at rupture does not pass through the origin unlike [βCD•Np]. It is likely that there is a certain threshold for the stress at rupture presumably because of the effect of multivalency.

It has been also demonstrated that γCD-gel discriminates the linker between aromatic guest moieties and the pAAm gel scaffold in macroscopic self-assembly [60]. We prepared eight samples of pAAm-based gel modified with aromatic moieties [i.e., benzyl (Bz), 2-naphthylmethyl (Np), 9-phenanthrylmethyl (Ph), and 1-pyrenylmethyl (Py)]

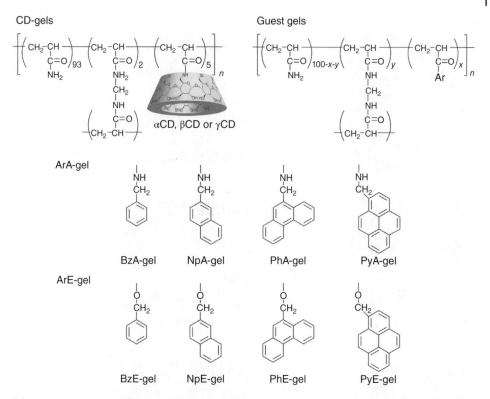

Scheme 5.4 Structures of gels modified with Bz, Np, Ph, and Py moieties, respectively, through ester and amide linkages.

through amide and ester linkages, respectively (Scheme 5.4). Interaction of CD-gels and guest-gels was investigated by mixing gel pieces in water. αCD-gel pieces did not form any gel assemblies with guest gels. βCD-gel pieces formed gel assemblies with gel pieces possessing smaller aromatics (i.e., Bz and Np) for both amide and ester linkages. These observations indicate that either αCD-gel or βCD-gel cannot discriminate the linkage between aromatic guest and gel scaffold. On the other hand, γCD-gel pieces formed gel assemblies with gel pieces modified with Np moieties through ester linkage and with those modified with Ph and Py through amide linkage, indicating that γCD-gel can discriminate the linkage for gel possessing these aromatic moieties (Figure 5.12). Since the cavity of γCD is large enough, γCD includes preferably dimeric aromatics [26]. Thus it is likely that the guest gels with which γCD-gel formed gel assemblies possess a larger fraction of dimer of aromatic moieties on the gel surface. However, since it was not possible to detect considerable differences in the fraction of dimeric aromatics by steady-state fluorescence, the difference in the fraction of dimer might be subtle. It is thus likely that the effect of multivalency on the gel interface is also responsible for the discrimination by γCD-gel of the linker between aromatic moieties and gel scaffold.

Utilizing the selectivity enhanced based on multivalency in macroscopic self-assembly, gel assemblies responsive to external stimuli, for example, temperature [61], pH [62], chemicals [63, 64], light [65], and redox [66], have been realized.

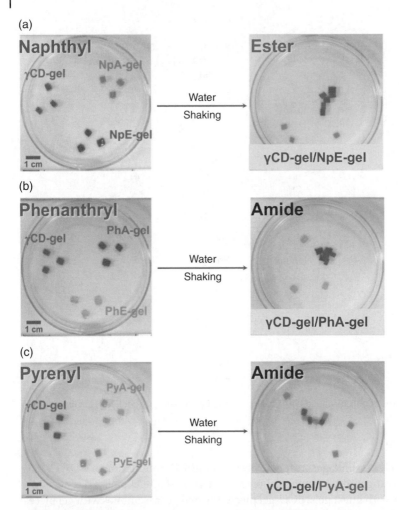

Figure 5.12 Photographs of the interaction of γCD-gel (dark green) with NpA-gel (green) and NpE-gel (blue) (a), with PhA-gel (purple) and PhE-gel (yellow) (b), and with PyA-gel (yellow) and PyE-gel (orange) (c). *Source*: Ref. [60]. Reproduced with permission of American Chemical Society. See color section.

5.3.3 Self-Healable Hydrogels Based on Multivalency

Recently, it has been reported that hydrogel samples of terpolymer formed from a CD-monomer, a guest-monomer, and a water soluble monomer show an excellent self-healing ability based on multivalency [67, 68].

Since guest-monomers are usually insoluble or only slightly soluble in water, it is difficult to carry out terpolymerization of a guest-monomer, a CD-monomer, and a water soluble monomer in water. Thus, our group prepared an inclusion complex from βCD-monomer with a monomer possessing an adamantyl moiety (Ad-monomer) by heating in water, and then conducted copolymerization of the inclusion complex with acrylamide (Figure 5.13). Since the inclusion complex of βCD-monomer and Ad-monomer

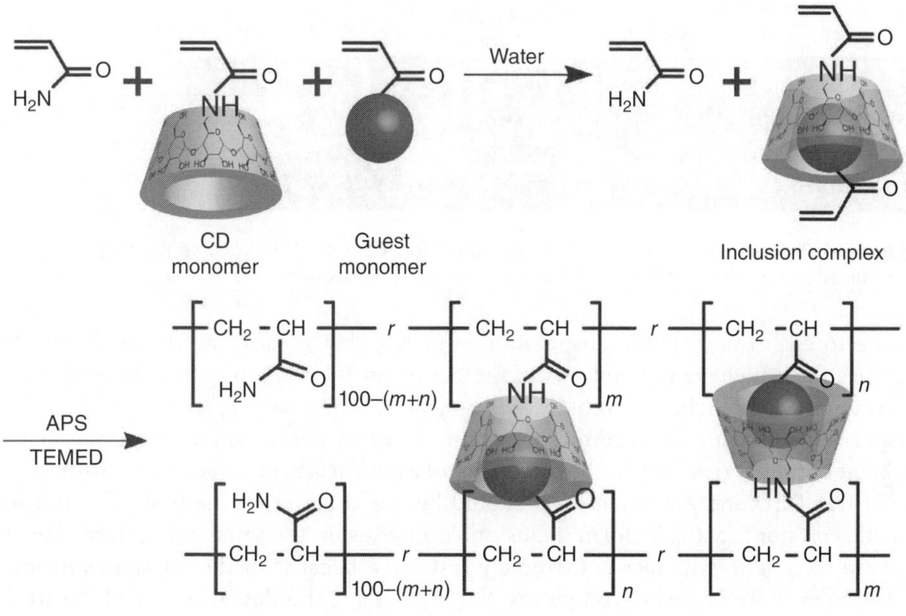

Figure 5.13 Preparation of self-healable hydrogel (βCD-Ad gel (m,n)). Source: Adapted from Refs [67, 68].

possesses two vinyl groups, the complex should act as a supramolecular crosslinker. The terpolymer samples obtained were hydrogels that exhibited turbidities depending on the mol% contents of βCD-monomer and Ad monomer; the hydrogel samples were rather transparent at lower CD and Ad contents (ca. 0.1 mol%) whereas highly turbid at higher CD and Ad contents (ca. 6–7 mol%). These hydrogel samples were not dissolved in good solvents for pAAm, that is, water and dimethyl sulfoxide (DMSO), even after washing repeatedly with these solvents, although the hydrogels should be a kind of physical gel. In a separate experiment, terpolymerization of βCD-monomer, Ad-monomer, and acrylamide in DMSO did not yield gel. It is thus concluded that the key in the formation of such hydrogel is copolymerization of the inclusion complex of monomers with acrylamide in water. Generally, in equilibria of inclusion complex formation of CDs in aqueous media, free and complexed species cannot be detected separately by ^{1}H NMR, indicative of exchange between the free and complexed species faster than the time scale of ^{1}H NMR (~ ms). Thus if an inclusion complex of CD and guest moieties act as a crosslinker, hydrogels should deform within a relatively short time because of fast rearrangement of the network structure. However, since the hydrogel samples obtained in this study maintain their shape even with immersing in solvent for a long time, it can be concluded that a crosslinking point is composed of two or more inclusion complexes.

The hydrogels show an excellent self-healing ability. When a gel piece was cut into half by a knife and then the cut halves were brought into contact on their cut surfaces, the gel pieces immediately adhered to each other. After standing for 24 h under humid conditions, the stress recovered to that of the virgin sample (Figure 5.14). It is noteworthy that the cut surface did not adhere to non-cut surfaces, and non-cut surfaces did not

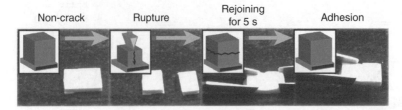

Figure 5.14 Photographs of a self-healing experiment for βCD-Ad gel (7,6). *Source*: Ref. [67]. Reproduced with permission of John Wiley and Sons. See color section.

adhere to each other. These observations indicate that most of all the βCD and Ad moieties form inclusion complexes in the gel or on the non-cut gel surfaces and the rearrrangement of inclusion complexes is very slow. When the gel piece is cut by a knife, dissociation of the inclusion complexes occurs more favorably than does the cleavage of covalent bonding, resulting in the formation of a cut surface on which there are a number of free βCD and Ad moieties. It should be noted here that the free βCD and Ad moieties do not practically form inclusion complexes on the same cut surface. This is because the chain dynamics is extremely restricted because of the crosslinks formed from two or more inclusion complexes. On the basis of the slow recovery of the stress after contact of the cut surfaces, the dynamics of polymer chains in the hydrogel occurs on a longer time scale (ca. 24 h).

Most recently, terpolymerization of βCD-monomer, Ad-monomer, and acrylamide at higher monomer concentrations (ca. $2 \, mol \, kg^{-1}$) yielded rather transparent hydrogels even at higher βCD and Ad contents (2–3 mol%) (Figure 5.15) [69]. The turbidity of the gel can be ascribable to light scattering based on the heterogeneous structure inside the gel. Thus transparent gel samples possess a less heterogeneous structure than that of the turbid gel samples. It is noteworthy that it was very difficult to pierce the film of transparent gel with sharps, indicating that the transparent gels prepared at higher monomer concentrations exhibited high flexibility and great toughness (Figure 5.15). The gel was also self-healable, indicating that a crosslink was still composed of two or more inclusion complexes.

5.4 Summary and Outlook

This chapter has described that multivalency provides unique properties for the CD/polymer systems to produce polymer and hydrogel systems which exhibit high selectivity or self-healing ability. If we combine the effect of multivalency with the steric and conformational effects of the main chain, environmentally benign functional materials comparable with biological systems will be realized for creation of sustainable societies.

Acknowledgments

The authors would like to express their gratitude to Professor Hiroyasu Yamaguchi and Associate Professor Yoshinori Takashima, Department of Macromolecular Science, Graduate School of Science, Osaka University.

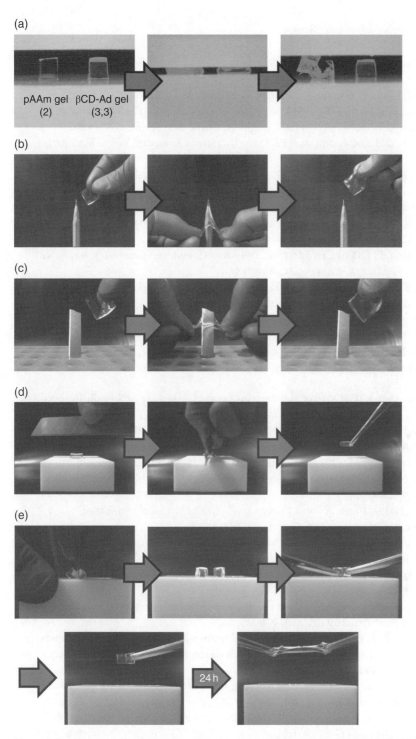

Figure 5.15 Photographs of a compression experiment of a cube-shaped βCD-Ad gel (3,3) compared with a cube-shaped pAAm gel chemically crosslinked with 2 mol% of *N,N*′-methylenebis(acrylamide) (a). Photographs of stab-resistant properties of βCD-Ad gel (2,2) using a pencil (b) and a cutter blade (c). Photographs of scission-resistant properties of a cuboid-shaped βCD-Ad gel (2,2) (d). Photographs of a self-healing experiment for βCD-Ad gel (3,3) (e). *Source*: Ref. [69]. Reproduced with permission of John Wiley and Sons.

References

1 Saenger, W., Jacob, J., Gessler, K., Steiner, T., Hoffmann, D., Sanbe, H., Koizumi, K., Smith, S. M., and Takaha, T. (1998) Structures of the Common Cyclodextrins and Their Larger Analogues Beyond the Doughnut, *Chem. Rev.* **98** (5), 1787–1802.

2 Bender, M. L. and Komiyama, M. (1978) *Cyclodextrin Chemistry*, Vol. 6, Springer, Berlin.

3 Szejtli, J. (1982) *Cyclodextrins and Their Inclusion Complexes*, Akadémiai Kiadó, Budapest.

4 Szejtli, J. and Osa, T. (Ed.) (1996) *Cyclodextrins*, Vol. 3, Pergamon, Oxford.

5 Harada, A. (1996) Cyclodextrins, In *Large Ring Molecules* (Semlyen, J. A., Ed.), pp. 407–432, John Wiley & Sons, Ltd, Chichester.

6 Dodziuk, H., (Ed.) (2006) *Cyclodextrins and Their Complexes: Chemistry, Analytical Methods, Applications*, Wiley-VCH, Weinheim.

7 Takata, T., Kihara, N., and Furusho, Y. (2004) Polyrotaxanes and Polycatenanes: Recent Advances in Syntheses and Applications of Polymers Comprising of Interlocked Structures, *Adv. Polym. Sci.* **171**, 1–75.

8 Wenz, G., Han, B.-H., and Müller, A. (2006) Cyclodextrin Rotaxanes and Polyrotaxanes, *Chem. Rev.* **106** (3), 782–817.

9 Wenz, G. (2009) Recognition of Monomers and Polymers by Cyclodextrins, *Adv. Polym. Sci.* **222**, 1–54.

10 Yuen, F. and Tam, K. C. (2010) Cyclodextrin-Assisted Assembly of Stimuli-Responsive Polymers in Aqueous Media, *Soft Matter* **6** (19), 4613–4630.

11 Zhou, J. and Ritter, H. (2010) Cyclodextrin Functionalized Polymers as Drug Delivery Systems, *Polym. Chem.* **1** (10), 1552–1559.

12 Chen, G. and Jiang, M. (2011) Cyclodextrin-Based Inclusion Complexation Bridging Supramolecular Chemistry and Macromolecular Self-Assembly, *Chem. Soc. Rev.* **40** (5), 2254–2266.

13 Harada, A., Hashidzume, A., and Takashima, Y. (2006) Cyclodextrin-Based Supramolecular Polymers, *Adv. Polym. Sci.* **201**, 1–43.

14 Harada, A., Hashidzume, A., and Miyauchi, M. (2006) Polymers Involving Cyclodextrin Moieties, In *Cyclodextrins and Their Complexes: Chemistry, Analytical Methods, Applications* (Dodziuk, H., Ed.), pp. 65–92, John Wiley & Sons, Inc., New York, NY.

15 Harada, A., Hashidzume, A., Yamaguchi, H., and Takashima, Y. (2009) Polymeric Rotaxanes, *Chem. Rev.* **109** (11), 5974–6023.

16 Schmidt, B. V. K. J., Hetzer, M., Ritter, H., and Barner-Kowollik, C. (2014) Complex Macromolecular Architecture Design via Cyclodextrin Host/Guest Complexes, *Prog. Polym. Sci.* **39** (1), 235–249.

17 Hashidzume, A. and Harada, A. (2011) Recognition of Polymer Side Chains by Cyclodextrins, *Polym. Chem.* **2** (10), 2146–2154.

18 Hashidzume, A. and Harada, A. (2015) Macromolecular Recognition: Recognition of Polymer Side Chains by Cyclodextrin, In *International Conference of Computational Methods in Sciences and Engineering 2015 (ICCMSE 2015)* (Simos, T. E., Kalogiratou, Z., and Monovasilis, T., Eds), p. 090016, AIP Publishing, Woodbury, NY.

19 Harada, A., Adachi, H., Kawaguchi, Y., and Kamachi, M. (1997) Recognition of Alkyl Groups on a Polymer Chain by Cyclodextrins, *Macromolecules* **30** (17), 5181–5182.

20 Hashidzume, A. and Harada, A. (2006) Macromolecular Recognition by Cyclodextrins. Interaction of Cyclodextrins with Polymethacrylamides Bearing Hydrophobic Amino Acid Residues, *Polymer* **47** (10), 3448–3454.

21 Oi, W., Hashidzume, A., and Harada, A. (2011) Macromolecular Recognition by Cyclodextrins. Interaction of Cyclodextrins with Poly(*N*-acryloyl-amino acids), *Polymer* **52** (3), 746–751.

22 Oi, W., Isobe, M., Hashidzume, A., and Harada, A. (2011) Macromolecular Recognition: Discrimination between Human and Bovine Serum Albumins by Cyclodextrins, *Macromol. Rapid Commun.* **32** (6), 501–505.

23 Taura, D., Hashidzume, A., and Harada, A. (2007) Macromolecular Recognition: Interactions of Cyclodextrins with Alternating Copolymer of Sodium Maleate and Dodecyl Vinyl Ether, *Macromol. Rapid Commun.* **28** (24), 2306–2310.

24 Taura, D., Hashidzume, A., Okumura, Y., and Harada, A. (2008) Cooperative Complexation of α-Cyclodextrin with Alternating Copolymers of Sodium Maleate and Dodecyl Vinyl Ether with Varying Molecular Weights, *Macromolecules* **41** (10), 3640–3645.

25 Taura, D., Taniguchi, Y., Hashidzume, A., and Harada, A. (2009) Macromolecular Recognition of Cyclodextrin: Inversion of Selectivity of β-Cyclodextrin toward Adamantyl Groups Induced by Macromolecular Chains, *Macromol. Rapid Commun.* **30** (20), 1741–1744.

26 Hashidzume, A., Zheng, Y., and Harada, A. (2012) Interaction of Cyclodextrins with Pyrene-Modified Polyacrylamide in a Mixed Solvent of Water and Dimethyl Sulfoxide as Studied by Steady-State Fluorescence, *Beilstein J. Org. Chem.* **8**, 1312–1317.

27 Hashidzume, A., Ito, F., Tomatsu, I., and Harada, A. (2005) Macromolecular Recognition by Polymer-Carrying Cyclodextrins. Interaction of a Polymer Bearing Cyclodextrin Moieties with Poly(acrylamide)s Bearing Aromatic Side Chains, *Macromol. Rapid Commun.* **26** (14), 1151–1154.

28 Ludden, M. J. W., Reinhoudt, D. N., and Huskens, J. (2006) Molecular Printboards: Versatile Platforms for the Creation and Positioning of Supramolecular Assemblies and Materials, *Chem. Soc. Rev.* **35** (11), 1122–1134.

29 Glass, J. E. (Ed.) (2000) *Associative Polymers in Aqueous Solutions*, Vol. 765, American Chemical Society, Washington, DC.

30 Hashidzume, A., Morishima, Y., and Szczubiałka, K. (2002) Amphiphilic Polyelectrolytes, In *Handbook of Polyelectrolytes and Their Applications* (Tripathy, S. K., Kumar, J., and Nalwa, H. S., Eds), pp. 1–63, American Scientific Publishers, Stevenson Ranch, CA.

31 Amiel, C., Sandier, A., Sebille, B., Valat, P., and Wintgens, V. (1995) Associations between Hydrophobically End-Capped Polyethylene Oxide and Water Soluble β-Cyclodextrin Polymers, *Int. J. Polym. Anal. Charact.* **1** (4), 289–300.

32 Amiel, C. and Sebille, B. (1996) New Associating Polymer Systems Involving Water-Soluble β-Cyclodextrin Polymers, *J. Inclusion Phenom. Mol. Recognit. Chem.* **25** (1–3), 61–67.

33 Weickenmeier, M., Wenz, G., and Huff, J. (1997) Association Thickener by Host Guest Interaction of a β-Cyclodextrin Polymer and Polymer with Hydrophobic Side-Groups, *Macromol. Rapid Commun.* **18** (12), 1117–1123.

34 Gosselet, N. M., Borie, C., Amiel, C., and Sebille, B. (1998) Aqueous Two Phase Systems from Cyclodextrin Polymers and Hydrophobically Modified Acrylic Polymers, *J. Dispersion Sci. Technol.* **19** (6 and 7), 805–820.

35 Amiel, C., Moine, L., Sandier, A., Brown, W., David, C., Hauss, F., Renard, E., Gosselet, M., and Sebille, B. (2001) Macromolecular Assemblies Generated by Inclusion Complexes between Amphipathic Polymers and β-Cyclodextrin Polymers in Aqueous Media, In *Stimuli-Responsive Water Soluble and Amphiphilic Polymers* (McCormick, C. L., Ed.), pp. 58–81, American Chemical Society, Washington, DC.

36 Galant, C., Amiel, C., Wintgens, V., Sébille, B., and Auvray, L. (2002) Ternary Complexes with Poly(β-cyclodextrin), Cationic Surfactant, and Polyanion in Dilute Aqueous Solution: A Viscometric and Small-Angle Neutron Scattering Study, *Langmuir* **18** (25), 9687–9695.

37 Galant, C., Amiel, C., and Auvray, L. (2004) Tailorable Polyelectrolyte Complexes Using Cyclodextrin Polymers, *J. Phys. Chem. B* **108** (50), 19218–19227.

38 Amiel, C., Galant, C., and Auvray, L. (2004) Ternary Complexes Involving a β-Cyclodextrin Polymer, a Cationic Surfactant and an Anionic Polymer, *Prog. Colloid Polym. Sci.* **126**, 44–46.

39 Gosselet, N. M., Layre, A. M., Wintgens, V., and Amiel, C. (2004) Physicochemical Study of Modified Dextrans with a β-Cyclodextrin Polymer, *Prog. Colloid Polym. Sci.* **126**, 21–24.

40 Galant, C., Amiel, C., and Auvray, L. (2005) Ternary Complex Formation in Aqueous Solution between a β-Cyclodextrin Polymer, a Cationic Surfactant and DNA, *Macromol. Biosci.* **5** (11), 1057–1065.

41 Wintgens, V. and Amiel, C. (2005) Surface Plasmon Resonance Study of the Interaction of a β-Cyclodextrin Polymer and Hydrophobically Modified Poly(*N*-isopropylacrylamide), *Langmuir* **21** (24), 11455–11461.

42 Wintgens, V., Charles, M., Allouache, F., and Amiel, C. (2005) Triggering the Thermosensitive Properties of Hydrophobically Modified Poly(*N*-isopropylacrylamide) by Complexation with Cyclodextrin Polymers, *Macromol. Chem. Phys.* **206** (18), 1853–1861.

43 Guo, X., Abdala, A. A., May, B. L., Lincoln, S. F., Khan, S. A., and Prud'homme, R. K. (2005) Novel Associative Polymer Networks Based on Cyclodextrin Inclusion Compounds, *Macromolecules* **38** (7), 3037–3040.

44 Li, L., Guo, X., Fu, L., Prud'homme, R. K., and Lincoln, S. F. (2008) Complexation Behavior of α-, β-, and γ-Cyclodextrin in Modulating and Constructing Polymer Networks, *Langmuir* **24** (15), 8290–8296.

45 Li, L., Guo, X., Wang, J., Liu, P., Prud'homme, R. K., May, B. L., and Lincoln, S. F. (2008) Polymer Networks Assembled by Host-Guest Inclusion between Adamantyl and β-Cyclodextrin Substituents on Poly(acrylic acid) in Aqueous Solution, *Macromolecules* **41** (22), 8677–8681.

46 Wintgens, V., Daoud-Mahammed, S., Gref, R., Bouteiller, L., and Amiel, C. (2008) Aqueous Polysaccharide Associations Mediated by β-Cyclodextrin Polymers, *Biomacromolecules* **9** (5), 1434–1442.

47 Koopmans, C. and Ritter, H. (2008) Formation of Physical Hydrogels via Host–Guest Interactions of β-Cyclodextrin Polymers and Copolymers Bearing Adamantyl Groups, *Macromolecules* **41** (20), 7418–7422.

48 Takashima, Y., Nakayama, T., Miyauchi, M., Kawaguchi, Y., Yamaguchi, H., and Harada, A. (2004) Complex Formation and Gelation between Copolymers Containing Pendant Azobenzene Groups and Cyclodextrin Polymers, *Chem. Lett.* **33**, 890–891.

49 Tomatsu, I., Hashidzume, A., and Harada, A. (2006) Contrast Viscosity Changes upon Photoirradiation for Mixtures of Poly(acrylic acid)-Based α-Cyclodextrin and Azobenzene Polymers, *J. Am. Chem. Soc.* **128** (7), 2226–2227.

50 Ogoshi, T., Takashima, Y., Yamaguchi, H., and Harada, A. (2007) Chemically-Responsive Sol-Gel Transition of Supramolecular Single-Walled Carbon Nanotubes (SWNTs) Hydrogel Made by Hybrids of SWNTs and Cyclodextrins, *J. Am. Chem. Soc.* **129** (16), 4878–4879.

51 Tamesue, S., Takashima, Y., Yamaguchi, H., Shinkai, S., and Harada, A. (2010) Photoswitchable Supramolecular Hydrogels Formed by Cyclodextrins and Azobenzene Polymers, *Angew. Chem. Int. Ed.* **49** (41), 7461–7464.

52 Tamesue, S., Takashima, Y., Yamaguchi, H., Shinkai, S., and Harada, A. (2011) Photochemically Controlled Supramolecular Curdlan/Single-Walled Carbon Nanotube Composite Gel: Preparation of Molecular Distaff by Cyclodextrin Modified Curdlan and Phase Transition Control, *Eur. J. Org. Chem.* **2011** (15), 2801–2806.

53 Lee, I. E. T., Hashidzume, A., and Harada, A. (2015) A Light-Controlled Release System Based on Molecular Recognition of Cyclodextrins, *Macromol. Rapid Commun.* **36** (23), 2055–2059.

54 Wenz, G., Weickenmeier, M., and Huff, J. (2000) Association Thickener by Host–Guest Interaction of β-Cyclodextrin Polymers and Guest Polymers, In *Associative Polymers in Aqueous Media* (Glass, J. E., Ed.), pp. 271–283, American Chemical Society, Washington, DC.

55 Nakahata, M., Takashima, Y., Yamaguchi, H., and Harada, A. (2011) Redox-Responsive Self-Healing Materials Formed from Host–Guest Polymers, *Nat. Commun.* **2** (10), 511.

56 Harada, A., Ito, F., Tomatsu, I., Shimoda, K., Hashidzume, A., Takashima, Y., Yamaguchi, H., and Kamitori, S. (2006) Spectroscopic Study on the Interaction of Cyclodextrins with Naphthyl Groups Attached to Poly(acrylamide) Backbone, *J. Photochem. Photobiol., A* **179** (1–2), 13–19.

57 Harada, A., Kobayashi, R., Takashima, Y., Hashidzume, A., and Yamaguchi, H. (2011) Macroscopic Self-Assembly through Molecular Recognition, *Nat. Chem.* **3** (1), 34–37.

58 Yamaguchi, H., Kobayashi, R., Takashima, Y., Hashidzume, A., and Harada, A. (2011) Self-Assembly of Gels through Molecular Recognition of Cyclodextrins: Shape Selectivity for Linear and Cyclic Guest Molecules, *Macromolecules* **44** (8), 2395–2399.

59 Zheng, Y., Hashidzume, A., Takashima, Y., Yamaguchi, H., and Harada, A. (2011) Macroscopic Observation of Molecular Recognition: Discrimination of the Substituted Position on Naphthyl Group by Polyacrylamide Gel Modified with β-Cyclodextrin, *Langmuir* **27** (22), 13790–13795.

60 Hashidzume, A., Zheng, Y., Takashima, Y., Yamaguchi, H., and Harada, A. (2013) Macroscopic Self-Assembly Based on Molecular Recognition: Effect of Linkage between Aromatics and the Polyacrylamide Gel Scaffold, Amide versus Ester, *Macromolecules* **46** (5), 1939–1947.

61 Zheng, Y., Hashidzume, A., Takashima, Y., Yamaguchi, H., and Harada, A. (2012) Temperature-Sensitive Macroscopic Assembly Based on Molecular Recognition, *ACS Macro Lett.* **1** (8), 1083–1085.

62 Zheng, Y., Hashidzume, A., and Harada, A. (2013) pH-Responsive Self-Assembly by Molecular Recognition on a Macroscopic Scale, *Macromol. Rapid Commun.* **34** (13), 1062–1066.

63 Zheng, Y., Hashidzume, A., Takashima, Y., Yamaguchi, H., and Harada, A. (2012) Switching of Macroscopic Molecular Recognition Selectivity Using a Mixed Solvent System, *Nat. Commun.* **3**, 831.

64 Nakamura, T., Takashima, Y., Hashidzume, A., Yamaguchi, H., and Harada, A. (2014) A Metal–Ion-Responsive Adhesive Material via Switching of Molecular Recognition Properties, *Nat. Commun.* **5**, 4622.

65 Yamaguchi, H., Kobayashi, Y., Kobayashi, R., Takashima, Y., Hashidzume, A., and Harada, A. (2012) Photoswitchable Gel Assembly Based on Molecular Recognition, *Nat. Commun.* **3**, 603.

66 Nakahata, M., Takashima, Y., and Harada, A. (2014) Redox-Responsive Macroscopic Gel Assembly Based on Discrete Dual Interactions, *Angew. Chem. Int. Ed.* **53** (14), 3617–3621.

67 Kakuta, T., Takashima, Y., Nakahata, M., Otsubo, M., Yamaguchi, H., and Harada, A. (2013) Preorganized Hydrogel: Self-Healing Properties of Supramolecular Hydrogels Formed by Polymerization of Host–Guest-Monomers that Contain Cyclodextrins and Hydrophobic Guest Groups, *Adv. Mater.* **25** (20), 2849–2853.

68 Kakuta, T., Takashima, Y., and Harada, A. (2013) Highly Elastic Supramolecular Hydrogels Using Host–Guest Inclusion Complexes with Cyclodextrins, *Macromolecules* **46** (11), 4575–4579.

69 Nakahata, M., Takashima, Y., and Harada, A. (2016) Highly Flexible, Tough, and Self-Healing Supramolecular Polymeric Materials Using Host–Guest Interaction, *Macromol. Rapid Commun.* **37** (1), 86–92.

6

Cucurbit[*n*]uril-Mediated Multiple Interactions

Zehuan Huang and Xi Zhang

Department of Chemistry, Tsinghua University, Beijing 100084, China

6.1 Introduction to Cucurbit[*n*]uril Chemistry

The cucurbit[*n*]urils (CB[*n*]s) constitute a family of water-soluble macrocyclic hosts, which have a hydrophobic cavity capable of binding one or two guest molecules (Figure 6.1) [1–3]. Three points contribute to the binding between CB[*n*] host and guest molecules [4–6]: (i) attraction between the host and guest; (ii) desolvation of the host cavity; and (iii) desolvation of the guest. In particular, the desolvation of the CB[*n*] host cavity releases high-energy water trapped in the cavity which is regarded as a non-classical hydrophobic effect as it has a favorable enthalpic signature. On account of their high affinities, CB[*n*]-mediated host–guest interactions are of great interest in fabricating supramolecular systems with well-defined composition and structure [7–11].

Among the cucurbit[*n*]uril familiy, cucurbit[8]uril (CB[8]) is a characteristic host which is capable of multiple binding with two guests (Figure 6.2). Based on whether binding different or the same guests, their complexation can be divided into heteroternary and homoternary complexations, respectively. CB[8]-mediated heteroternary and homoternary complexes have been exploited in the construction of supramolecular polymers [12–15], supramolecular hydrogels [16, 17], supra-amphiphiles [18–20] and other supramolecular functional self-assembled systems [21, 22].

6.2 Heteroternary Complexes

Cucurbit[8]uril-mediated heteroternary complexations were first discovered by Kim *et al.* in 2001 [23]. As shown in Figure 6.3, the authors found that methylviologen (MV) can form 1:1 host–guest complex with CB[8], as evidenced by ^{1}H NMR spectroscopy and mass spectrometry. As MV is an electron-poor guest, this 1:1 complex prompted the authors to study the inclusion of a second electron-rich guest into the MV-CB[8] complex. By adding one equivalent of an electron-rich aromatic guest (HB, HN), the

Multivalency: Concepts, Research & Applications, First Edition. Edited by Jurriaan Huskens,
Leonard J. Prins, Rainer Haag, and Bart Jan Ravoo.
© 2018 John Wiley & Sons Ltd. Published 2018 by John Wiley & Sons Ltd.

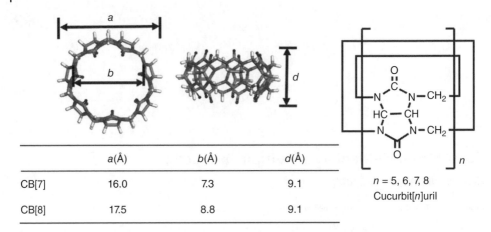

	a(Å)	b(Å)	d(Å)
CB[7]	16.0	7.3	9.1
CB[8]	17.5	8.8	9.1

n = 5, 6, 7, 8
Cucurbit[n]uril

Figure 6.1 Molecular structures and detailed information of cucurbit[n]urils. See color section.

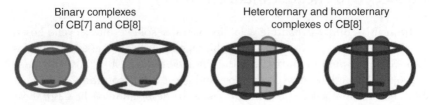

Binary complexes of CB[7] and CB[8]

Heteroternary and homoternary complexes of CB[8]

Figure 6.2 Schematic diagram of binary and ternary complexes based on cucurbit[n]urils.

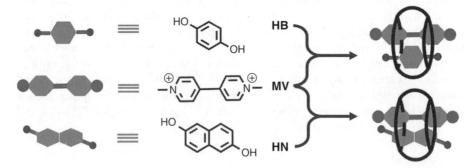

HB

MV

HN

Figure 6.3 Schematic diagram of heteroternary complexes of MV-CB[8]-HN and MV-CB[8]-HB.

heteroternary complexes were formed in a 1:1:1 ratio as indicated by ^{1}H NMR spectroscopy. Besides, the structures of the heteroternary complexes were also determined by X-ray crystallography.

Base on this pioneering work, Bush et al. [24] utilized this heteroternary complexation to realize the specific recognition between MV-CB[8] and aromatic α-amino acids in 2005. First, they discovered that tryptophan (Trp), tyrosine (Tyr), and phenylalanine (Phe) were all capable of binding with MV-CB[8] to form heteroternary complexes. Next, effects of electrostatic charge on peptides binding were investigated by employing N-terminal, internal and C-terminal peptides of tryptophan for demonstration.

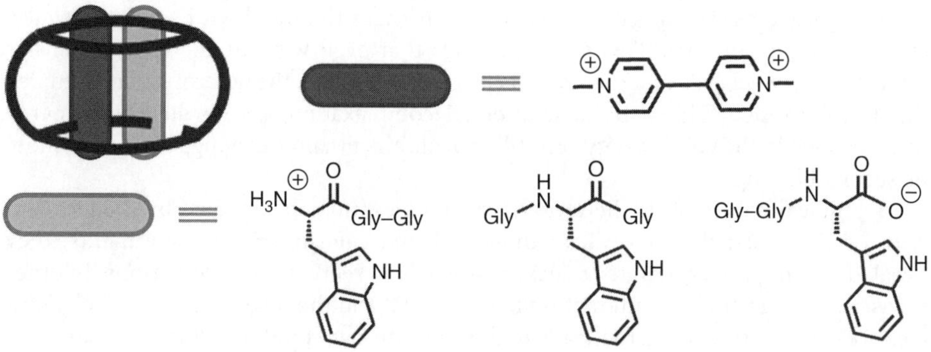

Figure 6.4 Schematic diagram of heteroternary complexes of MV-CB[8] binding with three kinds of tripeptides of tryptophan.

As shown in Figure 6.4, they found that MV-CB[8] bound Trp-Gly-Gly with high affinity, with 6-fold specificity over Gly-Trp-Gly, and with 40-fold specificity over Gly-Gly-Trp. These results showed that the positive charge can benefit the heteroternary complexation of MV-CB[8]-Trp while the negative charge may decrease the binding affinity. The electron-rich carbonyl rims of CB[8] prefer to bind with the positive charge to form an ion–dipole interaction. Therefore, the peptide recognition is strongly influenced by the electrostatic charge proximal to aromatic groups.

Jiao *et al.* [25] extended CB[8]-mediated heteroternary complexations from MV to imidazolium salts. As shown in Figure 6.5, three kinds of imidazolium salts were prepared and all of them formed 1:1 host–guest complexes with CB[8]. These complexes

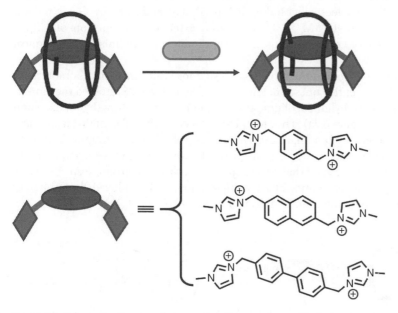

Figure 6.5 Schematic diagram of supramolecular cages formed by heteroternary complexes of three kinds of imidazolium salts and CB[8].

can be regarded as size-specific supramolecular cages that bind with small molecules or even organic solvents. This system reveals that even without the charge-transfer interaction between MV and the second electron-rich guest, the heteroternary complex can also be formed. This is because after 1:1 complexation, the residual high-energy water trapped in the CB[8] cavity can still provide an enthalpic driving force to promote the second binding.

The physical nature of the heteroternary complexation was studied by Biedermann *et al.* [5]. By molecular dynamics simulations and calorimetric measurements, they revealed that high-energy water release overwhelms electrostatic interactions. In other words, the charge-transfer interaction between MV and the second electron-rich guest does not contribute a lot energetically to the heteroternary binding, which also supports the previous research work of Jiao *et al.*

6.3 Homoternary Complexes

Besides heteroternary complexation, CB[8]-mediated homoternary complexation is also a powerful driving force in fabricating supramolecular systems, which have attracted lots of interest from chemists and biologists. The first example of homoternary complexation was found by Kim *et al.* in 2000 [26]. As shown in Figure 6.6, the protonated 2,6-bis(4,5-dihydro-1*H*-imidazol-2-yl)naphthalene formed a 2:1 host–guest complex with CB[8], as confirmed by ^{1}H NMR spectroscopy and mass spectrometry. Besides, the structure of this homoternary complex was also determined by X-ray crystallography.

An interesting sequence-specific recognition and cooperative dimerization of N-terminal aromatic peptides with CB[8] was discovered by Heitmann *et al.* [27]. Specifically, CB[8] can selectively bind with Trp-Gly-Gly and Phe-Gly-Gly with high affinity (ternary $K = 10^9 – 10^{11}\,\mathrm{M}^{-2}$). This research extended the horizon of CB[8]-mediated homoternary complexations from supramolecular chemistry to chemical biology.

What is the principle behind the molecular design for CB[8]-mediated homoternary complexation? To answer this question, Rauwald and Scherman [28] studied the dimerization of 1-phenyl-3-methylimidazolium bromide and 1-naphthyl-3-methylimidazolium bromide with CB[8] (Figure 6.7a). They found that the guest should contain an aromatic group as the homoternary binding motif and a positively charged group to bind with electron-rich carbonyl rims. Based on this principle, we utilized trimethylammonium and pyridium, containing a naphthalene group or an anthracene group, respectively, to realize the CB[8]-mediated homoternary complexation, and employed these interactions to fabricate water-soluble supramolecular polymers (Figure 6.7b) [29, 30].

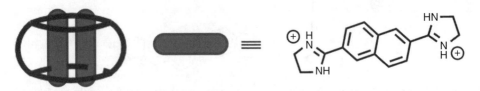

Figure 6.6 Schematic diagram of the first homoternary complex.

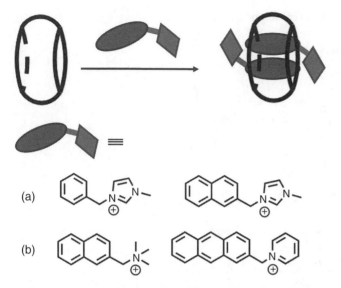

Figure 6.7 Schematic diagram of the homoternary complexes of CB[8] binding with (a) two imidazoliums with benzene and naphthalene and (b) trimethylammonium with naphthalene and pyridium with anthracene.

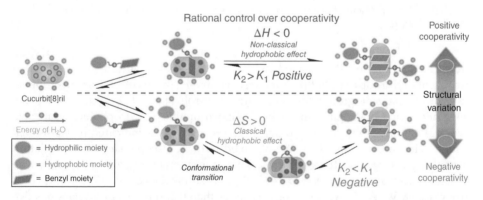

Figure 6.8 Schematic diagram of design principles on rationally controlling cooperativity of CB[8]-mediated π–π interactions by tuning classical and non-classical hydrophobic effects (blue balls represent bulk water, while orange and red balls represent high-energy water; the energy of the red balls is higher than that of the orange ones). See color section.

Although lots of effort has been made in designing new guests to realize the homoternary complexation with CB[8], there still remains the important question of how the cooperative effect influences this kind of ternary binding. We systematically investigated the cooperative binding of the CB[8]-mediated homoternary complexations (Figure 6.8). The binding cooperativity of CB[8]-mediated homoternary complexes is strongly influenced by the amphiphilicity of side groups of the guest molecule caused by an interplay between classical (entropy-driven) and non-classical (enthalpy-driven) hydrophobic effect.

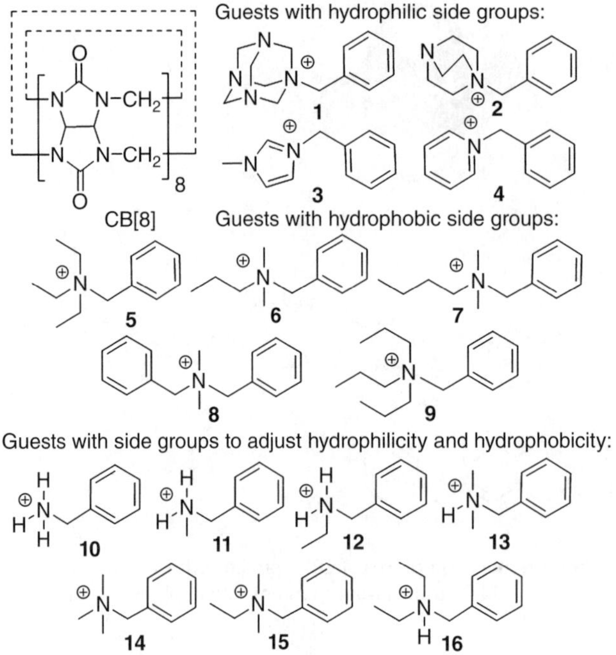

Figure 6.9 Molecular structures of a series of guest molecules (**1–16**) bearing various hydrophilic and hydrophobic side groups.

To this end, we rationally designed and prepared a series of guest molecules (**1–16**) bearing a benzyl group as the CB[8] homoternary binding motif with various hydrophilic and hydrophobic side groups for cooperative control (Figure 6.9) [31]. By gradually tuning side groups of the guest molecules from hydrophilic to hydrophobic, we are able to control binding from positive to negative cooperativity as indicated by isothermal titration calorimetry (Table 6.1). We obtained the macroscopic stepwise association constants by fitting the titration curves. Then, the interaction parameter (α) can be calculated based on the equation ($\alpha = 4 K_2/K_1$) to describe the cooperativity in a quantitative manner. When the value of log α is larger than 0.5, the system displays positive cooperativity, when log α is smaller than −0.5, the system displays negative cooperativity, and when log α is in between −0.5 and 0.5, the system is non-cooperative.

By introducing hydrophilic side groups on the guest molecules, the first binding event may release relatively few of the confined water molecules leaving more high-energy water in the CB[8] cavity for a more favorable second association. This leads to positive cooperativity as shown in Table 6.1 for four guest molecules **1–4** modified with hydrophilic side groups, which exhibited similar titration curves. Moreover, the mechanism behind their positive binding is corroborated by the stepwise enthalpy changes in Table 6.1. For guests **1–4**, all the second binding steps provide a greater change in enthalpy than the first binding step ($\Delta H_{12} - \Delta H_1 > \Delta H_2$), which means the positive cooperativity is dominant by the non-classical (enthalpy-driven) hydrophobic effect.

By contrast, the straightforward introduction of hydrophobic groups to the benzyl unit in the form of alkyl chains or additional aromatic rings can lead to simultaneous

Table 6.1 Thermodynamic parameters of all the host–guest complexes with CB[8] and interaction parameters for all guest molecules shown in Figure 6.9.

No.	Log α^a	$K_1{}^b$ $(10^5\,M^{-1})$	$K_2{}^b$ $(10^5\,M^{-1})$	$\Delta H_1{}^b$ $(kJ\,mol^{-1})$	$-T\Delta S_1{}^b$ $(kJ\,mol^{-1})$	$\Delta H_{12}{}^b$ $(kJ\,mol^{-1})$	$-T\Delta S_{12}{}^b$ $(kJ\,mol^{-1})$
1	2.32 ± 0.02	1.24 ± 0.04	64.60 ± 1.6	-36.4 ± 1.1	7.3 ± 1.2	-74.2 ± 2.2	6.2 ± 2.3
2	0.91 ± 0.02	11.70 ± 0.29	24.00 ± 0.32	-28.2 ± 0.2	-6.4 ± 0.2	-70.8 ± 0.3	-0.3 ± 0.4
3	1.28 ± 0.02	2.64 ± 0.05	12.60 ± 0.21	-26.6 ± 0.3	-4.3 ± 0.3	-69.4 ± 0.5	3.7 ± 0.6
4	0.87 ± 0.01	7.76 ± 0.12	14.40 ± 0.24	-30.0 ± 0.1	-3.6 ± 0.2	-69.1 ± 0.3	2.7 ± 0.4
5	-1.89 ± 0.02	26.00 ± 0.81	0.083 ± 0.001	-17.9 ± 0.1	-18.7 ± 0.1	-65.6 ± 0.5	6.6 ± 0.6
6	-1.11 ± 0.03	7.17 ± 0.30	0.139 ± 0.004	-22.5 ± 0.1	-11.0 ± 0.2	-68.7 ± 0.7	11.6 ± 0.8
7	-3.30 ± 0.05	193 ± 10	0.024 ± 0.001	-26.9 ± 0.1	-14.6 ± 0.2	-73.5 ± 2.4	12.6 ± 2.7
8	-3.02 ± 0.08	377 ± 50	0.089 ± 0.005	-29.4 ± 0.1	-13.9 ± 0.4	-72.1 ± 1.4	6.2 ± 1.9
9	$-^c$	363 ± 43	$-^c$	-26.0 ± 0.1	-17.2 ± 0.4	$-^c$	$-^c$
10	1.53 ± 0.02	1.29 ± 0.03	10.90 ± 0.20	-19.5 ± 0.3	-9.6 ± 0.4	-54.9 ± 0.7	-8.7 ± 0.8
11	1.00 ± 0.02	6.66 ± 0.20	16.50 ± 0.38	-16.4 ± 0.3	-16.9 ± 0.4	-58.7 ± 0.6	-10.1 ± 0.7
12	0.59 ± 0.11	10.9 ± 1.9	10.50 ± 0.82	-23.5 ± 0.7	-11.0 ± 1.1	-64.8 ± 1.3	-4.1 ± 2.0
13	0.06 ± 0.01	6.86 ± 0.09	1.99 ± 0.02	-16.9 ± 0.1	-16.5 ± 0.7	-50.9 ± 0.1	-12.7 ± 0.1
14	-0.38 ± 0.02	24.50 ± 0.63	2.58 ± 0.03	-20.9 ± 0.1	-15.6 ± 0.1	-68.5 ± 0.2	1.1 ± 0.3
15	-0.71 ± 0.01	35.30 ± 0.47	1.74 ± 0.02	-20.6 ± 0.1	-16.7 ± 0.1	-64.7 ± 0.1	-2.6 ± 0.2
16	-1.89 ± 0.02	31.00 ± 0.98	0.101 ± 0.004	-21.3 ± 0.1	-15.9 ± 0.1	-68.6 ± 0.5	8.6 ± 0.6

[a] Interaction parameters (log α) are calculated by microscopic stepwise binding constants (k_1, k_2).
[b] Mean value measured from ITC experiments at 25 °C in 50 mM sodium acetate buffer (pH 4.75), which are corrected by Sedphat and NIPIC and fitted by sequential binding model in Origin 7.0 within Microcal; K_1 and K_2 are macroscopic stepwise binding constants; ΔH_1 and ΔS_1 are the binding enthalpy and binding entropy, respectively, for the first step; ΔH_{12} and ΔS_{12} are the total binding enthalpy and binding entropy, respectively, for two steps.
[c] Not determined.

incorporation of both the aromatic benzyl moiety as well as a second hydrophobic group into the CB[8] cavity (Table 6.1). As evidenced by ^{1}H NMR spectroscopy and rotating-frame Overhauser effect spectroscopy, this very stable host–guest complex can be formed in a 1:1 ratio, which can inhibit the formation of the 2:1 complex thus resulting in negative cooperativity.

According to the enthalpy and entropy changes recorded in Table 6.1, a mechanism for negative cooperativity can be illustrated as in the following. For guests **5–8**, which carry hydrophobic side groups, the simultaneous incorporation of the benzyl group and another hydrophobic moiety in the initial binding step offers much greater entropic compensation ($T\Delta S_1 < -10\,kJ\,mol^{-1}$) than for guests **1–4**, which represents a classical hydrophobic driving force. Second guest molecule complexation with CB[8] to form the 2:1 homoternary complex yields a large entropy penalty ($\Delta S_{12}-\Delta S_1 > 20\,kJ\,mol^{-1}$), therefore suppressing the second association and decreasing the overall binding affinity compared with guests carrying a hydrophilic side group such as guests **1–4**. These two entropic effects can outweigh the non-classical hydrophobic effect making the first binding step substantially stronger than the second. It appears, therefore, that the stronger classical hydrophobic effect drives the occurrence of negatively cooperative events; thus negative cooperativity can be readily achieved in this system by selecting guests with hydrophobic side groups.

In an effort to achieve systems that displayed non-cooperativity, we attempted to find the right balance between hydrophilicity and hydrophobicity of the guest molecules. Thus we designed and prepared a series of guest molecules with increasing hydrophobicity of the side groups by systematically increasing the alkyl content one methylene group at a time. As shown in Table 6.1, three hydrophilic guests, ammonium (**10**), *N*-methylammonium (**11**) and *N*-ethylammonium (**12**) still exhibited considerable positive cooperativity on account of their abundant hydrogen-bonding with bulk water. However, by gradually increasing the hydrophobic alkyl chains of guest molecules to *N,N*-dimethylammonium (**13**) and trimethylammonium (**14**), they displayed smaller and similar interaction parameters (α). Their log α values are in the range of −0.5 to 0.5, which is close to 0 (Table 6.1), indicating non-cooperativity. These data confirm that non-cooperativity can be readily achieved by balancing the hydrophilicity and hydrophobicity of the guest molecules.

In summary, cooperativity of multiple non-covalent interactions with CB[8] can be rationally controlled by tuning the interplay between the classical and non-classical hydrophobic effects. Specifically, positive cooperativity, non-cooperativity and negative cooperativity can all be achieved for structurally similar guest molecules with the same host by varying the amphiphilicity of side groups present on the guest molecules from hydrophilic to hydrophobic. This line of research represents an in-depth study of modulating classical and non-classical hydrophobic effects, which will enrich the field of supramolecular chemistry leading to important advances and the realization of controlling cooperativity of non-covalent interactions.

6.4 Conclusions

CB[8] with rigid skeleton and eight electron-rich carbonyl rims in each portal side allows the formation of CB[8]-mediated heteroternary and homoternary complexes with high binding affinities. Cooperativity can be fine-tuned by balancing the classical and non-classical hydrophobic effect. It is highly anticipated that CB[n]-mediated heteroternary and homoternary complexations can be used as the driving forces for fabricating supramolecular polymers and assemblies with controlled architectures and functions.

References

1 Barrow SJ, Kasera S, Rowland MJ, Del Barrio J, Scherman OA. Cucurbituril-based molecular recognition. Chem. Rev. 2015;115(22):12320–406.
2 Lee JW, Samal S, Selvapalam N, Kim H, Kim K. Cucurbituril homologues and derivatives: new opportunities in supramolecular chemistry. Acc. Chem. Res. 2003;36(8):621–30.

3 Lagona J, Mukhopadhyay P, Chakrabarti S, Isaacs L. The cucurbit[n]uril family. Angew. Chem. Int. Ed. 2005;44(31):4844–70.

4 Biedermann F, Uzunova VD, Scherman OA, Nau WM, De Simone A. Release of high-energy water as an essential driving force for the high-affinity binding of cucurbit[n]urils. J. Am. Chem. Soc. 2012;134(37):15318–23.

5 Biedermann F, Vendruscolo M, Scherman OA, De Simone A, Nau WM. Cucurbit[8]uril and blue-box: high-energy water release overwhelms electrostatic interactions. J. Am. Chem. Soc. 2013;135(39):14879–88.

6 Biedermann F, Nau WM, Schneider HJ. The hydrophobic effect revisited – studies with supramolecular complexes imply high-energy water as a noncovalent driving force. Angew. Chem. Int. Ed. 2014;53(42):11158–71.

7 Das D, Scherman OA. Cucurbituril: at the interface of small molecule host–guest chemistry and dynamic aggregates. Isr. J. Chem. 2011;51(5–6):537–50.

8 Isaacs L. Stimuli responsive systems constructed using cucurbit[n]uril-type molecular containers. Acc. Chem. Res. 2014;47(7):2052–62.

9 Assaf KI, Nau WM. Cucurbiturils: from synthesis to high-affinity binding and catalysis. Chem. Soc. Rev. 2015;44(2):394–418.

10 Isaacs L. Cucurbit[n]urils: from mechanism to structure and function. Chem. Commun. 2009(6):619–29.

11 Masson E, Ling X, Joseph R, Kyeremeh-Mensah L, Lu X. Cucurbituril chemistry: a tale of supramolecular success. RSC Adv. 2012;2(4):1213–47.

12 Liu Y, Yu Y, Gao J, Wang Z, Zhang X. Water-soluble supramolecular polymerization driven by multiple host-stabilized charge-transfer interactions. Angew. Chem. Int. Ed. 2010;49(37):6576–9.

13 del Barrio J, Horton PN, Lairez D, Lloyd GO, Toprakcioglu C, Scherman OA. Photocontrol over cucurbit[8]uril complexes: stoichiometry and supramolecular polymers. J. Am. Chem. Soc. 2013;135(32):11760–3.

14 Huang Z, Yang L, Liu Y, Wang Z, Scherman OA, Zhang X. Supramolecular polymerization promoted and controlled through self-sorting. Angew. Chem. Int. Ed. 2014;53(21):5351–5.

15 Yang L, Tan X, Wang Z, Zhang X. Supramolecular polymers: historical development, preparation, characterization, and functions. Chem. Rev. 2015;115(15):7196–239.

16 Appel EA, del Barrio J, Loh XJ, Scherman OA. Supramolecular polymeric hydrogels. Chem. Soc. Rev. 2012;41(18):6195–214.

17 Li C, Rowland MJ, Shao Y, Cao T, Chen C, Jia H, Zhou X, Yang Z, Scherman OA, Liu D. Responsive double network hydrogels of interpenetrating DNA and CB[8] host-guest supramolecular systems. Adv. Mater. 2015;27(21):3298–304.

18 Jeon YJ, Bharadwaj PK, Choi S, Lee J, Kim K. Supramolecular amphiphiles: spontaneous formation of vesicles triggered by formation of a charge-transfer complex in a host. Angew. Chem. Int. Ed. 2002;41(23):4474–6.

19 Wang G, Kang Y, Tang B, Zhang X. Tuning the surface activity of gemini amphiphile by the host-guest interaction of cucurbit[7]uril. Langmuir 2015;31(1):120–4.

20 Mondal JH, Ghosh T, Ahmed S, Das D. Dual self-sorting by cucurbit[8]uril to transform a mixed micelle to vesicle. Langmuir 2014;30(39):11528–34.

21 Mohanty J, Nau WM. Ultrastable rhodamine with cucurbituril. Angew. Chem. Int. Ed. 2005;44(24):3750–4.

22 Joseph R, Nkrumah A, Clark RJ, Masson E. Stabilization of cucurbituril/guest assemblies via long-range Coulombic and CH...O interactions. J. Am. Chem. Soc. 2014;136(18):6602–7.

23 Kim H-J, Heo J, Jeon WS, Lee E, Kim J, Sakamoto S, Yamaguchi K, Kim K. Selective inclusion of a hetero-guest pair in a molecular host: formation of stable charge-transfer complexes in cucurbit[8]uril. Angew. Chem. Int. Ed. 2001;40(8):1526–9.

24 Bush ME, Bouley ND, Urbach AR. Charge-mediated recognition of N-terminal tryptophan in aqueous solution by a synthetic host. J. Am. Chem. Soc. 2005;127(41):14511–7.

25 Jiao D, Biedermann F, Scherman OA. Size selective supramolecular cages from aryl-bisimidazolium derivatives and cucurbit[8]uril. Org. Lett. 2011;13(12):3044–7.

26 Kim J, Jung I-S, Kim S-Y, Lee E, Kang J-K, Sakamoto S, Yamaguchi K, Kim K. New cucurbituril homologues: synthesis, isolation, characterization, and X-ray crystal structure of cucurbit[n]uril (n = 5, 7, and 8). J. Am. Chem. Soc. 2000;122(3):540–1.

27 Heitmann LM, Taylor AB, Hart PJ, Urbach AR. Sequence-specific recognition and cooperative dimerization of N-terminal aromatic peptides in aqueous solution by a synthetic host. J. Am. Chem. Soc. 2006;128(38):12574–81.

28 Rauwald U, Scherman OA. Supramolecular block copolymers with cucurbit[8]uril in water. Angew. Chem. Int. Ed. 2008;47(21):3950–3.

29 Liu Y, Liu K, Wang Z, Zhang X. Host-enhanced pi-pi interaction for water-soluble supramolecular polymerization. Chem. Eur. J. 2011;17(36):9930–5.

30 Liu Y, Fang R, Tan X, Wang Z, Zhang X. Supramolecular polymerization at low monomer concentrations: enhancing intermolecular interactions and suppressing cyclization by rational molecular design. Chem. Eur. J. 2012;18(49):15650–4.

31 Huang Z, Qin K, Deng G, Wu G, Bai Y, Xu J, Wang Z, Yu Z, Scherman OA, Zhang X. Supramolecular chemistry of cucurbiturils: controlling multiple non-covalent interactions from positive to negative cooperativity. Langmuir 2016;32(47):12352–60.

7

Multivalency as a Design Criterion in Catalyst Development

Paolo Scrimin, Maria A. Cardona, Carlos M. León Prieto, and Leonard J. Prins

Department of Chemical Sciences, University of Padova, Via Marzolo 1, 35131 Padova, Italy

7.1   Introduction

Multivalency is most often associated with the binding interaction between molecular partners through the simultaneous occurrence of multiple binding events. The aim of this chapter is to illustrate how multivalency relates to catalysis, in particular referring to those cases in which multivalency is purposely used as a design criterion to develop catalysts. The attachment of homogenous catalysts to multivalent scaffolds such as dendrimers, nanoparticles, or macroscopic resins has received tremendous attention in the past decades [1–5]. This interest is predominantly caused by the possibility to create hybrid systems that combine the advantages of heterogeneous and homogeneous catalysts [6]. Anchoring of a catalyst on a solid support creates the possibility of catalyst separation and, thus, recycling, which leads to a potential cost reduction. Although of obvious importance, multivalent catalysts that have been prepared for this purpose will not be discussed here and the reader is referred to the numerous reviews cited above that provide overviews of such systems. Rather, in this chapter the focus will be on systems in which multivalency is an essential prerequisite for observing or enhancing catalytic activity. This will involve a discussion of synthetic systems that express cooperative catalysis. This implies that, just as what happens in the active site of an enzyme, catalytic activity originates from the interplay between two (or more) functional groups that are in close proximity in the multivalent scaffold. Indeed, special attention will be paid to the interpretation of the Michaelis–Menten parameters for such multivalent enzyme mimics. Additional topics that will be treated are the ability of the multivalent system to alter the local reaction conditions and induce different reaction mechanisms. Finally, a section is dedicated to what happens in the special case where a multivalent catalyst acts on a multivalent substrate. It is not the purpose of this chapter to provide an exhaustive overview of all examples that have appeared in the literature. Examples have been selected based on the insight they can provide in discussing the relationship between multivalency and catalysis.

Multivalency: Concepts, Research & Applications, First Edition. Edited by Jurriaan Huskens, Leonard J. Prins, Rainer Haag, and Bart Jan Ravoo.
© 2018 John Wiley & Sons Ltd. Published 2018 by John Wiley & Sons Ltd.

7.2 Formation of Enzyme-Like Catalytic Pockets

The close proximity of multiple functional groups in a single molecular structure provides multivalent systems with an excellent opportunity to create catalytic sites similar
to the active site of enzymes. The emergence of dendrimers led rapidly to the realization
that the dendritic shell may have a strong effect on the performances of a catalytic core.
As illustration we report an example by Javor *et al.* [7], who have extensively studied
peptide dendrimers as enzyme-like catalysts [8,9]. The attractiveness of this system is
the simplicity of the catalytic site in the core and the enormous structural variety
inserted in the branches of the third generation dendrimer (Figure 7.1). The fact that
amino acids are used to construct the structure strongly enhances the analogy with
enzymes. A library of over 65 000 peptide dendrimers was prepared following a combinatorial approach. Variable amino acids at the catalytic core included nucleophilic (His,
Cys) and cationic (Arg) residues for substrate binding and catalysis. In the outer regions,
aromatic residues were chosen between aromatic residues (Tyr, Phe, Trp) to assist in
binding. Finally, polar, negatively charged residues and small hydrophobic residues
were distributed throughout the entire structure. The library was screened on-bead for
esterase activity using a fluorogenic butyrate ester. Strongly fluorescent beads were then
selected for sequencing and analysed for the presence of consensus-sequences. Active
sequences were found to contain at least one histidine or arginine in the catalytic
core and predominantly aromatic residues at the outer positions. Representative hits
were resynthesized and were found to catalyse the hydrolysis of activated esters with
saturation kinetics and multiple turn-overs. An important contribution to catalysis by
the apolar outer layers was observed attributed to an increase in substrate binding. The
importance of this study is that it demonstrates the possibility to create multivalent
enzyme-like structures in which the properties of the active site are altered because of
the surrounding structure.

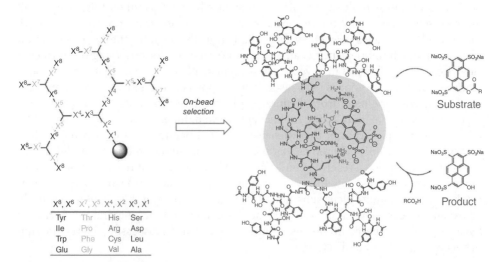

Figure 7.1 On-bead selection of a catalytic dendrimer from a combinatorial library of 65 536 different
dendrimers. The yellow circle highlights the reactive core with a substrate. *Source*: Ref. [7]. Reproduced
with permission of American Chemical Society. See color section.

The attractive feature of dendrimers is their monodisperse structure, which is well-defined on the molecular level. This allows for a precise determination of the effect of structural changes on the catalytic performances of the system. Yet, the multi-step covalent synthesis of dendrimers poses challenges in terms of yields and purification. For that reason, self-assembled monolayers (SAMs) on gold nanoparticles have emerged as attractive alternatives for the formation of multivalent catalytic systems, because they form spontaneously via a strong sulfur–gold (Au) interaction [10,11]. The potential of these systems to create enzyme-like catalysts was nicely illustrated by Belser *et al.* [12] in a study on catalytic SAMs on gold nanoparticles. They distinguished two different situations for a mixed monolayer system composed of two different thiols, one bearing a catalytic group and the other one with an inert head group (Figure 7.2). A convex catalytic site is formed in the case where the catalytic thiol contains a longer spacer compared with the surrounding inert thiol. In that case, the catalytic group extends out of the monolayer surface. In principle, the activity should resemble that of the monomeric reference catalyst, although the neighbouring head groups of inert thiols may affect the catalytic activity because of interactions with the substrate or an alteration of the local chemical environment (see examples below). Alternatively, in the case where the catalytic thiol has a shorter spacer compared with the surrounding thiol a concave catalytic site is formed, which has more similarities with the active site of an enzyme. In an initial study, they explored the first case by embedding thiolates with chiral rhodium-PYRPHOS head groups in monolayers of *n*-alkanethiolates of different length and head group polarity on Au nanoparticles (NPs) with a diameter of around 3 nm. The catalytic activity of these systems was evaluated in the hydrogenation of methyl α-acetamido-cinnamate. Scanning tunnelling microscopy measurements of analogously prepared two-dimensional (2D) monolayers revealed a statistical distribution of the catalytic head groups, which is of crucial importance when the synergetic effect between the two thiols is studied (see Section 7.5.2). As expected, the catalytic activity was similar to that of the analogous homogeneous catalyst both in terms of enantioselectivity and conversion, but only in the case where apolar head groups were present on the neighbouring thiols. The presence of polar amino and hydroxy end groups caused a decisive decrease in yield and enantioselectivity. Although this study did not provide an explanation for this effect, it clearly shows that neighbouring end groups play an important role in governing the performance of the catalytic system.

A more complete demonstration was provided by Paluti and Gawalt [13,14] in a study on the activity of aza-bis(oxazoline) copper complexes embedded in 2D SAMs composed of alkane thiols (Figure 7.3). Apart from the convex situation analysed by Belser *et al.*, systems were also analysed in which the catalyst was embedded within the monolayer or at an equal distance compared with the surrounding head groups. All systems were tested in the cyclopropanation reaction of ethyl diazoacetate and styrene, thus permitting an analysis of the product distribution both in terms of cis/trans ratio as well as the enantioselectivity of each diastereoisomer. As a reference, the monomeric aza-bis(oxazoline) catalyst with bulky *t*-butyl substituents gave a cis/trans ratio of 20/80 with respective enantiomeric excess (*ee*) values of 80 and 87%. Nearly identical results were obtained in the case where the catalyst was positioned well above the surrounding monolayer (cis/trans 23/77; $ee_{cis} = 81$; $ee_{trans} = 85$) indicating that this construction provides supported catalysts with homogeneous properties. Changes were observed upon positioning the catalyst at an even distance compared with the monolayer surface.

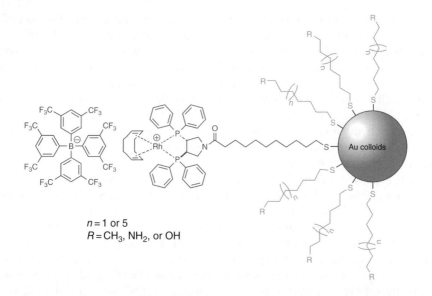

$n = 1$ or 5
$R = CH_3$, NH_2, or OH

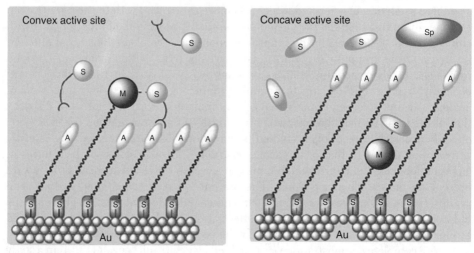

Figure 7.2 Formation of convex and concave catalytic sites on a monolayer depending on the relative spacer length of the catalytic and surrounding inert thiols. *Source*: Ref. [12]. Reproduced with permission of American Chemical Society.

The cis/trans ratio changed slightly to 16/84, but a significant increase in the *ee* of the trans product was observed (from 87 to 93%). For the cis isomer no change in enantioselectivity was reported (82 versus 80), but, remarkably, the opposite enantiomer was favoured. Finally, embedding the catalyst within the monolayer caused a drop both in the cis/trans ratio (28/72), but also in the *ee* values of both products ($ee_{cis} = 37$; $ee_{trans} = 44$). The latter results were tentatively ascribed to the occurrence of steric interactions between the alkyl chains and the catalysts. From this comparative study it emerges that the most advantageous situation occurs when the catalyst is levelled with the monolayer surface.

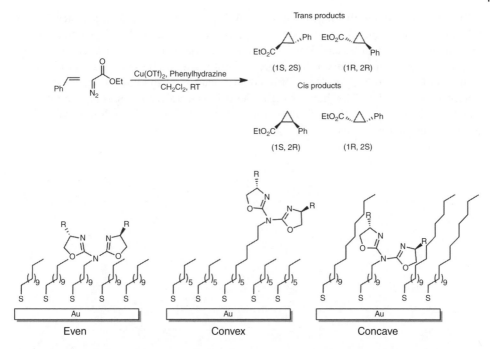

Figure 7.3 Three catalytic 2D SAMs used to study the cyclopropanation reaction between styrene and ethyl diazoacetate. *Source*: Ref. [13]. Reproduced with permission of Elsevier.

7.3 Cooperativity Between Functional Groups

An important step towards artificial enzymes requires the exploitation of multivalent scaffolds for the formation of catalytic sites in which multiple functional groups cooperatively act on the substrate [15]. This way the scaffold assumes a fundamental role in preorganizing the chemical functionalities. Typically, this results in very strong rate accelerations over the background, because the monomeric units by themselves are not or hardly active. A classic example of cooperative catalysis is the imidazole-catalysed hydrolysis of carboxylic esters around pH 7 in which different imidazoles provide both nucleophilic and a general acid/base contribution. Delort *et al.* [16] explored the occurrence of this mechanism in a series of peptide dendrimers of different generation (up till the fourth) containing His-residues in every generation (Figure 7.4). An additional Ser-residue was also presented in each generation as previous studies had shown an enhanced activity when this residue was present. Importantly, all dendrimers exhibited enzyme-like saturation kinetics in the hydrolysis of pyrene trisulfonate esters and Michaelis–Menten parameters could be determined for each generation. Comparison of the k_{cat}, K_M and k_{cat}/K_M values for each generation gave a valuable insight in the cooperativity between functional groups in the dendrimer. A systematic study of the dendritic effect in peptide dendrimer catalysis revealed that the catalytic rate constant k_{cat} and substrate binding constant $1/K_M$ both increased with increasing generation number. The dendrimers showed rate accelerations up to $k_{cat}/k_{uncat} = 20\ 000$ and K_M values around 0.1 mM. The experiments showed thus a strong positive dendritic effect resulting from cooperative binding and catalysis [17–20]. A very strong indication for the effective occurrence of cooperative interactions between imidazoles came from the very different rate profiles as a function of pH compared with the monomeric reference catalyst 4-methylimidazole. For the reference catalyst, an increase in rate was observed as the pH increased in agreement with a deprotonation of the imidazole (creating the nucleophile). On the other hand, for the dendrimer-catalysed reactions the rate was slightly bell-shaped over the pH range studied (4.5–7.5) indicating a double role of the imidazole in the mechanism.

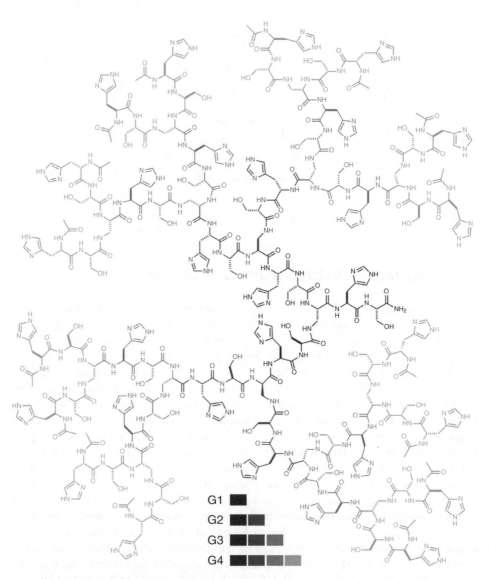

Figure 7.4 Catalytic dendrimers containing His-residues in every generation. Observed positive dendritic effect in the hydrolysis of activated esters. *Source*: Ref. [16]. Reproduced with permission of American Chemical Society. See color section.

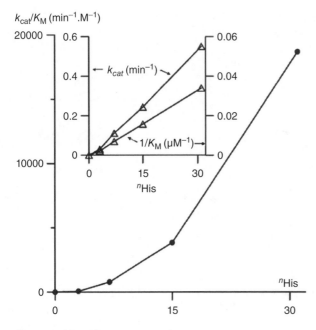

k_{cat}/K_M (min^{-1}.M^{-1})

Figure 7.4 (Cont'd)

Another class of reactions that generally requires the joined action of two catalytic groups is the hydrolytic cleavage of phosphodiesters [21,22]. This reaction is biologically very relevant as phosphodiesters constitute the backbone of DNA and RNA, the polymers that carry genetic information. The high stability of this bond is reflected by the estimated half-life of 10^{10} years for the cleavage of dimethyl phosphate. Not far from this low reactivity is the time required for hydrolytically cleaving the P-O bond of DNA. On the other hand, RNA is more labile because the nucleophilic attack on phosphorus is performed intramolecularly by the -O(H) in the 2′-position of the ribose. This reduces the half-life of RNA to roughly 10^4 years. Enzymes involved in the DNA or RNA cleavage typically have multiple transition metal ions (Zn^{2+}, Mg^{2+}) in the active site. These metal ions work in a concerted manner through nucleophile activation and stabilization of both the transition state and leaving group. Accordingly, this reaction provides an excellent test case to determine the occurrence of cooperativity in multivalent catalysts. Manea et al. [23] prepared Au NPs with a mixed monolayer composed of thiols terminating with a catalytic 1,4,9-triazacyclononane (TACN)·Zn^{2+} head group and inert octane thiols (Figure 7.5). NMR spectroscopic analysis revealed that the thiols were present in a 1.2:1 ratio in the monolayer. The catalytic activity of the NPs was tested on the transphosphorylation of 2-hydroxypropyl-p-nitrophenyl phosphate (HPNPP), which is typically used as an RNA model compound. Importantly, Zn^{2+} is fundamental to catalysis, because hardly any activity over the background reaction is observed in the absence of metal ion. This offers an attractive possibility to correlate the presence of active catalysts on the surface to the overall activity by measuring the reaction rate as a function of the amount of $Zn(NO_3)_2$ added to a NP solution. The observed sigmoidal profile is a strong indication that the catalytic activity originates from the cooperative

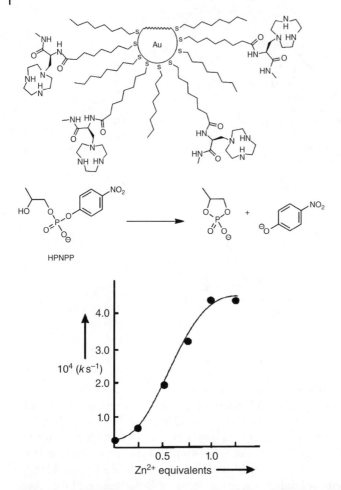

HPNPP

Figure 7.5 Catalytic NPs for the transphosphorylation of HPNPP, an RNA model substrate. A sigmoidal curve is observed when the reaction rate is measured as a function of the number of equivalents of Zn^{2+} (relative to the head groups). *Source*: Ref. [23]. Reproduced with permission of John Wiley and Sons.

action of two neighbouring $TACN \cdot Zn^{2+}$ complexes (Figure 7.5). At low Zn^{2+} concentrations isolated complexes are formed in the monolayer with a low catalytic activity. However, after a $TACN:Zn^{2+}$ ratio of around 1:0.3 has been reached a strong increase in reactivity is observed, because the gradual saturation of the monolayer with Zn^{2+} causes the formation of dinuclear catalytic sites. A maximum reactivity of the NPs is reached when the monolayer is fully saturated with Zn^{2+}. In subsequent studies, this observation of cooperativity was used to understand in detail the origin of the dendritic effect in multivalent catalysts (see Section 7.5).

The induction of cooperativity between neighbouring groups in multivalent systems is not limited to biomimetic reactions, but is an attractive strategy for any reaction that has an order higher than one in the concentration of catalyst. A well-known example is the asymmetric ring opening of epoxides catalysed by chiral salen $\cdot$ Co^{3+} complexes.

Substantial mechanistic evidence is in support of a mechanism involving the cooperative action of two complexes through the simultaneous activation of both the nucleophile and the epoxide (Figure 7.6a and b) [24,25]. Early on, this led to the speculation that the incorporation of these complexes in dendritic structures might lead to highly efficient catalytic systems because of an enforced cooperativity [26]. Commercially available polyamidoamine (PAMAM) dendrimers containing 4, 8 or 16 head groups were functionalized at the periphery with chiral salen·Co^{3+} complexes and tested for catalytic activity in the hydrolytic kinetic resolution of a terminal epoxide (Figure 7.6c). Not only did the catalytic dendrimers exhibit a significantly enhanced activity (normalized for the number of Co^{3+} complexes) compared with the monomeric complex, but, surprisingly also compared with the reference dimeric complex. Interestingly, a maximum reactivity per cobalt was attained for the dendrimer containing four complexes. The positive dendritic effect was ascribed to restricted conformations imposed by the dendrimer structure, creating a higher effective molarity of salen·Co^{3+} complexes. An alternative explanation relied on the occurrence of higher order interactions between the catalytic centres in dendrimers of higher generation. Later on (Section 7.4), an example will be discussed in which the multivalent structure indeed induces a different mechanistic pathway, which is one of the hallmarks of enzymatic catalysis. The same catalytic units were also exploited in a Au NP-based multivalent catalyst by inserting salen-terminated thiols in an n-octanethiol covered monolayer through the place exchange reaction [27]. A final 3:1 ratio of catalytic and inert thiolates ensured the possibility of forming dinuclear catalytic pockets. The observation that this NP exhibited a complete kinetic resolution of racemic hexane-1-oxide within just 5 h (as compared with 52 h required for the monomeric reference catalyst at the same loading) confirmed the efficacy of embedding this catalyst on a NP.

7.4 Mechanistic Effects

Apart from a direct control over activity by creating catalytic sites through the precise positioning of functional groups on a multivalent scaffold, it has also been demonstrated that the scaffold itself can exert an indirect effect on catalysis by creating a local chemical environment that is different from the bulk. This is exemplified by a study of Au NPs terminating with a HisPhe-OH dipeptide (Figure 7.7) [28]. The monomeric peptide itself is a modest catalyst for the hydrolysis of 2,4-dinitrophenylbutanoate. Its incorporation in the nanosystem led to a significant increase in activity of at least one order of magnitude. Yet, the most interesting difference was the observed catalytic activity as a function of pH. For the monomeric peptide, an increase in activity was observed upon an increase in pH, consistent with the deprotonation of imidazole ($pK_a = 6.6$), which is the catalytically relevant nucleophile. On the other hand, the profile observed for the NPs indicates the formation of the first nucleophilic species with pK_a 4.2 and a second one with pK_a 8.1. These pK_a values were assigned to the carboxylic acid and the imidazolium, respectively. The reason for the higher value of the pK_a of the imidazolium in the NP is due to the anionic nature of the NP that disfavours the deprotonation of the imidazolium cation. The confinement of the catalytic units in the monolayer covering the NPs triggers a cooperative hydrolytic mechanism operative at pH < 7 in which a carboxylate and an imidazolium ion act as a general base and general acid, respectively.

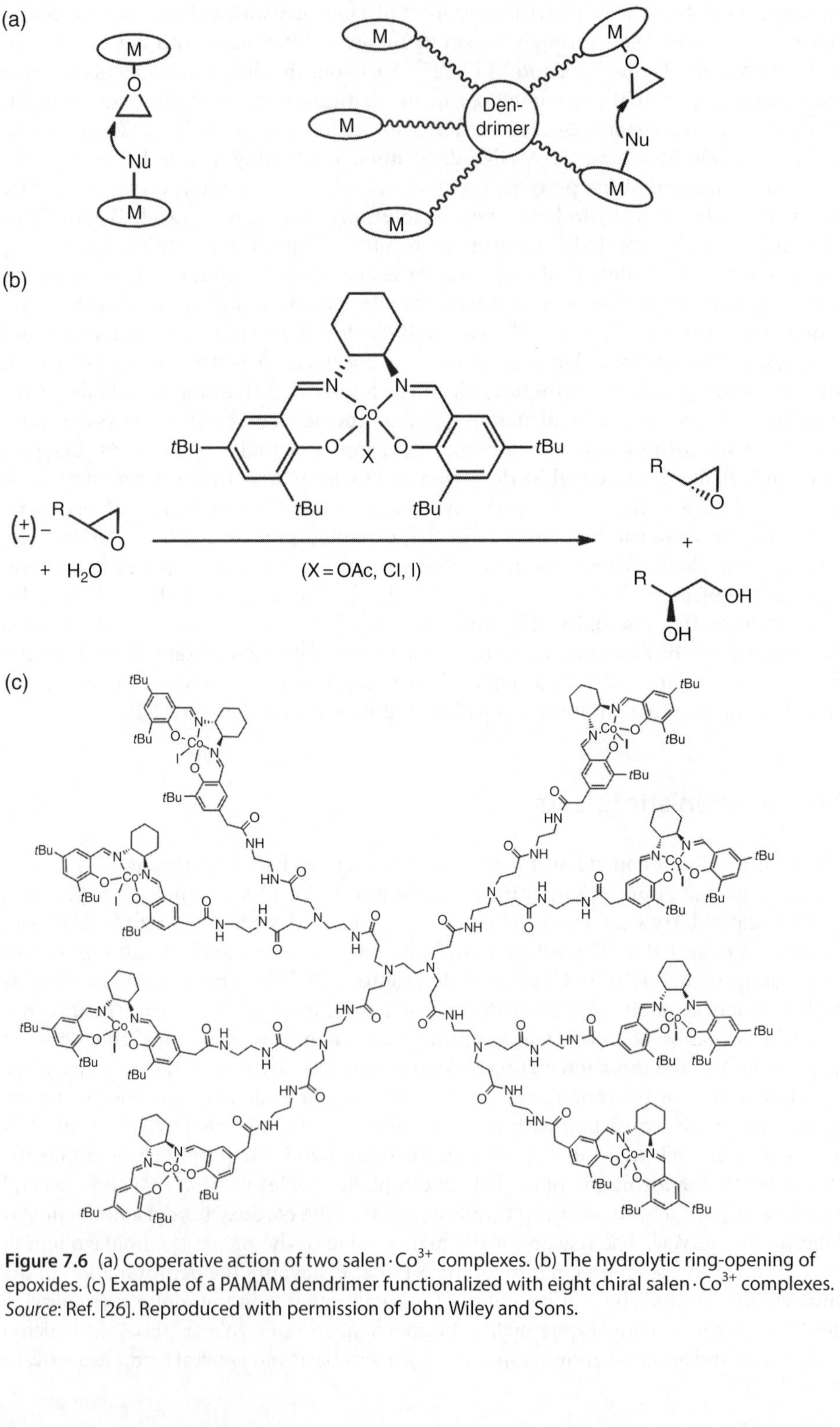

(a)

(b)

(c)

Figure 7.6 (a) Cooperative action of two salen·Co^{3+} complexes. (b) The hydrolytic ring-opening of epoxides. (c) Example of a PAMAM dendrimer functionalized with eight chiral salen·Co^{3+} complexes. *Source*: Ref. [26]. Reproduced with permission of John Wiley and Sons.

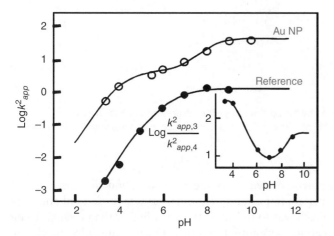

Reference: CH₃CO-His-Phe-OH

Figure 7.7 Catalytic activity of Au NPs functionalized with a catalytic dipeptide as a function of pH compared with that of the monomeric reference catalyst. *Source*: Ref. [28]. Reproduced with permission of American Chemical Society.

The absence of this mechanism in the monomeric catalyst results in a 300-fold rate acceleration at acidic pH for the NP-based catalyst.

In an entirely different system enhanced catalytic activity was also found to originate from a local pH effect in combination with other factors. Zaramella *et al.* [29,30] self-assembled small negatively charged peptides on the surface of Au NPs passivated with thiols containing positively charged quaternary ammonium salts (Figure 7.8). These peptides were equipped with one or more His-residues as catalytic units for the hydrolysis of esters. When bound to the surface, the peptides accelerated the cleavage of the *p*-nitrophenyl ester of *N*-Cbz-protected phenylalanine by more than two orders of magnitude. However, this rate enhancement did not originate from the cooperative action between two His-residues on the same peptide because a linear correlation was observed between the number of His-residues present in the peptides and the second-order rate constant. Yet, cooperativity was not observed between His-residues on neighbouring peptides, because in that case the rate should have exponentially increased upon saturating the monolayer surface with peptides. A detailed analysis showed that the main

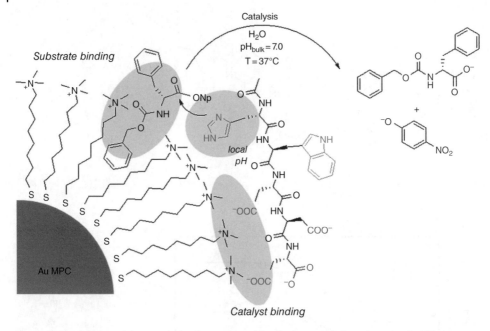

Figure 7.8 Catalytic activity caused by the co-assembly of substrate and catalyst on a Au NP surface aided by an increase in local pH. *Source*: Ref. [29]. Reproduced with permission of American Chemical Society.

reason for the enhanced catalytic activity was the co-localization of substrate and catalyst on the multivalent surface. Importantly, the catalysis was further enhanced by the local pH at the surface which was found to be 0.7 units higher than the pH of the bulk solvent, caused by the cationic ammonium groups. The pH tuning by the charge of the surface is very similar to what is observed with cationic micelles or vesicles. The higher local pH increased the concentration of unprotonated imidazole which acted as the nucleophile during catalysis.

7.5 The Dendritic Effect in Multivalent Nanozymes

Synthetic multivalent catalysts frequently display enzyme-like Michaelis–Menten reaction kinetics and have been coined nanozymes also for that reason [23,31]. The basis of Michaelis–Menten kinetics is a model that assumes that the substrate (S) is bound by an enzyme (E) yielding the complex E·S, after which the chemical transformation into product (P) takes place. Complex formation is determined by the dissociation constant K_M, whereas the efficiency of the catalytic process is determined by the first-order rate constant k_{cat}. This leads then to the following expression for the initial rate v_{init}.

$$v_{init} = \frac{k_{cat}[E]_0[S]_0}{K_M + [S]_0} \qquad (7.1)$$

Single site Michaelis–Menten kinetics

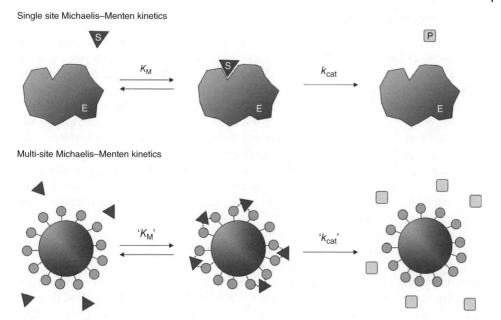

Multi-site Michaelis–Menten kinetics

Figure 7.9 Michaelis–Menten saturation kinetics for an enzyme and a multivalent nanozyme.

The values for K_M and k_{cat} are obtained by fitting a plot of the initial rate as a function of the initial substrate concentration to Equation 7.1. As mentioned, multivalent catalysts frequently display similar saturation kinetics and, consequently, the catalytic properties of such systems are typically evaluated in terms of the parameters k_{cat} and K_M. Comparison of these values with those of the monomeric reference catalyst or comparison of these values within a series of structurally multivalent catalysts, such as dendrimers of different generations, has often led to the observation that multivalent catalysts display a strong *dendritic effect*. This implies that the catalytic performances progressively increase as a function of the valency of the scaffold.

Yet, an intrinsic difference exists between an enzyme and the multivalent catalysts described in this chapter. The Michaelis–Menten model at the basis of Equation 7.1 explicitly refers to an enzyme with a single active site (Figure 7.9). This implies that the enzyme forms a 1:1 complex with the substrate and that, consequently, at saturation each enzyme is saturated with a single substrate. Evidently, this is not the case for multivalent catalysts. Here, multiple binding sites are present and at saturation the single multivalent catalyst is saturated with multiple substrates (Figure 7.9). Thus, the k_{cat} and K_M values obtained from fitting the saturation curve to Equation 7.1 are macroscopic values composed of all microscopic binding and catalytic events that occur simultaneously within the multivalent system. For two multivalent systems, a dendritic catalyst and a Au NP-based catalyst, our group has developed theoretical models in order to determine the relation between the microscopic values for the individual catalytic sites and the macroscopic ones measured for the entire system [32,33]. These studies give a surprising insight into the origin of the dendritic effect.

7.5.1 Peptide-Based Dendrimers for the Cleavage of Phosphodiesters

A series of dendrimers of different generation with a peptidic Lys-backbone and a varying number of peripheral 1,4,9-triazacyclononane (TACN)$\cdot$Zn^{2+} complexes (ranging from 4 to 32) were analysed for their ability to catalyse the transphosphorylation of HPNPP (Figure 7.5 and Figure 7.10) [32]. Michaelis–Menten-like saturation kinetics was observed and values for k_{cat} and K_M were obtained by fitting the data to Equation 7.1. Plots of the obtained values as a function of the valency indicate an increase of k_{cat} and a decrease in K_M (stronger binding) as the valency of the dendrimer increases (Figure 7.10). At first glance, this suggests that a positive dendritic effect originates from both an improved catalytic efficiency *and* a higher affinity of the substrate for the catalyst. The combined effect leads to a significant increase in the apparent second-order rate constant k_{cat}/K_M, often taken as a measure to quantify the dendritic effect. Yet, how do these results need to be interpreted?

Obviously, because of the presence of multiple catalytic head groups in the dendrimers the value for k_{cat} is artificially inflated. A correct comparison of k_{cat} between the different generations requires a normalization on the actual number of catalytic head groups present. Indeed, after correction it is observed that the k_{cat} value increases only from dendrimer D4 to D8 after which it remains constant and even drops for the D32 (D4, D8 and D32 refer to dendrimers with 4, 8 or 32 TACN head groups, respectively). This indicates that from an efficiency point of view, the catalytic process does not improve upon increasing the valency of the dendrimer. A less obvious issue that needs to be considered in evaluating k_{cat} is the maximum saturation level that can be reached for a multivalent system. It is reminded that the maximum rate for an enzyme is obtained when all of the enzyme is saturated with substrate. For a multivalent system, the question is whether all catalytic sites can be simultaneously saturated with substrates because of geometric constraints. For example, consider a system, like the example discussed here, in which the catalytic site is composed of two neighbouring head groups. In the case where the multivalent system contains an odd number of catalytic head groups, this implies that at saturation not all head groups participate in catalysis. This leads to an intrinsic underestimation of the value for k_{cat}. Alternatively, a non-optimal saturation level may result from geometric constraints. Although it may be the cause for the observed lower activity of dendrimer D4 compared with D8, it is evident that this intrinsic effect is much harder to quantify. Yet, its importance emerges in a clear manner from the analysis of catalytic SAMs (see Section 7.5.2).

Even after normalization of the k_{cat} values, a (now linear) increase of k_{cat}/K_M is observed as a function of valency driven by a decrease in K_M. Although less strong, this still points to a positive dendritic effect. However, the K_M value must also be interpreted with caution because of the multivalent nature of the catalyst. This is exemplified with a simple case in which catalysis by a dimeric and a tetrameric catalyst is compared (Figure 7.11). As before, catalysis requires the cooperative interaction of two head groups. In the case for example where 12 head groups are clustered in dimers, a total number of 6 binding sites are present in the catalyst. However, the same number of head groups can create 18 potential binding sites in the tetrameric catalyst. This points to an important aspect of multivalent catalysts: clustering of catalytic units in a multivalent system leads to a significant increase in the apparent number of binding sites. The consequence is an increase in the apparent affinity of the substrate for the multivalent catalysts, leading towards an apparent decrease in K_M upon fitting the saturation profile to Equation 7.1.

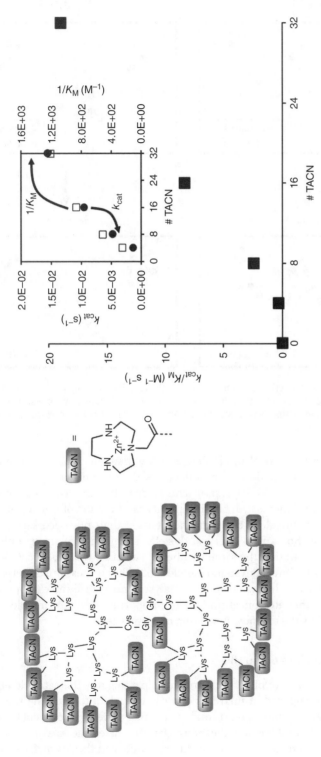

Figure 7.10 Representative structure of a peptidic Lys-based dendrimer with 32 catalytic TACN·Zn^{2+} head groups and plots of the overall Michaelis–Menten parameters for the cleavage of HPNPP (see Figure 7.5) as a function of the number of catalytic units present in the dendrimer. *Source*: Ref. [32]. Reproduced with permission of American Chemical Society.

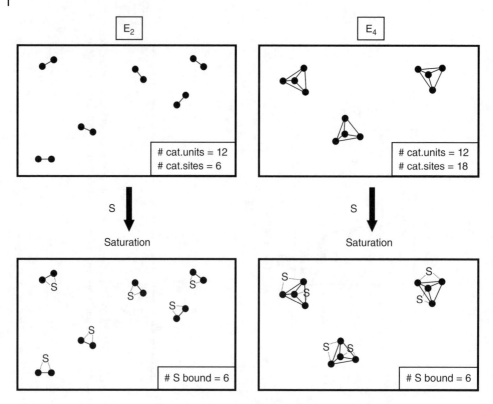

Figure 7.11 Effect of the clustering of catalytic units on the creation of catalytic sites in multivalent models containing two or four catalytic units. Each catalytic unit is depicted by a black circle and each catalytic site by a line connecting two catalytic units. *Source*: Ref. [32]. Reproduced with permission of American Chemical Society.

Simulations using a theoretical model indeed show that the apparent affinity of the substrate increases as a function of the valency of the system (with the same microscopic binding constant K_M for each site). This leads to the important conclusion that the (generally) observed increase in binding affinity as a function of valency is an intrinsic phenomenon of multivalent catalysts and by itself does not necessarily require a chemical explanation. So, how to interpret this effect? Evidently, it affects only K_M and not k_{cat}, because at saturation the same number of substrate molecules can be accommodated by the multivalent catalyst, independent of whether these are dimeric or tetrameric. Basically, it tells us that a multivalent catalyst is intrinsically able to operate more efficiently at lower substrate concentrations. An important design principle for multivalent catalysts is thus the efficacy in creating catalytic sites.

7.5.2 Catalytic 3D SAMs on Au NPs

In a follow-up study, we exploited the same principle to study the catalytic efficacy of SAMs on Au NPs composed of mixtures of a catalytic thiol (containing the identical TACN $\cdot$ Zn^{2+} head group) and an inert thiol (containing a triethylene glycol head group) (Figure 7.12a) [33]. A fundamental question regarding mixed monolayer protected Au NPs regards the spatial distribution of the different thiols on the Au surface, which can

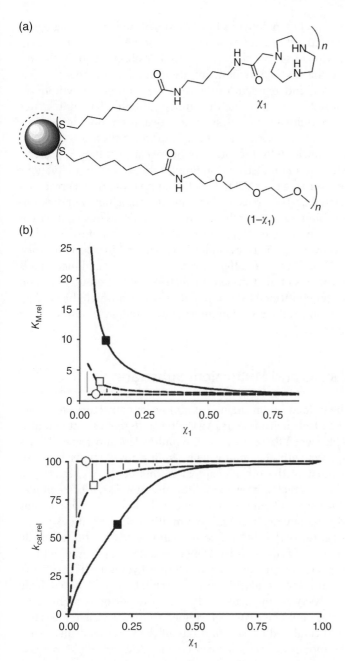

Figure 7.12 (a) Au NPs functionalized with catalytic and inert thiols. (b) Simulated correlations between K_M and k_{cat} as a function of the mole fraction χ_1. The solid line marked with (■) represents the correlation in the case where the catalytic thiols are randomly distributed in the monolayer. The dashed line marked with (□) represents homodomain formation of the catalytic thiol after monolayer formation. The dashed line marked with (○) represents the self-sorting of thiols on different NPs during monolayer formation (extreme case of phase separation). *Source:* Ref. [33]. Reproduced with permission of John Wiley and Sons.

be either random or phase-separated in domains [34,35]. Model simulations analogous to those described before indicated that a different behaviour of the 'overall' Michaelis–Menten parameters as a function of surface thiol ratio would be observed for both situations (Figure 7.12b) [33]. In the case where domains are formed, the model predicts that both K_M and k_{cat} are nearly independent of the mole fraction of catalytic thiols in the monolayer. This is caused by the fact that the number of potential binding sites and the saturation level of the multivalent catalyst are identical for a monolayer composed of just catalytic head groups and a mixed monolayer in which these are clustered together, obviously after normalization for the total number of catalytic groups present. On the other hand, a statistical random distribution implies that as the mole fractions of catalytic groups decreases (starting from 0.3 to 0.4), the number of groups present in small clusters or even as isolated units increases. Consequently, the efficacy of the multivalent catalyst in generating binding sites for the substrate diminishes (lower apparent binding affinity, thus observed increase in K_M) and the saturation level of the multivalent catalyst is also reduced (lower k_{cat}). This was indeed experimentally observed when the catalytic activity of a series of NPs with different ratios of catalytic and inert thiols was studied. This led to the conclusion that these were distributed in a random fashion in the monolayer. This was later confirmed by an alternative study in which the binding affinity of fluorescent oligoanions to the mixed monolayer was studied as a function of the monolayer composition [36].

7.6 Multivalent Catalysts and Multivalent Substrates

The discussion so far has been focused on the use of multivalent structures to create catalytic sites or different chemical environments. An entirely different aspect of multivalent catalysts comes into play when these interact with multivalent substrates. This is the closest analogue compared with the multivalent binding interactions between two partners discussed in most of the other chapters in this book. Yet, related to catalysis it is an argument that has received relatively little attention [37,38]. Recently, a first attempt was made by McKay and Finn [39] to qualitatively describe what happens when both the catalyst and substrate are attached to a multivalent dendritic support. The reaction under investigation is the hydrazone formation between the aldehyde groups attached to the periphery of PAMAM-dendrimers of different generation (with 16, 31 and 61 end groups, respectively) and nitrobenzoxadiazole hydrazine (Figure 7.13). This reaction is catalysed by anthranilic acid which forms a Schiff base with the aldehyde as intermediate. Also this catalyst was supported on PAMAM-dendrimers of the same generations. The reaction rates could be easily determined from UV-visible measurements. Compared with the uncatalysed background rate of the two monovalent substrates, it was found that the monovalent catalyst caused a modest three-fold increase in activity. Incorporation of the substrate in the dendrimer caused a slight increase in background reactivity, but did not affect the three-fold increase in rate acceleration induced by the monovalent catalyst. In contrast, dendrimer-supported catalysts caused much faster reaction rates with polyvalent substrates, but not with monovalent ones. Depending on the valency of both systems, rate accelerations ranging from 90 to 1300 were measured relative to the strictly monovalent reference catalyst. The authors suggest two reasons that may be at the origin of this effect. The first is an enhanced initial

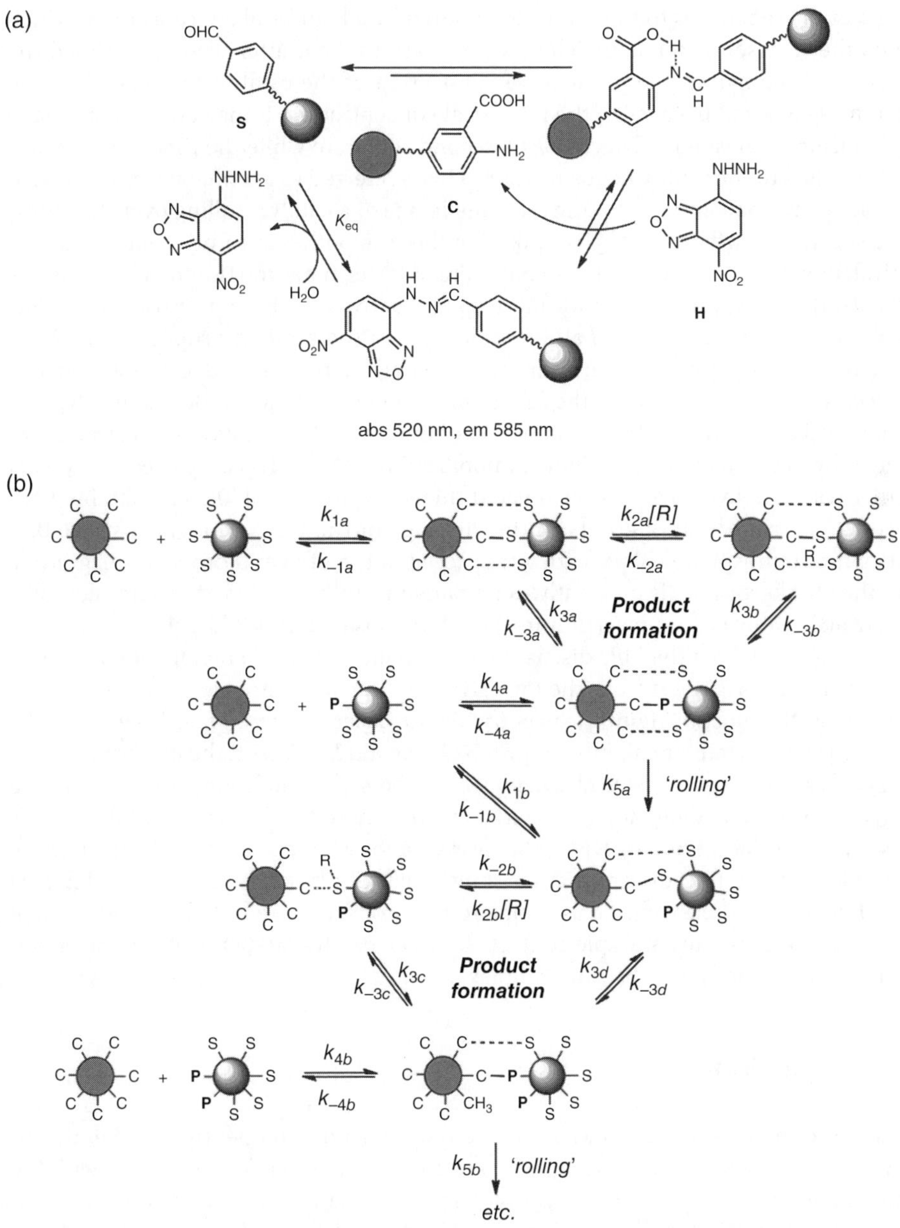

Figure 7.13 (a) Intermediates in the nucleophilic catalysis of hydrazone formation by monovalent or polyvalent anthranilic acid. The equilibrium on the bottom left shows the concomitant anthranilate-catalysed hydrolysis of the hydrazone that establishes K_{eq}. (b) Overall catalytic scheme. C, catalyst; S, substrate; R, capturing reagent, such as nitrobenzoxadiazole hydrazine in the present case; P, product; k_1, k_{-1}, initial association/dissociation rate constant, respectively; k_3, product-forming step; k_{-3}, catalysed product-decomposition step; k_{-4}, escape $\sim k_{-1}$; k_5, polyvalent association rate constant ('rolling' rate). Letter designations (k_{3a}, k_{3b}, etc.) denote the rates of the same fundamental step; these rates may differ in different cycles, probably only slightly in the early stages of the reaction. Accelerated substrate transformation may occur if 'rolling' (k_5) is much faster than diffusion-controlled steps (k_{-4}, k_1). *Source:* Ref. [39]. Reproduced with permission of John Wiley and Sons.

binding of the substrates to the catalysts favoured by a high local concentration (which favours the re-association step). The second reason is that after having catalysed the formation of the hydrazone product, re-connection of the catalyst to a neighbouring substrate is favoured because of the high local concentration. It may also well be that a neighbouring catalyst already activates a second substrate while the first catalytic unit is still attached to the multivalent substrate. This is referred to as enhanced processivity, or a 'rolling' mechanism, which transmits the image of a catalyst rolling over the multivalent substrate surface. Strong support for these hypotheses came from a study of multivalent substrates and catalysts containing a lower mole fraction of active species on the dendrimer surface. A significant drop in reactivity for both multivalent systems was observed when the functional group density on the scaffold dropped from 75 to 50%. However, the simultaneous variation of both substrate and catalyst loadings produced a dramatic drop when the functional group density was reduced to 75%.

A practical application of the combination of multivalent catalysts and substrates was provided by Bonomi *et al.* [40]. They immobilized BAPA•Zn(II) complexes on Au NPs together with triethylene glycol-terminated thiols (Figure 7.14) [40]. The obtained NP showed an extraordinary activity in the cleavage of the DNA model substrate bis-*p*-nitrophenyl phosphate (BNP) with a rate acceleration of over 5 orders of magnitude over the background. Importantly, comparison with the monomeric complex BAPA•Zn(II) showed that insertion in the SAM caused a 100-fold gain in reactivity. Similar as observed for the NPs discussed above (Figure 7.5), the major source for this rate acceleration originates from the formation of dinuclear catalytic sites in the SAM. Nonetheless, the most intriguing aspect of this NP-based catalyst appeared when DNA was used as a substrate. Incubation of pBR 322 plasmid DNA with the catalyst resulted in a significant phosphodiester cleavage. Under the same conditions, the monomeric catalyst showed no activity at all. Remarkably, the amount of linear DNA formed was 50% larger than that of nicked DNA, the formation of which is statistically much more favourable. This result indicates that the multivalent NP-based catalyst preferably performs double strand cleavage. This seems to be a consequence of the multivalency of the NPs, which generates multiple contacts between catalyst and cleavable bonds upon formation of the catalyst–substrate complex.

7.7 Conclusions

As discussed in Section 7.1, multivalent catalysts are most frequently associated with numerous homogeneous catalysts attached to a multivalent scaffold, prepared with the scope to facilitate recovery and re-usage of the precious catalysts. A good design implied that no loss of catalytic performance was observed compared with the reference homogeneous catalyst and/or during subsequent cycles. Occasional improvements of the catalytic performance (e.g. in rate or enantioselectivity) was a welcome additional bonus not sought for. Yet, the examples discussed in this chapter illustrate that the potential of multivalent catalysts goes far beyond this. Design principles are emerging aimed at exploiting the ability of multivalent systems to induce cooperativity between functional groups. At difference with traditional multivalent catalysts, this implies that multivalency becomes a prerequisite for observing catalytic activity. The possibility to alter in a controlled manner the local chemical environment in a multivalent catalyst becomes an

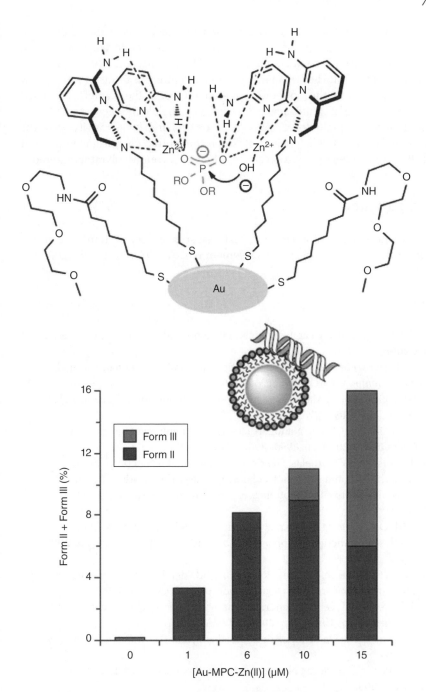

Figure 7.14 Catalytic Au NPs for the cleavage of phosphodiesters. The result of the cleavage of plasmid DNA and a cartoon representing the interaction between the catalyst and the biopolymer are also shown. *Source*: Ref. [40]. Reproduced with permission of American Chemical Society. See color section.

attractive strategy to control actively the catalytic performance. It is further shown that multivalent catalysts bear intrinsic advantages compared with monovalent ones when substrate binding to the catalytic system plays a role. The enhanced number of potential binding sites in a multivalent catalyst makes it perform better (in terms of rate) at low concentration compared with the monomeric reference. Overall, the examples discussed here mark the importance of multivalency as a design principle for catalysts. The next challenge is to develop methodology aimed at maximizing cooperativity in these catalysts, create local enzyme-like pockets of which the (chiral) chemical environment can be controlled, and widen the reaction scope, which is currently still rather limited.

Acknowledgements

Funding from the European Union Horizon 2020 research and innovation programme under the Marie Sklodowska-Curie grant agreement No. 642793 is acknowledged.

References

1 D. Mery, D. Astruc, Dendritic catalysis: Major concepts and recent progress, *Coord. Chem. Rev.* **2006**, *250*, 1965–1979.

2 S. Roy, M. A. Pericas, Functionalized nanoparticles as catalysts for enantioselective processes, *Org. Biomol. Chem.* **2009**, *7*, 2669–2677.

3 J. M. Fraile, J. I. Garcia, J. A. Mayoral, Noncovalent immobilization of enantioselective catalysts, *Chem. Rev.* **2009**, *109*, 360–417.

4 A. Schatz, O. Reiser, W. J. Stark, Nanoparticles as semi-heterogeneous catalyst supports, *Chem. Eur. J.* **2010**, *16*, 8950–8967.

5 D. Astruc, E. Boisselier, C. Ornelas, Dendrimers designed for functions: From physical, photophysical, and supramolecular properties to applications in sensing, catalysis, molecular electronics, photonics, and nanomedicine, *Chem. Rev.* **2010**, *110*, 1857–1959.

6 A. Thomas, M. Driess, Bridging the materials gap in catalysis: Entrapment of molecular catalysts in functional supports and beyond, *Angew. Chem. Int. Ed.* **2009**, *48*, 1890–1892.

7 S. Javor, E. Delort, T. Darbre, J. L. Reymond, A peptide dendrimer enzyme model with a single catalytic site at the core, *J. Am. Chem. Soc.* **2007**, *129*, 13238–13246.

8 T. Darbre, J. L. Reymond, Peptide dendrimers as artificial enzymes, receptors, and drug-delivery agents, *Acc. Chem. Res.* **2006**, *39*, 925–934.

9 J. Kofoed, J. L. Reymond, Dendrimers as artificial enzymes, *Curr. Opin. Chem. Biol.* **2005**, *9*, 656–664.

10 L. Pasquato, P. Pengo, P. Scrimin, Nanozymes: Functional nanoparticle-based catalysts, *Supramol. Chem.* **2005**, *17*, 163–171.

11 G. Pieters, L. J. Prins, Catalytic self-assembled monolayers on gold nanoparticles, *New J. Chem.* **2012**, *36*, 1931–1939.

12 T. Belser, M. Stohr, A. Pfaltz, Immobilization of rhodium complexes at thiolate monolayers on gold surfaces: Catalytic and structural studies, *J. Am. Chem. Soc.* **2005**, *127*, 8720–8731.

13 C. C. Paluti, E. S. Gawalt, Immobilized aza-bis(oxazoline) copper catalysts on SAMs: Selectivity dependence on catalytic site embedding, *J. Cat.* **2009**, *267*, 105–113.

14 C. C. Paluti, E. S. Gawalt, Immobilized aza-bis(oxazoline) copper catalysts on alkanethiol self-assembled monolayers on gold: Selectivity dependence on surface electronic environments, *J. Cat.* **2010**, *275*, 149–157.

15 C. Guarise, F. Manea, G. Zaupa, L. Pasquato, L. J. Prins, P. Scrimin, Cooperative nanosystems, *J. Pept. Sci.* **2008**, *14*, 174–183.

16 E. Delort, T. Darbre, J. L. Reymond, A strong positive dendritic effect in a peptide dendrimer-catalyzed ester hydrolysis reaction, *J. Am. Chem. Soc.* **2004**, *126*, 15642–15643.

17 L. Ropartz, R. E. Morris, D. F. Foster, D. J. Cole-Hamilton, Increased selectivity in hydroformylation reactions using dendrimer based catalysts; a positive dendrimer effect, *Chem. Commun.* **2001**, 361–362.

18 A. Dahan, M. Portnoy, Remarkable dendritic effect in the polymer-supported catalysis of the Heck arylation of olefins, *Org. Lett.* **2003**, *5*, 1197–1200.

19 Y. Ribourdouille, G. D. Engel, M. Richard-Plouet, L. H. Gade, A strongly positive dendrimer effect in asymmetric catalysis: Allylic aminations with Pyrphos-palladium functionalised PPI and PAMAM dendrimers, *Chem. Commun.* **2003**, 1228–1229.

20 B. Helms, J. M. J. Frechet, The dendrimer effect in homogeneous catalysis, *Adv. Synth. Cat.* **2006**, *348*, 1125–1148.

21 D. E. Wilcox, Binuclear metallohydrolases, *Chem. Rev.* **1996**, *96*, 2435–2458.

22 F. Mancin, P. Scrimin, P. Tecilla, U. Tonellato, Artificial metallonucleases, *Chem. Commun.* **2005**, 2540–2548.

23 F. Manea, F. B. Houillon, L. Pasquato, P. Scrimin, Nanozymes: Gold-nanoparticle-based transphosphorylation catalysts, *Angew. Chem. Int. Ed.* **2004**, *43*, 6165–6169.

24 K. B. Hansen, J. L. Leighton, E. N. Jacobsen, On the mechanism of asymmetric nucleophilic ring-opening of epoxides catalyzed by (salen)Cr^{III} complexes, *J. Am. Chem. Soc.* **1996**, *118*, 10924–10925.

25 L. P. C. Nielsen, C. P. Stevenson, D. G. Blackmond, E. N. Jacobsen, Mechanistic investigation leads to a synthetic improvement in the hydrolytic kinetic resolution of terminal epoxides, *J. Am. Chem. Soc.* **2004**, *126*, 1360–1362.

26 R. Breinbauer, E. N. Jacobsen, Cooperative asymmetric catalysis with dendirmeric Co(salen) complexes, *Angew. Chem. Int. Ed.* **2000**, *39*, 3604–3607.

27 T. Belser, E. N. Jacobsen, Cooperative catalysis in the hydrolytic kinetic resolution of epoxides by chiral (salen)Co(III) complexes immobilized on gold colloids, *Adv. Synth. Cat.* **2008**, *350*, 967–971.

28 P. Pengo, S. Polizzi, L. Pasquato, P. Scrimin, Carboxylate – Imidazole cooperativity in dipeptide-functionalized gold nanoparticles with esterase-like activity, *J. Am. Chem. Soc.* **2005**, *127*, 1616–1617.

29 D. Zaramella, P. Scrimin, L. J. Prins, Self-Assembly of a catalytic multivalent peptide-nanoparticle complex, *J. Am. Chem. Soc.* **2012**, *134*, 8396–8399.

30 D. Zaramella, P. Scrimin, L. J. Prins, Catalysis of transesterification reactions by a self-assembled nanosystem, *Int. J. Mol. Sci.* **2013**, *14*, 2011–2021.

31 H. Wei, E. K. Wang, Nanomaterials with enzyme-like characteristics (nanozymes): next-generation artificial enzymes, *Chem. Soc. Rev.* **2013**, *42*, 6060–6093.

32 G. Zaupa, P. Scrimin, L. J. Prins, Origin of the dendritic effect in multivalent enzyme-like catalysts, *J. Am. Chem. Soc.* **2008**, *130*, 5699–5709.

33 G. Zaupa, C. Mora, R. Bonomi, L. J. Prins, P. Scrimin, Catalytic self-assembled monolayers on Au nanoparticles: The source of catalysis of a transphosphorylation reaction, *Chem. Eur. J.* **2011**, *17*, 4879–4889.

34 C. Gentilini, L. Pasquato, Morphology of mixed-monolayers protecting metal nanoparticles, *J. Mater. Chem.* **2010**, *20*, 1403–1412.

35 D. Rodriguez-Fernandez, L. M. Liz-Marzan, Metallic Janus and patchy particles, *Part. Part. Syst. Charact.* **2013**, *30*, 46–60.

36 R. Bonomi, A. Cazzolaro, L. J. Prins, Assessment of the morphology of mixed SAMs on Au nanoparticles using a fluorescent probe, *Chem. Commun.* **2011**, *47*, 445–447.

37 X. L. Liao, R. T. Petty, M. Mrksich, A spatially propagating biochemical reaction, *Angew. Chem. Int. Ed.* **2011**, *50*, 706–708.

38 R. A. Pavlick, S. Sengupta, T. McFadden, H. Zhang, A. Sen, A polymerization-powered motor, *Angew. Chem. Int. Ed.* **2011**, *50*, 9374–9377.

39 C. S. McKay, M. G. Finn, Polyvalent catalysts operating on polyvalent substrates: A model for surface-controlled reactivity, *Angew. Chem. Int. Ed.* **2016**, DOI: 10.1002/anie.201602797.

40 R. Bonomi, F. Selvestrel, V. Lombardo, C. Sissi, S. Polizzi, F. Mancin, U. Tonellato, P. Scrimin, Phosphate diester and DNA hydrolysis by a multivalent, nanoparticle-based catalyst, *J. Am. Chem. Soc.* **2008**, *130*, 15744.

8

Multivalent Molecular Recognition on the Surface of Bilayer Vesicles

Jens Voskuhl[1],*, Ulrike Kauscher[2,3],*, and Bart Jan Ravoo[2]

[1] Institute of Organic Chemistry, University of Duisburg-Essen, Universitätsstrasse 7, 45117 Essen, Germany
[2] Organic Chemistry Institute, Westfälische Wilhelms-Universität Münster, Corrensstrasse 40, 48149 Münster, Germany
[3] Present address: Department of Materials and Department of Bioengineering, Imperial College London, Prince Consort Rd, SW7 2AZ London, UK

8.1 Introduction

The current state of the art of supramolecular chemistry and self-assembly of soft matter includes the mimicking of biological processes such as enzyme catalysis, dynamic self-assembly, and molecular recognition. Nature is an important source of inspiration since it shows the chemist how to use weak and non-covalent interactions to build complex supramolecular structures such as proteins, virus capsids and membranes from simple components such as amino acids, carbohydrates, lipids, and nucleotides. Since each of these structures are held together by non-covalent interactions, they are inherently dynamic and respond to even small changes in their environment. Especially membranes are of high interest since these complex and dynamic structures play a crucial role in biology and medicine. Membranes separate the "outside" and the "inside" of every living cell, membranes mediate the transfer of small and large molecules between organelles, and membranes are crucial in neurotransmission.

Nearly all properties of biological membranes can be conveniently investigated in liposomes or vesicles. These structures consist of a bilayer membrane of phospholipids which encapsulates an aqueous compartment. Vesicles also occur inside and outside cells ("exovesicles"), for example during neurotransmission in the synapse. Ever since the discovery that it is not only phospholipids that can form bilayer vesicles, the synthesis of a plethora of artificial amphiphiles with vesicle forming properties became an important branch of supramolecular chemistry [1]. Due to their analogy in morphology and properties compared with natural liposomes and membranes, the discovery of synthetic vesicles opened a new and burgeoning field called biomimetic membrane chemistry. Artificial vesicles can be used as versatile model systems to design, investigate and understand complex biological processes such as delivery of

---

* These authors contributed equally to this chapter.

Multivalency: Concepts, Research & Applications, First Edition. Edited by Jurriaan Huskens, Leonard J. Prins, Rainer Haag, and Bart Jan Ravoo.
© 2018 John Wiley & Sons Ltd. Published 2018 by John Wiley & Sons Ltd.

neurotransmitters between synapses, infections, tissue growth and ion transport, since the variation of important parameters such as composition and size of the vesicles is rather straightforward.

The molecular structure of typical phospholipids features a hydrophilic head group and two hydrophobic tails. The hydrophilic head group consists of a negatively charged phosphate group substituted with hydrophilic units such as amines or carbohydrates. The tails consist of saturated or unsaturated alkyl chains with up to 24 carbon atoms (Figure 8.1). Due to the balance between the hydrophobic part (which minimizes contact with water) and the hydrophilic part (which is strongly hydrated) of the lipid, these amphiphilic molecules form stable bilayer membranes in aqueous media.

In 2000, Ravoo and Darcy reported synthetic vesicles composed of amphiphilic cyclo-dextrins (Figure 8.2) [2]. Cyclodextrins are host molecules for hydrophobic molecules and cyclodextrin vesicles combine the remarkable ability of cyclodextrins for molecular recognition with the formation of stable bilayer vesicles in aqueous media. It has been shown that cyclodextrins embedded in vesicles form host–guest complexes with typical guests such as adamantane, azobenzene, and t-butylbenzene [3]. As a result, cyclodextrin vesicles are very easy to functionalize with any desired molecule that bears a hydrophobic anchor fitting into the cavity of the cyclodextrins. Functionalization is straightforward since addition of this guest molecule to the aqueous solution of the vesicle will result in spontaneous decoration of the host vesicle surface due to the formation of an inclusion complex.

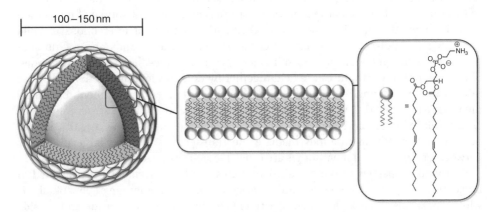

Figure 8.1 Molecular structure of phospholipids and their assembly into liposomes.

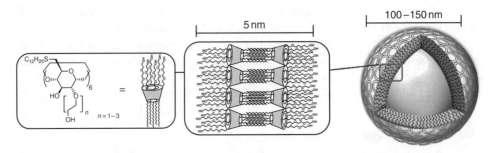

Figure 8.2 Molecular structure of amphiphilic cyclodextrins and their assembly into vesicles.

Molecular recognition and intermolecular reactions at membrane surfaces are influenced by the local environment at the membrane surface, which is in essence an interface between a hydrocarbon nanolayer and an aqueous solution. The confinement and organization of molecules in a membrane results in high local concentrations (up to 1 M) that are very unusual in homogeneous solutions. In addition, molecular recognition at biological membranes, such as proteins binding to carbohydrate residues of glycolipids, is usually mediated by multivalent instead of monovalent interactions. Multivalent interactions at membranes are not only stronger, but usually also more selective than monovalent interactions. Moreover, they can be attenuated or amplified by competing ligands or inhibitors. Multivalent interactions benefit from the high local concentration of interacting molecules at the membrane–water interface. It should be emphasized here that a biological membrane is not a simple two-dimensional fluid, since lipids and proteins may be present in domains [4]. In biological membranes, multivalent interactions can be strengthened by receptor clustering or the formation of "lipid rafts" in fluid membranes. Hence, receptor clustering and raft formation are intimately linked not only to membrane fluidity but also to the efficiency of multivalent interactions. In synthetic vesicles, the surface density of interacting molecules can be tuned by mixing with inert amphiphiles, and lateral diffusion in the membrane is a function of temperature and chain length of membrane components. A fluid bilayer can behave as an adaptable matrix in which interacting species will find optimal multivalent binding modes, exclusively accessible by clustering of binding units on the membrane surface.

In this chapter we will review the topic of multivalent molecular recognition at membrane surfaces. We will focus on chemical systems and synthetic vesicles, since multivalent effects at biological membranes are treated separately in Chapters 9 and 13. Using selected examples from the recent literature, we will show how multivalency can be used to induce selective interaction ("recognition") between vesicles. Furthermore, we will discuss biomimetic vesicles that either act as a multivalent platform or serve as a versatile model system for biological membrane fusion. Finally, we will describe a number of examples of supramolecular materials based on multivalent interactions of vesicles.

8.2 Molecular Recognition of Vesicles

The molecular recognition of vesicles is a highly interesting biomimetic phenomenon since recognition and interaction of biological membranes through non-covalent interactions plays a crucial role in a multitude of cellular processes such as endocytosis and exocytosis. Also fusion and fission of membranes, which are crucial processes in cell–vesicle interaction can easily be mimicked by artificial binding motifs which leads to a higher degree of understanding of these highly complex biological phenomena. In this section, we will describe how synthetic recognition motifs (such as metal coordination, host–guest interaction, and hydrogen bonding) can be embedded in bilayer vesicles and how interaction of vesicles can be induced by multivalent interactions. Key issues involve the balance of intravesicular and intervesicular interactions as well as the distribution of recognition units in the membrane (homogeneous distribution versus clustering). Since many recognition motifs are rather sensitive to changes in the environment, also the interaction of vesicles by multivalent molecular recognition can be stimulus responsive. Biomimetic recognition motifs involving for example carbohydrates, peptides, proteins, and nucleic acids are discussed in the next section.

8.2.1 Metal Coordination

Early studies on the recognition of metal ions on the surface of bilayer vesicles were conducted by Constable *et al.* [5]. The introduction of terpyridine (Ter) ligands on the surface of vesicles leads to membranes with artificial "receptors" which respond to the addition of Fe^{2+} ions leading to a change in color based on coordination and to a spontaneous vesicle aggregation induced by the formation of multiple $(Ter)_2Fe^{2+}$ complexes between the bilayers. Addition of an ion-capturing agent (EDTA) leads to a disaggregation of the vesicle, so that vesicle interaction is fully reversible. Further studies were performed by Marchi-Artzner *et al.* [6] who investigated the coordination of lanthanides to vesicles bearing diketone ligands. It was found that vesicle aggregation and luminescence of the lanthanide increases starting at a lanthanide-to-diketone ratio higher than 1:10 due to an energy transfer from the Eu^{3+} to the ligand [6].

First investigations concerning the metal ion recognition properties of cyclodextrin vesicles were reported in 2007 using heterobifunctional linkers composed of adamantane and ethylenediamine separated by a tetraethyleneglycol spacer (**1**) [7]. Upon mixing this monovalent guest with cyclodextrin vesicles in the presence of Ni^{2+} a spontaneous aggregation of the vesicles was observed based on the formation of intervesicular Ni^{2+} complexes, whereas Cu^{2+} ions form intravesicular complexes only (Figure 8.3). Interestingly, the strongest coordination complex is formed by Cu^{2+}, resulting in a saturation of the vesicle surface with $(1)_2Cu^{2+}$ and no significant intervesicular binding. In contrast, Ni^{2+} forms much weaker complexes with **1**, resulting in many free coordination sites and ligands on the vesicle surface, leading to the formation of multiple intervesicular coordination complexes and rapid aggregation of the vesicles. The interaction can be reversed by the addition of EDTA.

Inspired by the results describes above, Nalluri *et al.* reported the use of metal ions to induce a conformational change of a divalent guest molecule which leads to a switching between inter- and intravesicular recognition of cyclodextrin vesicles [8]. When divalent guest **2** is added to a dilute vesicle solution a spontaneous aggregation is observed due to the formation of multiple non-covalent cross-links between the cyclodextrin vesicles (Figure 8.3). However, aggregation is disrupted upon addition of metal ions, since metal ion coordination results in a bent conformation of the divalent guest and intervesicular binding is no longer possible due to the restricted conformation of the coordination complex. If the metal ion is extracted with EDTA, the linker can relax and the vesicles aggregate again. In contrast, divalent guest **3** can act as a cross-linker, but it is not flexible enough to bend upon metal coordination, so that aggregation persists even in the presence of metal ions.

Webb *et al.* introduced an interesting artificial binding motif based on the interaction of an amphiphilic Cu^{2+}-iminodiacetate (CuIDA) complex with polyhistidines [9]. This study elegantly showed the role of ligand clustering in membranes for the induction of aggregation and fusion (Figure 8.4). In this case the multivalent coordination of CuIDA and histidine was used to force liposomes to fuse. The relatively weak interaction of CuIDA and histidine ($K_a = 10^3\,M^{-1}$) is strongly amplified by multivalent binding of polyhistidines. The concomitant change in the distribution of the lipids was measured using pyrene moieties attached to the IDA amphiphile. In the absence of Cu^{2+} only pyrene

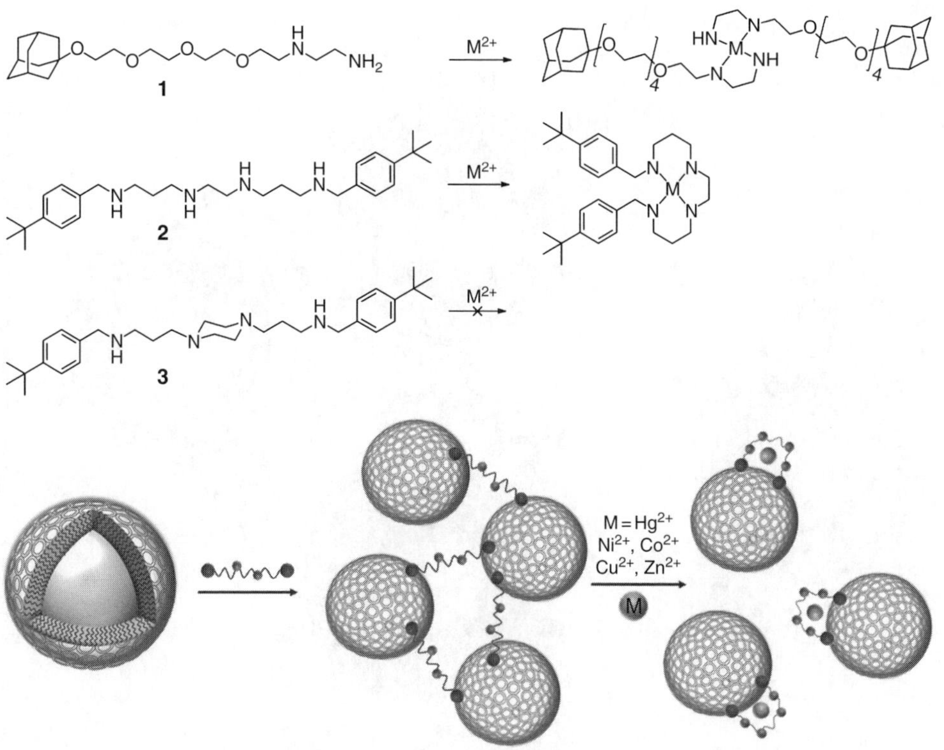

Figure 8.3 Molecular structures of heterobifunctional linkers that bind metals and cyclodextrins and schematic presentation of the molecular recognition of cyclodextrin vesicles mediated by metal ions.

monomer emission was observed. Upon addition of Cu^{2+}, the pyrene monomer emission decreased due to quenching by Cu^{2+}. Clustering of CuIDA resulted in an increased pyrene excimer emission. Upon addition of polyhistidine to egg yolk phosphatidylcholine (EPC) liposomes, an increase in turbidity indicated liposome aggregation followed by fusion, whereas aggregation (but no fusion) was observed for distearoyl phosphatidylcholine (DSPC) liposomes. Further studies showed that a match between valency of the oligohistidines and CuIDA is necessary to induce efficient liposome aggregation [10] and that clustering of the CuIDA ligand can be induced by fluorophobic interactions in the membrane [11].

In a comparable approach, Turkyilmaz *et al.* prepared liposomes containing Zn^{2+} bis(dipicolylamine) (Zn_2BDPA) as an affinity unit for anionic cell membranes [12]. The Zn_2BDPA coordination complex is known to associate with phosphate polyanions. They synthesized a new kind of amphiphile consisting of the metal recognition unit and long aliphatic chains. This amphiphile was then incorporated into liposomes. They showed that these multivalent liposomes selectively and efficiently target bacteria even in the presence of mammalian cells causing agglutination of the bacteria. Moreover, these vesicles also showed as well a high affinity for dead or dying human cancer cells that had been treated with a chemotherapeutic agent.

Gruber *et al.* [13] showed that effective cleavage of phosphodiesters can be achieved with vesicles bearing Zn^{2+}-cyclen complexes at the liposomal surface due to clustering

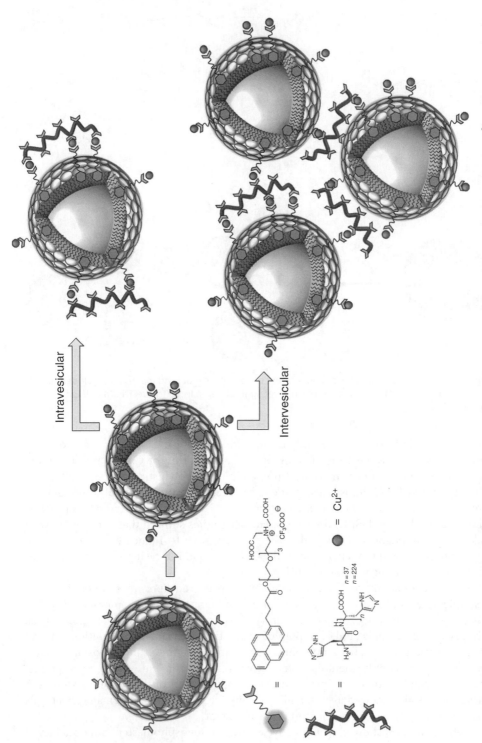

Figure 8.4 Schematic representation of polyhistidine induced aggregation of vesicles functionalized with a pyrene containing Cu^{2+}-iminodiacetate (IDA) ligand. See color section.

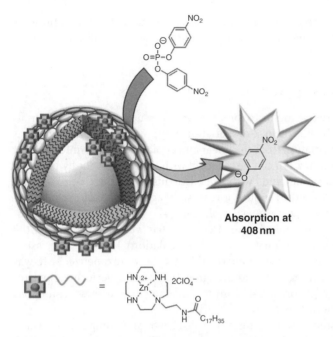

Figure 8.5 Vesicles bearing Zn^{2+}-cyclen moieties induce efficient dephosphorylation due to clustering of the active sites on the membrane surface.

of the active sites in the membrane (Figure 8.5). This system can be seen as a model system for membrane restricted enzymes that bear Zn^{2+}-fingers. To date, this supramolecular system is one of the most active artificial metal catalysts for the cleavage of phosphodiesters reported [13]. Receptor clustering can also be observed when a coumarin moiety is attached to a biscyclen receptor as the hydrophobic part of an amphiphile. The effect of clustering can directly be monitored via fluorescence quenching of the fluorophore. Moreover, molecular recognition of phosphorylated proteins can be achieved when this amphiphile is used [14].

Receptor clustering induced by metal complexes on the surface of vesicles was recently shown by Tomas and Milanesi who used Zn^{2+}-porphyrins attached to cholesterol groups which can be embedded into lipid membranes [15]. Upon addition of positively charged N-heterocycles an effective clustering of receptor can be achieved due to an interaction with Zn^{2+} ions in the axial position, since the ligand neutralizes the negative charge of the porphyrin bearing amphiphile. If, however, negatively charged ligands are used an increased repulsion of the receptors is observed which can be monitored by UV-vis spectroscopy.

Finally, it was shown that binding information can be transported through bilayer membranes in the presence of Cu^{2+} ions [16]. To this end a bolaamphiphile (i.e., an amphiphile consisting of two hydrophilic head groups linked by a hydrophobic unit) was synthesized bearing two cholesterol units which are attached to dansyl moieties via an ethylenediamine bridge. Addition of Cu^{2+} triggered a dimerization on the outer bilayer membrane of the artificial receptor unit due to the formation of a 2:1 complex between the ethylenediamines and a Cu^{2+} ion, which could be directly observed as a decrease in fluorescence of the dansyl units. Thus, receptor clustering on the surface enhances the binding of Cu^{2+} on the inside of the vesicle. This model system represents one of the first examples of signal transduction across a bilayer membrane using an artificial binding motif.

8.2.2 Light Responsive Interactions

Light is a particularly attractive stimulus to direct responsive supramolecular materials since light can be applied with a high degree of spatial and temporal control. Azobenzenes isomerize under UV light irradiation from the linear and apolar trans form into the bent and polar cis form. The isomerization can be reversed with visible light (or by heating). Importantly, the *trans*-azobenzene binds with cyclodextrins while the *cis*-azobenzene does not. This behavior was used to induce a reversible aggregation of cyclodextrin vesicles using a divalent azobenzene guest molecule (**4**) [17]. The divalent *trans*-azobenzene forms multiple cross-links between cyclodextrin vesicles resulting in the formation of vesicles clusters (and ultimately, vesicle flocculation). However, the vesicle clusters are readily dispersed by photoisomerization to the *cis*-azobenzene. It was shown that light responsive aggregation and deaggregation can be performed by irradiation at 365 nm for deaggregation and 455 nm for reaggregation over several cycles without significant fatigue of the system.

A more sophisticated system was investigated in 2011 where a competitive ternary system was used to induce a specific recognition between cyclodextrin vesicles [18]. In this system α-cyclodextrin vesicles were mixed with divalent guest molecules (**5**) with a rather weak affinity to cyclodextrins. The formation of multiple cross-links leads to aggregation of vesicles, which however can be dispersed using a *trans*-azobenzene competitor (**6**) since it binds stronger to the cyclodextrin host (Figure 8.6). Under UV light irradiation the *cis*-azobenzene is released from the cavity and a reaggregation was observed due to the presence of divalent guest. This system nicely shows the interplay and competition between weak and strong binding partners in a multivalent membrane mimetic system. An elegant combination of different stimuli in a supramolecular system was described by Samanta and Ravoo using guest molecules with light, redox and metal ion sensitive properties (**7**) to show an orthogonal switching between aggregated and deaggregated cyclodextrin vesicles due to stimulus responsive multivalent cross-linking of the cyclodextrin vesicles [19]. It was shown that all three possible stimuli can be triggered in an orthogonal way giving a multiresponsive heteroternary system.

Jin *et al.* applied the host–guest recognition between cyclodextrins and azobenzenes on branched polymersomes [20] (Figure 8.7). To this end two different types of

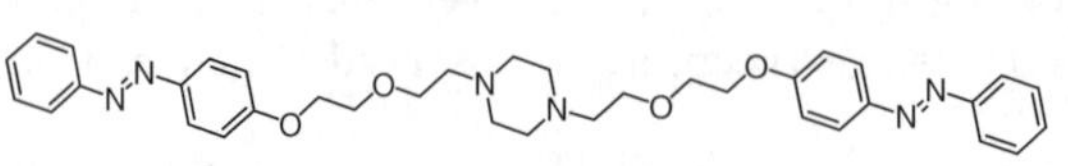

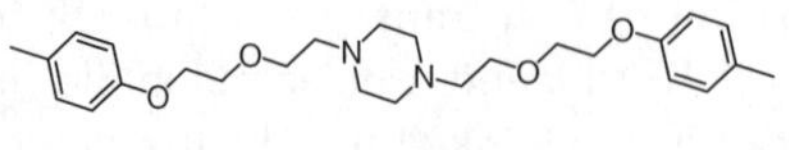

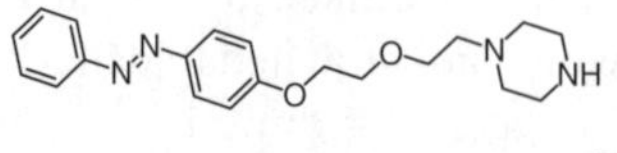

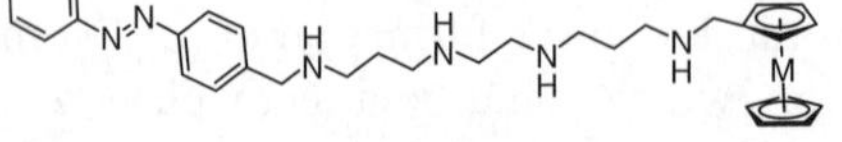

Figure 8.6 Light responsive guest molecules for the molecular recognition of cyclodextrin vesicles.

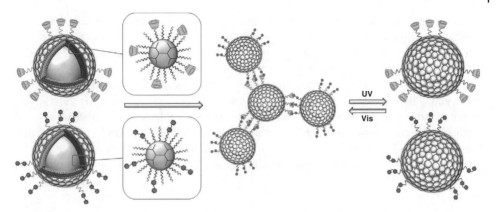

Figure 8.7 Molecular recognition of complementary polymersomes bearing either cyclodextrins or azobenzenes and UV light induced disaggregation.

hyperbranched polymers were synthesized bearing cyclodextrins or azobenzenes which assemble into large polymersomes. By mixing these two types, large aggregates of polymersomes were obtained which can be dispersed under UV light irradiation. The light-responsive clustering of the polymersomes can be attributed to the light-responsive interaction of the cyclodextrin and azobenzene recognition units.

In an alternative light-responsive system, phtalocyanines were immobilized on cyclodextrin vesicles. Phthalocyanines are well known dyes which act as potent photo-sensitizers to produce highly reactive necrotic singlet oxygen under red light irradiation. In 2012 Voskuhl *et al.* reported the immobilization of a water soluble phthalocyanine on cyclodextrin vesicles [21]. To this end, the phthalocyanine was functionalized with two adamantane units. Due to the fact that the adamantanes are positioned in close proximity on the phthalocyanine, exclusively intravesicular binding to the cyclodextrin vesicles was observed. Moreover, the divalent immobilization and the negative charge of the phthalocyanine prevent the aggregation of the dye on the vesicle surface, leading to significantly enhanced production of singlet oxygen.

8.2.3 Hydrogen Bonding and Electrostatic Interactions

A classic binding motif based on hydrogen bonding in combination with electrostatic interaction is the guanidinium–phosphate interaction. Paleos and coworkers made several contributions in which this motif was used to merge and fuse liposomes. Poly(arginine) coated vesicles aggregate and fuse with dihexadecylphosphate bearing liposomes, which was demonstrated by a calcein leakage assay [22]. More recently, Pons and coworkers showed that arginine-based surfactants form stable cationic vesicles in aqueous media and are able to interact with anionic phosphate-based vesicles under the formation of catanionic hybrid vesicles based on the combination of electrostatic interactions and hydrogen bonding between the guanidinium moiety and the phosphate residue [23]. Bong and coworkers used the well-known binding between cyanuric acid and melamine, which is able to form heterodimers via three hydrogen bonds, to form large vesicle aggregates. To this end two novel amphiphiles containing these binding motifs were synthesized. More recently the same group synthesized a trivalent variation of these amphiphiles which are able to fuse liposomes instead of docking based on multivalent interaction between the complementary liposomes. It was proposed that the formation of hydrogen bonding driven rosettes induce the stalk opening between the liposomes which is a crucial part for membrane fusion [24].

The combination of peptidic binding motifs such as β-sheet forming "leucine zippers" (Leu-Glu)$_4$ in combination with multivalent molecular recognition via host–guest complexation was described by Versluis *et al.* in 2009 [25]. When guest **8** was mixed with cyclodextrin vesicles at pH 5.0 the attached guest molecules underwent a formation of intervesicular β-sheets leading to a remarkable transition of the bilayer vesicles into nanotubes (Figure 8.8). This morphological change is fully reversible upon pH

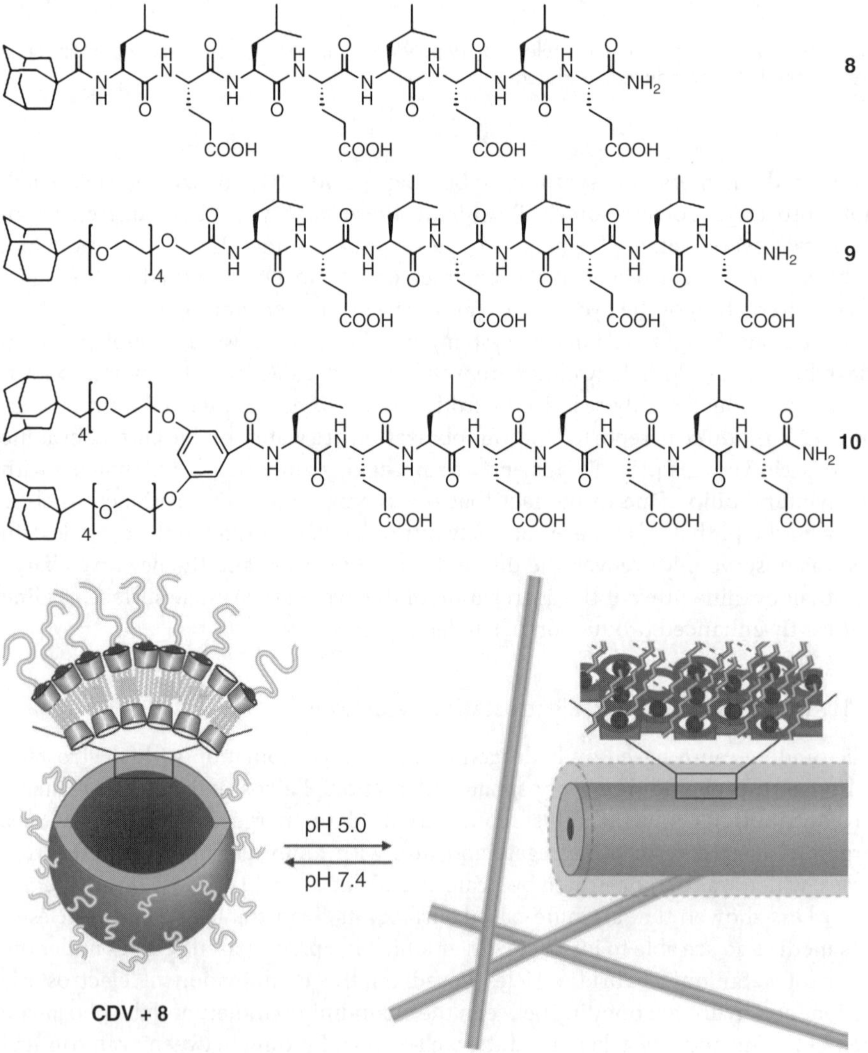

Figure 8.8 Adamantane-leucine zipper conjugates for pH responsive morphological changes of cyclodextrin vesicles and schematic presentation of the morphological change of cyclodextrin vesicles. *Source*: Ref. [25]. Reproduced with permission of American Chemical Society.

increase since the glutamic acid groups are partially protonated and electrostatic repulsion under formation of random coils is achieved leading to a reassembly of the vesicles. A follow up study by Versluis *et al.* used variations of guest **8** to get a deeper insight in the influence of binding valency and spacer length on the morphology [26]. It was found that increasing the spacer length leads to bundled fibres (**9**) whereas a divalent guest (**10**) disrupts the vesicles and gives wormlike micelles.

Another approach to pH-sensitive supramolecular systems was described by Voskuhl *et al.* in 2011 [27]. In this study a fully synthetic system was based on a guanidiniumcarbonylpyrrole carboxylate in combination with host–guest interactions between cyclodextrin vesicles and an adamantane moiety (Figure 8.9). The guest (**11**) exists as a monomeric species at higher or lower pH values since here a charged form is predominant and repulsive forces hinder the molecules from dimerizing, whereas at neutral pH guest **11** forms dimers with a weak interaction ($K_a = 10^2\,M^{-1}$). Due to the fact that multivalent interactions can occur between two colliding vesicles, a strong but pH dependent aggregation of the vesicles is observed.

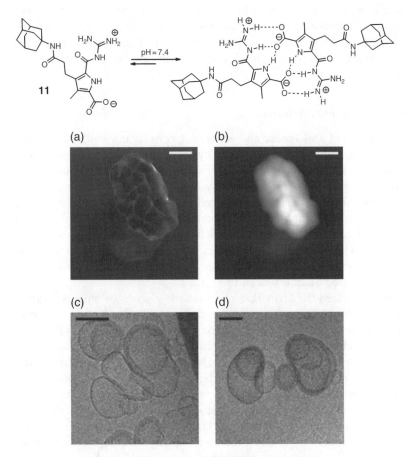

Figure 8.9 pH responsive adamantane-substituted guanidiniumpyrroles and the pH dependent dimerization as well as the corresponding atomic force microscopy (a and b) and cryo-transmission electron microscopy (cryo-TEM; c and d) images of the aggregates. *Source*: Ref. [27]. Reproduced with permission of John Wiley and Sons. See color section.

8.3 Biomimetic Vesicles

The construction of well-defined, self-assembled systems with adaptive properties comparable with living organisms is a prime challenge within supramolecular chemistry. This biomimetic research area benefits from the application of biological components and principles in synthetic materials and medicine, and at the same time aids the understanding of complex assembly mechanisms found throughout nature. As outlined above, vesicles occur in all types of cells and synthetic vesicles formed by synthetic amphiphiles are biomimetic nanostructures in their own right. Moreover, vesicles are also often used as platforms to investigate biomimetic processes. This is not surprising, since a large number and variety of complex proteins and carbohydrates are found embedded within or confined at the surface of biological lipid membranes to regulate a multitude of recognition, transport and signaling events that are critical to living cells and organisms. Since vesicles are prepared rather easily and many functional molecules can easily be incorporated within the membrane or the internal compartment of the vesicles, biological processes are often studied with vesicles. In this section, we will describe a number of examples of biomimetic vesicles. On one hand, we will look at vesicles as model platforms for multivalent interactions. On the other hand, we will discuss how multivalent recognition can lead to membrane fusion. Throughout this section, we will focus on biomimetic recognition motifs involving carbohydrates, peptides, proteins, and nucleic acids.

8.3.1 Vesicles as Multivalent Platforms

The extracellular surfaces of prokaryotic and eukaryotic cells are covered with a dense layer of complex carbohydrates, called the glycocalyx. The glycocalyx has many functions ranging from the simple protection and stabilization of the plasma membranes to complex tasks like the regulation of the immune system. One very important property of the glycocalyx is the adhesion of cells through cell–cell and cell–matrix interactions of the carbohydrates. An important family of proteins that bind to carbohydrates are called lectins. The binding constants for the interaction of a carbohydrate with a single lectin binding site are in the range of 10^3–$10^4 M^{-1}$. Lectins usually have more than one binding site for carbohydrates, so that multivalency strongly enhances their overall affinity to the glycocalyx (see Chapters 13 and 14).

The Ravoo group described the formation of a versatile glycocalyx on cyclodextrin vesicles [28]. Adamantane-functionalized carbohydrates (including lactose, maltose, and mannose) were added to a solution of cyclodextrin vesicles (Figure 8.10). The adamantanes bind to the cyclodextrin cavities, while the carbohydrates are exposed to the water phase. It was demonstrated by using dynamic light scattering that the vesicles stay intact while effective binding to the lectins could be shown via an agglutination assay. The addition of lectins resulted in multivalent intervesicular binding and aggregation of the vesicles. This process was followed by the increase of the optical density of the solution. It was shown that these binding processes are highly selective: agglutination occurs exclusively if lectins are added to vesicles decorated with a "matching" carbohydrate that binds to that particular lectin [e.g., concanavalin A (ConA) binds to mannose-decorated vesicles while peanut agglutinin (PNA) binds to lactose-decorated vesicles].

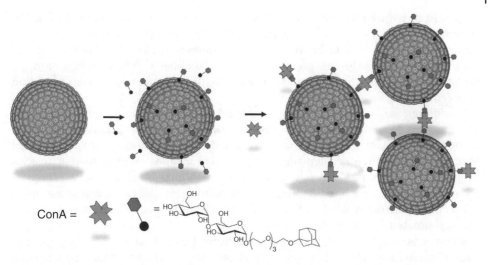

Figure 8.10 Artificial glycocalyx formed by decoration of cyclodextrin vesicles with glycoside-conjugated adamantanes. *Source*: Ref. [28]. Reproduced with permission of John Wiley and Sons.

Also, it was shown that agglutination is determined by selective molecular recognition: if excess of cyclodextrin (competing with the cyclodextrin vesicles) or excess of carbohydrate (competing with the adamantine-carbohydrate conjugate) is added, the agglutination is completely reversed.

In a more detailed study of the binding process between cyclodextrin vesicles, adamantane containing carbohydrate ligands and lectins, Ravoo and coworkers investigated the influence of valency on vesicle agglutination [29]. To this end different ratios of maltose and lactose appended guests were immobilized on cyclodextrin vesicles and the rate and extent of vesicle clustering in the presence of either ConA or PNA was investigated (Figure 8.11). It was observed that at least a divalent binding towards lectins

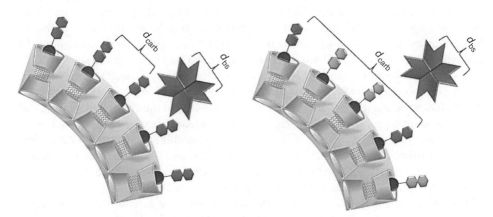

Figure 8.11 Relationship between distance of the binding sites (d_{bs}) and the distance of carbohydrate ligands (d_{carb}) on the surface of vesicles. *Source*: Ref. [29]. Reproduced with permission of American Chemical Society.

is needed to induce efficient binding since a single carbohydrate binding to a lectin pocket is too weak ($K_a \sim 1000\,\mathrm{M}^{-1}$) to stabilize the vesicle cluster. The critical factor for agglutination is the distance between binding carbohydrates (d_{carb}) and the distance of the binding sites in the protein (d_{bs}). Only if the distance between binding ligands (d_{carb}) is equal to or less than the binding site distance (d_{bs}) an effective multivalent interaction between the vesicles is obtained.

In addition, the influence of carbohydrate and guest valency on the lectin binding of cyclodextrins decorated with carbohydrates was investigated [30]. To this end different carbohydrate and adamantane containing ligands were synthesized. It was observed that guests with three adamantane units and one mannose unit cross-link vesicles even in the absence of ConA based on an intermolecular binding process. The best lectin binder is a compound bearing two carbohydrates and two adamantane units which binds inherently divalent both to the vesicle surface and to lectin and induces fast agglutination at low concentrations.

Webb and coworkers used vesicles as a multivalent platform for biocatalytic glycosidation to produce sialylated liposomes that may target cells or be masked from the immune response [31]. They synthesized a pyrene-perfluoroalkyl membrane anchor that forms microdomains in phopholipid bilayers. They modified this anchor with lactose as an acceptor glycolipid to study the *Trypanosoma cruzi trans*-sialidase (TcTS) mediated transfer of sialic acid onto the lactose head group (Figure 8.12). This system mimics the biosynthesis of GM3 from glycosylceramide *in vivo* and provides vesicles with a synthetic sialylated glycocalyx.

The Ravoo group used cyclodextrin vesicles as a multivalent platform to capture biomolecules such as DNA, peptides, and proteins and facilitated a release as a response to UV light irradiation to develop the vesicles as stimulus-responsive delivery systems. Ligands that bind to biomolecules were conjugated to azobenzene units, which are suitable guest molecules for cyclodextrins. As described in the previous section, the *trans*-azobenzene that binds cyclodextrin can be isomerized with irradiation at 365 nm to the *cis*-azobenzene, which does not bind cyclodextrin. The isomerization can be reversed with irradiation at 455 nm. Thus, irradiation of cyclodextrin vesicles decorated with azobenzenes enables a switch from a multivalent, high affinity state to a monovalent, low affinity state. In this way, it was possible to trap biomolecules reversibly in clusters of cyclodextrin vesicles, which were decorated with ligands conjugated to azobenzenes. On the one hand, they utilized a modular strategy that was based on light-reversible host-guest chemistry between the azobenzene unit and the cyclodextrin vesicles and electrostatic interactions between the linker on the cyclodextrin vesicles and the biomolecules [32, 33]. In this way they demonstrated the controlled binding and release of four types of DNA and six different proteins proving the large scope of possible targets (Figure 8.13). The selectivity and reversibility was tested with optical density measurements, dynamic light scattering (DLS) and ζ-potential measurements and isothermal titration calorimetry (ITC). Using carbohydrates as ligands, it was also possible to capture and release lectins in response to light [34].

To combine the stability of liposomes and the receptor function of amphiphilic cyclodextrins in membranes, Kauscher *et al.* developed hybrid vesicles composed of lipids and cyclodextrins [35]. It was shown that lipids such as 1,2-dioleoyl-*sn*-glycero-3-phosphocholine (DOPC), 1,2-dioleoyl-*sn*-glycero-3-phosphoethanolamine (DOPE), and cholesterol can be mixed with amphiphilic cyclodextrins in every ratio. The result

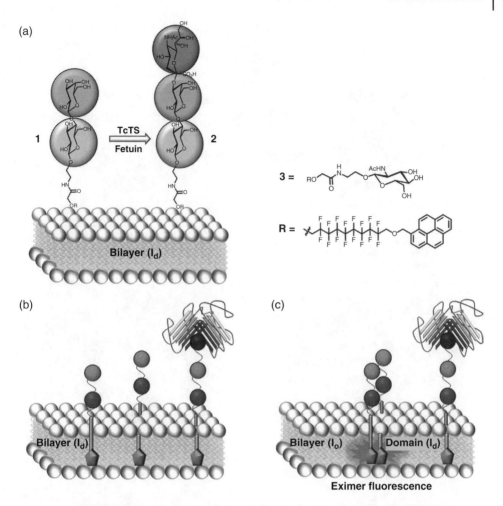

Figure 8.12 (a) Transfer of a sialic acid residue to phase separating Lac lipid **1** mediated by TcTS, producing Neu5Ac(α2-3)Gal(β1-4)Glc capped lipid **2**. GlcNAc lipid **3** is a phase-separating non-substrate control. (b) Lac lipid **1** is dispersed across the dimyristoyl phosphatidylcholine (DMPC) in a liquid ordered (l_d) bilayer (DMPC) membrane at 37 °C. (c) Microdomains are formed in liquid ordered (l_o, DMPC–cholesterol) membranes at 37 °C.

is a mixed vesicle that combines the properties of all components (Figure 8.14). Cyclodextrin vesicles have very leaky membranes and hydrophilic molecules that are encapsulated in the inner compartment will leak out over very short times. Vesicles formed by a mixture of lipids and amphiphilic cyclodextrins were demonstrated to have much less permeable membranes that are able to retain cargo inside the vesicle compartments. Moreover, the receptor function of the amphiphilic cyclodextrins is retained in the membranes, even if only a very small amount (10 mol%) is mixed with lipids in the membranes. Using a fluorescent guest, it could be shown that the cyclodextrins are distributed homogeneously in the lipid membrane (Figure 8.14). In this way a new type of biomimetic vesicle was obtained, which can recognize molecules on their external membrane and transport molecules effectively.

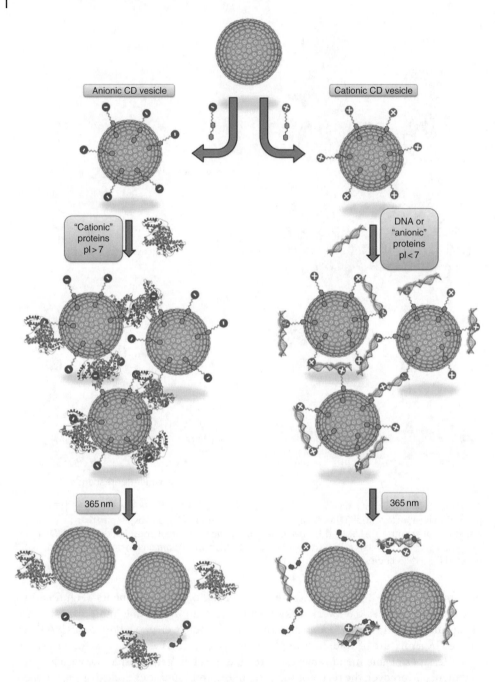

Figure 8.13 Light-responsive capture and release of DNA and proteins from a multivalent ternary complex with cyclodextrin vesicles and azobenzene conjugated to polyvalent cations or anions. Photoisomerization of the azobenzene results in a switch from a multivalent, high affinity state to a monovalent, low affinity state.

(a)

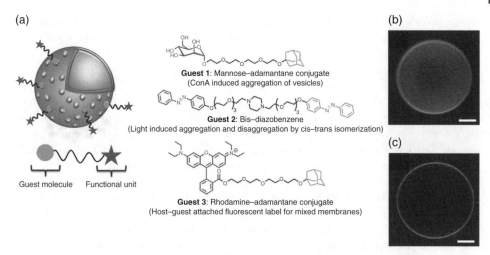

(b)

(c)

Figure 8.14 (a) Multivalent host–guest interaction of mixed vesicles of cyclodextrins and phospholipids. (b) and (c) Giant unilamellar vesicles prepared from a lipid film containing 30% cyclodextrin. Fluorescent staining was obtained by the addition of guest **3**. (b) Confocal laser scanning microscopy (CLSM) image of a reconstructed three-dimensional image of a vesicle hemisphere from confocal image sections. (c) CLSM single image slice through the vesicle equator. *Source*: Ref. [35]. Reproduced with permission of American Chemical Society. See color section.

8.3.2 Membrane Fusion

Fusion is a biological process, which is vital for the transport of hydrophilic molecules across membranes. Processes like the release of hormones into the bloodstream or the release of digestive enzymes into the gastrointestinal tract require the fusion of membranes [36]. Fusion describes the process in which two membranes are coming into close proximity so that they join in the end while the content of the former two compartments are mixed. This process is entirely dependent on the multivalent interaction between recognition units embedded in the membrane. The SNARE complex is a well-known protein based regulation system for fusion processes between synaptic vesicles [37]. The proteins anchored in different membranes interact via their helices, which fold around each other and result in a supercoiled structure [38]. These supercoiled structures form via the multivalent interaction of at least two or more amino acid strands [39]. The specificity of this binding process is determined by the amino acid sequence of the peptide strands. The majority of coiled-coil forming peptides are characterized by heptad repeats (a-b-c-d-e-f-g) in which apolar amino acids occur at most of the a and d positions, while charged amino acids are positioned at e and g. This results in an amphiphilic helix. To facilitate fusion this coil–coil interaction needs to overcome the initial energy barrier of the membrane fusion [35].

Tareste and coworkers explored the highly dynamic association process of SNARE folding during the fusion process between bilayer membranes [40]. They found that SNARE folding starts when the opponent liposome membranes are 8 nm apart. The membranes are then pulled together, storing an increasing amount of energy within the SNAREpin fold until a plateau is reached at about 4 nm membrane distance. This is followed by the unstructuring of the SNAREpin forming a strong membrane bridging complex.

Kros and coworkers designed two oligopeptide hybrids which include all important functional properties of membrane bound SNARE proteins [41]. Their recognition motif consists of three heptad repeats and is connected by a flexible poly(ethylene glycol) (PEG) chain to a phospholipid tail, which anchors the mimicry protein into the membrane of liposomes. They then showed that their reduced SNARE model with the shortest known hetero coiled-coil sequence meets all criteria of a natural SNARE protein and allows for selective membrane fusion between liposomes.

The thermodynamics and kinetics of the coiled-coil formation of the peptide recognition motifs used by Kros and coworkers were investigated by Janshoff and coworkers [42]. They connected the peptide sequences via a cysteine linker to maleimide-functionalized lipids which were incorporated into lipid bilayers. They also bound the inverted peptide sequences to these lipid anchors to compare these peptides with different superhelical macrodipoles and a predominantly parallel or antiparallel orientation. They found that a parallel coiled-coil formation is needed for significant lipid mixing and attributed this result to the difference in proximity needed to overcome the hydration barrier.

Light-responsive biomimetic fusion was recently reported by Kong [43]. A steric shielding and rapid, photoinduced deshielding of complementary fusogenic peptides was achieved on the surface of liposomal membranes. A light-degradable PEG linker was added to liposomes via a cholesterol anchor and covered the fusogenic peptides. UV light triggered a cleavage between the PEG and its cholesterol anchor leading to the degradation of the shield around the fusogenic peptides. As a result, liposome fusion was induced. In a more biological context, they also showed that this mechanism leads to an accumulation of liposomes at prefunctionalized cellular membranes.

The coiled-coil driven fusion between natural phospholipids and synthetic cyclodextrin vesicles has also been investigated [44]. A pair of fusogenic coiled-coil forming peptides was used for this study. One peptide was modified with a cholesterol anchor to bind into lipid membranes while the other was conjugated to an adamantane, which can bind to the hydrophobic cavity of cyclodextrins (Figure 8.15). Both anchors were separated from their functional peptide parts via a PEG spacer to verify the effective binding without any repulsive strains. When the cyclodextrin vesicles decorated with adamantane-modified peptide **K** (APK) and the liposomes decorated with peptide **E** (CPE) were added to each other a fusion process occurred. The different stages of the fusion process were studied using circular dichroism spectroscopy, light scattering, lipid and content mixing assays as well as cryo-electron tomography. The fusion of lipid and non-lipid vesicles had to that date not been reported in the literature and thus provided new valuable detailed information regarding the fusion processes that happen in nature. An increase of the vesicle sizes from ca. 190 nm to 900 nm was recorded via dynamic light scattering 10 min after the fusion process was allowed to start by mixing all components together. Content mixing assays, as well as cryo-TEM and cryo-electron tomography verified a successful fusion process in which hybrid lipid cyclodextrin vesicles are formed. However, lipid mixing assays showed that it is not a full fusion that occurs but an outer leaflet mixing, called hemi-fusion. This is shown to have two reasons: (1) the peptide **K** does not interact strongly enough with a cyclodextrin

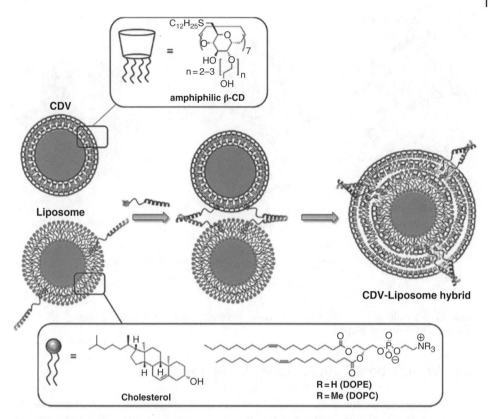

Figure 8.15 Fusion triggered by multivalent peptide ligands between cyclodextrin vesicles and liposomes. *Source*: Ref. [44]. Reproduced with permission of Royal Society of Chemistry.

membrane; and (2) the curvature of cyclodextrin vesicles is close to zero due to the macrocyclic structure of cyclodextrin amphiphiles, which is known to impede full fusion. The study showed that it was possible to prepare complex hybrid cyclodextrin lipid vesicles using a SNARE protein mimic.

Kimoon Kim and coworkers have carried out further studies on SNARE proteins by single vesicle content mixing assays [45]. They introduced a novel fluorescence resonance energy transfer (FRET) sensor that is based on the host–guest chemistry between cucurbit[7]uril (CB[7]) and adamantylamine (Ad) (Figure 8.16). The sensor was then used to monitor content mixing assays. To this end CB[7] was conjugated with a donor dye Cy3 and Ad was conjugated with an acceptor dye Cy5. The fusion of vesicles containing either CB[7]-Cy3 or Ad-Cy5 in their internal compartments results in the mixing of the content and so to the formation of the host–guest complex between CB[7] and Ad. The FRET signal then emerged, which was monitored by total internal reflection fluorescence microscopy (TIRF). With this system they managed to observe for the first time the flickering events, that is, a repetitive opening and closure of the fusion pore in an *in vitro* content mixing assay. Moreover, it was possible to measure the pore opening rate and the dwell time of the pore closure state. In this way it was possible to study short-lived events in complex processes as typically found in nature.

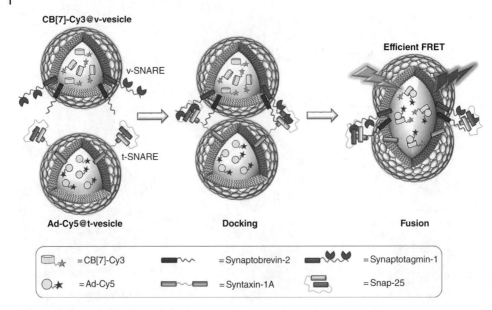

Figure 8.16 SNARE-protein induced vesicle fusion and fluorescence readout using host–guest induced FRET. See color section.

8.4 Vesicle-based Supramolecular Materials

A novel approach for the use of vesicles involves their use as multivalent platforms and dynamic building blocks for the construction of soft nanomaterials. Since the production of functional vesicles from various amphiphiles is rather straightforward, this strategy opens a new field for the assembly of novel soft materials such as hydrogels, immobilized vesicles, polymer capsules, and supported bilayers. In this section we will highlight how multivalent interactions can aid the design and construction of supramolecular materials based on vesicles.

8.4.1 Hydrogels

Hao *et al.* described temperature responsive chitosan-based vesicle hydrogels [46]. To this end chitosan was modified with n-dodecyl chains which can interact with vesicles composed of n-dodecyltrimethyl ammonium bromide (DTAB) and 5-methyl salicylic acid (MSA) via insertion of the hydrophobic alkyl chains in the bilayer (Figure 8.17). Due to the multivalent cross-linking of the polymer and the vesicles, a hydrogel results that exhibits excellent stability at room temperature. However, an increase in temperature transforms the hydrogel into a solution due to a phase transition of DTAB/MSA vesicles into wormlike micelles. A hydrogel is obtained if the temperature is decreased. The sol–gel transition can be repeated for several cycles and the gelation temperature can be controlled by variation of the ratio of DTAB and MSA in the vesicles.

A recent example for the use of cyclodextrin vesicles as cross-linker in a hydrogel was reported by Himmelein *et al.* [47]. When an adamantane-modified hydroxyethylcellulose (HEC) is mixed with cyclodextrin vesicles in dilute aqueous solution (<2% organic material)

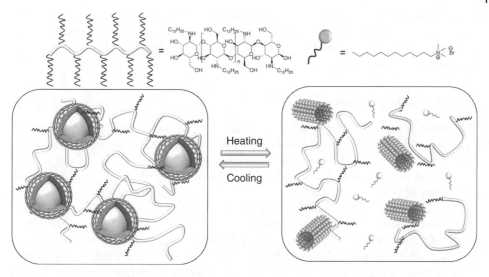

Figure 8.17 Formation of a hydrogel by multivalent hydrophobic anchoring of n-dodecyl substituted chitosan into vesicles composed of DTAB and MSA.

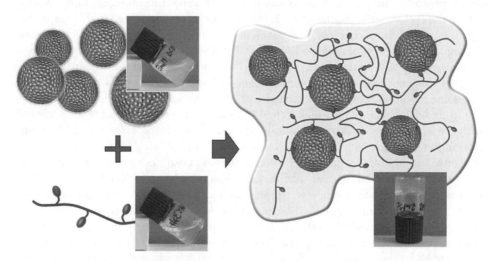

Figure 8.18 Supramolecular hydrogel formed by multivalent interaction of cyclodextrin host vesicles and adamantane guest polymers.

a spontaneous gelation process is observed due to the multivalent interaction of the adamantanes on the HEC and the cyclodextrins in the vesicles (Figure 8.18). The vesicles remain intact upon gelation which was shown by cryo-TEM and SAXS. Furthermore, it was observed that this hydrogel dissolves upon addition of competitive guest molecules such as adamantane. The gel is shear-thinning, which can be explained by a rupture of the host–guest complexes under shear stress. Interestingly, the gel is restored nearly instantaneously when shear is relieved, indication that this supramolecular gel is an injectable gel. A follow up study by Lange *et al.* [48] described an Fmoc-protected tripeptide sequence with an adamantane moiety which is readily available for inclusion at the surface of cyclodextrin vesicles. When these components were mixed a hydrogel was obtained due to multivalent interaction of the adamantanes on the self-assembled peptide scaffold and the cyclodextrins in the vesicles.

8.4.2 Immobilization of Vesicles

The immobilization of intact vesicles on solid supports is a challenging task since these highly dynamic structures tend to rupture on surfaces. Several approaches to stabilize vesicles on surfaces have been conducted. The multivalent recognition of complementary binding motifs on the vesicle and on the substrate is a mild and specific approach to immobilize vesicles.

Very recently a novel approach for the immobilization of liposomes and vesicles using multivalent coiled-coil peptide interactions was described. When the well-known binding motif (KIAALKE)$_3$ was attached to a solid support such as gold or silicon via click chemistry on a self-assembled monolayer (SAM), liposomes bearing the complementary binding motif (EIAALEK)$_3$ via a membrane anchor (cholesterol) were able to bind to the surface without rupture [49]. This was demonstrated by dye encapsulation in the liposomes. In another experiment cyclodextrin vesicles instead of liposomes were used. In this case adamantane was used to attach the peptide binding motif in a non-covalent way to the cyclodextrin vesicles. Even in this case intact vesicles were immobilized. This example shows how multiple non-covalent supramolecular weak interactions can orthogonally be assembled on solid supports. Another approach was recently described by Roling *et al.* [50]. In that case, either mannose or biotin was immobilized on a solid support using click chemistry in combination with microcontact printing. To this end, an alkene modified SAM was patterned with mannose or biotin thiols using a PDMS stamp under UV light irradiation. When the mannose-terminated SAM is incubated with the lectin ConA an immobilization of proteins is observed based on a multivalent interaction of ConA with mannose. Since some of the binding sites on ConA (most likely two out of four) remain unoccupied a layer of liposomes containing amphiphilic cyclodextrins decorated with an adamantane-mannose conjugate can be deposited on the ConA adlayer. In the same way, streptavidin can be deposited on a biotin-modified substrate, and liposomes contained cyclodextrins decorated with an adamantane-biotin conjugate can be deposited on the streptavidin layer. Interestingly, the supramolecular immobilization of protein and vesicles can be repeated several times leading to intact vesicle multilayers (Figure 8.19). However, stable multilayers are obtained only in the case of the biotin-streptavidin pair.

8.4.3 Nanoparticles and Nanocontainers

An innovative approach to the preparation of tailor made nanocontainers is the use of vesicles and liposomes as template for polymer capsules. A first example was reported by Lee *et al.*[51] who used a cholesterol-modified polyacrylate. Upon insertion into the bilayer membrane, a cross-linking of the polyacrylate shell was induced by addition of a diamine in the presence of a peptide coupling agent. The liposomes entrapped in a multivalent polyacrylate shell have a higher stability and lower permeability than the template liposomes. A recent paper by Samanta *et al.* [52] used surface immobilized

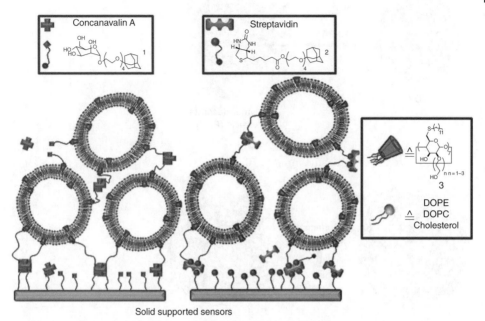

Figure 8.19 Layer-by-layer deposition of vesicle multilayers mediated by multivalent interactions. Cyclodextrin vesicles decorated with guest molecules, either **1** or **2**, display a high density of mannose or biotin, respectively, on their surface. Addition of concanavalin A or streptavidin, respectively, leads to the formation of non-covalent intervesicular links. If applied in succession on a suitable surface, vesicle multilayers can be obtained by layer-by-layer deposition of vesicles and protein. *Source*: Ref. [50]. Reproduced with permission of American Chemical Society.

adamantane-terminated polyacrylates on cyclodextrin vesicles which could be cross-linked by diamines (C) leading to a robust polymer shell which is stable even after removal of the cyclodextrin vesicle template (Figure 8.20). The size of the resulting polymer capsule is determined by the size of the template vesicle, whereas the thickness of the polymer shell is determined by the length of the polyacrylate. Also in this case it was found that the polymer-shelled vesicle has a higher stability and a much lower permeability than the template vesicle.

Finally, an interesting example of multivalent interactions of vesicles was reported by Schenkel *et al.* [53] when superparamagnetic nanoparticles (MNP) were incorporated into the bilayer membranes of cyclodextrin vesicles (CDV). These vesicles can be manipulated by an external magnetic field to form linear aggregates parallel to the field lines of the external field (Figure 8.21). The linear aggregates instantaneously disassemble when the magnetic field is switched off. However, multivalent cross-linking with a divalent azobenzene linker (Figure 8.6) that binds to the cyclodextrins on the surface of the vesicles results in stabilization of the linear aggregates even when the magnetic field is switched off. However, since only *trans*-azobenzenes (not *cis*-azobenzenes) bind to the cyclodextrin vesicles, the linear aggregates disassemble upon UV light irradiation. Thus, a unique self-organization process results from the alignment of superparamagnetic nanoparticles in a magnetic field and light-responsive multivalent interaction of cyclodextrins and azobenzenes.

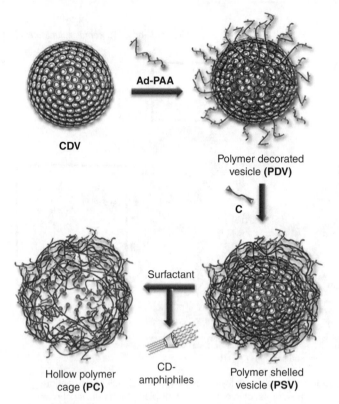

Figure 8.20 Polymer capsules formed by self-assembly and cross-linking of polyacrylate on a cyclodextrin vesicle. *Source*: Ref. [52]. Reproduced with permission of American Chemical Society.

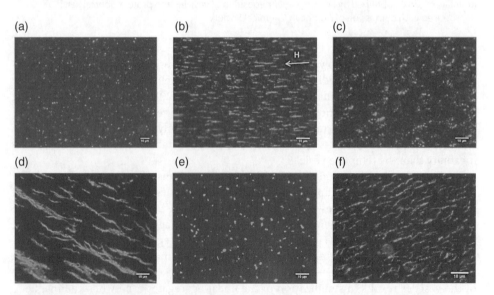

Figure 8.21 Fluorescence microscopy images of sulforhodamine B labeled magnetic vesicles. (a) MNP-CDV in the absence of an external magnetic field. (b) Formation of linear aggregates of MNP-CDV in a magnetic field. (c) Spontaneous disassembly of linear aggregates of MNP-CDV after switching off the magnetic field. (d) Linear aggregates of MNP-CDV formed in an external magnetic field and stabilized by the cross-linker (**4**). The aggregates deform but persist after switching off the external field. (e) Disaggregation of MNP-CDV and *cis*-**4** after UV irradiation for 30 min in the absence of an external magnetic field. (f) Re-aggregation of MNP-CDV and *trans*-**4** in a magnetic field after visible light irradiation for 30 min.

8.5   Conclusion

Multivalency is a key principle in the chemical and biological properties and interactions of vesicles. Vesicles can be seen as a biomimetic multivalent platform equipped with either synthetic or biological recognition units that lead to multivalent interactions at the surface of vesicles. Such multivalent interactions can result in the selective and high affinity binding of smaller and larger molecules to vesicles as well as in highly specific binding between vesicles. In some cases, the interaction responds to external stimuli such as metal ions, pH or light. In other cases, tight multivalent interactions can result in fusion of vesicles, similar to fusion of biological membranes. Multivalent interactions of vesicles can also give rise to soft materials such as hydrogels or nanocontainers. In all cases, the density and dynamics of the binding partners is critical to high affinity, multivalent interaction.

Acknowledgment

Financial support from the DFG EXC 1003 *Cells in Motion–Cluster of Excellence* at the Westfälische Wilhelms-Universität Münster (Germany) is gratefully acknowledged.

References

1 B. J. Ravoo, Vesicles in supramolecular chemisty. In: Supramolecular Chemistry: from Molecules to Nanomaterials (Eds P. A. Gale and J. W. Steed), Wiley-VCH, 2012, Vol. 2, p. 501.

2 B. J. Ravoo, R. Darcy, Cyclodextrin bilayer vesicles. Angew. Chem. Int. Ed. 2000, 39, 4324.

3 P. Falvey, C. W. Lim, R. Darcy, T. Revermann, U. Karst, M. Giesbers, A. T. M. Marcelis, A. Lazar, A. W. Coleman, D. N. Reinhoudt, B. J. Ravoo, Bilayer vesicles of amphiphilic cyclodextrins: Host membranes that recognize guest molecules. Chem. Eur. J., 2005, 11, 1171.

4 W. H. Binder, V. Barragan, F. M. Menger, Domains and rafts in lipid membranes. Angew. Chem. Int. Ed., 2003, 42, 5802.

5 E. C. Constable, W. Meier, C. Nardin, S. Mundwiler, Reversible metal-directed assembly of clusters of vesicles. Chem. Commun., 1999, 1483.

6 V. Marchi-Artzner, M. J. Brienne, T. Gulik-Krzywicki, J. C. Dedieu, J. M. Lehn, Selective complexation and transport of europium ions at the interface of vesicles. Chem. Eur. J., 2004, 10, 2342.

7 C. W. Lim, O. Crespo-Biel, M. C. A. Stuart, D. N. Reinhoudt, J. Huskens, B. J. Ravoo, Intravesicular and intervesicular interaction by orthogonal multivalent host–guest and metal–ligand complexation. Proc. Natl Acad. Sci. USA, 2007, 104, 6986.

8 S. K. M. Nalluri, J. B. Bultema, E. J. Boekema, B. J. Ravoo, Metal ion responsive adhesion of vesicles by conformational switching of a non-covalent linker. Chem. Sci. 2011, 2, 2383.

9 S. J. Webb, L. Trembleau, R. J. Mart and X. Wang, Membrane composition determines the fate of aggregated vesicles. Org. Biomol. Chem. 2005, 3, 3615.

10 X. Wang, R. J. Mart, S. J. Webb, Vesicle aggregation by multivalent ligands: relating crosslinking ability to surface affinity. Org. Biomol. Chem. 2007, 5, 2498.

11 R. J. Mart, K. P. Liem, X. Wang, S. J. Webb, The effect of receptor clustering on vesicle – vesicle adhesion. J. Am. Chem. Soc. 2006, 128, 14462.

12 S. Turkyilmaz, D.R. Rice, R. Palumbo, B. D. Smith, Selective recognition of anionic cell membranes using targeted liposomes coated with zinc(II)-bis(dipicolylamine) affinity units. Org. Biomol. Chem., 2014, 5645.

13 B. Gruber, E. Kataev, J. Aschenbrenner, S. Stadlbauer, B. König, Vesicles and micelles from amphiphilic zinc(II)–cyclen complexes as highly potent promoters of hydrolytic DNA cleavage. J. Am. Chem. Soc. 2011, 133, 20704.

14 B. Gruber, S. Stadlbauer, K. Woinaroschy, B. König, Luminescent vesicular receptors for the recognition of biologically important phosphate species. Org. Biomol. Chem. 2010, 8, 3704.

15 S. Tomas, L. Milanesi, Mutual modulation between membrane-embedded receptor clustering and ligand binding in lipid membranes. Nat. Chem. 2010, 2, 1077.

16 H. Dijkstra, J. Hutchinson, C. Hunter, H. Qin, S. Tomas, S. Webb, N. Williams, Transmission of binding information across lipid bilayers. Chem. Eur. J. 2007, 13, 7215.

17 S. K. M. Nalluri, B. J. Ravoo, Light-responsive molecular recognition and adhesion of vesicles. Angew. Chem. Int. Ed. 2010, 49, 31.

18 S. K. M. Nalluri, J. B. Bultema, E. J. Boekema, B. J. Ravoo, Photoresponsive molecular recognition and adhesion of vesicles in a competitive ternary supramolecular system. Chem. Eur. J. 2011, 17, 10297.

19 A. Samanta, B. J. Ravoo, Metal ion, light, and redox responsive interaction of vesicles by a supramolecular switch. Chem. Eur. J. 2014, 20, 4966.

20 H. Jin, Y. Zheng, Y. Liu, H. Cheng, Y. Zhou, D. Yan, Reversible and large-scale cytomimetic vesicle aggregation: light-responsive host–guest interactions. Angew. Chem. Int. Ed. 2011, 50, 10352.

21 J. Voskuhl, U. Kauscher, M. Gruener, H. Frisch, B. Wibbeling, C. A. Strassert, B. J. Ravoo, A soft supramolecular carrier with enhanced singlet oxygen photosensitizing properties. Soft Matter 2013, 9, 8, 2453.

22 I. Tsogas, D. Tsiourvas, G. Nounesis, C. M. Paleos, Modeling cell membrane transport: Interaction of guanidinylated poly(propylene imine) dendrimers with a liposomal membrane consisting of phosphate-based lipids. Langmuir 2006, 22, 11323.

23 N. Lozano, A. Pinazo, C. La Mesa, L. Perez, P. Andreozzi, R. Pons, Catanionic vesicles formed with arginine-based surfactants and 1,2-dipalmitoyl-*sn*-glycero-3-phosphate monosodium salt. J. Phys. Chem. B 2009, 113, 6312.

24 M. Ma, Y. Gong, D. Bong, Lipid membrane adhesion and fusion driven by designed, minimally multivalent hydrogen-bonding lipids. J. Am. Chem. Soc. 2009, 131, 16919.

25 F. Versluis, I. Tomatsu, S. Kehr, C. Fregonese, A. W. J. W. Tepper, M. C. A. Stuart, B. J. Ravoo, R. I. Koning, A. Kros, Shape and release control of a peptide decorated vesicle through ph sensitive orthogonal supramolecular interactions. J. Am. Chem. Soc. 2009, 131, 13186.

26 F. Versluis, J. Voskuhl, M. C. A. Stuart, J. B. Bultema, S. Kehr, B. J. Ravoo, A. Kros, Power struggles between oligopeptides and cyclodextrin vesicles. Soft Matter 2012, 8, 33, 8770.

27 J. Voskuhl, T. Fenske, M. C. A. Stuart, B. Wibbeling, C. Schmuck, B. J. Ravoo, Molecular recognition of vesicles: Host–guest interactions combined with specific dimerization of zwitterions. Chem. Eur. J. 2010, 16, 8300.

28 J. Voskuhl, M. C. A. Stuart, B. J. Ravoo, Sugar-decorated sugar vesicles: Lectin–carbohydrate recognition at the surface of cyclodextrin vesicles. Chem. Eur. J. 2010, 16, 2790.

29 R. V. Vico, J. Voskuhl, B. J. Ravoo, Multivalent interaction of cyclodextrin vesicles, carbohydrate guests, and lectins: A kinetic investigation. Langmuir 2011, 27, 1391.

30 U. Kauscher, B. J. Ravoo, Mannose-decorated cyclodextrin vesicles: The interplay of multivalency and surface density in lectin–carbohydrate recognition. Beilstein J. Org. Chem. 2012, 8, 1543.

31 G. T. Noble, F. L. Craven, M. D. Segarra-Maset, J. E. R. Martínez, R. Sardzík, S. L. Flitsch, S. J. Webb, Sialylation of lactosyl lipids in membrane microdomains by *T. cruzi trans*-sialidase. Org. Biomol. Chem. 2014, 9272.

32 S. K. M. Nalluri, J. Voskuhl, J. L. Bultema, E. J. Boekema, B. J. Ravoo, Light-responsive capture and release of DNA in a ternary supramolecular complex. Angew. Chem. Int. Ed. 2011, 50, 9747.

33 J. Moratz, A. Samanta, J. Voskuhl, S. K. M Nalluri, B. J. Ravoo, Light-triggered capture and release of DNA and proteins by host–guest binding and electrostatic interaction. Chem. Eur. J. 2015, 21, 3271.

34 A. Samanta, M. C. A. Stuart, B. J. Ravoo, Photoresponsive capture and release of lectins in multilamellar complexes. J. Am. Chem. Soc. 2012, 134, 19909.

35 U. Kauscher, M. C. A. Stuart, P. Drücker, H. Galla, B. J. Ravoo, Incorporation of amphiphilic cyclodextrins into liposomes as artificial receptor units. Langmuir 2013, 29, 7377.

36 H. Robson Marsden, A. Kros, Self-assembly of coiled coils in synthetic biology: Inspiration and progress. Angew. Chem. Int Ed. 2010, 49, 2988.

37 N. C. Collins, H. Thordal-Christensen, V. Lipka, S. Bau, E. Kombrink, J.L. Qiu, R. Hückelhoven, M. Stein, A. Freialdenhoven, S. C. Somerville, P. Schulze-Lefert, SNARE-protein-mediated disease resistance at the plant cell wall. Nature 2003, 425, 973.

38 T. Y. Yoon, B. Okumus, F. Zhang, Y. K. Shin, T. Ha, Multiple intermediates in SNARE-induced membrane fusion. Proc. Natl Acad. Sci. USA 2006, 103, 19731.

39 L. K. Tamm, J. Crane, V. Kiessling, Membrane fusion: a structural perspective on the interplay of lipids and proteins. Curr. Opin. Struct. Biol. 2003, 13, 453.

40 F. Li, F. Pincet, E. Perez, W. S. Eng, T. J. Melia, J. E. Rothman, D. Tareste, Energetics and dynamics of SNAREpin folding across lipid bilayers. Nat. Struct. Mol. Biol. 2007, 14, 890.

41 H. Robson Marsden, N. A. Elbers, P. H. H. Bomans, N. A. J. M. Sommerdijk, A. Kros, A reduced SNARE model for membrane fusion. Angew. Chem. Int Ed. 2009, 48, 2330.

42 G. Pähler, C. Panse, U. Diederichsen, A. Janshoff, Coiled-coil formation on lipid bilayers—Implications for docking and fusion efficiency. Biophys. J. 2012, 103, 2295.

43 L. Kong, S. H. C. Askes, S. Bonnet, A. Kros, F. Campbell, Temporal control of membrane fusion through photolabile PEGylation of liposome membranes. Angew. Chem. Int. Ed. 2016, 55, 1396.

44 F. Versluis, J. Voskuhl, H. Friedrich, B.J. Ravoo, P. H. H. Bomans, M. C. A. Stuart, N. A. J. M. Sommerdijk, A. Kros, Coiled coil driven membrane fusion between cyclodextrin vesicles and liposomes. Soft Matter 2014, 10, 9746.

45 B. Gong, B.-K. Choi, J.-Y. Kim, D. Shetty, Y. H. Ko, N. Selvapalam, N. K. Lee, K. Kim, High affinity host–guest FRET pair for single-vesicle content-mixing assay: Observation of flickering fusion events. J. Am. Chem. Soc. 2015, 137, 8908.

46 X. Hao, H. Liu, Y. Xie, C. Fang, H. Yang, Thermal-responsive self-healing hydrogel based on hydrophobically modified chitosan and vesicle. Colloid. Polym. Sci. 2013, 291, 1749.

47 S. Himmelein, V. Lewe, M. C. A. Stuart, B. J. Ravoo, A carbohydrate-based hydrogel containing vesicles as responsive non-covalent cross-linkers. Chem. Sci. 2014, 5, 1054.

48 S. C. Lange, J. Unsleber, P. Drücker, H.-J. Galla, M. P. Waller, B. J. Ravoo, pH response and molecular recognition in a low molecular weight peptide hydrogel. Org. Biomol. Chem. 2015, 13, 561.

49 J. Voskuhl, C. Wendeln, F. Versluis, E. C. Fritz, O. Roling, H. Zope, C. Schulz, S. Rinnen, H. F. Arlinghaus, B. J. Ravoo, A. Kros, Immobilization of liposomes and vesicles on patterned surfaces by a peptide coiled-coil binding motif. Angew. Chem., Int. Ed. 2012, 51, 12616.

50 O. Roling, C. Wendeln, U. Kauscher, P. Seelheim, H. J. Galla, B. J. Ravoo, Layer-by-layer deposition of vesicles mediated by supramolecular interactions. Langmuir 2013, 29, 10174.

51 S. M. Lee, H. Chen, C. M. Dettmer, T. V. O'Halloran, S. T. J. Nguyen, Polymer-caged liposomes: A pH-responsive delivery system with high stability. J. Am. Chem. Soc. 2007, 129, 15096.

52 A. Samanta, M. Tesch, U. Keller, J. Klingauf, A. Studer, B. J. Ravoo, Fabrication of hydrophilic polymer nanocontainers by use of supramolecular templates. J. Am. Chem. Soc. 2015, 137, 1967.

53 J. H. Schenkel, A. Samanta, B. J. Ravoo, Self-assembly of soft hybrid materials directed by light and a magnetic field. Adv. Mater. 2014, 26, 1076.

Part III

Multivalent Systems in Biology

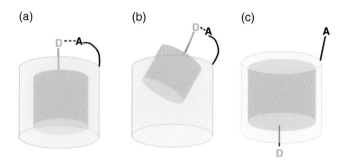

Figure 1.7 Schematic consequences of mismatch: (a) similar interaction in- and outside and sufficient matching (e.g. Case I); (b) stronger interaction outside (Case II); (c) stronger interaction inside cavity (e.g. Case III). *Source*: Ref. [40]. Reproduced with permission of Royal Society of Chemistry.

Multivalency: Concepts, Research & Applications, First Edition. Edited by Jurriaan Huskens, Leonard J. Prins, Rainer Haag, and Bart Jan Ravoo.
© 2018 John Wiley & Sons Ltd. Published 2018 by John Wiley & Sons Ltd.

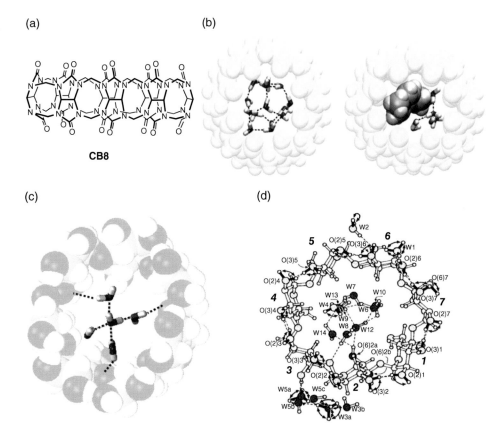

Figure 1.9 Examples of high energy water. (a) Cucurbit[8]uril (CB8) and 14 water molecules. (b) CB8 with viologen as guest and 6 water molecules in the cavity. (c) ß-Cyclodextrin with 5 water molecules, all from molecular dynamics simulations. *Source*: Ref. [48]. Reproduced with permission of John Wiley and Sons. (d) ß-Cyclodextrin dodecahydrate structure derived from neutron diffraction. *Source*: Ref. [49]. Reproduced with permission of American Chemical Society.

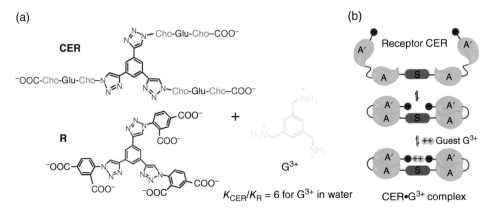

Figure 1.10 (a) Receptor CER with and R without steroidal arms, tricationic guest G^{3+}. (b) Hydrophobic interactions between the steroidal arms of CER preorganizes the CER for binding of guest G^{3+}. *Source*: Ref. [73]. Reproduced with permission of American Chemical Society.

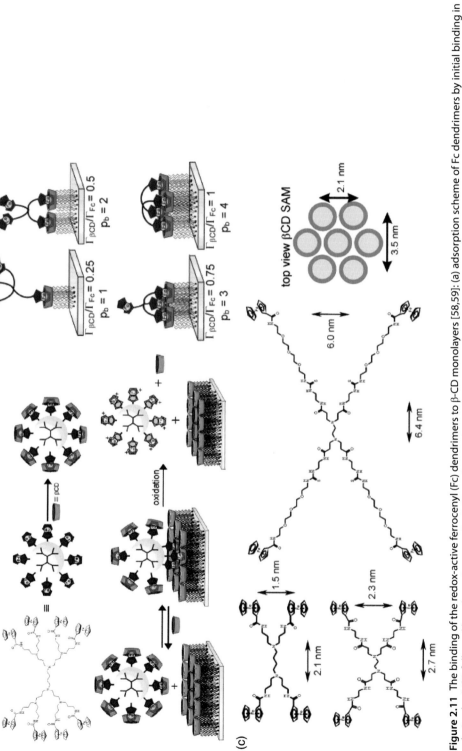

Figure 2.11 The binding of the redox-active ferrocenyl (Fc) dendrimers to β-CD monolayers [58,59]: (a) adsorption scheme of Fc dendrimers by initial binding in solution by β-CD followed by adsorption onto the β-CD surface, with subsequent oxidation-induced desorption; (b) schematic representation of the four possible binding modes of the generation-1 dendrimer with the numbers, p_b, of bound sites and the predicted coverage ratios of β-CD and Fc depicted below; (c) comparison between dimensions of the tetra-topic PPI (left top), PAMAM (left bottom), and oligo(ethylene glycol)-extended PAMAM (center) on the one hand and the β-CD lattice period (right) on the other hand, predicting the (confirmed) binding valencies of 2, 3, and 4, respectively. *Source:* Adapted from Refs [58,59]. Reproduced with permission of American Chemical Society.

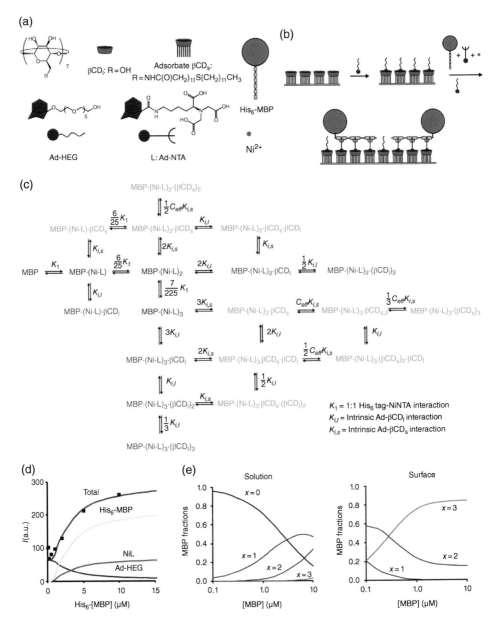

Figure 2.13 Multivalent assembly of His-tagged proteins on a receptor surface [77]. (a) Compounds used in this study (MBP: mono-His$_6$-tagged maltose binding protein). (b) Adsorption scheme for the assembly of MBP·(Ni·L)$_3$ at a β-CD SAMs in the presence of Ad-HEG. (c) Equilibria for all species (solution and surface) for the attachment of His$_6$-MBP at the molecular printboard (charges are omitted for clarity). Subsequent complexation steps of Ni·L to MBP in solution are shown in red, other solution species in blue, and all surface species are given in green. (d) Equilibrium values of the SPR intensities (markers) of a titration of MBP onto the β-CD SAM and corresponding fit to the model and contributions by different components to the signal (solid lines). (e) Speciation modeling showing fractions of His$_6$-MBP·(Ni·L)$_x$ (i.e., complexed to different numbers, x, of Ni·L) as a function of [His$_6$-MBP] in solution (left) and at the surface (right). *Source*: Adapted from Ref. [77]. Reproduced with permission of John Wiley and Sons.

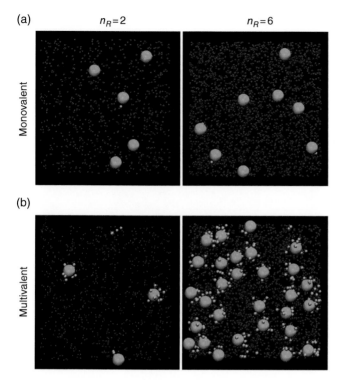

Figure 3.2 Simulation snapshots comparing the targeting selectivity of monovalent and multivalent guest nanoparticles. We compare the adsorption onto two host surfaces with receptor concentrations (n_R) that differ by a factor of three. (a) The monovalent guests provide little selectivity: increasing by three times the receptor coverage just increases the average number of bound guests by 1.8 (i.e. from 5.4 to 9.7 bound particles on average). (b) The multivalent nanoparticles behave super selectively: an increase of the receptor coverage by a factor three causes a 10-fold increase in the average number of adsorbed particles. The multivalent guests have ten ligands per particle. The individual bonds of the multivalent case (b) are weaker than those in the monovalent case (a). *Source*: Ref. [8]. Reproduced with permission of National Academy of Sciences of the United states of America.

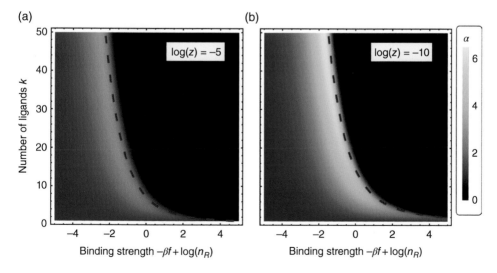

Figure 3.11 The selectivity landscape as function of the valency k and the rescaled binding strength $-\beta f + \log(n_R)$. The landscape was obtained by calculating the selectivity α using Eq. (3.16). The activity of multivalent particles was chosen as: (a) $z = \exp(-5)$; and (b) $z = \exp(-10)$. Both plots use the same colour scale. The dashed curves represent the approximate optimal selectivity relation given by Eq. (3.17), which rather accurately fits the maximum selectivity region.

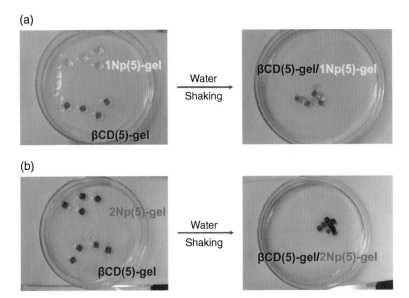

Figure 5.8 Photographs for gel assemblies formed from βCD(5)-gel/1 Np(5)-gel (a) and βCD(5)-gel/2 Np(5)-gel (b). *Source*: Ref. [59]. Reproduced with permission of American Chemical Society.

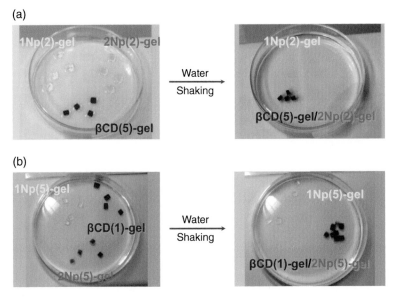

Figure 5.9 Photographs for gel assemblies formed from βCD(5)-gel/2 Np(2)-gel in the presence of 1 Np(2)-gel (a) and βCD(1)-gel/2 Np(5)-gel in the presence of 1 Np(5)-gel (b). *Source*: Ref. [59]. Reproduced with permission of American Chemical Society.

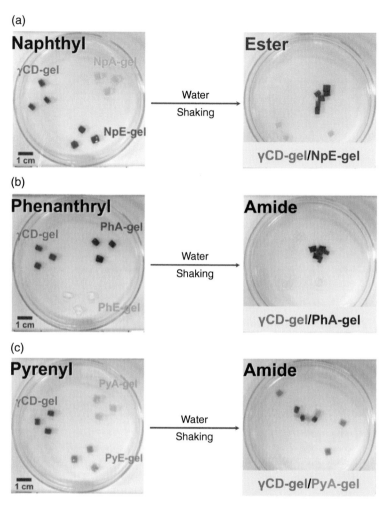

Figure 5.12 Photographs of the interaction of γCD-gel (dark green) with NpA-gel (green) and NpE-gel (blue) (a), with PhA-gel (purple) and PhE-gel (yellow) (b), and with PyA-gel (yellow) and PyE-gel (orange) (c). *Source*: Ref. [60]. Reproduced with permission of American Chemical Society.

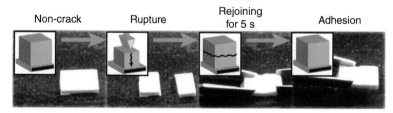

Figure 5.14 Photographs of a self-healing experiment for βCD-Ad gel (7,6). *Source*: Ref. [67]. Reproduced with permission of John Wiley and Sons.

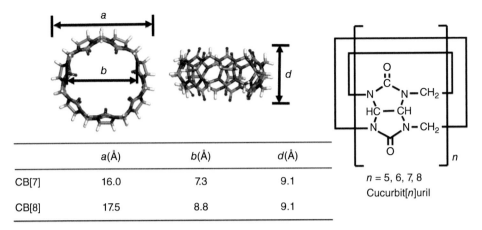

	a(Å)	b(Å)	d(Å)
CB[7]	16.0	7.3	9.1
CB[8]	17.5	8.8	9.1

n = 5, 6, 7, 8
Cucurbit[n]uril

Figure 6.1 Molecular structures and detailed information of cucurbit[n]urils.

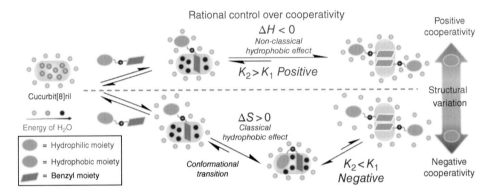

Figure 6.8 Schematic diagram of design principles on rationally controlling cooperativity of CB[8]-mediated π–π interactions by tuning classical and non-classical hydrophobic effects (blue balls represent bulk water, while orange and red balls represent high-energy water; the energy of the red balls is higher than that of the orange ones).

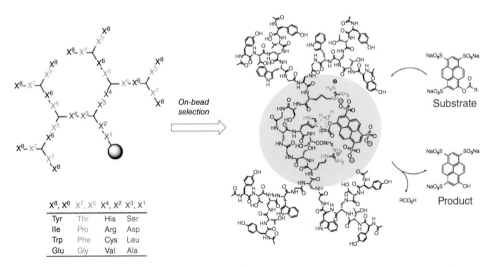

X8, X6	X7, X5	X4, X2	X3, X1
Tyr	Thr	His	Ser
Ile	Pro	Arg	Asp
Trp	Phe	Cys	Leu
Glu	Gly	Val	Ala

Figure 7.1 On-bead selection of a catalytic dendrimer from a combinatorial library of 65 536 different dendrimers. The yellow circle highlights the reactive core with a substrate. *Source*: Ref. [7]. Reproduced with permission of American Chemical Society.

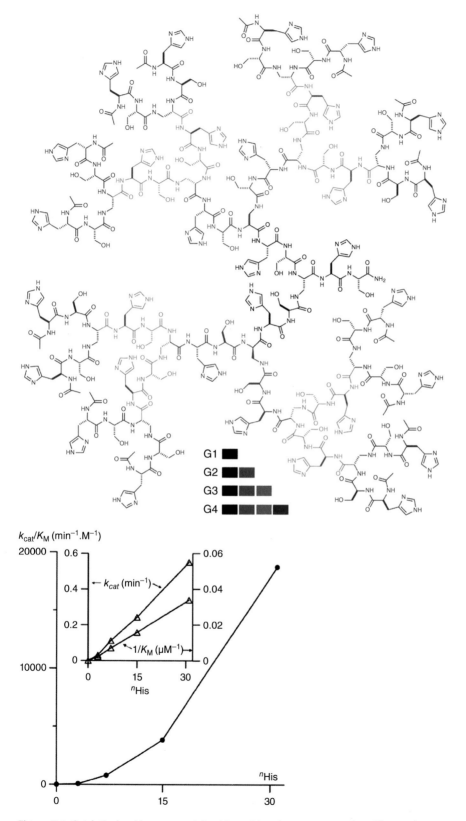

Figure 7.4 Catalytic dendrimers containing His-residues in every generation. Observed positive dendritic effect in the hydrolysis of activated esters. *Source*: Ref. [16]. Reproduced with permission of American Chemical Society.

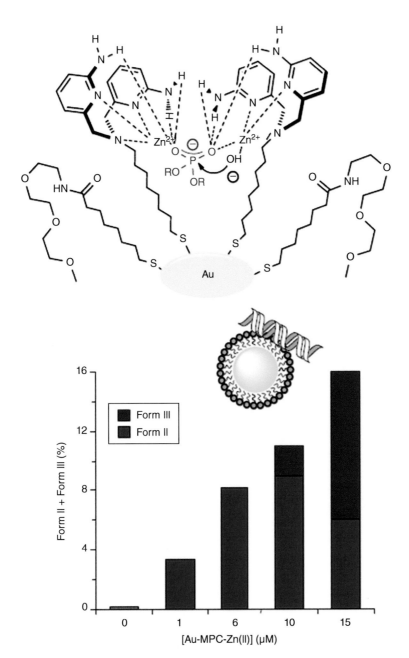

Figure 7.14 Catalytic Au NPs for the cleavage of phosphodiesters. The result of the cleavage of plasmid DNA and a cartoon representing the interaction between the catalyst and the biopolymer are also shown. *Source*: Ref. [40]. Reproduced with permission of American Chemical Society.

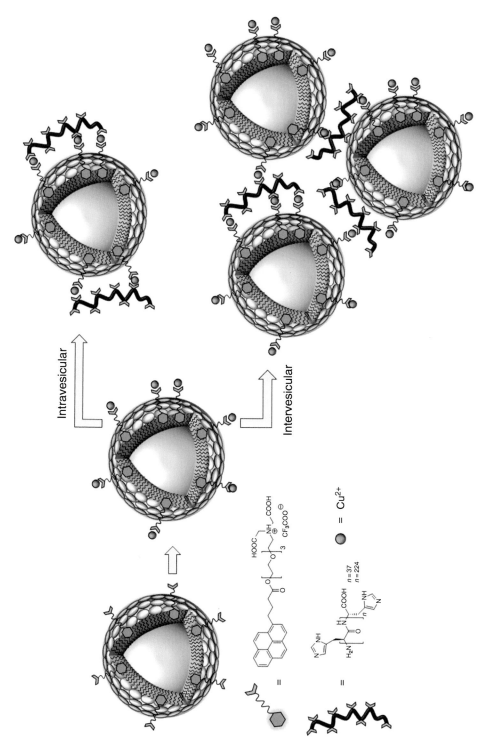

Figure 8.4 Schematic representation of polyhistidine induced aggregation of vesicles functionalized with a pyrene containing Cu^{2+}-iminodiacetate (IDA) ligand.

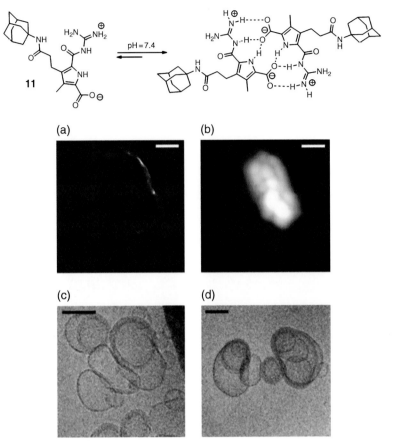

Figure 8.9 pH responsive adamantane-substituted guanidiniumpyrroles and the pH dependent dimerization as well as the corresponding atomic force microscopy (a and b) and cryo-transmission electron microscopy (cryo-TEM; c and d) images of the aggregates. *Source*: Ref. [27]. Reproduced with permission of John Wiley and Sons.

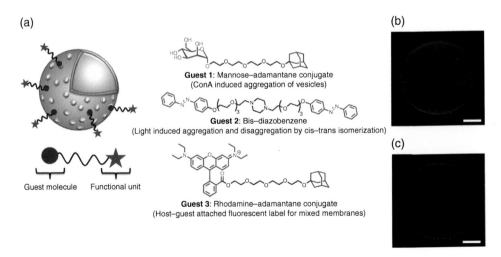

Figure 8.14 (a) Multivalent host–guest interaction of mixed vesicles of cyclodextrins and phospholipids. (b) and (c) Giant unilamellar vesicles prepared from a lipid film containing 30% cyclodextrin. Fluorescent staining was obtained by the addition of guest **3**. (b) Confocal laser scanning microscopy (CLSM) image of a reconstructed three-dimensional image of a vesicle hemisphere from confocal image sections. (c) CLSM single image slice through the vesicle equator. *Source*: Ref. [35]. Reproduced with permission of American Chemical Society.

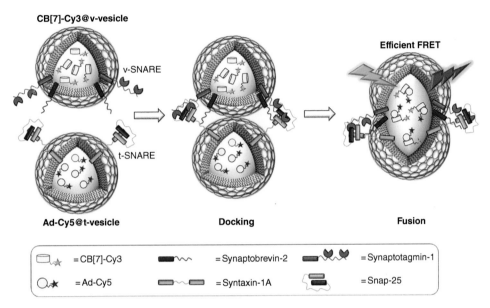

Figure 8.16 SNARE-protein induced vesicle fusion and fluorescence readout using host–guest induced FRET.

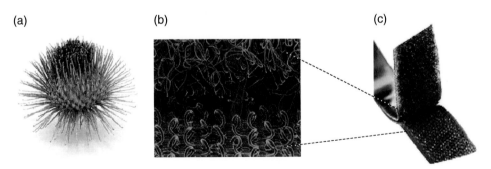

Figure 9.1 (a) A natural burr. (b) The binding mode of velcro on a molecular level. (c) A velcro fastener. *Source*: Adapted from Ref. [2]. Reproduced with permission of John Wiley and Sons.

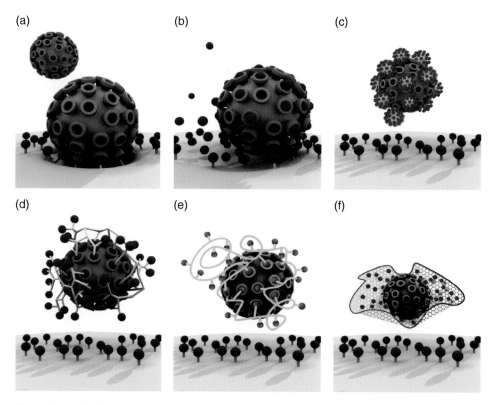

Figure 9.2 (a) Multivalent interaction of a pathogen particle to the cell surface. (b) A standard, highly dosed monovalent drug cannot efficiently prevent pathogen adhesion by blocking single pathogen receptors. (c) A globular multivalent inhibitor decorated with multiple ligands is able to bind several receptors simultaneously and additionally influence the operability of further receptors due to its shape and size. (d) In contrast to the more rigid globular architectures, dendritic or starlike polymer-based structures can be disposed as highly adaptive inhibitors to enhance the binding and shielding efficiency. (e) Linear polymeric inhibitors offer a more flexible and higher surface area compared with their globular equivalents. Due to this fact, they are capable of stretching and coiling to access more receptor sites simultaneously, and additionally can sterically shield further receptors. (f) Since a flexible two-dimensional architecture functionalized by numerous multivalent ligands is not only able to strongly interact with the pathogen and shield several binding sites, it should ideally be able to wrap the whole infectious particle. *Source*: (a and b) Adapted from Ref. [2]. Reproduced with permission of John Wiley and Sons. (c–f) Adapted from Ref. [9]. Reproduced with permission of American Chemical Society.

(a)

(b)

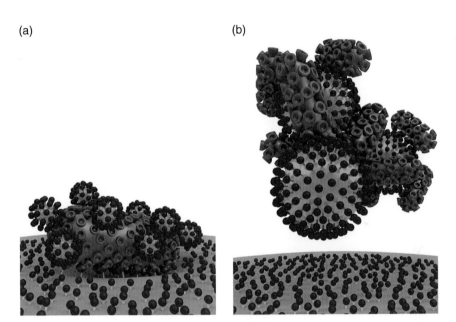

Figure 9.11 Schematic representation of the size-dependent virus inhibition by ligand functionalized gold nanoparticles according to the TEM data. (a) Although smaller sized gold nanoparticles decorate virions, the inhibition of virus–cell binding turned out to be inefficient. (b) Larger virus-sized gold nanoparticles induced the formation of virus-inhibitor clusters, thus inhibiting the virus–cell binding more efficiently. *Source*: Adapted from Ref. [72]. Reproduced with permission of Royal Society of Chemistry.

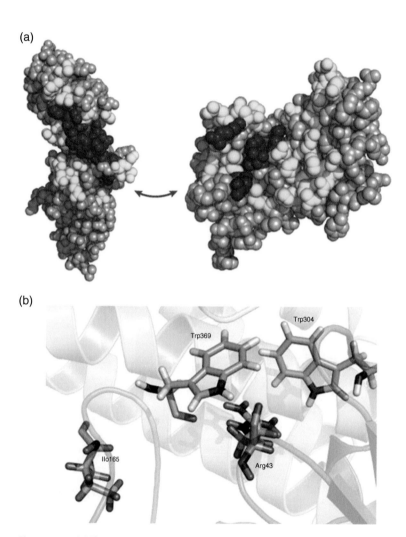

Figure 10.1 (a) The protein–protein complex of human growth hormone binding protein (hGHbp) and growth hormone (hGH) (PDB ID: 3HHR) is separated to expose the interacting surface. The proteins are colored either purple (hGH) or gray (hGHbp) except for the "hot spot" (red) and "rim" (yellow) residues. (b)The hGH (in cyan) complexed with its receptor, hGHbp (in blue). The four hot spot residues are highlighted by a stick representation and the two tryptophan residues are colored in pink. The PDB ID: 1A22. *Source*: Refs [9,17]. Reproduced with permission of John Wiley and Sons and Nature Publishing Group.

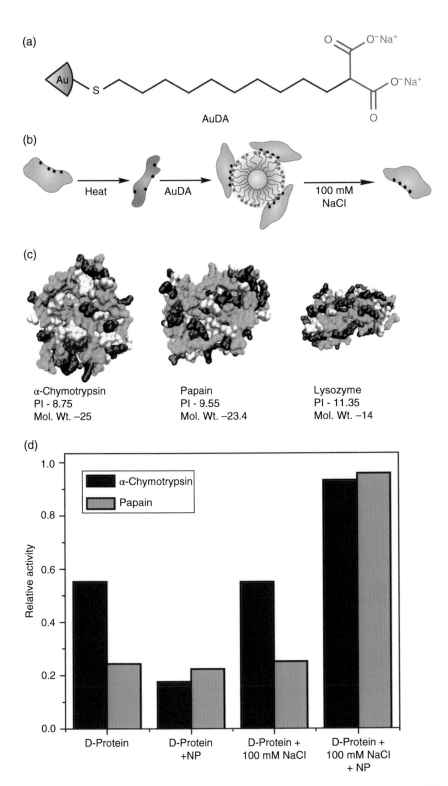

Figure 10.18 (a) Schematic representation of the structure of the AuDA (2 nm core) and (b) thermal denaturation followed by NP mediated refolding of proteins. (c) Surface structural features of three positively charged proteins used in the refolding study. Color scheme for the proteins: basic residues (blue), acidic residues (red), polar residues (green) and nonpolar residues (gray). (d) Enzymatic activity of thermally denatured ChT and papain in the presence of AuDA and 100 mM NaCl solution in 5 mM sodium phosphate buffer (pH = 7.4). *Source*: Ref. [64]. Reproduced with permission of Royal Society of Chemistry.

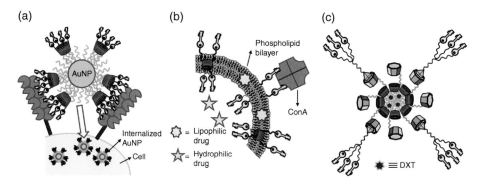

Figure 11.6 A clustered multivalent interaction (a) of the AuNPs functionalized with mannosylated calixarenes **19** targeting the human macrophage mannose receptors. (b) Section of a DOPC liposome functionalized with the bolaamphiphile **21** and exposing glucose units that can efficiently interact with concanavalin A (ConA): the functionalized liposomes can also be loaded with a lipophilic or hydrophilic cargo for targeted drug delivery. Self-assembled nanospheres (c) of calix[4] arene-β-cyclodextrin heterodimers **22** with a cargo of docetaxel (DXT) in the central core and noncovalently functionalized with mannosyl dendrons for targeted drug delivery.

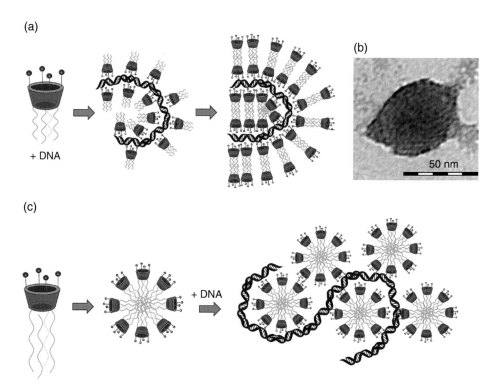

Figure 11.8 The two proposed mechanisms for DNA condensation with calixarenes **25a** and **26**. (a) Sequential binding and condensation: amphiphile **25a** binds to DNA as monomer or in small aggregates and then hydophobic interaction leads to condensation; (b) the snake-like ultra-thin TEM image of the **25a**-DNA aggregate; (c) hierarchical self-assembly, with micelles of amphiphile **26** interacting with a DNA filament. *Source*: Ref. [12]. Reproduced with permission of Royal Society of Chemistry.

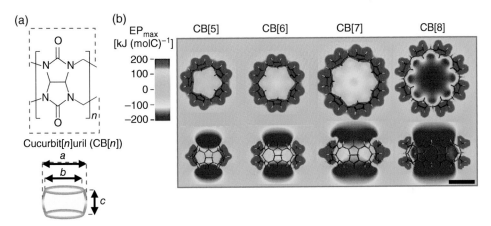

(a)

O

‑N   N‑

N   N‑

O n

Cucurbit[n]uril (CB[n])

a

b

c

(b)

EP_{max}
[kJ (molC)$^{-1}$]

200 —
100 —
0 —
−100 —
−200 —

CB[5] CB[6] CB[7] CB[8]

Figure 12.1 (a) Molecular and schematic structure of cucurbit[n]uril macrocycles. (b) Calculated electrostatic potential (EP) maps for CB[5], CB[6], CB[7], and CB[8] as cross-sectional and side views. Scale bar: approximately 4 Å. *Source*: Adapted from Ref. [16]. Reproduced with permission of American Chemical Society.

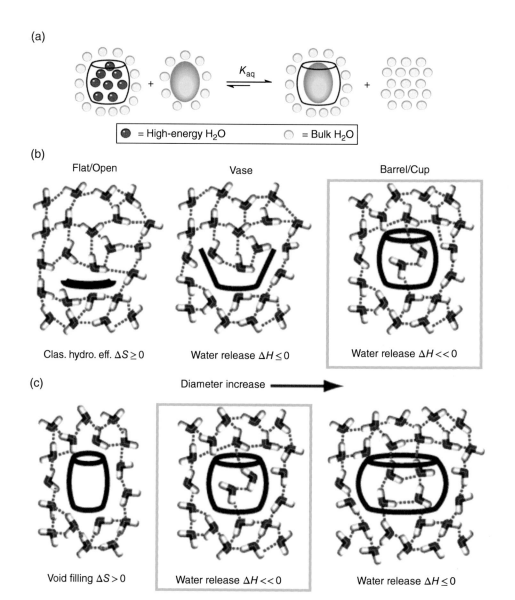

(a)

= High-energy H_2O = Bulk H_2O

(b)

Flat/Open

Vase

Barrel/Cup

Clas. hydro. eff. $\Delta S \geq 0$

Water release $\Delta H \leq 0$

Water release $\Delta H \ll 0$

(c)

Diameter increase ➝

Void filling $\Delta S > 0$

Water release $\Delta H \ll 0$

Water release $\Delta H \leq 0$

Figure 12.4 Schematic representations of (a) the release of high energy water molecules upon complex formation and of the effect of the host shape (b) and size (c) on the water molecules network. *Source*: (a) Adapted from Ref. [17]. Reproduced with permission of American Chemical Society. (b and c) Ref. [3]. Reproduced with permission of John Wiley and Sons.

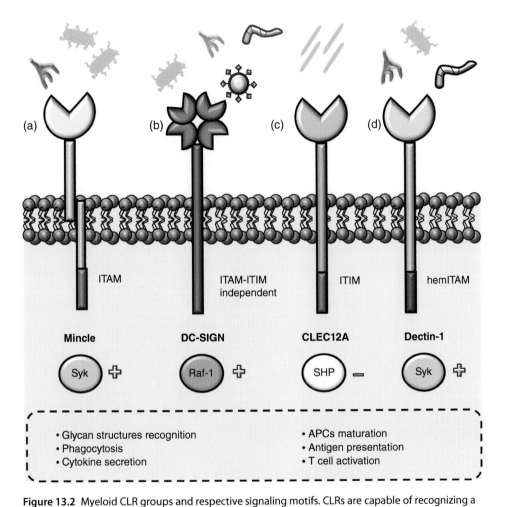

Figure 13.2 Myeloid CLR groups and respective signaling motifs. CLRs are capable of recognizing a vast array of antigens from either pathogens, such as bacteria, viruses, parasites, fungi, or self-antigens. Upon CLR engagement, a signaling cascade is triggered through binding of early adaptors and the recruitment of kinases or phosphatases. Myeloid CLRs can be grouped into four groups based on the cytoplasmic signaling motifs and early adaptors: (a) immunoreceptor tyrosine-based activating motif (ITAM)-coupled CLRs, like Mincle or Dectin-2, signal via spleen tyrosine kinase, Syk, by forming an association with ITAM-bearing adaptors, like DAP12 or FcRγ chain; (b) ITAM-immunoreceptor tyrosine-based inhibitory motif (ITIM) independent CLRs, such as DC-SIGN and mannose receptor, possess tyrosine-based motifs involved in endocytosis, however signal independently from Syk or phosphatases; (c) ITIM-coupled CLRs dampen immune responses by the recruitment of phosphatases SHP-1 and SHP-2 via the ITIM motif; and (d) hemITAM-coupled CLRs, such as Dectin-1 and SIGNR3, modulate immune responses by transducing the signal by a single tyrosine-based motif in their tail and Syk recruitment.

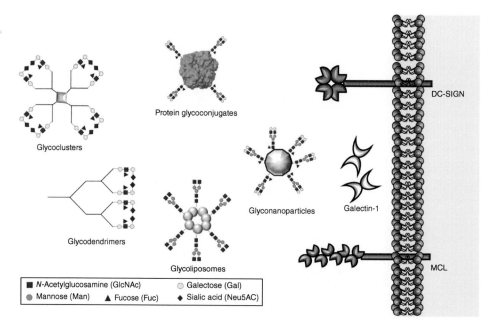

Protein glycoconjugates

Glycoclusters

Glycodendrimers

Glycoliposomes

Glyconanoparticles

Galectin-1

DC-SIGN

MCL

■ N-Acetylglucosamine (GlcNAc)   ◯ Galectose (Gal)
● Mannose (Man)   ▲ Fucose (Fuc)   ◆ Sialic acid (Neu5AC)

Figure 13.3 Schematic representation of the multivalent glycan carriers employed to target lectins. Glycan-carrier systems that enable multivalent presentation of glycans include glycoclusters, glycan-modified antigens (proteins or peptides), liposomes, dendrimers, and nanoparticles.

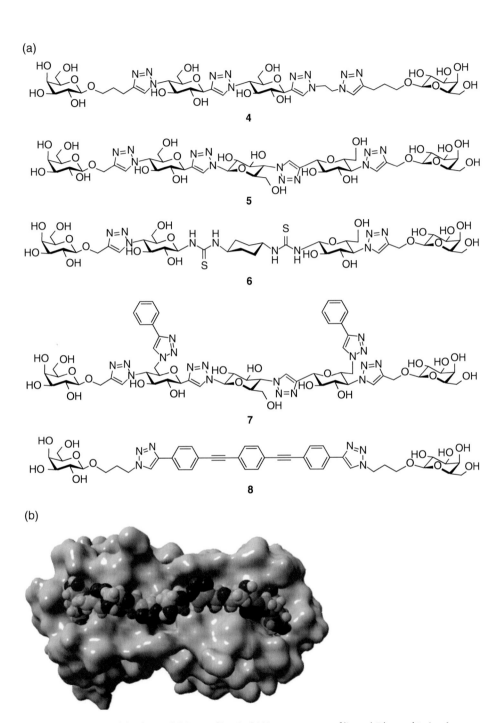

Figure 14.4 (a) Rigid divalent inhibitors of LecA. (b) X-ray structure of ligand **5** bound to LecA.

9

Blocking Pathogens by Multivalent Inhibitors

Sumati Bhatia, Benjamin Ziem, and Rainer Haag

Institute of Chemistry and Biochemistry, Freie Universität Berlin, Takustraße 3, 14195 Berlin, Germany

9.1 Introduction

Multivalency is a unique concept in nature to achieve strong interfacial reversible interactions between m-valent ligands and n-valent receptors of the participating binding partners (with m, $n > 1$) to increase the binding strength and the kinetic stability. Furthermore, multivalency plays an essential role in biological systems for recognition, adhesion, and signaling processes involving antibodies, membranes, molecules, cells, and pathogens such as viruses and bacteria [1–4]. Understanding the mode of action at the molecular level is the first priority for designing multivalent scaffolds, which can play a huge role in the fields of medicine, bio- and supramolecular chemistry, or materials science. The velcro fastener represents one example of these decoding and developing processes, which are necessary to achieve a marketable system. Inspired by the natural burr (Figure 9.1a), the velcro fastener (Figure 9.1b) also uses hooks and loops for reversible multivalent ligand–receptor interactions.

Another example for natural multivalency is the pathogen adhesion, for example, of bacteria, or virus particles to the host cells. This adsorption process is based on non-covalent bindings towards the favored glycan or protein structure on the extra cellular matrix or directly on the cell surface (Figure 9.2a). Influenza virus, for example, initially binds to sialic acid in the mucous while herpesviridae-beta prefers heparansulfate as an interaction partner in the extracellular matrix [1]. After cellular uptake, the genetic material is released to initiate pathogen reproduction followed by the spreading of new virus particles. So far, only small monovalent drugs (Figure 9.2b), for example, Tamiflu as neuraminidase inhibitor against influenza infections and other antibiotics for different bacterial and fungal infections, have been applied in high doses for medical treatment, which can lead to resistance of the pathogens [5,6]. A new approach is, therefore, required, which not only binds with several receptors on the target but also prevents cell adhesion through steric shielding (Figure 9.2c–f). This strategy can lower the dose of multivalent drugs to achieve an improved efficacy in therapies as compared with monovalent analogs.

Multivalency: Concepts, Research & Applications, First Edition. Edited by Jurriaan Huskens,
Leonard J. Prins, Rainer Haag, and Bart Jan Ravoo.
© 2018 John Wiley & Sons Ltd. Published 2018 by John Wiley & Sons Ltd.

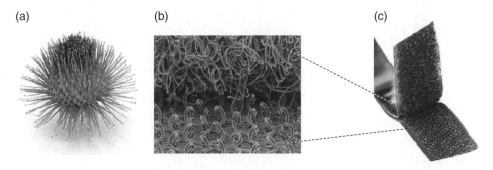

Figure 9.1 (a) A natural burr. (b) The binding mode of velcro on a molecular level. (c) A velcro fastener. *Source*: Adapted from Ref. [2]. Reproduced with permission of John Wiley and Sons. See color section.

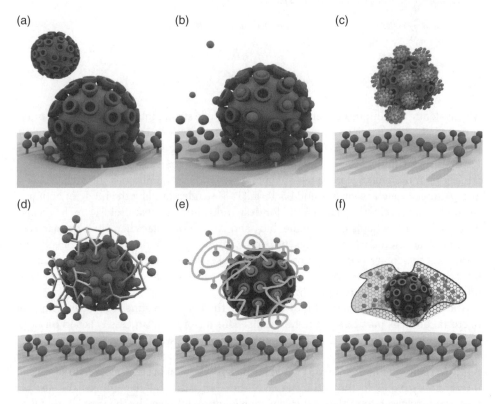

Figure 9.2 (a) Multivalent interaction of a pathogen particle to the cell surface. (b) A standard, highly dosed monovalent drug cannot efficiently prevent pathogen adhesion by blocking single pathogen receptors. (c) A globular multivalent inhibitor decorated with multiple ligands is able to bind several receptors simultaneously and additionally influence the operability of further receptors due to its shape and size. (d) In contrast to the more rigid globular architectures, dendritic or starlike polymer-based structures can be disposed as highly adaptive inhibitors to enhance the binding and shielding efficiency. (e) Linear polymeric inhibitors offer a more flexible and higher surface area compared with their globular equivalents. Due to this fact, they are capable of stretching and coiling to access more receptor sites simultaneously, and additionally can sterically shield further receptors. (f) Since a flexible two-dimensional architecture functionalized by numerous multivalent ligands is not only able to strongly interact with the pathogen and shield several binding sites, it should ideally be able to wrap the whole infectious particle. *Source*: (a and b) Adapted from Ref. [2]. Reproduced with permission of John Wiley and Sons. (c–f) Adapted from Ref. [9]. Reproduced with permission of American Chemical Society. See color section.

Developing an efficient multivalent pathogen inhibitor is a complex process. First of all, the binding mode of the target pathogen has to be decrypted using thermodynamic and kinetic simulations. An accurate theoretical modeling [7,8] is, however, very difficult, due to the complexity of biological systems, which often prevents good theoretical predictions. Furthermore, toxicity as well as biocompatibility also play a major role in the development process, since the intended inhibitors should not be harmful to the treated organisms. Although constant efforts have brought up a number of multivalent inhibitors with great potential, even against drug-resistant pathogens, the pharmaceutical industry has not yet translated this concept. The reason might be that the production of polymeric multivalent ligand architectures is more challenging in terms of polydispersity and reproducible therapeutic efficacy than that of monovalent drug molecules [9]. In spite of these challenges a few systems [10,11] have been successfully tested *in vivo* and further studies are required to explore the full potential of multivalency for pathogen blocking.

9.2 Design of Multivalent Ligand Architectures

The monovalent ligand–receptor interaction is mostly an enthalpy driven process where a ligand diffuses in solution to the target and binds a receptor with a free energy of interaction $\Delta G = \Delta H - T\Delta S$, where ΔG is the free energy of binding and is the sum of an enthalpic (ΔH) and an entropic ($-T\Delta S$) contributions. Only two states (bound and unbound) exist in a monovalent system from which the corresponding free enthalpy difference can be calculated. In the multivalent system, significant entropic penalties are incorporated by the conformational flexibility of the scaffold itself and the spacer linking the multiple ligands to the backbone. An additional entropic contribution is imposed by the release of water molecules from the binding site during hydrophobic interactions.

To assess the multivalent binding effect, Whitesides and coworkers [1] proposed an enhancement factor β, which is the ratio of the binding constant for the multivalent binding (K_{multi}) of a multivalent ligand to a multivalent receptor and the binding constant for the monovalent binding (K_{mono}) of a monovalent ligand to a multivalent receptor. Multivalent binding interactions are complex and dynamic in nature, which makes the kinetic evaluation of the binding process inevitable. The first binding event between a multivalent ligand and a multivalent receptor produces the spatial proximity at the interface that leads to high local concentration. The preorganization of ligands makes the following binding events faster. The overall association rate (k_{on}) of several binding events during multivalent interactions is expressed kinetically. It is a diffusion limited parameter and a weighted average quantity for many elementary binding process. The k_{on} is not always significantly affected in multivalent interactions. In contrast, the lower dissociation rates k_{off} in multivalent systems usually determine the large differences in dissociation constant (K_d) values of the multivalent and the corresponding monovalent systems. Kim *et al.* [12] have recently reported a binding kinetic study for monovalent and bivalent 15-base aptamers (15-Apt) for thrombin binding. The relative k_{off} value for bivalent 15-Apt was 50 times lower than that for monovalent 15-Apt. The relative k_{on} of bivalent 15-Apt was found to be similar to the free 15-Apt. The bivalent 15-Apt exhibited ~60 times higher association constant (K_a) value than the single 15-Apt for thrombin binding [12]. An efficient design of a multivalent inhibitor requires the optimization of the scaffold size and flexibility, spacer length, and ligand density to reduce the thermodynamic penalties and to match the receptor's topological distribution. An optimized multivalent inhibitor should ideally increase its on-rate due to a faster binding to the multivalent receptor surface and decrease its off-rate through continuous rebinding of the ligands in close proximity. Consequently, the multivalent ligand construct will possess a longer life-time during the binding with the receptor surface than the monovalent ligand.

The first step in desiging a multivalent ligand inhibitor is knowledge about the receptor's surface. This includes an intrinsic affinity of the monovalent ligand–receptor pair, size, charge, and mechanical property of the pathogen and inhibitor architecture, and topological distribution profile of receptors on the target/pathogen surface. Multivalent ligand architecture needs to be optimized in terms of ligand density. This can be achieved by varying the ratio of the active functional groups to the non-active functional groups on the carrier backbone. Too much deviation from the optimum ligand density can significantly affect the inhibitory potential of the scaffold. For example, the most abundant trimeric hemagglutinin (HA) glycoprotein (400–500 copies, 10–12 nm apart from each other per virus) each having three binding sites for sialic acid (SA), on the influenza virus surface mediates virus attachment with the cell surface [13–16] (Figure 9.3). X-ray crystallographic studies [17] showed that SA binding sites are located in shallow pockets on the globular head domain of HA trimers. An extensive study was carried out by Roy and coworkers and Whitesides and coworkers [18–20] to inhibit the erythrocyte agglutination by the influenza A virus by different multivalent polyacrylamide sialosides (PAA-SA). The multivalent PAA-o-SA ligand architectures bearing intermediate mole fractions of SA residues ($\chi_{SA} = 0.2$–0.6) [19] were found to be more potent inhibitors than the others. Similarly, a study on sialic acid-conjugated nanogels [21] showed that 50-nm-sized SA conjugated nanogel with only 12% SA conjugation was much more potent, inhibiting 80% of the influenza A virus, than the 70-nm-sized nanogel with 80% SA, which inhibited only 60% of the virus from binding to the cells (Figure 9.4).

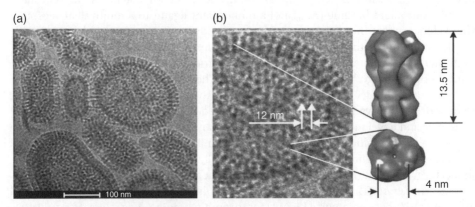

Figure 9.3 (a) The cryo-transmission electron microscopy (TEM) image of influenza A virus (X31/H$_3$N$_2$) shows an abundant distribution of hemagglutinin (HA) glycoprotein trimers on the virus surface. (b) Inter trimeric distances and the length of the stem domain of HA$_3$ are indicated as observed in the TEM image. Patches on the globular domain of HA trimers show the receptors for the sialic acid sugar residues. *Source*: (a) Adapted from Ref. [9]. Reproduced with permission of American Chemical Society. (b) Adapted from Ref. [71]. Reproduced with permission of John Wiley and Sons.

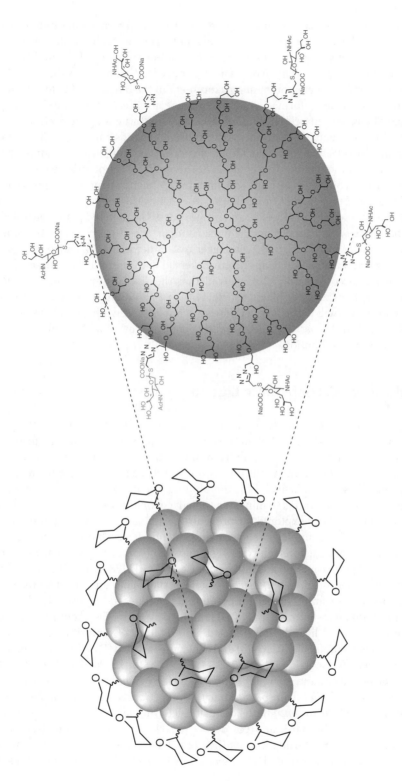

Figure 9.4 Sialic acid-functionalized polyglycerol nanogels consisting of a number of chemically crosslinked dendritic polyglycerol (dPG) units. *Source:* http://pubs.rsc.org/en/content/articlehtml/2014/md/c4md00143e. Used under CC-By 3.0 https://creativecommons.org/licenses/by-nc/3.0/.

Ligands can be linked via optimized spacers to the carrier backbone not only to increase the local ligand density for rebinding but also to reach the less exposed receptors on the target surface. Furthermore, the inhibitory potency of the multivalent compound should be expressed in terms of ligand concentration to assess the impact of multivalent presentation. Zeta potential measurement of the target and the multivalent inhibitor provides further information about non-specific electrostatic interaction at the interface. The pathogen-cell binding inhibition by multivalent ligand architectures is the result of not only enhanced affinities of multivalent ligands but an overall effect of other additional effects like steric shielding, clustering, or statistical rebinding [22], and so on. Although the quantitative analysis of these effects is not straightforward, a careful comparison of binding constants and inhibitory concentrations of the multivalent inhibitor may unravel the contributions of other effects towards prevention of pathogen adhesion to cells. Depending on the size and properties of the targeted pathogen, different carrier backbones, for example, dendrimers, nanoparticles, linear or hyperbranched polymers, nanogels or microgels, and even bigger two-dimensional (2D) adaptable platforms, can be selected. Studies by Haag and coworkers [23], Kiessling and coworkers [22], and Whitesides and coworkers [1,24] showed that the size, geometry, and also the inhibitor-to-pathogen size ratio play a crucial role in the binding mechanism. In the following sections, we will describe the different scaffold types with suitable examples from the literature to discuss the role of multivalency and other additional effects in pathogen–cell binding inhibition.

9.3 Multivalent Carbohydrate Ligands

Carbohydrate–protein interactions play an important role in the binding of pathogens to a cell surface. Monovalent carbohydrates are typically weakly bound to proteins with their dissociation constants ranging from mM to μM. When presented in a multivalent fashion, multiple simultaneous interactions of these carbohydrate ligands can lead to high affinities, which is useful for the binding of pathogens with cell surfaces. Various natural and unnatural glyconjugates (neoglycoconjugates) with different spatial arrangements have been explored for preventing pathogen–cell binding inhibition and thus infection [25]. Examples of such ligands include P^k saccharide for Shiga-like toxins (SLT) [26], monosialotetrahexosyl ganglioside (GM1 oligosaccharides) for cholera toxin (CT) [27], α-mannoside and galabiose for FimH [28] and P-fimbriae [29], respectively, expressed on *Escherichia coli* (*E. coli*), sialic acid and silayl lactose for influenza, α-D-galactose and L-fucose for cytototoxic lectin A and B, respectively, produced from *Pseudomonas aeruginosa* [30,31], and many more. Different multivalent carbohydrate-conjugated ligands can be used as inhibitors of dendritic cell-specific intercellular adhesion molecule (ICAM 3) grabbing non-integrin (DC-SIGN)-mediated binding of different pathogens [32] like HIV-1, HIV-2, SIV, Ebola virus, hepatitis C virus, cytomegalo virus, and dengue virus, bacteria (*Mycobacterium tuberculosis*, *Helicobacter pylori*, and *Klebsiella pneumonia*), yeasts (*Candida albicans*), and parasites *Leishmania* and *Schistosoma*) to the dendritic cells on mucosal surfaces [3]. In the recent investigation by Bernardi and coworkers [33,34], a tetravalent dendron with four copies of linear trimannoside was synthesized. This tetravalent derivative inhibited the DC-SIGN-mediated trans HIV infection process of CD4 + T lymphocytes in

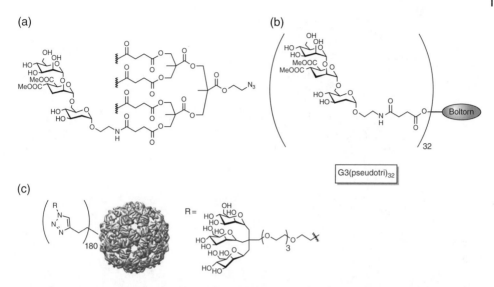

Figure 9.5 (a) Tetravalent linear trimannoside conjugate for HIV infection inhibition. (b) Trimannoside conjugates of G3 Boltorn-type dendrimers for Ebola pseudotyped virus infection inhibition. (c) Mannose conjugated proteic scaffold Qβ-(Man₉)₁₈₀ bearing 1620 mannose residues for Ebola pseudovirus infection inhibition, prepared by click reaction of propargylated Qβ phage with pentaerythritol-based azide terminated glycodendron. *Source*: (c) Adapted from Ref. [36]. Reproduced with permission of Nature Publishing Group.

the presence of elevated viral loads in cellular and cervical explants models (Figure 9.5a). The multivalent linear di- and trimannosides on the G3 Boltron-type dendrimers are also known to strongly inhibit cell infection by Ebola pseudo-typed viral particles by blocking DC-SIGN [35] receptor in the nanomolar concentration range (Figure 9.5b). Virus-like glycol-dendri-protein-nanoparticles up to the size of 32 nm and bearing up to 1620 mannose residues based on a proteic scaffold Qβ phage were prepared [36]. These glycodendrinonanoparticles were capable of mimicking pathogens both in size and valency, blocked a model infection of T-lymphocytes and human dendritic cells by Ebola virus in the picomolar range (Figure 9.5c).

Both Shiga and cholera toxin infections cause millions of deaths per year. These toxins belong to the family of bacterial AB₅ toxins, where the enzymatically cleavable A subunit is connected to the B-pentamer. The B-pentamer contains an oligosaccharide recognizing domain which strongly binds to the receptors on the outer leaflet of many cell membranes. Several multivalent oligosaccharides have been explored that target the B₅ subunit of both cholera and Shiga-like toxins to block its cellular uptake. Excellent multivalent binders that match the topology of the receptor B₅ protein structure were designed and explored. The 'starfish'-like inhibitors initially designed by Bundle and coworkers [37,38] with their planar and radially distributed binding units have shown their potential in blocking the cellular uptake of bacterial cholera and Shiga toxins [39,40]. This concept was further expanded from pentavalent to octa- and decavalent inhibitors of AB₅ toxins (Figure 9.6). The higher avidity of the octavalent system compared with the pentavalent was explained for a situation where several statistical options for binding were available [8].

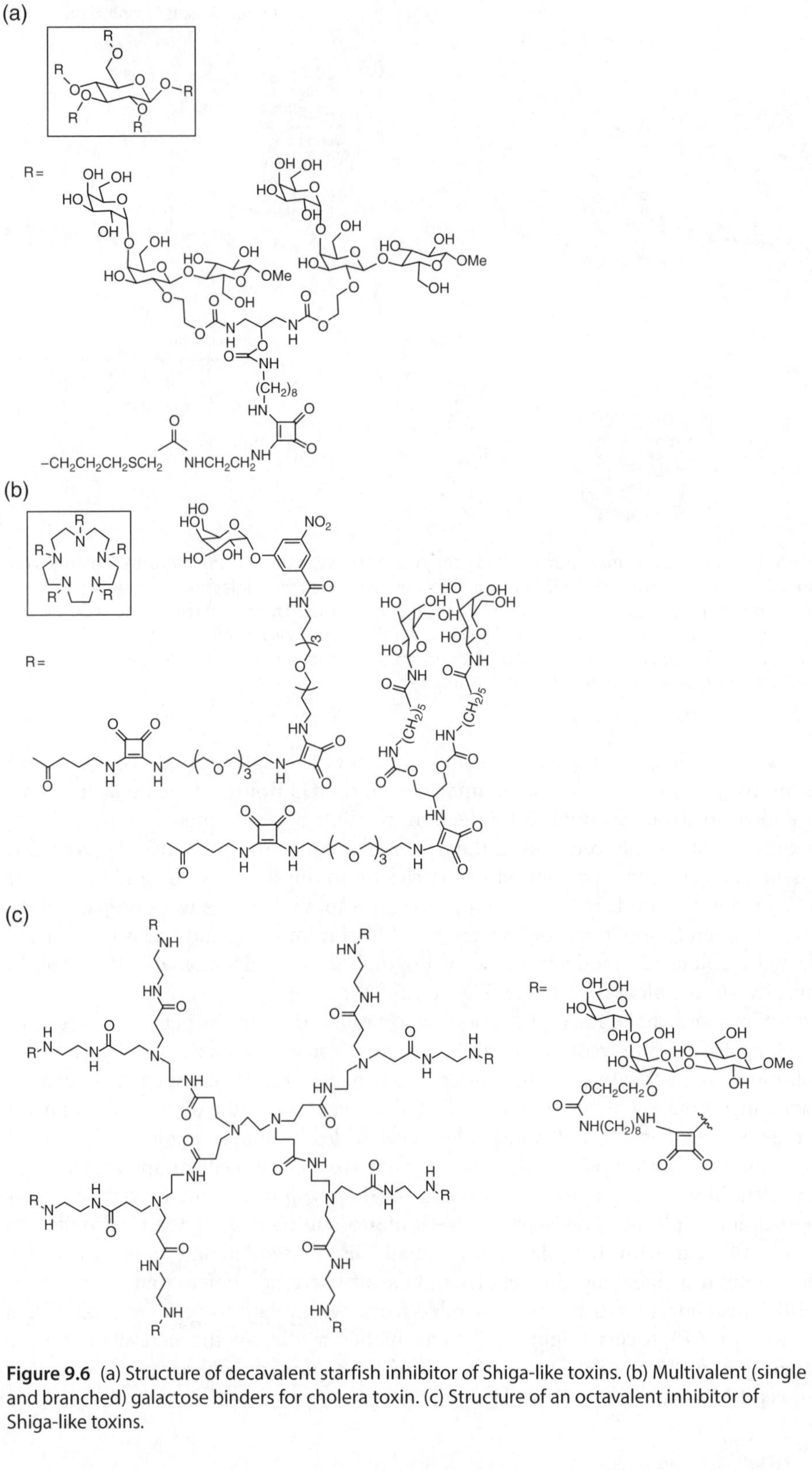

Figure 9.6 (a) Structure of decavalent starfish inhibitor of Shiga-like toxins. (b) Multivalent (single and branched) galactose binders for cholera toxin. (c) Structure of an octavalent inhibitor of Shiga-like toxins.

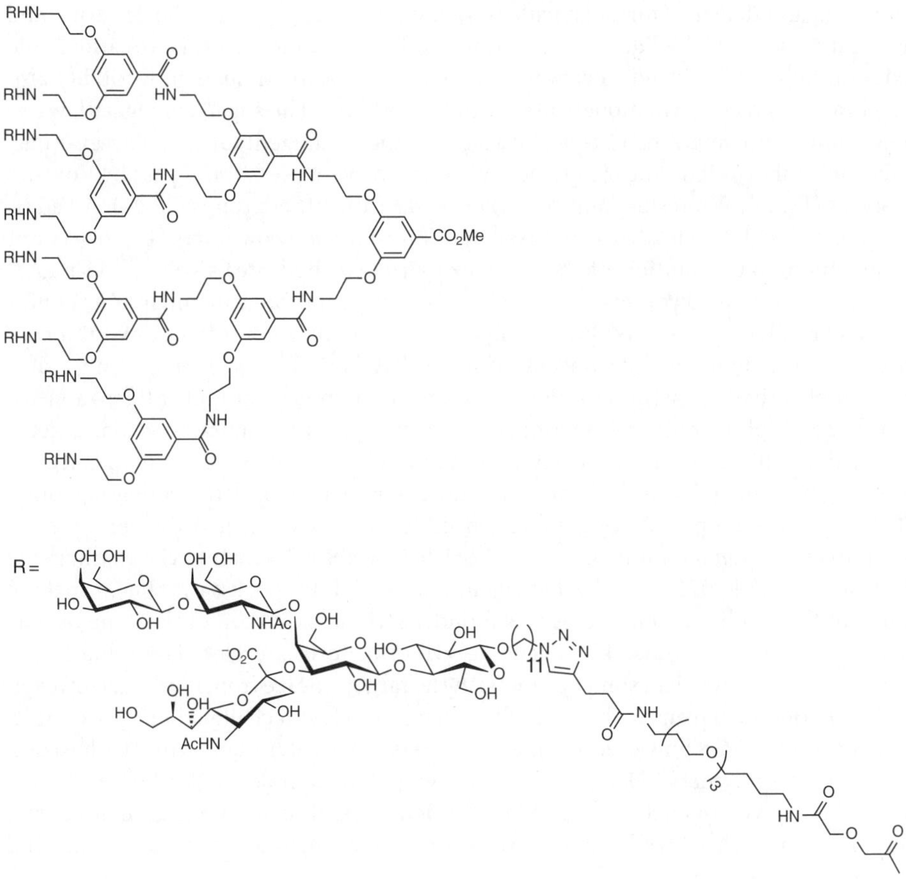

Figure 9.7 Octavalent GM1os inhibitor of the CTB$_5$-subunit.

The multivalent version of the oligoscaccharide of ganglioside (GM1os) using glycodendrimers synthesized by Pieters and coworkers [40] strongly inhibits the cholera toxin B5-subunit (CTB$_5$). The oligovalent glycodendrimer, for example, was 380 000-fold more potent with IC$_{50}$ = 50 pM than the monovalent GM1-mimic (Figure 9.7). The tetravalent and octavalent galactose with long spacer arms, prepared as simplified versions of GM1os, competed well with the natural GM1os [41].

9.4 Scaffold Architecture

9.4.1 Linear and Dendritic Scaffolds

Different linear, branched, and globular carriers can be used as a scaffold to prepare multivalent ligand inhibitors of infection by different pathogens. The optimal choice of the scaffold for the preparation of a multivalent ligand architecture is difficult. Ligands are presented more rigidly and exposed on the dendritic carriers than the relatively flexible linear backbones, which determine the affinity of ligands. Therefore, the optimum ligand densities for a particular target might be different for the dendritic and linear carriers. Flexible ligands can attain different conformations to afford the maximum number of ligand–receptor pairs but, at the same time, high flexibility also incorporates high conformational entropy to the binding. The sugar-conjugated, water swollen linear polymers can sterically stabilize the pathogen surface. For example, higher molecular weight linear polymers were extensively used in the past by Roy and coworkers (18,20), Whitesides and coworkers (1,42), and other groups (43,44). An additional steric shielding effect was proposed by Whitesides and coworkers [19] to account for the nanomolar inhibition of the influenza virus by the water swollen linear PAA sialosides. Chen and coworkers [45] have also recently reported zanamivir (ZA) conjugated linear poly-L-glutamine (PGN) polymers for inhibition of influenza viral fusion and release at subnanomolar concentrations of ZA [46]. These PGN-ZA conjugates were also effective against monovalent ZA- and oseltamivir resistant influenza virus. Recently amphiphilic antiviral peptides that form high order supramolecular assemblies for the multivalent display of inhibitors have also gained a lot of importance for preventing binding of influenza virus to the host cells. The stearylated antiviral peptide B (PeBGF), for example, showed enhanced inhibitory effect against the serotypes of both human pathogenic influenza virus A/Aichi/2/1968 H3N2, and avian pathogenic A/FPV/Rostock/34 H7N1 in the hemagglutination inhibition assay. The unspecific binding to the cell lipid membranes was found a serious limitation of this type of self-assembling antiviral peptide [47]. The large size of dendritic or globular architectures [48–52] can sterically shield some receptors at the target surface from binding. However, affording perfect dendrons with size greater than 10 nm is still challenging, for example, the diameter of the G8 polyamidoamine dendron is not greater than 8 nm. Synthesizing hyperbranched polymers [53,54] is less challenging, for example, 2–20 nm sized hyperbranched polyglycerol backbones can be afforded depending on the degree of polymerization. Even larger sized globular backbones may include nanogels or metallic nanoparticles.

Also, fullerenes have emerged as carbon-based, globular carriers for the preparation of multivalent ligand architectures. The mannosylated fullerenes [55] with different linker lengths were investigated to inhibit the FimH of *E. coli*. The lipopolysaccharide (LPS) protects and stabilizes the bacterial membrane. The glycosylated fullerenes **1, 2**, and **3** displaying the mannopyranose core structure of bacterial L,D-heptosides have shown the inhibition of the LPS heptosyltransferase WaaC in the low micromolar range ($IC_{50} = 11$, 47, and 6.7 μM) and thus inhibit the bacterial LPS synthesis (Figure 9.8). The monomeric glycosides, on the other hand, displayed IC_{50} values above 400 mM [55]. Recently Muñoz *et al.* [56] reported the synthesis of the giant tridecafullerenes decorated with 120 peripheral carbohydrate (mannose or galactose) units. The mannosylated "superballs" have shown great potential in inhibiting cell infection by artificial Ebola virus particles with IC_{50} in the subnanomolar range.

Dendritic or linear backbones also have different biological impacts. Heparin, for example, is a natural linear sulfated polysaccharide, which has been used in the last few decades as a viral entry inhibitor, as well as for prevention and treatment of thrombosis [57–61]. However, its therapeutic application is limited due to its anticoagulant effect [62]. Further drawbacks are associated with the preparation of heparin, since it has to be isolated from mammalian organs. This way of extraction always bears the potential risk of contamination by pathogens and leads to a highly polydisperse heparin [63].

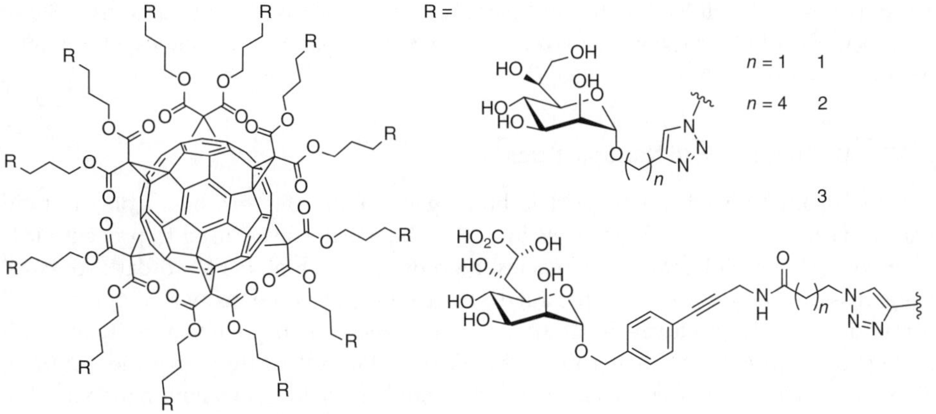

R =

n = 1   1
n = 4   2

3

Figure 9.8 Dodecavalent glycoclusters based on fullerene hexa-adducts to inhibit the LPS heptosyl-transferase WaaC with nanomolar affinity.

On the basis of these disadvantages dendritic polyglycerol sulfate (dPGS) has been developed as a synthetic heparin mimetic, which has a similar activity profile but not showing the characteristic limitations of heparin (Figure 9.9). In contrast to heparin, dPGS can be fully synthesized with a straightforward protocol that results in a low polydisperse product, which shows only a low anticoagulant effect but is able to

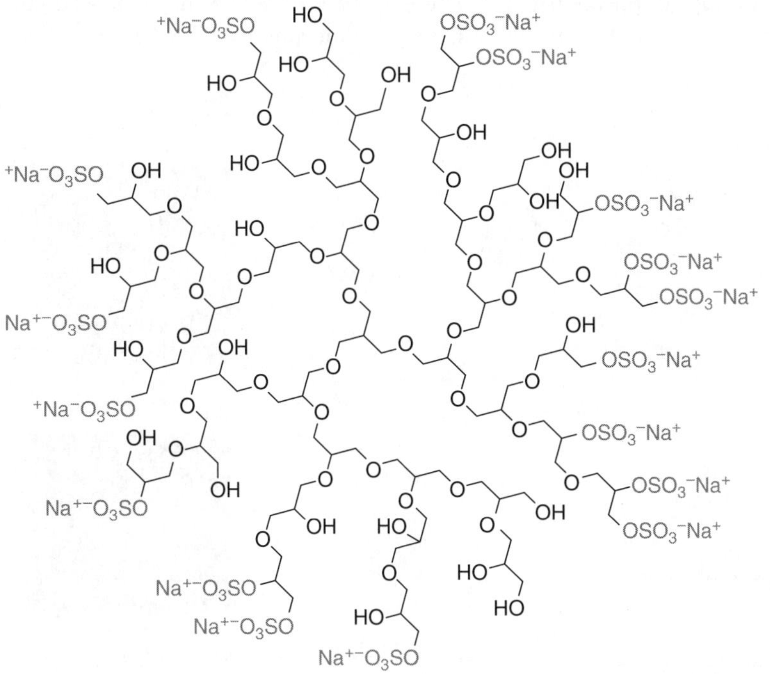

Figure 9.9 Schematic representation of an idealized dendritic polyglycerol sulfate structure. The negatively charged sulfate end groups and the corresponding counterions are shown in gray.

strongly interact with leukocytes and pathogens via electrostatic interactions [23,64]. Furthermore, dPGS is tunable in size, is highly biocompatible, and shows a high anti-inflammatory effect *in vivo* [63,65,66].

9.4.2 Multivalent Gold Nanoparticles

Competitive, multivalent nanoparticle binding inhibitors that are based on the rigid core, for example, gold nanoparticles [67], have been extensively used to prevent virus infections. The use of gold nanoparticles, moreover, gives high contrast "dark spots" in transmission electron microscopy, due to their high electron density, which is helpful for visualizing the multivalent attachment of ligand-coated gold nanoparticles with pathogens (Figure 9.10). Furthermore, the possible controlled size preparation of these gold nanoparticles allows for a size dependent study of rigid multivalent nanoparticles. In the recent past, various gold nanoparticles functionalized [67] with multivalent ligands have been studied for the biological interactions and for biolabels. For example, mannose-conjugated gold nanoparticles were used to visualize the position of individual FimH proteins on the type 1 pilus of *E. coli* with electron microscopy. The quantitative analysis [68] of multivalent interaction of different mannose-conjugated gold nanoparticles with tetrameric plant lectin, concanavalin A (ConA), shows the dependence of affinity on nanoparticle size and linker of mannose ligands. Martínez-Avila *et al.* [69] synthesized a series of gold nanoparticles with different spacers and densities of oligomannoside. Surface plasmon resonance-based competitive assays were applied to evaluate selected mannose- conjugated gold nanoparticles as inhibitors of DC-SIGN binding to HIV envelop glycoprotein gp120. The gp120 was direcctly immobilized on the sensor chip surface and binding measurements of fluid-phase DC-SIGN at a fixed

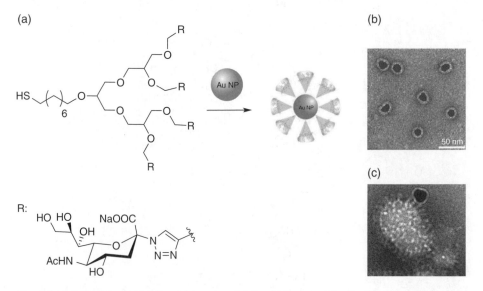

Figure 9.10 (a) Synthesis of sialic acid-functionalized gold nanoparticles (Au NPs). (b) Electron microscopic visualization of 14-nm-sized gold nanoparticles with a sialic acid corona. (c) Preparation showing multiple binding of individually functionalized Au NPs to viral HAs. *Source:* Adapted from Ref. [71]. Reproduced with permission of John Wiley and Sons.

concentration in the presence of mannose-conjugated gold nanoparticles or free alkyl amino (oligo)mannosides were performed at varying stoichiometric ratios. The disaccharide Manα1-2Manα containing gold nanoparticles was found to be the best inhibitor, showing nanomolar inhibitory concentrations (100% inhibition at 115 nM). This was 20 000 fold higher than the corresponding monomeric disaccharide (100% inhibition at 2.2 mM). Furthermore, these multivalent manno-glyco nanoparticles [70] also inhibited DC-SIGN-mediated HIV-1 trans-infection of human T cells at nanomolar concentrations in experimental settings that mimicked the natural route of virus transmission. Papp *et al.* [71] chemically conjugated 2- and 14-nm-sized gold nanoparticles with sialic acid-terminated dendron and investigated for binding with influenza virus. The larger sized gold nanoparticle inhibitors had high affinity for HA on the influenza virus surface as observed by electron microscopy techniques and biochemical analysis (Figure 9.10). This result pointed out the relevance of the contact area between an inhibitor and its target. Vonnemann *et al.* [72] studied different sized polysulfated gold nanoparticles for the vesicular stomatitic virus (VSV)–cell binding inhibition. The polysulfated nanoparticles smaller than the size of the VSV with diameters ≤50 nm inhibited VSV–cell binding only to a small extent, whereas the VSV-sized nanoparticles (≥52 nm) formed VSV/nanoparticle clusters and effectively inhibited virus–cell binding and thus infection (Figure 9.11). For the virus-sized nanoparticle inhibitors, the increased contact area between the virus and inhibitor enhanced the polyvalent effect and thus led to cluster formation. This effect was not observed for the nanoparticle inhibitors, which were much bigger than the virus itself.

(a)   (b)

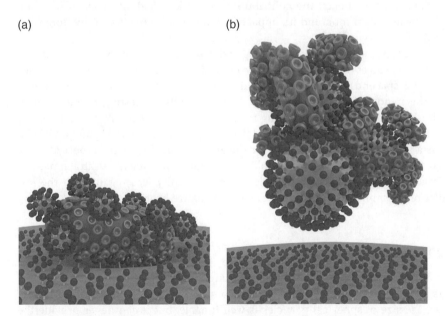

Figure 9.11 Schematic representation of the size-dependent virus inhibition by ligand functionalized gold nanoparticles according to the TEM data. (a) Although smaller sized gold nanoparticles decorate virions, the inhibition of virus–cell binding turned out to be inefficient. (b) Larger virus-sized gold nanoparticles induced the formation of virus-inhibitor clusters, thus inhibiting the virus–cell binding more efficiently. *Source*: Adapted from Ref. [72]. Reproduced with permission of Royal Society of Chemistry. See color section.

Vonnemann *et al.* [23] explored the quantitative impact of virus concentration, inhibitor size, steric shielding, and multivalency in the inhibition process by spherical inhibitors for both strong and weak ligand/receptor pairs. Different sized streptavidin-coated silica nanoparticles were used as strong binding inhibitors for 10-µm-sized biotinylated silica particles. The dPG and dPGS-coated gold nanoparticles for leukocyte-selectin-coated binders were taken as an example for weak ligand/receptor pairs. The impact of the varying sizes and ligand densities on the evaluation of IC_{50} was systematically investigated during inhibition. A modified version of the Cheng–Prusoff-equation was adapted to account for the multivalent inhibitors covering the surface of the binder. In this equation, $IC_{50} = k_d^{multi} + 0.5P[B]$, the first term was the contribution to multivalency, and the prefactor of the binder concentration [B] accounted for the steric shielding contribution. Their analysis concluded that the optimal size of the globular inhibitor was smaller than the binder's in most cases. The multivalent dissociation constant K_d^{multi} for large globular inhibitors exponentially depended on the contact area and therefore was much lower than the virus (pathogen) experimental/biological concentration, which shows the predominance of the steric shielding effect over multivalency in body fluids. The impact of steric shielding [23], furthermore, was only noticeable when all the inhibitors were bound to the pathogen.

Their analysis showed that the multivalent association constant for large globular inhibitors exponentially increased with inhibitor/virus contact area and ligand density. The particle and volume normalized IC_{50} value of an inhibitor predominantly depended on its multivalent association constant at a very low virus concentration. Compared with the multivalency effects, the contribution of steric shielding to the IC_{50} values of inhibitors was only minor, and its impact was only noticeable if all inhibitors were bound to the binder (virus).

Inorganic nanoparticles are usually conjugated with ligands via poly(ethylene glycol) (PEG) spacer to protect them from serum protein adsorption [73]. The attachment of a ligand to a PEG spacer can significantly impair its receptor affinity, which is counterbalanced by the nanoparticle's multivalency. Therefore, affinity testing of monovalent ligand with PEG spacer before attaching to the nanoparticle is highly recommended. A recent study by Hennig *et al.* [74] showed that PEGylation of a model ligand EXP3174, which was a small molecule antagonist for the angiotensin II receptor type 1 (AT1R), decreased the receptor affinity by 58-fold to the native ligand. After attachment of the PEGylated ligand to the nanoparticle, receptor affinity was regained in the low nanomolar range similar to the affinity range of the native ligand.

9.4.3 2D Platforms

Globular inhibitors decorated with multiple functionalities are able to bind several receptors at once but they are limited in their contact area due to their shape and size. Of course, the size of spherical particles is well tunable, but it can never provide the same contact area as its weight or size comparable sheet-like analog. Since interactions of virus particles or bacteria are based on dynamic processes at the interfaces, a huge surface area with optimum ligand density is advantageous.

Furthermore, flexible 2D platforms coated with numerous multivalent ligands are not only able to bind several pathogen receptors at once but can ideally wrap the whole

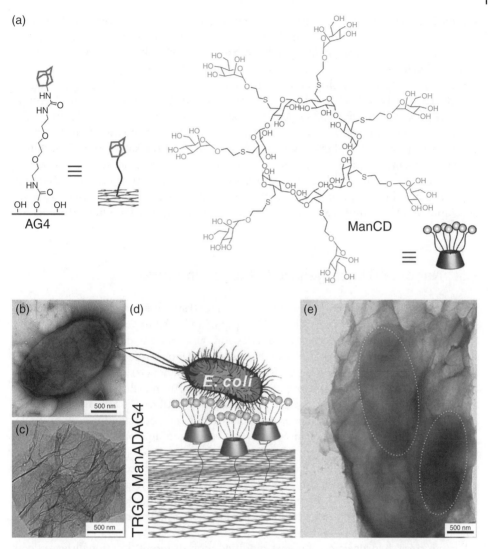

Figure 9.12 Schematic of (a) chemical structures of adamantyl-functionalized graphene derivative (AG4) and heptamannosylated ß-cyclodextrin (ManCD). TEM images of the (b) ORN178 *E. coli* and (c) ManCD-AG4 hybrid. (d) Schematic representation for capturing *E. coli* by mannosylated cyclodextrin, adamantly-functionalized (ManCD@AG4) graphene sheets. (e) *E. coli* agglutination incubated with ManCD@AG4. The dashed gray circles outline the captured bacteria. *Source*: Adapted from Ref. [4]. Reproduced with permission of American Chemical Society.

infectious particle. To satisfy all these expectations, a supramolecular host–guest construct was developed on the basis of adamantyl-functionalized, thermally reduced graphene oxide (TRGO) decorated with multivalent sugar ligands (Figure 9.12a).

The formation of host–guest inclusions on the carbon surface provides a versatile strategy, for not only increasing the intrinsic water solubility of graphene-based materials, but also promote binding of the biofunctional groups with the sheet-like architecture. The combination of the unique 2D surface area of the graphene and the specific binding ability of carbohydrates results in a flexible platform that is able to wrap selectively and agglutinate *E. coli* (Figure 9.12b–e). Through the powerful thermal IR-absorption properties of the graphene-based material, it is possible to kill the wrapped bacteria by short IR-laser irradiation with high efficiency [4].

Functionalization, flexibility, and size are essential criteria for embedding pathogens into the carbon-based 2D architecture. Additionally, implementing functional groups partially convert hybridized honeycomb-like structures from sp^2 to sp^3 and thus results in increased flexibility of the sheet [75]. Due to these changes, the modified sheets can bend, extensively fold, and even wrap smaller particles. If the size of the 2D architecture is much smaller than that of the bacteria or the virus particle, several sheets will adhere on the pathogen and can fully cover the whole surface to isolate the infectious particle. Since the inhibition of cell–pathogen is still an urgent and unsolved problem, development of new multivalent 2D platforms is a new strategy for efficiently blocking pathogens.

9.5 Nano- and Microgels for Pathogen Inhibition

Hydrogels are highly hydrated three-dimensional (3D) systems, which offer an excellent biocompatibility as well as an adjustable stiffness and swelling behavior [76]. Due to these facts, hydrogels are perfect candidates for various biomedical applications [77] such as targeted drug delivery [78–80], biosensing [81], or tissue engineering [82–84]. Furthermore, the gel size can be varied from 20 nm (nanogels) [21,85] up to 400 μm (microgels) [86] by different techniques. The commonly used techniques are miniemulsion, microfluidic templating, or inverse nanoprecipitation [87–89]. In the case of miniemulsion, the gel formation is based on a polymer crosslinking, for example, partially acrylated polyglycerol in a surfactant-stabilized aqueous nanodroplet. This technique, however, is limited to nanometer-sized gels on its own, but in combination with droplet-based microfluidic and macro crosslinkers, for example, poly(ethylene glycol) diacrylate, the gel size can be augmented to microgels (Figure 9.13). By embedding cleavable bonds or linker into the polymer matrix, such as acetals, disulfides, ketals, or phosphate esters, a high biodegradation and an efficient degradation can be guaranteed [85,87].

Furthermore, hydrogels can be post modified with different functionalities such as sulfates, amines, or specific glycan structures to build up a new type of multivalent pathogen inhibitor. One example for such an inhibitor is the polyglycerol-based glycoarchitecture coated with multiple sialic acids for pathogen interaction (e.g., Figure 9.4). On the basis of influenza virus, Papp *et al.* [21] demonstrated the high inhibition potential of modified nanogels depending on the gel size and the degree of sialic acid functionalization.

Until now, nano- and microgels have been frequently used as transporters for encapsulation and targeted release of cells or pharmaceutical biomacromolecules triggered by external stimuli such as pH changes or reductive environments [88,89]. However, in the future, degradable hydrogels could be designed not only as a direct pathogen inhibitor but also to develop 3D pathogen traps. Additionally, microgels [90] are easily able to mimic the natural extracellular matrix, which is known to interact with various types of pathogens. Ideally, after adhesion of the infectious particle to the gel surface, diffusion

(a)

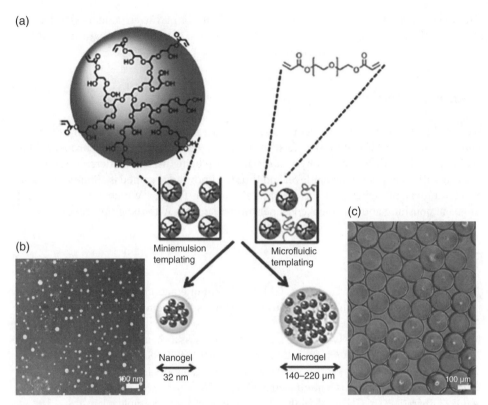

(b)

(c)

Miniemulsion templating

Microfluidic templating

Nanogel
←→
32 nm

Microgel
←→
140–220 μm

100 nm

100 μm

Figure 9.13 (a) Schematic representation of a nano- or microgel formation via miniemulsion and microfluidic templating. The nanogel (on the left) consists of crosslinked acrylated polyglycerol, while the microgel (on the right) is the result of combining the above-mentioned nanogel and a poly(ethylene glycol) diacrylate macro crosslinker. (b) TEM micrograph of nanogel particles. (c) Optical micrograph of water-swollen microgel particles. *Source*: Adapted from Ref. [86]. Reproduced with permission of Elsevier.

into the gel network should take place so that the pathogen is trapped inside. There is still much work to be done for developing biocompatible or biodegradable 3D gel networks for efficient pathogen inhibition in biological systems.

9.6 Conclusion

Prevention of pathogen inhibition is one of the major goals for human health. Different multivalent ligand scaffolds have been synthesized and explored in the area of multivalent pathogen inhibition. The design of efficient multivalent inhibitors is a complex process and involves optimizing ligand density, flexibility, size, and charge. The larger sized inhibitor architectures such as 2D platforms and 3D gel networks can be extremely useful to combat pathogens like micrometer-sized bacteria. However, the biocompatibility, biodegradability, and systemic clearance are inevitable for their application in biological systems. Many problems have still to be solved in this area. No reports are available yet on the fate of bound inhibitor–pathogen complexes in biological systems.

Questions have also been raised on the applicability of pathogen binders for the late stages of infection. In spite of these challenges, the area of multivalent pathogen inhibitors has a great potential for future research.

Acknowledgments

We would like to thank the Deutsche Forschungsgemeinschaft (DFG) for financial support within the Collaborative Research Centre 765 as well as the European ITN "Multiapp" where different biophysical methods are being used to understand multivalency quantitatively at molecular level and thus the knowledge gained is applied to design novel multivalent interfaces. We are grateful to Dr Pamela Winchester for careful language polishing and Dr Wiebke Fischer for her support with the graphical design.

References

1 Mammen M, Choi SK, Whitesides, GM. Polyvalent interactions in biological systems: Implications for design and use of multivalent ligands and inhibitors. Angew. Chem. Int. Ed. 1998;37(20):2754–94.
2 Fasting C, Schalley CA, Weber M, *et al*. Multivalency as a chemical organization and action principle. Angew. Chem. Int. Ed. 2012;51(42):10472–98.
3 Bhatia S, Dimde M, Haag R. Multivalent glycoconjugates as vaccines and potential drug candidates. Med. Chem. Commun. 2014;(5):862–78.
4 Qi Z, Bharate P, Lai CH, *et al*. Multivalency at interfaces: Supramolecular carbohydrate-functionalized graphene derivatives for bacterial capture, release,.and Disinfection. Nano Lett. 2015;15(9):6051–7.
5 Jiang W., Kim BY, Rutka JT, Chan WC. Nanoparticle-mediated cellular response is size-dependent. Nat. Nanotechnol. 2008;3:145–50.
6 Moscona A. Medical management of influenza infection. Annu. Rev. Med. 2008;59:397–413.
7 Kane RS. Thermodynamics of multivalent interactions: Influence of the linker Langmuir 2010;26(11):8636–40.
8 Kitov PI, Bundle DR. On the nature of the multivalency effect: A thermodynamic model. J. Am. Chem. Soc. 2003;125(52):16271–84.
9 Bhatia S, Camacho LC, Haag R. Pathogen inhibition by multivalent ligand architectures. J. Am. Chem. Soc. 2016. DOI: 10.1021/jacs.5b12950.
10 Karlson KA. Meaning and therapeutic potential of microbial recognition of host glycoconjugates. Mol. Microbiol. 1998;29(1):1–11.
11 Landers JJ, Cao Z, Lee I, Piehler LT, *et al*. Prevention of influenza pneumonitis by sialic acid-conjugated dendritic polymers. JID 2002;186(9):1222–30.
12 Kim Y, Cao Z, Tan W. Molecular assembly for high-performance bivalent nucleic acid inhibitor. PNAS 2008;105(15):5664–9.
13 Yamaguchi M, Danev R, Nishiyama K, *et al*. Zernike phase contrast electron microscopy of ice-embedded influenza A virus. J. Struct. Biol. 2008;162(2):271–6.
14 Ruigrok RWH, Calder LJ, Wharton SA. Electron microscopy of the influenza virus submembranal structure. Virol. 1989;173(1):311–16.

15 Li S, Sieben C, Ludwig K, *et al.* pH-Controlled two-step uncoating of influenza virus. Biophys. J. 2014;106(7):1447–56.

16 Boettcher C, Ludwig K, Herrmann A, *et al.* Structure of influenza haemagglutinin at neutral and at fusogenic pH by electron cryo-microscopy. FEBS 1999;463(3):255–9.

17 Weis W, Brown JH, Cusack S, *et al.* Structure of the influenza virus haemagglutinin complexed with its receptor, sialic acid. Nature 1998;333(6172):426–31.

18 Roy R, Pon RA, Tropper FD, Andersson FO. Michael addition of poly-L-lysine to N-acryloylated sialosides. Syntheses of influenza A virus haemagglutinin inhibitor and group B meningococcal polysaccharide vaccines. J. Chem. Soc. Chem. Commun. 1993;(3):264–65.

19 Lees WJ, Spaltenstein A, Kingery-Wood JE, Whitesides GM. Polyacrylamides bearing pendant alpha-sialoside groups strongly inhibit agglutination of erythrocytes by influenza A virus: Multivalency and steric stabilization of particulate biological systems. J. Med. Chem. 1994;37(20):3419–33.

20 Roy R, Andersson FO,.Harms G, *et al.* Synthesis of esterase-resistant 9-O-acetylated polysialoside as inhibitor of influenza C virus hemagglutinin. Angew. Chem. Int. Ed. Engl. 1992;31(11):1478–81.

21 Papp I, Sieben C, Sisson AL, *et al.* Inhibition of influenza virus activity by multivalent glycoarchitectures with matched sizes. ChemBioChem 2011;12(6):887–95.

22 Gestwicki JE, Cairo CW, Strong LE, *et al.* Influencing receptor – ligand binding mechanisms with multivalent ligand architecture. J. Am. Chem. Soc. 2002;124(50):14922–33.

23 Vonnemann J, Liese S, Kuehne C, *et al.* Size dependence of steric shielding and multivalency effects for globular binding inhibitors. J. Am. Chem. Soc. 2015;137(7):2572–9.

24 Choi SK, Mammen M, Whitesides GM. Monomeric inhibitors of influenza neuraminidase enhance the hemagglutination inhibition activities of polyacrylamides presenting multiple C-sialoside groups. Chem. Biol. 1996;3(2):97–104.

25 Bernardi A, Jimenez-Barbero J, Casnati A, *et al.* Multivalent glycoconjugates as anti-pathogenic agents. Chem. Soc. Rev. 2013;42:4709–27.

26 Ling H, Boodhoo A, Hazes B, *et al.* Structure of the shiga-like toxin I B-pentamer complexed with an analogue of its receptor Gb3. Biochemistry 1998;37(7):1777–88.

27 Fan E, Merritt EA, Verlinde CLMJ, Hol WGJ. AB_5 toxins: structures and inhibitor design. Curr. Opin. Struct. Biol. 2000;10(6):680–6.

28 Heidecke CD, Lindhorst TK. Iterative synthesis of spacered glycodendrons as oligomannoside mimetics and evaluation of their antiadhesive properties. Chem. Eur. J. 2007;13(32):9056–67.

29 Lane MC, Mobley HLT. Role of P-fimbrial-mediated adherence in pyelonephritis and persistence of uropathogenic *Escherichia coli* (UPEC) in the mammalian kidney. Kidney Int. 2007;72(1):19–25.

30 Gilboa-Garber N. *Pseudomonas aeruginosa* lectins. Methods Enzymol. 1982;83:378–85.

31 Imberty A, Wimmerova M, Mitchell EP, Gilboa-Garber N. Structures of the lectins from *Pseudomonas aeruginosa*: insights into the molecular basis for host glycan recognition. Microbes Infect. 2004;6(2):221–8.

32 van Kooyk Y, Geijtenbeek TB. DC-SIGN: escape mechanism for pathogens. Nat. Rev. Immunol. 2003;3(9):697–709.

33 Sattin S, Daghetti A, Thepaut M, *et al*. Inhibition of DC-SIGN-mediated HIV infection by a linear trimannoside mimic in a tetravalent presentation. ACS Chem. Biol. 2010;5(3):301–12.

34 Berzi A, Reina JJ, Ottria R, *et al*. A glycomimetic compound inhibits DC-SIGN-mediated HIV infection in cellular and cervical explant models. AIDS 2012;26(2):127–37.

35 Luczkowiak J, Sattin S, Sutkeviciute I, *et al*. Pseudosaccharide functionalized dendrimers as potent inhibitors of DC-SIGN dependent Ebola pseudotyped viral infection. Bioconjug. Chem. 2011;22(7):1354–65.

36 Ribeiro-Viana R, Sanchez-Navarro M, Luczkowiak J, *et al*. Virus-like glycodendrinanoparticles displaying quasi-equivalent nested polyvalency upon glycoprotein platforms potently block viral infection. Nat. Commun. 2012;3:1303.

37 Kitov PI, Sadowska JM, Mulvey G, *et al*. Shiga-like toxins are neutralized by tailored multivalent carbohydrate ligands. Nature 2000;403(6670):669–72.

38 Jacobson JM, Yin J, Kitov PI, *et al*. The crystal structure of shiga toxin type 2 with bound disaccharide guides the design of a heterobifunctional toxin inhibitor. J. Biol. Chem. 2014;289(2):885–94.

39 Tsutsuki K, Watanabe-Takahashi M, Takenaka Y, *et al*. Identification of a peptide-based neutralizer that potently inhibits both shiga toxins 1 and 2 by targeting specific receptor-binding regions. Infect. Immunol. 2013;81(6):2133–8.

40 Pukin AV, Branderhorst HM, Sisu C, *et al*. Strong inhibition of cholera toxin by multivalent GM1 derivatives. ChemBioChem 2007;8(13):1500–3.

41 Branderhorst HM, Liskamp RM, Visser GM, Pieters RJ. Strong inhibition of cholera toxin binding by galactose dendrimers. Chem. Commun. (Camb.) 2007;(47):5043–5.

42 Sigal GB, Mammen M, Dahmann G, Whitesides GM. Polyacrylamides bearing pendant α-sialoside groups strongly inhibit agglutination of erythrocytes by influenza virus: The strong inhibition reflects enhanced binding through cooperative polyvalent interactions. J. Am. Chem. Soc. 1996;118(16):3789–800.

43 Kiessling LL, Gestwicki JE, Strong LE. Synthetic multivalent ligands in the exploration of cell-surface interactions. Curr. Opin. Chem. Biol. 2000;4(6):696–703.

44 Kanai M, Mortell KH, Kiessling LL. Varying the size of multivalent ligands: The dependence of Concanavalin A binding on neoglycopolymer length. J. Am. Chem. Soc. 1997;119(41):9931–2.

45 Lee CM, Weight AK, Haldar J, *et al*. Polymer-attached zanamivir inhibits synergistically both early and late stages of influenza virus infection. PNAS 2012;109(50):20385–90.

46 Weight AK, Haldar J, Alvarez de Cienfuegos L, *et al*. Attaching zanamivir to a polymer markedly enhances its activity against drug-resistant strains of influenza a virus. J. Pharm. Sci. 2011;100(3):831–5.

47 Lauster D, Pawolski D, Storm J, *et al*. Potential of acylated peptides to target the influenza A virus. Beilstein J. Org. Chem. 2015;11:589–95.

48 Mansfield ML, Klushin LI. Intrinsic viscosity of model starburst dendrimers. J. Phys. Chem. 1992;96(10):3994–8.

49 Murat M, Grest GS, Molecular dynamics study of dendrimer molecules in solvents of varying quality. Macromolecules 1996; 29(4):1278–85.

50 Stechemesser S, Eimer W. Solvent-dependent swelling of poly(amido amine) starburst dendrimers. Macromolecules 1997;30(7):2204–6.

51 Tomalia DA. Architecturally driven properties based on the dendritic state. High Perform. Polym. 2001;13(2):S1–S10.

52 Newkome GR, Shreiner CD. Poly(amidoamine), polypropylenimine, and related dendrimers and dendrons possessing different $1 \to 2$ branching motifs: An overview of the divergent procedures. Polymer 2008;49(1):1–173.

53 Staedtler AM, Hellmund M, Sheikhi Mehrabadi F, *et al.* Optimized effective charge density and size of polyglycerol amines leads to strong knockdown efficacy in vivo. J. Mater. Chem. B 2015;3(46):8993–9000.

54 Kainthan RK, Muliawan EB, Hatzikiriakos SG, Brooks DE. Synthesis, characterization, and viscoelastic properties of high molecular weight hyperbranched polyglycerols. Macromolecules 2006;39(22):7708–17.

55 Durka M, Buffet K, Iehl J, et al. The functional valency of dodecamannosylated fullerenes with *Escherichia coli* FimH-towards novel bacterial antiadhesives. Chem. Commun. (Camb.) 2011;47(4):1321–3.

56 Munoz A, Sigwalt D, Illescas BM, *et al.* Synthesis of giant globular multivalent glycofullerenes as potent inhibitors in a model of Ebola virus infection. Nat. Chem. 2016;8(1):50–7.

57 Bengali Z, Satheshkumar PS, Moss B. Orthopoxvirus species and strain differences in cell entry. Virol. 2012;433(2):506–12.

58 Bengali Z, Townsley AC, Moss B. Vaccinia virus strain differences in cell attachment and entry. Virol. 2009;389(1–2):132–40.

59 Carter GC, Law M, Hollinshead M, Smith GL. Entry of the vaccinia virus intracellular mature virion and its interactions with glycosaminoglycans. J. Gen. Virol. 2005;86(Pt 5):1279–90.

60 Chung CS, Hsiao JC, Chang YS, Chang W. A27L protein mediates vaccinia virus interaction with cell surface heparan sulfate. J. Virol. 1998;72(2):1577–85.

61 Whitbeck JC, Foo CH, Ponce de Leon M, *et al.* Vaccinia virus exhibits cell-type-dependent entry characteristics. Virol. 2009;385(2):383–91.

62 Rabenstein DL. Heparin and heparan sulfate: structure and function. Nat. Prod. Rep. 2002;19(3):312–331.

63 Turk H, Haag R, Alban S. Dendritic polyglycerol sulfates as new heparin analogues and potent inhibitors of the complement system. Bioconjug. Chem. 2004;15(1):162–7.

64 Calderon M, Quadir MA, Sharma SK, Haag R. Dendritic polyglycerols for biomedical applications. Adv. Mater. 2010;22(2):190–218.

65 Kainthan RK, Janzen J, Levin E, *et al.* Biocompatibility testing of branched and linear polyglycidol. Biomacromolecules 2006;7(3):703–9.

66 Dernedde J, Rausch A, Weinhart M, *et al.* Dendritic polyglycerol sulfates as multivalent inhibitors of inflammation. PNAS 2010;107(46):19679–84.

67 Niemeyer CM. Nanoparticles, proteins, and nucleic acids: Biotechnology meets materials science. Angew. Chem. Int. Ed. 2001;40(22):4128–58.

68 Lin CC, Yeh YC, Yang CY, *et al.* Quantitative analysis of multivalent interactions of carbohydrate-encapsulated gold nanoparticles with concanavalin A. Chem. Commun. 2003;(23):2920–1.

69 Martinez-Avila O, Hijazi K, Marradi M, *et al.* Gold manno-glyconanoparticles: Multivalent systems to block HIV-1 gp120 binding to the lectin DC-SIGN. Chem. Eur. J. 2009;15(38):9874–88.

70 Martinez-Avila O, Bedoya LM, Marradi M, *et al.* Multivalent manno-glyconanoparticles inhibit DC-SIGN-mediated HIV-1 trans-infection of human T cells. ChemBioChem 2009;10(11):1806–9.

71 Papp I, Sieben C, Ludwig K, *et al*. Inhibition of influenza virus infection by multivalent sialic-acid-functionalized gold nanoparticles. Small 2010;6(24):2900–6.

72 Vonnemann J, Sieben C, Wolff C, *et al*. Virus inhibition induced by polyvalent nanoparticles of different sizes. Nanoscale 2014;6(4):2353–60.

73 Khan S, Gupta A, Verma NC, Nandi CK. Kinetics of protein adsorption on gold nanoparticle with variable protein structure and nanoparticle size. J. Chem. Phys. 2015;143(16):164709.

74 Hennig R, Pollinger K, Veser A, *et al*. Nanoparticle multivalency counterbalances the ligand affinity loss upon PEGylation. J. Control. Release 2014;194:20–7.

75 Stankovich S, Dikin DA, Dommett GH, *et al*. Graphene-based composite materials. Nature 2006;442(7100):282–6.

76 Richter M, Steinhilber D, Haag R, von Klitzing R. Visualization of real-time degradation of pH-responsive polyglycerol nanogels via atomic force microscopy. Macromol. Rapid Commun. 2014;35(23):2018–22.

77 Seliktar D. Designing cell-compatible hydrogels for biomedical applications. Science 2012;336(6085):1124–8.

78 Oh JK, Drumright R, Siegwart DJ, Matyjaszewski K. The development of microgels/ nanogels for drug delivery applications. Prog. Polym. Sci. 2008;33(4):448–77.

79 Sisson AL, Steinhilber D, Rossow T, *et al*. Biocompatible functionalized polyglycerol microgels with cell penetrating properties. Angew. Chem. Int. Ed. 2009;48(41):7540.

80 Hamidi M, Azadi A, Rafiei P. Hydrogel nanoparticles in drug delivery. Adv. Drug Deliv. Rev. 2008;60(15):1638–49.

81 Richter A, Paschew G, Klatt S, *et al*. Review on hydrogel-based pH sensors and microsensors. Sensors 2008;8(1):561–81.

82 Nguyen MK, Lee DS. Injectable biodegradable hydrogels. Macromol. Biosci. 2010;10(6):563–79.

83 Drury JL, Mooney DJ. Hydrogels for tissue engineering: scaffold design variables and applications. Biomaterials 2003;24(24):4337–51.

84 Fedorovich NE, Alblas J, de Wijn JR, *et al*. Hydrogels as extracellular matrices for skeletal tissue engineering: state-of-the-art and novel application in organ printing. Tissue Eng. 2007;13(8):1905–25.

85 Sisson AL, Haag R. Polyglycerol nanogels: highly functional scaffolds for biomedical applications. Soft Matter 2010;6(20):4968–75.

86 Steinhilber D, Seiffert S, Heyman JA, *et al*. Hyperbranched polyglycerols on the nanometer and micrometer scale. Biomaterials 2011;32(5):1311–6.

87 Steinhilber D, Witting M, Zhang X, *et al*. Surfactant free preparation of biodegradable dendritic polyglycerol nanogels by inverse nanoprecipitation for encapsulation and release of pharmaceutical biomacromolecules. J. Control. Release 2013;169(3):289–95.

88 Fleige E, Quadir MA, Haag R. Stimuli-responsive polymeric nanocarriers for the controlled transport of active compounds: concepts and applications. Adv. Drug Deliv. Rev. 2012;64(9):866–84.

89 Steinhilber D, Sisson AL, Mangoldt D, *et al*. Synthesis, reductive cleavage, and cellular interaction studies of biodegradable, polyglycerol nanogels. Adv. Funct. Mater. 2010;20(23):4133–8.

90 Rossow T, Heyman JA, Ehrlicher AJ, *et al*. Controlled synthesis of cell-laden microgels by radical-free gelation in droplet microfluidics. J. Am. Chem. Soc. 2012;134(10):4983–9.

10

Multivalent Protein Recognition Using Synthetic Receptors

Akash Gupta, Moumita Ray, and Vincent M. Rotello

Department of Chemistry, University of Massachusetts, Amherst, 710 North Pleasant Street, Amherst, MA 01002, USA

10.1 Introduction

Protein–protein interactions (PPIs) play vital roles in numerous biological processes [1]. The supramolecular nature of these interactions differs based on the composition and affinity of the involved peptide strands and in the dynamics of the proteins [2]. Rational approaches towards the recognition of protein surfaces such as co-crystallization of proteins along with their counterparts or mutational analysis of the protein–protein interfaces provides better insight into the structural epitopes involved in PPIs, providing important clues for biomimetic strategies. In the following section, we will review physical and chemical properties of protein surfaces involved in PPI, in the context of designing synthetic receptors discussed in the later sections.

10.2 Structural Properties of Protein Surfaces

In nature, there are distinct families of protein–protein complexes. There are heterodimeric protein–protein complexes, such as hormone–receptor or antigen–antibody complexes, in which each protein has its own stable identity upon isolation. Homo-dimeric/oligomeric protein complexes, on the other hand, consist of multiple identical proteins that do not have a stable identity on their own.

10.2.1 Protein–Protein Interfacial Areas

Protein–protein interfaces rely on inter-protein proximity coupled with concomitant changes in solvent accessibility upon binding. Protein–protein associations are formed by the electrostatic and hydrophobic interactions established between functionalities on the constituent proteins. The flexibility of the peptides involved in the interactions is likewise crucial. The standard size of the protein interfacial area varies from 1200 to 2000 Å^2. Finally, size matters: smaller interfacial areas in the range of 1150–1200 Å^2 are

Multivalency: Concepts, Research & Applications, First Edition. Edited by Jurriaan Huskens, Leonard J. Prins, Rainer Haag, and Bart Jan Ravoo.
© 2018 John Wiley & Sons Ltd. Published 2018 by John Wiley & Sons Ltd.

predominant in low-stability complexes [3]. Larger interfacial areas in the range of $2000-4660\,\text{Å}^2$ are prevalent within enzyme–inhibitor complexes. However, $600\,\text{Å}^2$ is considered as a minimum area required for a "water-tight" energetically favorable interaction [4].

10.2.2 Chemical Nature of the Protein–Protein Interface

Cooperative networks of non-covalent forces provide stability to PPIs. Protein–protein interfaces are generally hydrophobic and bury a large extent of their non-polar surface area upon interaction. Therefore, hydrophobicity is one of the key non-covalent interactions found in protein–protein interfaces [5]. Van der Waals contact among non-polar amino acids is a major component in hydrophobic interactions, coupled with the entropically favorable release of water upon binding. Protein residues at the interface are organized as hydrophobic patches. These patches vary from 1 to 15 amino acids, with their sizes ranging from 200 to $400\,\text{Å}^2$ [6]. It is important to note that while hydrophobic interactions are the major driving force in PPIs, they lack directionality, requiring additional interactions to impart selectivity/specificity to PPIs [7].

Electrostatic interactions are the other major non-covalent force involved in PPI formation [8]. These interactions are distance dependent (inversely proportional to the square of the distance), and significantly enhance the diffusion-controlled association of the proteins. Hydrogen bonds (H-bonds) are likewise important, and tend to be proportional to the area of protein interface, with ~1 H-bond per $100-200\,\text{Å}^2$ of protein interface [9]. Water molecules surrounding the protein–protein interface form $\sim24\%$ of the H-bonds with the protein surfaces, while the rest is formed by the amino acids at the protein interfaces.

10.2.3 "Hot Spots"

Amino acid residues residing at the core of the protein–protein interface, known as the "hot spot", are vital for PPI stability [10]. Alanine-scanning mutagenesis is one of the most popular methods for mapping hot spots associated with PPIs. In this method, an alanine is substituted at specific sites to probe the role of individual amino acid residues [11,12]. This methodical approach determines the contribution of the side chain to the binding energy at the protein–protein interface. Hot spots are multipurpose – the same hot spots can be used by proteins to bind to multiple partners. Therefore, it is necessary to understand the structural features of a hot spot to elucidate the chemical nature of PPIs, which in turn is helpful in designing complementary binding partners for the proteins.

Hot spots have distinctive amino acid compositions, with Trp (21%), Arg (13.3%), Tyr (12.3%) and Ile (9.6%) being very regularly identified in hot spots [9]. Conversely, Leu, Phe, Ser and Thr are rarely found in hot spots. The abundance of Trp in hot spots arises from the large surface area and aromatic nature of the side chain [13]. Trp mutation to Ala generates huge cavities, resulting in destabilization of protein–protein complexes [14]. Moreover, Trp protects weak H-bonds from bulk solvent [15]. The abundance of Tyr is also due to its hydrophobic surface. Arg, on the other hand, forms H-bonded salt bridges due to the presence of the positive charge on the guanidinium group. This concept is further illustrated in the following example. There are 29 interfacial amino acid residues that are involved in the complexation of human growth hormone and the growth hormone protein [16,17]. Among them only four are hot spots, including two Trp, one Arg and one Ile residue as shown in Figure 10.1. As expected, Ala substitution at any of these sites reduces the binding affinity between human growth hormone and the growth hormone protein drastically.

(a)

(b)

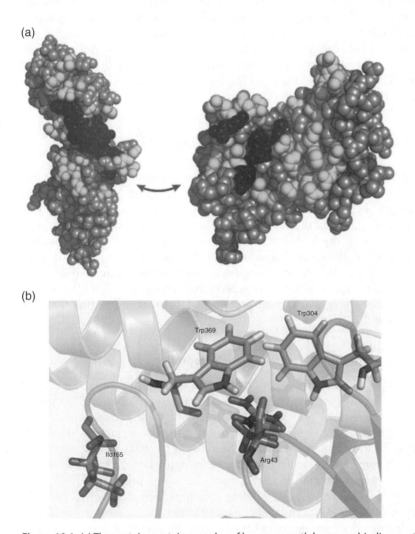

Figure 10.1 (a) The protein–protein complex of human growth hormone binding protein (hGHbp) and growth hormone (hGH) (PDB ID: 3HHR) is separated to expose the interacting surface. The proteins are colored either purple (hGH) or gray (hGHbp) except for the "hot spot" (red) and "rim" (yellow) residues. (b) The hGH (in cyan) complexed with its receptor, hGHbp (in blue). The four hot spot residues are highlighted by a stick representation and the two tryptophan residues are colored in pink. The PDB ID: 1A22. *Source*: Refs [9,17]. Reproduced with permission of John Wiley and Sons and Nature Publishing Group. See color section.

10.2.4 O-Ring Structure

Along with the hot spot, the O-ring is another structural feature of PPIs worth mentioning. The O-ring is a rim of amino acids surrounding the hot spots in protein–protein complexes. This ring of amino acids excludes water from the interface, providing stability to the complex. Tyr, Trp and Asn residues are abundant at the O-ring motifs. These amino acids form hydrophobic and H-bonding interactions to stabilize the protein–protein complex. Therefore, at the protein–protein interface, there is a core and a rim of amino acid residues differing in composition. Upon complex formation, the core region is buried, while the rim of amino acids is partly solvent accessible.

Complex formation between lysozyme HEL and FVD1.3 antibody provides a demonstration of O-ring dynamics [18]. Simulation studies show that at the HEL-FVD1.3 interface interfacial water is shielded from the bulk solvent surrounding the protein complex, illustrating the concept of the O-ring.

10.3 Synthetic Receptors for Protein Surface Recognition

Synthetic receptors that can selectively bind to the biomacromolecule surface serve as useful tools for competing with, and hence regulating, PPIs. Artificial receptors provide a means to observe and control cellular processes for diagnostic and therapeutic applications [19]. One of the key challenges in the design of synthetic receptors for proteins is providing the large surface area required for high affinity binding with the protein surface. Hence, multivalent, expansive and structurally well-defined scaffolds are required for efficient and selective interaction. Synthetic scaffolds such as porphyrins, calixarenes, molecular tweezers, and nanoparticles (NPs) exhibit selective interaction with highly irregular protein surfaces that can be fine-tuned by surface functionalization.

10.3.1 Porphyrin Scaffolds for Protein Surface Recognition

In pioneering studies, Jain and Hamilton demonstrated [20] surface recognition of cytochrome c (cyt c) using porphyrin-based artificial receptors. Cyt c is highly cationic (pI = 10) and interacts with its protein partners chiefly through a hydrophobic patch surrounded by cationic Arg and Lys residues. In their studies, tetraphenylporphyrin (TPP) derivatives bearing anionic side chains were used to study the importance of electrostatic interaction between surface and receptor (Figure 10.2a). On increasing the number of free acid groups on the side chains from four in Receptor 1 (Figure 10.2b, **1**) to eight in Receptor 2 (Figure 10.2b, **2**), binding constants increased by five-fold. The binding affinity of the scaffolds was further enhanced by decorating the porphyrin scaffolds with Tyr-Asp dipeptide sequences in Receptor 3 (Figure 10.2b, **3**) due to increased hydrophobic interactions. In subsequent studies [21], TPP-scaffolds with larger hydrophobic cores and increased anionic charge (e.g., Receptor 4; Figure 10.2b, **4**) exhibited the strongest affinity with cyt c. These receptors demonstrated good selectivity in binding to cyt c ($K_d = 0.67$ nM) when compared with some closely related proteins such as ferredoxin ($K_d = 17 \mu M$) and cyt c551 ($K_d = 180$ nM). These findings provide insight on the importance of charge and size complementarity for efficient binding between surface and receptor. The thermal stability of cyt c upon binding with TPP derivatives

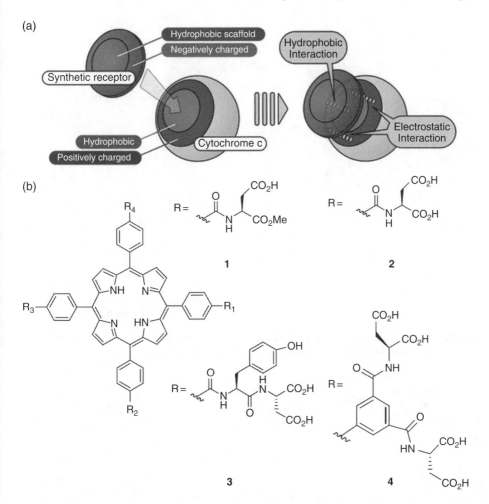

Figure 10.2 (a) Schematic representation of cytochrome c binding by synthetic receptors. (b) Structures of compounds employed in the denaturation studies. *Source*: (a) Ref. [21]. Reproduced with permission of Elsevier. (b) Ref. [25]. Reproduced with permission of John Wiley and Sons.

was also studied [22] and in the presence of 1.2 eq of Receptor 3 a reduction in the melting point of cyt c from 85 to 64 °C was observed. However, no changes were observed in the native structure of the protein at room temperature.

The above studies suggested that the modulation of functionalities on porphyrin scaffolds could yield agents capable of denaturing cyt c in physiological conditions [23]. Copper(II) porphyrin derivatives undergo extensive π–π stacking interaction in water, leading to stronger interaction with the hydrophobic surface of the folded protein. These copper–porphyrin dimers resulted in a strong denaturing effect on cyt c by inducing substantial conformational changes. Apart from unfolding, metallo-porphyrin derivatives also assist in the proteolytic degradation of cyt c. Receptor 5 (Figure 10.3) enables digestion of cyt c by trypsin in physiological conditions within 15 min (Figure 10.4). These receptors bind to α-helices and other folded parts of cyt c, allowing trypsin to access the active site of the protein. Further studies were performed to

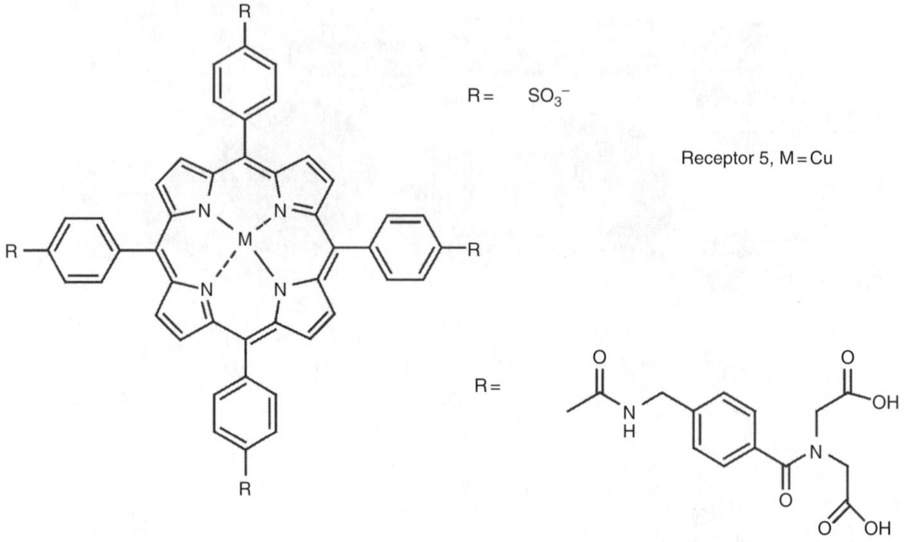

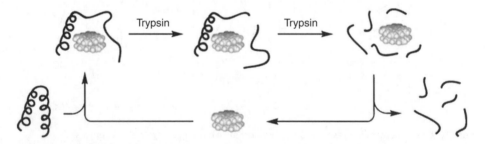

R = SO$_3^-$

Receptor 5, M = Cu

Receptor 6, M = Cu

Figure 10.3 Chemical structures employed in the denaturation and digestion studies of cytochrome c. *Source*: Ref. [22]. Reproduced with permission of American Chemical Society.

Figure 10.4 Proposed catalytic cycle for the accelerated digestion of cytochrome c by porphyrin dimer. *Source*: Ref. [23]. Reproduced with permission of American Chemical Society.

ascertain the selectivity of receptors towards protein surfaces. The binding affinity of anionic copper porphyrins was examined against larger heme proteins such as hemoglobin (64.5 kDa, pI = 6.9) and myoglobin (17.2 kDa, pI = 7.2) [24]. Receptor 6 (Figure 10.3) selectively activated the proteolytic digestion of the heme proteins, whereas no effect was observed on the stability of azurin and α-lactalbumin (Figure 10.5).

TPP derivatives are promising candidates for protein surface recognition owing to their large hydrophobic surface area (>300 Å^2) and ease of surface functionalization [25]. Moreover, they are highly fluorescent and display intensity changes upon binding to target protein. Exploiting these characteristic features, Hamilton and coworkers designed an array of TPP derivatives for pattern-based detection of proteins. TPP derivatives were decorated with a range of amino acids and dipeptides, resulting in the synthesis of 35 different molecules with different physico-chemical properties [26].

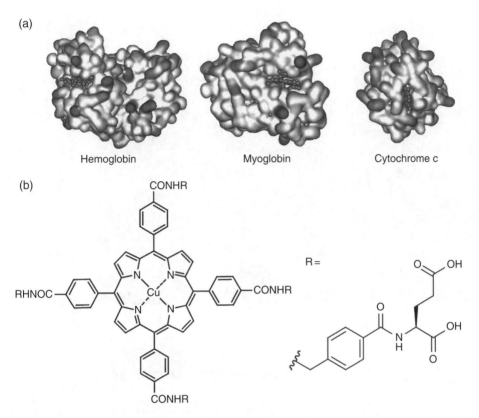

Figure 10.5 (a) The relative sizes of cytochrome c (MW = 12 kDa, PDB: 3CYT), myoglobin (MW = 17.2 kDa, PDB: 1MBN) and hemoglobin (MW = 64.5 kDa, PDB: 2HCO). (b) Receptor structure. *Source*: Ref. [24]. Reproduced with permission of Royal Society of Chemistry).

The molecules comprised 4 to 8 hydrophobic groups with charges ranging from +8 to −8. The array of TPP derivatives was used to differentiate ferredoxin (pI 2.75), cyt c551 (pI 4.7), myoglobin (pI 6.8) and cyt c (pI 10.6) (Figure 10.6). Differential binding of receptors with target proteins generated distinct fluorescent responses that could be easily identified with the naked eye. The response of receptors could be attributed to their charge complementarity with proteins (i.e., stronger binding yields greater fluorescence quenching). Anionic receptors exhibited stronger fluorescence quenching with cationic cyt c, whereas cationic receptors exhibited increased quenching with ferredoxin. Subsequently, Hamilton and coworkers coupled porphyrin-based arrays with pattern recognition techniques for identification of both metal and non-metal containing proteins [27]. A library of TPP-based receptors with different surface functionality was titrated against target proteins to generate different degrees of receptor fluorescence quenching. α-Lactalbumin and lysozyme exhibited reduced fluorescence quenching when compared with other metal containing proteins (cyt c, ferredoxin) (Figure 10.7). The interaction of porphyrin receptors with protein surfaces yielded distinct fluorescence patterns for each protein. These fluorescence patterns were analyzed using principal component analysis, generating unique fingerprints for every sample.

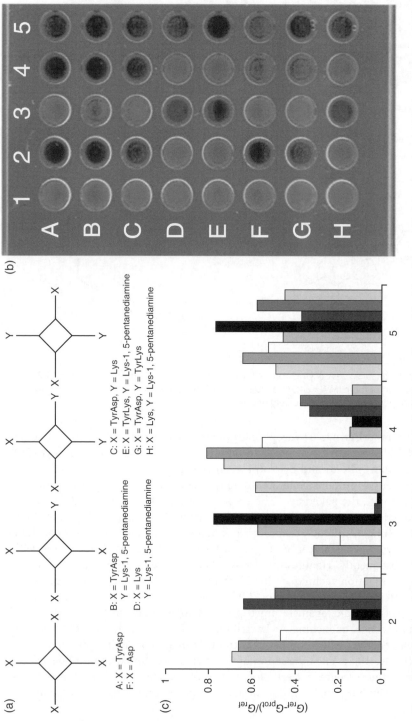

Figure 10.6 (a) Schematic structure of porphyrin receptors A–H. (b) Picture of part of the 96 well quartz plate irradiated with UV light (302 nm) and producing red fluorescence. Each well contains a solution of porphyrins A–H; in the wells of columns 2–5 were added 3 eq of: 2, cytochrome c (pI = 10.6, MW ≈ 12 500); 3, ferredoxin (pI = 2.75, MW ≈ 5500); 4, cytochrome c551 (pI = 4.7, MW ≈ 9000); 5, myoglobin (pI = 6.8, MW ≈ 18 000). (c) Fingerprints of cytochrome c (2), ferredoxin (3), cytochrome c551 (4), and myoglobin (5) based on the 8-porphyrin array. Each bar quantifies the extent of fluorescence attenuation measured as the quantity $(G_{ref} - G_{prot})/G_{ref}$, where G_{ref} is the average gray value for the blank wells and G_{prot} is the average gray value for the protein-containing wells. *Source:* Ref. [26]. Reproduced with permission of American Chemical Society.

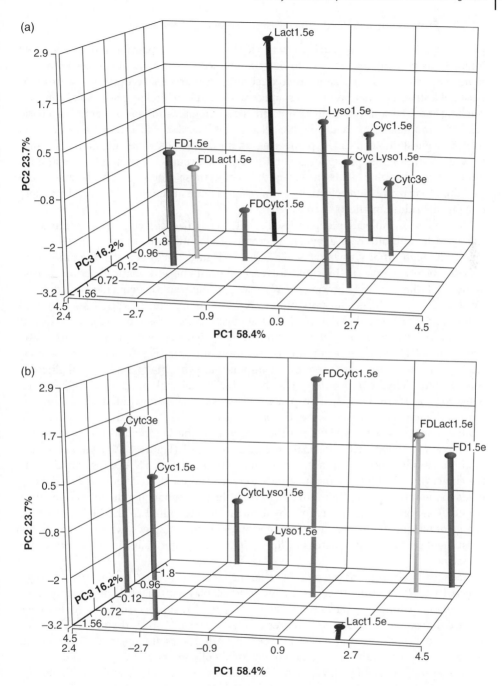

Figure 10.7 Principal component analysis mapping of protein samples against an 8-porphyrin array (a) or a 16-porphyrin array (b). The samples were tested on a 16-porphyrin array, while (a) is extracted from (b) by taking the data matrix of the first eight porphyrin receptors. *Source*: Ref. [27]. Reproduced with permission of American Chemical Society.

10.3.2 Protein Surface Recognition Using Molecular Tweezers

Non-covalent interactions such as hydrogen bonding, ion pairing and arene–arene interactions play important roles in determining PPIs. Synthetic molecules known as molecular "clips" and "tweezers" are frequently used as artificial receptors and are composed of a structure that provides a shaped cavity that can bind to guest molecules by π stacking and CH–π interactions [28]. Further functionalization of the central aromatic unit can modulate the hydrophobic, electrostatic and steric interactions between host and guest molecules. In pioneering studies, Fokkens *et al.* created molecular tweezers with anionic phosphonate and phosphate groups for recognition of ammonium ions and amino acids in water [29]. Later these electron rich moieties were exploited to provide high selectivity towards lysine and arginine residues and a variety of *N*-alkylpyridinium guests such as NAD$^+$.

The ability of molecular clips to selectively recognize cofactors and amino acids was used for inhibition of enzymatic activity of alcohol dehydrogenase (ADH) [30]. Due to its high affinity towards NAD$^+$, Clip 1 (Figure 10.8a) depletes the amount of cofactor below a critical threshold for carrying out the enzymatic reaction. Excess amount of the scaffold, however, irreversibly denatured the enzyme. In contrast, Tweezer 2 (Figure 10.8a) does not bind to NAD$^+$, but is instead strongly associated to lysine residues proximal to the ADH site thereby blocking substrate entry. Just 0.6 eq of Tweezer 2 was able to completely shut down the enzyme activity. However, enzymatic activity of ADH could be restored upon addition of lysine (Figure 10.8b). In a similar study, molecular clips were used for the inhibition of enzymatic oxidation of glucose-6-phosphate (G6P) with NADP$^+$ catalyzed by glucose-6-phosphate dehydrogenase (G6PD) [31]. Phosphate and phosphonate substituted clips inhibited enzymatic activity of G6PD by cofactor inclusion, both demonstrating comparable binding constants with NADP$^+$. Further analysis of the kinetic and thermodynamic parameters revealed that phosphate-substituted Clip 3 (Figure 10.9a) follows a dual inhibition mode. Clip 3 forms a stable ternary complex in which NADP$^+$ is bound inside the clip cavity, blocking the nicotinamide to reach the substrate. Additionally, the clip binds to the substrate binding sites on enzymes, further enhancing the inhibition effect (Figure 10.9b and c). This dual mechanism can be attributed to the higher negative charge of phosphate groups that promotes binding with enzymes, whereas the methyl substituent on the phosphonates in Clip 4 (Figure 10.9a) sterically hinders enzyme-clip association.

Molecular tweezers developed by Klärner and coworkers demonstrate a high selectivity towards lysine residues. Lysine residues play a crucial role in the early steps of protein misfolding, causing several neurological disorders such as Alzheimer's disease. Phosphate substituted Tweezer 5 (Figure 10.9a) displayed strong affinity for Lys-16 and Lys-28 on the amyloid β-protein (Aβ) 15–29 fragment [32]. Moreover, Tweezer 5 prevented the aggregation of non-toxic oligomers of Aβ and even dissolved preformed Aβ, critical factors for Alzheimer's disease. A similar activity of Tweezer 5 was observed against other amyloidogenic proteins involved in various human diseases. Finally, aggregation-induced cell toxicity could be reduced by the molecular tweezer against a wide range of amyloidogenic proteins. Subsequent studies were performed using Tweezer 5 which improved spatial memory of AD transgenic mice, suggesting that these mechanisms also operate *in vivo* [33].

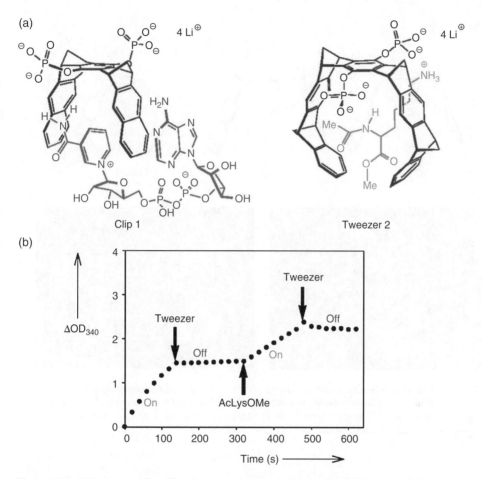

Figure 10.8 (a) Structures of a molecular tweezer and a molecular clip. (b) Enzyme switching: alternating addition of tweezer and lysine derivative turns ADH enzyme activity on and off. *Source*: Ref. [30]. Reproduced with permission of American Chemical Society.

In a study by Ottomann and coworkers, molecular tweezers were used to inhibit PPIs between adapter protein 14-3-3 and two partner proteins, kinase C-Raf and exoenzyme S [34]. The high binding of tweezer with 14-3-3 protein was attributed to strong hydrophobic interactions between the concave aromatic scaffold and the surface-exposed Lys-214 residue (Figure 10.10). Additionally, one phosphate group of the tweezer forms an ion pair with the amino group of a Lys residue, blocking the site for association of 14-3-3 proteins with exoenzyme S and phosphorylated C-Raf. These observations were corroborated by isothermal titration calorimetry (ITC), X-ray crystallography and computational simulation (Figure 10.10). ITC studies revealed that this interaction is an exothermic process comprising two independent binding events with different affinities, while crystallographic studies suggested that the molecular tweezers primarily target certain surface-exposed lysine residues, which are a hotspot for 14-3-3σ binding with its partner proteins.

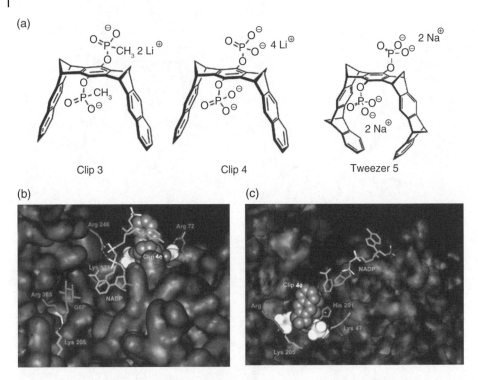

Clip 3   Clip 4   Tweezer 5

Figure 10.9 (a) Structures of a molecular tweezer and molecular clips. Molecular dynamics simulations illustrating the two modes of inhibition of G6PD by clip (b) ternary complex of enzyme, clip and cofactor, and (c) binary enzyme complex of clip occupying the substrate-binding site. *Source*: Refs [31,32]. Reproduced with permission of John Wiley and Sons and American Chemical Society.

10.3.3 Calixarene Scaffolds for Protein Surface Recognition

Calixarenes are widely used as receptors for biomolecule recognition owing to the ability to vary the size and shape of their cavities for molecular recognition [35]. These features have enabled calixarene-based systems to be used in the development of biosensors, biotechnology and drug discovery (see also Chapter 11) [36]. Hamilton and coworkers have designed calixarene-based receptors for protein surface recognition of cyt c [37]. These receptors are composed of a calixarene core functionalized with four cyclic constrained peptide loops. These loops mimic the hypervariable loops in an antibody, while the acidic calixarene strongly binds with the positively charged lysine groups in the protein near its heme region. Affinity chromatography and gel filtration indicated that longer alkyl chain substituents at the lower rim interfered with the binding at the upper rim. In subsequent studies, the affinity and selectivity of calixarene receptors was tuned by changing the amino-acid sequence [38]. By changing the sequence of loop regions, receptors with diverse functionalities were generated covering approximately $450\,\text{Å}^2$ in area. The most potent inhibitor from the library of receptors exhibited slow binding kinetics with α-chymotrypsin (ChT), similar to that of natural proteinase inhibitors (Figure 10.11) [39].

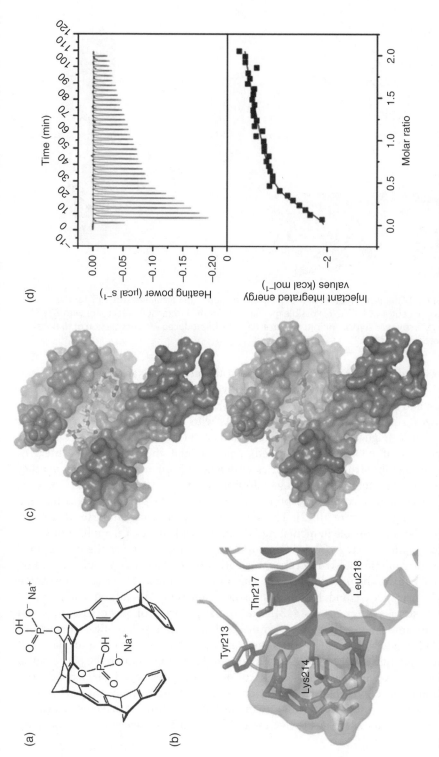

Figure 10.10 (a) Structure of the tweezer. (b) Detailed view of tweezers binding with Lys 214 from 14-3-3σ. (c) Binding site competition between molecular tweezers and C-RafpS259 or ExoS. Top, superimposition of C-RafpS259 (stick structure) binding to the complex of tweezers and 14-3-3σ. Bottom, superimposition of ExoS interaction motif (stick structure) binding to the complex of tweezers and 14-3-3σ. (d) ITC of 14-3-3σ titrated with tweezers shows exothermic binding. *Source*: Ref. [34]. Reproduced with permission of Nature Publishing Group.

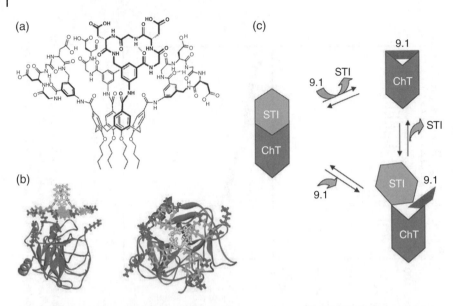

Figure 10.11 (a) Chemical structure of the calixarene receptor. (b) Calculated structure for the interaction of receptor with ChT. (c) Possible mechanisms for the interaction of receptor with ChT-STI complexes. STI, soybean trypsin inhibitor. *Source*: Ref. [39]. Reproduced with permission of National Academy of Sciences, USA.

Mendoza and coworkers have demonstrated the use of calixarene-based synthetic receptors for binding and stabilizing negatively charged α-helices, abundant in glutamates and aspartates [40]. A conical calix[4]arene functionalized with four guanidinomethyl residues at the upper rim interacts with negatively charged residues at the p53 surface (Figure 10.12). The interaction of the calixarene receptors bridged the monomers of p53-R337H (mutated) and retained their tetrameric integrity, providing means for structural recoveries of damaged protein assemblies. In another study, the crystal structure of a protein-calixarene complex was reported [41]. Binding studies indicate that calixarene receptor binds to two or more binding sites by forming a dynamic complex (Figure 10.12). The above-mentioned studies indicate that calixarenes can play a major role in mediating PPIs for valuable biological applications.

Another popular strategy for protein surface recognition involves the use of small molecules that mimic the structural conformation of the key amino acid residues on the original helix of the protein [42]. The Hamilton group reported that trisubstituted terphenyl derivatives could be used as α-helix mimetics to inhibit calmodulin–phosphodiesterase interactions [43]. Additionally, series of terephthalamide and 4,4-dicarboxamide derivatives were synthesized to inhibit Bcl-x_L–Bak interactions [(44,45]. Subsequently, oligamide-based helix mimetics were designed to regulate islet amyloid polypeptide (IAPP) aggregation, a protein that is important in the pathology of type II diabetes [46]. Similarly, Wilson and coworkers have developed oligoamide derivatives as inhibitors of p53–hDM2 interactions [47]. The use of α-helix mimics for regulating PPIs provides a detailed insight into the structural features of proteins that can be used for designing broad range of selective and therapeutically relevant inhibitors [48].

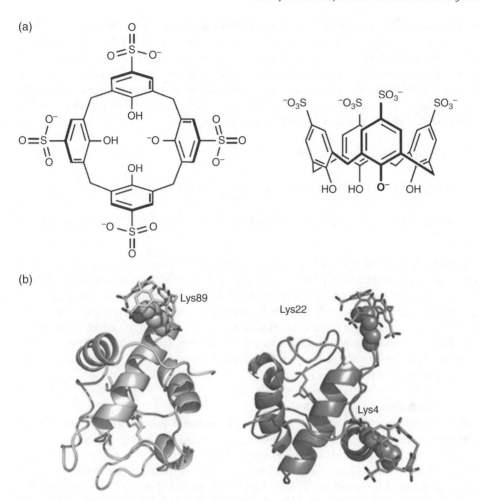

Figure 10.12 (a) Structure of p-sulfonatocalix[4]arene (sclx$_4$). (b) Crystal structure of the cytochrome c-sclx$_4$ complex. The asymmetric unit comprises two molecules of cytochrome c and three molecules of sclx$_4$. *Source*: Ref. [41]. Reproduced with permission of Nature Publishing Group.

10.3.4 Recognition of Protein Surfaces Using Nanoparticles

NPs provide versatile platforms for protein surface recognition owing to their large surface area and structurally well-defined architectures for high affinity binding. There are a wide variety of core materials available, featuring different optical, electronic and magnetic properties. Such inherent physical properties make them suitable constructs for designing biosensors, clinical diagnostics and therapeutics [49–52]. Additionally, the tunable core-size of NPs, comparable with the size of proteins allows efficient interactions with biomacromolecules. Most importantly, the surface of the NPs can be tailored with a wide range of functionalities, offering versatility in creating protein surface receptors [53].

It is important to understand the nature of interaction between NPs and proteins to design suitable receptors for protein surface recognition. Studies on NP–protein interactions suggest that electrostatics is the most significant non-covalent force acting on the NP–protein interface. Thermodynamic studies provide a more fundamental understanding of NP–protein interactions [54]. Water molecules surrounding the NPs and proteins play a significant role in their association. The process of NP-protein complex formation can be described by the following equations:

$$\text{Protein} + \text{NP} \rightleftharpoons \text{Protein-NP} \tag{10.1}$$

$$x\left(H_2O\right)_p + y\left(H_2O\right)_n \rightleftharpoons \left(x+y-z\right)\left(H_2O\right)_{p\text{-}n} + z\left(H_2O\right) \tag{10.2}$$

$$\text{Protein}\cdot x\left(H_2O\right)_p + \text{NP}\cdot y\left(H_2O\right)_n \rightleftharpoons \text{Protein-NP}\cdot\left(x+y-z\right)\left(H_2O\right)_{p\text{-}n} + z\left(H_2O\right) \tag{10.3}$$

where $(H_2O)_p$, $(H_2O)_n$ and $(H_2O)_{p\text{-}n}$ are the water molecules associated with protein, NPs, and the protein-NP complex, respectively. Equation 10.1 describes the formation of the protein-NP complex which is enthalpically favored (formation of new bonds) and entropically opposed (losing degrees of freedom). However, Equation 10.2 describing the solvent reorganization process is entropically favored due to release of water molecules from the solvation shells of both NPs and proteins. The combined Equation 10.3 indicates that just like PPIs, NP-protein complex formation can be enthalpically or entropically favored depending upon the forces behind the protein–NP interaction.

10.3.4.1 Nanoparticles as Protein Mimics

The ability to tailor a NP surface with diverse chemical functionalities makes them excellent candidates for mimicking protein surfaces. Additionally, the hydrodynamic radius of small NPs functionalized with ligands is commensurate in size to that of mid-size globular proteins. These similarities were further corroborated by thermodynamic studies performed by Rotello and coworkers which revealed that NP–protein interactions closely resembled PPIs (Figure 10.13). Structurally diverse amino acid terminated anionic NPs were used as receptors to study the process of complex formation with three positively charged proteins (histone, cyt c and ChT) (Figure 10.13a). [55] Binding of NPs with ChT (bearing 17 positively charged residues) was enthalpically driven, while binding to cyt c and histone (20 and 62 positively charged residues, respectively) was entropically driven. These studies indicate that besides electrostatic attraction, other factors also play a crucial role in the complexation process. Enthalpy–entropy compensation studies on NP–protein interactions demonstrate an excellent linear correlation between the two thermodynamic quantities. Thermodynamically, this behavior is more similar to PPIs than any other known synthetic system (Figure 10.13b and c). Further studies were conducted with a series of cationic NPs bearing different functionalities against a panel of proteins with different sizes and geometries. It was observed that ΔH was the driving force when the proteins were large enough to accommodate several NPs, whereas ΔS was the driving force in the case of smaller proteins. This behavior can be attributed to the large number of water molecules per NP that were released in complexation with smaller proteins [54]. In the following section, we will discuss the role of surface functionality in multivalent recognition elements for proteins.

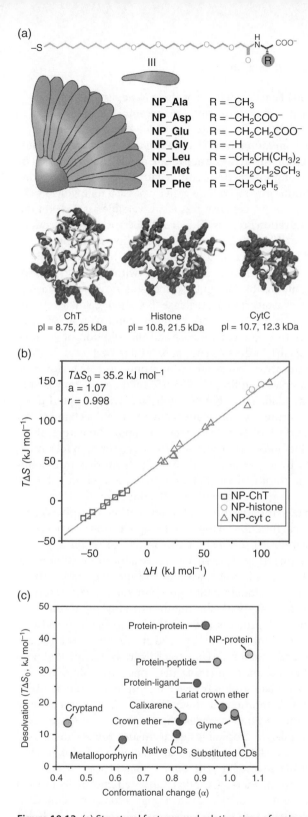

(a)

III

NP_Ala	R = –CH$_3$
NP_Asp	R = –CH$_2$COO$^-$
NP_Glu	R = –CH$_2$CH$_2$COO$^-$
NP_Gly	R = –H
NP_Leu	R = –CH$_2$CH(CH$_3$)$_2$
NP_Met	R = –CH$_2$CH$_2$SCH$_3$
NP_Phe	R = –CH$_2$C$_6$H$_5$

ChT
pl = 8.75, 25 kDa

Histone
pl = 10.8, 21.5 kDa

CytC
pl = 10.7, 12.3 kDa

(b)

$T\Delta S_0 = 35.2$ kJ mol^{-1}
a = 1.07
r = 0.998

$T\Delta S$ (kJ mol^{-1})

ΔH (kJ mol^{-1})

□ NP-ChT
○ NP-histone
△ NP-cyt c

(c)

Desolvation ($T\Delta S_0$, kJ mol^{-1})

Protein-protein
NP-protein
Protein-peptide
Protein-ligand
Lariat crown ether
Cryptand
Calixarene
Crown ether
Glyme
Metalloporphyrin
Native CDs   Substituted CDs

Conformational change (α)

Figure 10.13 (a) Structural features and relative sizes of amino acid-functionalized gold nanoparticles and proteins. The overlapping spheres in the proteins represent the positively charged residues on their surface. (b) Plots of entropy ($T\Delta S$) versus enthalpy (ΔH) for nanoparticle–protein interactions. (c) Slope (α) and intercept ($T\Delta S_0$) values for various host–guest systems. Protein–ligand interactions have been divided into protein–peptide and protein–other (protein–ligand) interactions. *Source*: Ref. [55]. Reproduced with permission of7 American Chemical Society.

10.3.4.2   Regulating the Structure and Function of Proteins Using Nanoparticles

Selective and specific binding of proteins with synthetic receptors enables them to modulate protein structure and activity. In early studies of NP–protein interactions, Rotello and coworkers designed mixed monolayer NPs bearing anionic charges for surface recognition and activity inhibition of ChT. The active site of ChT is surrounded by cationic residues that promote strong interaction with negatively charged receptors. Alkanethiol functionalized NPs inhibit the enzymatic activity of ChT following a two-step inactivation mechanism [56]. This process features fast reversible association followed by a slower irreversible denaturation process. The strong electrostatic interactions resulted in an inhibition constant of $K_i = 10.4 \pm 1.3$ nM and a binding stoichiometry of five molecules of ChT to one NP (Figure 10.14). Denaturation of ChT at the interface was attributed to hydrophobic interactions of the protein with the NP interior (Figure 10.14). The electrostatic interaction is highly selective as no interaction was observed for other proteins such as elastase, β-galactosidase (β-Gal), or cellular retinoic acid binding protein. In subsequent studies, the interaction between ChT and NPs was tuned by changing the ionic strength of aqueous media and the addition of cationic surfactants [57]. Enzyme activity of ChT in the presence of NPs increased from 5 to 97% of the native activity as the salt concentration was increased from 0 to 1.5 M. However, variation in ionic strength after complete binding of NP-ChT could restore only part of the enzymatic activity (<35%). The association of ChT with NPs was disrupted using cationic surfactants, partially restoring its enzymatic activity. The mechanism for release of proteins bound to the surface of NPs was dependent upon the nature of the surfactant [58]. Alkyl surfactant liberated ChT from the NP surface through formation of a bilayer structure as indicated by a ~4.5 nm increase in the hydrodynamic radius of the complex. However, cationic thiol and alcohol were incorporated into the NP monolayers, thereby releasing protein molecules without affecting the size of the NPs.

NP–protein interactions can change the structure and conformation of proteins causing them to lose important biological functions. However, for applications such as protein delivery and enzyme stabilization it is critical for the protein to retain its native structure upon protein binding. Engineering of NP surfaces with appropriate functional groups can regulate their interaction with proteins without altering their native structure. The oligo(ethylene glycol) (OEG) functionality provides such a characteristic where the anti-fouling properties increase with the number of ethylene glycol (EG) units. To prevent the denaturation of proteins, Rotello and coworkers introduced tetra(ethylene glycol) spacers before anionic carboxylate head groups on CdSe NPs [59]. These anionic NPs exhibited strong affinity for ChT, whereas hydroxyl terminated neutral NPs did not interact with the protein, confirming the important role played by electrostatic attraction. Circular dichroism studies revealed retention of the structure of NP-bound proteins and complete restoration of their activity was observed upon increasing the ionic strength of the solution. Besides charge complementarity, NPs bearing chemical motifs featuring polar, hydrophobic and aromatic groups can be used to probe NP–protein interactions. In a related study, a library of amino-acid functionalized NPs was used to systematically investigate the interactions at the NP–protein interface (Figure 10.15a). Gold nanoparticles (AuNPs) inhibited the hydrolysis of *N*-succinyl-L-phenylaniline-*p*-nitroanilide (SPNA) by forming a complex with ChT. The degree of inhibition was attributed to the electrostatic and hydrophobic interactions between the hydrophobic patches of the protein and receptor surface [60].

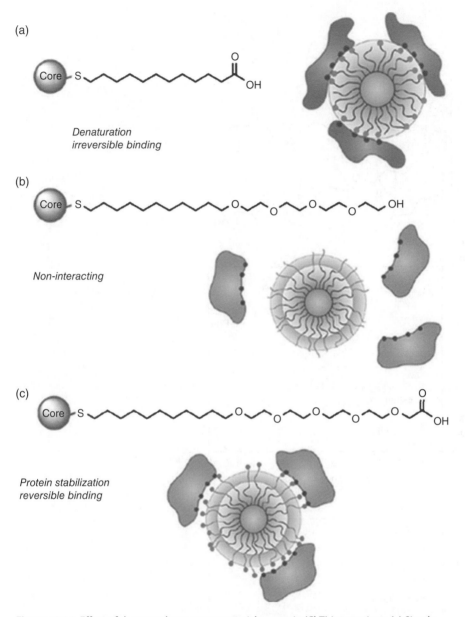

Figure 10.14 Effect of the monolayer on nanoparticle–protein (ChT) interactions. (a) Simple alkanethiol-based monolayer results in protein denaturation. (b) Tetra(ethylene glycol)-functionalized particles are non-interacting. (c) Termination of the tetra(ethylene glycol) layer with carboxylate groups results in reversible binding to ChT that stabilizes the protein against denaturation. *Source*: Ref. [53]. Reproduced with permission of American Chemical Society.

Binding constants (10^6–$10^7 M^{-1}$) for NP-protein complexes were comparable with naturally occurring protein–ChT inhibitor interactions. The degree of inhibition and denaturation rate constants of ChT could be correlated to the hydrophobicity of the side chains. For example, aspartic acid (Asp), asparagine (Asn) and glutamic acid (Glu)

(a)

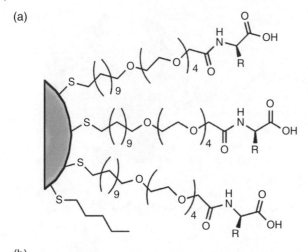

(b)

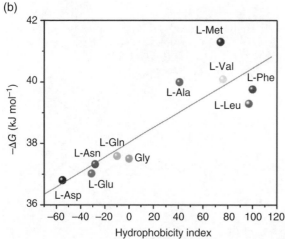

(c)

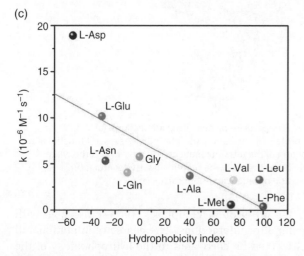

Figure 10.15 (a) Amino acid-decorated gold nanoparticles. (b) Correlation between the ΔG of NP–ChT interaction and the hydrophobicity index of amino acid side chains. (c) Correlation between the denaturation rate constants (k) of ChT and the hydrophobicity index of amino acid side chains in NPs. *Source*: Ref. [60]. Reproduced with permission of American Chemical Society.

with polar side chains showed 80% suppression of ChT activity, whereas methionine (Met) and alanine (Ala) with hydrophobic side chain showed an inhibition of 60% (Figure 10.15b and c). Moreover, hydrophobic side chains did not induce any changes in the conformation of the protein, while hydrophilic amino acid NPs destabilized the ChT structure. Further studies revealed that the denaturation rate constants of ChT were also depended upon the OEG tethers [61]. Tunable denaturation of ChT could be achieved by varying the combination of amino acid side chain and OEG tethers, with four repeat units [i.e., tetra (ethylene glycol)] providing essentially no denaturation. The reduced denaturation is attributed to the micelle-like stabilization induced by the OEG segment that prevents non-specific hydrophobic interactions between the interior alkyl chain and protein.

The binding of NPs to a protein surface can not only alter the secondary structure of the protein but it can also be used to modulate the enzyme activity. For example, Rotello and coworkers investigated the activity of NP bound ChT towards substrates with different net charges [62] (Figure 10.16a). NP-ChT complexes exhibited very low activities towards negatively charged substrates, whereas no change in activity was observed towards cationic substrates. This difference in selectivity of the NP-protein complex towards the substrates can be attributed to the electrostatic repulsion/attraction between substrates and the negatively charged monolayer that restricts/allows the entry of substrate in the active pocket of surface bound enzyme. Moderate inhibition in the activity of the neutral substrate was caused due to the presence of steric hindrance (Figure 10.16b and c). Further kinetic studies revealed that ChT-NP activity increased 3-fold compared with ChT, whereas it decreased by ~50 and ~95% against neutral and negatively charged substrate, respectively (Figure 10.17). Moreover, the catalytic behavior of surface bound enzymes is correlated to the properties of amino acid side chains. These studies suggest that chemical functionalities on the NP surface can be exploited to fine-tune enzymatic activity [63].

In contrast to the general belief that NPs tend to denature the native protein, appropriate functionalization of the NP surface can enable refolding of thermally denatured proteins. For example, Rotello and coworkers designed dicarboxylate capped AuNPs (AuDA) to restore the enzymatic activity of thermally denatured ChT and papain by ~93 and ~97%, respectively. The protein folding process can be described as a two-step process. First, AuDA binds to the unfolded proteins through complementary electrostatic interactions, facilitating refolding and preventing aggregation (Figure 10.18) [64]. Partial refolding is followed by subsequent release of proteins by increasing the ionic strength of the solution.

Understanding the association of NPs with proteins plays an important role in biosensing and biomedical applications. Avoiding interactions leading to the formation of protein coronas is important for designing "stealth" particles for delivery applications. The undesirable adsorption of proteins on the NP surface not only reduces their targeting capabilities but also promotes their elimination from the body [65,66]. An appropriate surface functionalization of NPs can reduce the interactions of NPs with proteins by providing them with non-fouling characteristics. For instance, Rotello and coworkers designed a library of zwitterionic AuNPs that did not adsorb proteins and hence did not form hard coronas at physiological serum concentrations [67]. Additionally, the hydrophobic motifs present on the NP surfaces dictated their biological response such as hemolytic activity and cellular uptake [68,69]. Furthermore,

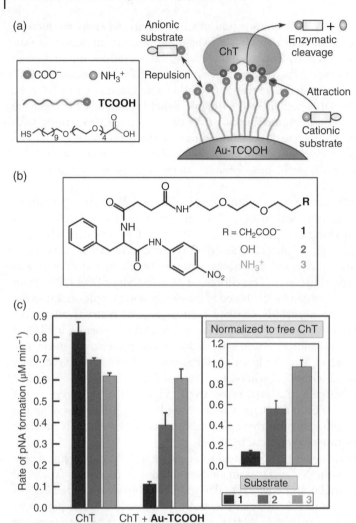

Figure 10.16 (a) Chemical structure of the TCOOH ligand and schematic depiction of substrate–monolayer-interaction-induced enzyme selectivity. (b) Structures of modified SPNA substrates. (c) initial rates of ChT hydrolysis of these modified substrates. (Inset) Normalized activity of Au-TCOOH-bound ChT toward substrates. *Source*: Ref. [62]. Reproduced with permission of American Chemical Society.

zwitterionic carboxy-betaine ligands were reported to provide high stability of larger AuNPs (15, 25 and 35 nm in diameter) to physical and chemical changes (e.g., ionic strength gradients, contact with biofluids and freeze–drying cycles). These NPs could be further functionalized with biomolecules such as avidin to provide a plug and play system for biological applications [70].

10.3.4.3 Nanoparticle-based Protein Sensors

An effective protein sensor, like any biosensor system, features a recognition element and a transduction component for signaling the binding event. NPs can act as recognition

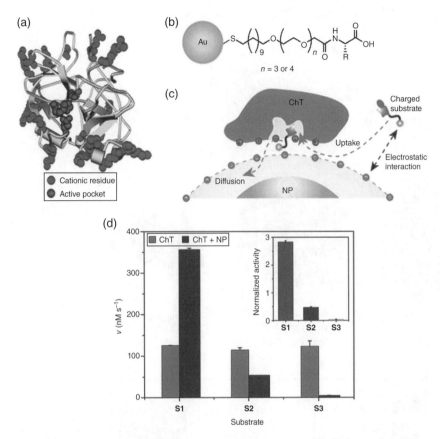

Figure 10.17 (a) Molecular structure of α-chymotrypsin. (b) Chemical structure of amino-acid-functionalized AuNPs and SPNA-derived substrates. (c) Schematic representation of monolayer-controlled diffusion of the substrate into and the product away from the active pocket of NP-bound ChT. (d) Generation rate (v) of 4-nitroaniline from different substrates under catalysis by ChT or ChT – NP-Glu complex. Inset shows the normalized activity of ChT-NP complex to ChT toward different substrates. *Source*: Ref. [63]. Reproduced with permission of American Chemical Society.

elements owing to their versatile structural diversity and upon binding with fluorescent molecules/macromolecules they can form entities with unique optical properties [49].

Differential affinity between sensor and analytes can produce distinct responses that can be used as fingerprint regions for the identification of each analyte. Based on this strategy, Rotello and coworkers designed a sensor array comprising six cationic AuNP elements and one anionic poly(*p*-phenylene ethynylene) (PPE) polymer for the identification of protein analytes (Figure 10.19). The fluorescence of the PPE polymer quenches upon binding to cationic AuNPs due to the high surface plasma resonance of the gold core. Subsequent binding of NPs with protein analytes displaces the polymer, generating signature patterns for each analyte (Figure 10.19). Using this approach seven different proteins with different sizes and isoelectric points were identified with an accuracy of 94% using linear discriminant analysis (LDA) [71]. Further studies were conducted for the identification of unknown proteins in serum. Serum is a highly complex biological fluid that is composed of more than 20 000 proteins, with an overall concentration of ~1 mM.

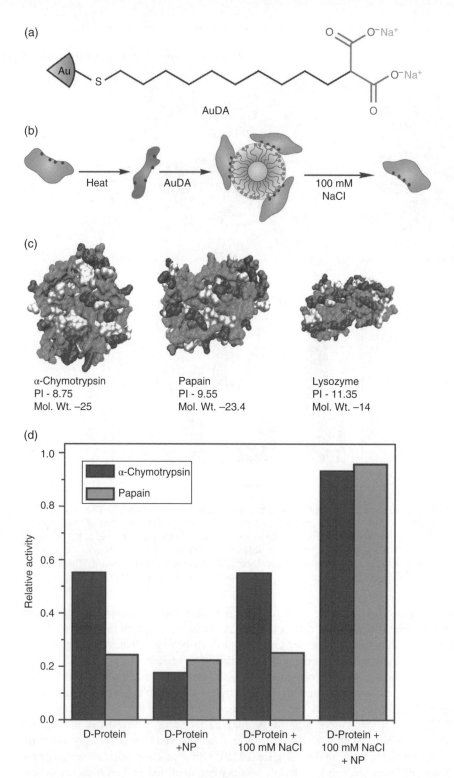

Figure 10.18 (a) Schematic representation of the structure of the AuDA (2 nm core) and (b) thermal denaturation followed by NP mediated refolding of proteins. (c) Surface structural features of three positively charged proteins used in the refolding study. Color scheme for the proteins: basic residues (blue), acidic residues (red), polar residues (green) and nonpolar residues (gray). (d) Enzymatic activity of thermally denatured ChT and papain in the presence of AuDA and 100 mM NaCl solution in 5 mM sodium phosphate buffer (pH = 7.4). *Source*: Ref. [64]. Reproduced with permission of Royal Society of Chemistry. See color section.

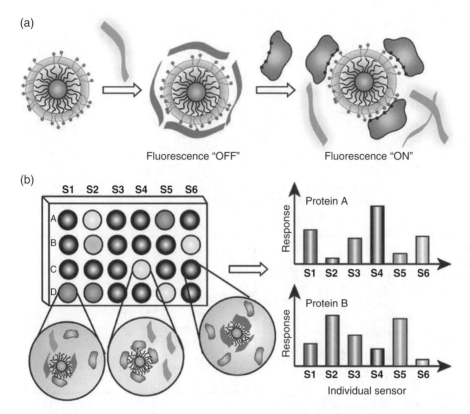

Figure 10.19 (a) Displacement of quenched fluorescent polymer by the protein analyte with the concomitant restoration of fluorescence. (b) Pattern generation through the differential release of fluorescent polymers from gold nanoparticles.

In this study, cationic NPs complexed with green fluorescent proteins (GFP) were used to detect different proteins spiked in undiluted human serum at 500 nM (Figure 10.20). The NP–GFP-based sensor array could differentiate the five most abundant human serum proteins with an accuracy of 97% [72]. The sensitivity of the NP-based sensor systems was further improved by using an enzyme amplified array sensing. In this approach, the ability of cationic AuNPs to inhibit the enzymatic activity of β-Gal was coupled with an analyte displacement strategy. Upon addition of analyte proteins, the activity of β-Gal was restored, resulting in an amplified fluorescent signal. Proteins at concentrations as low as 1 nM were identified with 100% accuracy in desalted human urine, with an overall concentration of protein of $\sim$1.5 μM (Figure 10.21) [73].

NPs have also been used for protein sensing by modifying enzyme-linked immunoabsorbent assays (ELISA). In this study, biotinylated NPs were used as mimics for biotinylated antibodies that selectively bind target proteins. After washing the unbound NPs/antibodies, HRP-conjugated avidin is added that immobilizes the enzyme on the ELISA plate due to avidin–biotin interaction. Finally, upon addition of HRP-substrate the observed color change can be used for the identification of the corresponding protein (Figure 10.22). Shae and coworkers designed a library of polymeric NPs with different charge, hydrophobicity, and H-bonding ability. These NPs were screened for their affinity towards histone and fibrinogen proteins. These ELISA-mimic systems have been used for a rapid search of synthetic polymers with high or low affinities for target biological macromolecules [74].

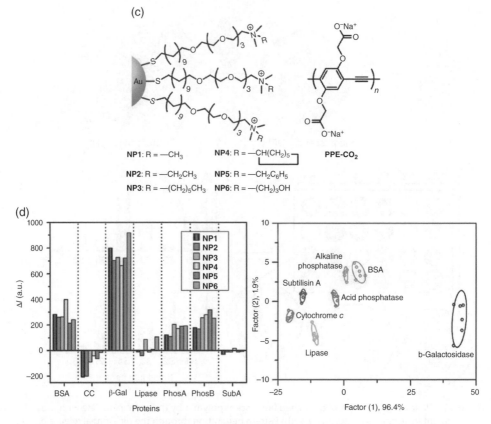

Figure 10.19 (Cont'd) (c) Chemical structure of cationic gold nanoparticles and anionic fluorescent polymer PPE-CO₂ ($n \approx 12$). (d) Fluorescence response (ΔI) patterns of the NP-PPE sensor array (NP1–NP6) against various proteins (CC, cytochrome c; β-Gal, β-galactosidase; PhosA, acid phosphatase; PhosB, alkaline phosphatase; SubA, subtilisin); and canonical score plot for the first two factors of simplified fluorescence response patterns. The 95% confidence ellipses for the individual proteins are also shown. *Source*: Ref. [71]. Reproduced with permission of Nature Publishing Group.

10.4 Future Perspective and Challenges

Efforts directed towards identifying and targeting PPIs for therapeutics has gathered pace in recent years. "Conformational drugs" have emerged as new therapeutics [75,76]. These drugs have the potential to alter protein structure and hence, regulate protein function. Besides, the structural rationale of these receptors contributes to new designs of sensing arrays to be used for medical diagnostics.

However, certain issues need to be addressed before using these molecular receptors or NPs efficiently as drugs. The performance and pharmacological activity of these receptors as drugs are principally dependent on ADMET (adsorption, distribution, metabolism, excretion, and toxicity) factors. For NP-based systems, these factors are distinct from those of small molecules because of their large size and multivalency, which gives rise to a new set of ADMET challenges. One such obstacle is delivering the agents at the disease site, even though enhanced permeability and retention time of the

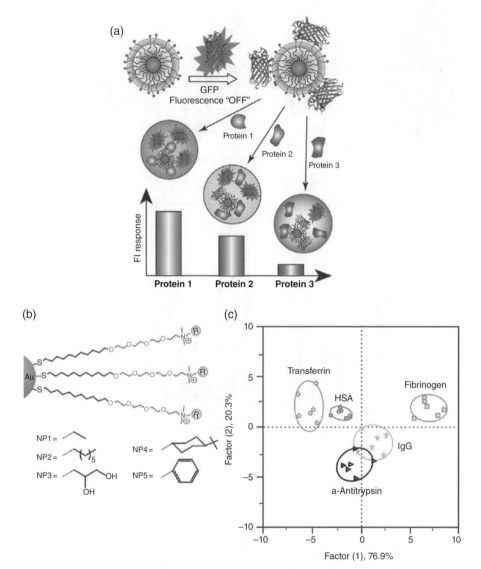

Figure 10.20 (a) Schematic illustration of the competitive binding between protein and quenched NP-GFP complexes and protein aggregation leading to fluorescence light up or further quenching. (b) Chemical structure of cationic AuNPs. (c) Canonical score plot for the fluorescence patterns as obtained from LDA against five protein analytes at a fixed concentration (500 nM) with 95% confidence ellipses. *Source*: Ref. [72]. Reproduced with permission of Nature Publishing Group.

agents can be added to our advantage. Additionally, the recognition of the surface functionalized NPs by the reticuloendothelial system is another issue that needs to be solved. Finally, protein corona formation on the surface of the large molecules, polymers, and NPs masks their targeting efficiency.

There has been significant progress in the field of designing and developing synthetic receptors for protein recognition in the last decade. However, more work in

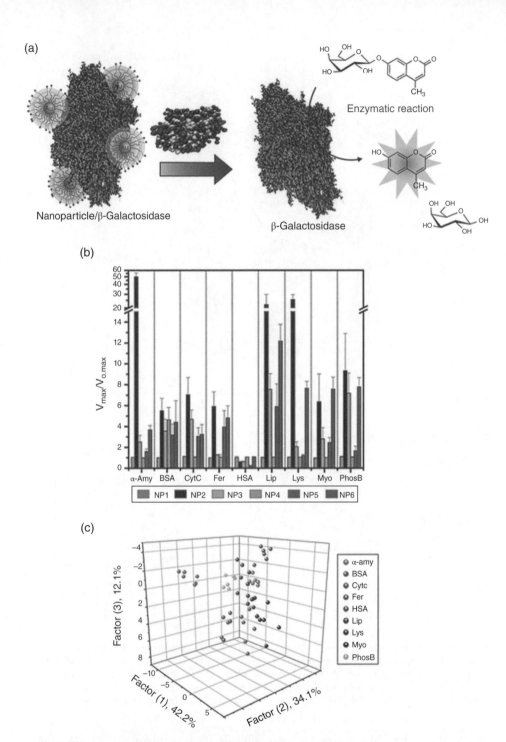

Figure 10.21 A schematic representation of a sensor element in the sensor composed of β-Gal and cationic AuNPs and differentiation of proteins in three dimensions. (a) As shown, β-Gal is displaced from the β-Gal–AuNP complex by protein analytes, restoring the catalytic activity of β-Gal toward the fluorogenic substrate 4-methylumbelliferyl-β-ᴅ-galactopyranoside, resulting in an amplified signal for detection. (b) Differential protein pattern of the nine proteins at 1 nM. (c) Canonical score plot of the first three factors of fluorescence response patterns obtained through β-Gal–AuNP sensor array against nine target proteins in 1 nM concentration. *Source*: Ref. [73]. Reproduced with permission of American Chemical Society.

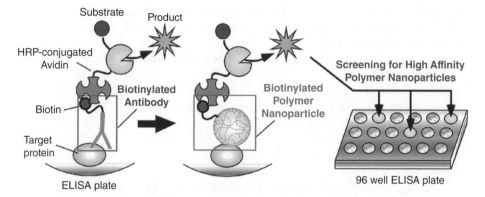

Figure 10.22 Concept of an ELISA-mimic screen (middle and right). A standard ELISA using a biotinylated antibody is also illustrated (left). *Source*: Ref. [74]. Reproduced with permission of American Chemical Society.

the years ahead will enormously expand our knowledge about the complex fundamental aspects of protein complexation by synthetic systems. This understanding will in turn improve our ability to develop these systems into therapeutics and diagnostics for use in the clinic.

Acknowledgment

This work was supported by the NIH (GM077173).

References

1 Jones S, Thornton JM. Principles of protein-protein interactions. Proc. Natl Acad. Sci. USA 1996;93(1):13–20.
2 Milroy LG, Grossmann TN, Hennig S, *et al.* Modulators of protein-protein interactions. Chem. Rev. 2014;114:4695–748.
3 Janin J, Chothia C. The structure of protein-protein recognition sites. J. Biol. Chem. 1990;265(27):16027–30.
4 Horton N, Lewis M. Calculation of the free energy of association for protein complexes. Protein Sci. 2008;1(1):169–81.
5 Young L, Jernigan RL, Covell DG. A role for surface hydrophobicity in protein-protein recognition. Protein Sci. 1994;3(5):717–29.
6 Lijnzaad P, Argos P. Hydrophobic patches on protein subunit interfaces: Characteristics and prediction. Proteins Struct. Funct. Genet. 1997;28(3):333–43.
7 Fernández A, Scheraga HA. Insufficiently dehydrated hydrogen bonds as determinants of protein interactions. Proc. Natl Acad. Sci. USA 2003 7;100(1):113–8.
8 Sheinerman FB, Norel R, Honig B. Electrostatic aspects of protein-protein interactions. Curr. Opin. Struct. Biol. 2000;10(2):153–9.
9 Moreira IS, Fernandes PA, Ramos MJ. Hot spots – A review of the protein-protein interface determinant amino-acid residues. Proteins Struct. Funct. Genet. 2007;68(4):803–12.

10 Keskin O, Ma B, Nussinov R. Hot regions in protein-protein interactions: The organization and contribution of structurally conserved hot spot residues. J. Mol. Biol. 2005 4;345(5):1281–94.

11 Wells JA. Additivity of mutational effects in proteins. Biochemistry 1990;29(37):8509–17.

12 Cunningham BC, Wells JA. High-resolution epitope mapping of hGH-receptor interactions by alanine-scanning mutagenesis. Science 1989;244(4908):1081–5.

13 Samanta U, Pal D, Chakrabarti P. Environment of tryptophan side chains in proteins. Proteins 2000;38(3):288–300.

14 Bogan AA, Thorn KS. Anatomy of hot spots in protein interfaces. J. Mol. Biol. 1998;280(1):1–9.

15 Fernández A. Desolvation shell of hydrogen bonds in folded proteins, protein complexes and folding pathways. FEBS Lett. 2002;527(1–3):166–70.

16 de Vos AM, Ultsch M, Kossiakoff AA. Human growth hormone and extracellular domain of its receptor: crystal structure of the complex. Science 1992;255(5042):306–12.

17 Nero TL, Morton CJ, Holien JK, *et al.* Oncogenic protein interfaces: Small molecules, big challenges. Nat. Rev. Cancer 2014;14(4):248–62.

18 Moreira IS, Fernandes PA, Ramos MJ. Hot spot occlusion from bulk water: A comprehensive study of the complex between the lysozyme HEL and the antibody FVD1.3. J. Phys. Chem. B 2007;111(10):2697–706.

19 Ray M, Gupta A, Rotello VM. Synthetic receptors for protein surfaces. In: Synthetic Receptors for Biomolecules: Design Principles and Applications. The Royal Society of Chemistry; 2015; pp. 369–403.

20 Jain RK, Hamilton AD. Designing protein denaturants: Synthetic agents induce cytochrome c unfolding at low concentrations and stoichiometries. Angew. Chem. Int. Ed. 2002;41(4):641–3.

21 Aya T, Hamilton AD. Tetrabiphenylporphyrin-based receptors for protein surfaces show sub-nanomolar affinity and enhance unfolding. Bioorg. Med. Chem. Lett. 2003;13(16):2651–4.

22 Wilson AJ, Groves K, Jain RK, *et al.* Directed denaturation: Room temperature and stoichiometric unfolding of cytochrome c by a metalloporphyrin dimer. J. Am. Chem. Soc. 2003;125(15):4420–1.

23 Groves K, Wilson AJ, Hamilton AD. Catalytic unfolding and proteolysis of cytochrome C induced by synthetic binding agents. J. Am. Chem. Soc. 2004;126(40):12833–42.

24 Fletcher S, Hamilton AD. Denaturation and accelerated proteolysis of sizeable heme proteins by synthetic metalloporphyrins. New J. Chem. 2007;31(5):623.

25 Yin H, Hamilton AD. Strategies for targeting protein-protein interactions with synthetic agents. Angew. Chem. Int. Ed. 2005;44(27):4130–63.

26 Baldini L, Wilson AJ, Hong J, Hamilton AD. Pattern-based detection of different proteins using an array of fluorescent protein surface receptors. J. Am. Chem. Soc. 2004;126(18):5656–7.

27 Zhou H, Baldini L, Hong J, *et al.* Pattern recognition of proteins based on an array of functionalized porphyrins. J. Am. Chem. Soc. 2006;128(7):2421–5.

28 Klärner FG, Schrader T. Aromatic interactions by molecular tweezers and clips in chemical and biological systems. Acc. Chem. Res. 2013 16;46(4):967–78.

29 Fokkens M, Schrader T, Klärner FG. A molecular tweezer for lysine and arginine. J. Am. Chem. Soc. 2005;127(41):14415–21.

30 Talbiersky P, Bastkowski F, Klärner FG, Schrader T. Molecular clip and tweezer introduce new mechanisms of enzyme inhibition. J. Am. Chem. Soc. 2008;130(30):9824–8.

31 Kirsch M, Talbiersky P, Polkowska J, *et al*. A mechanism of efficient G6PD inhibition by a molecular clip. Angew. Chem. Int. Ed. 2009;48(16):2886–90.

32 Sinha S, Lopes DHJ, Du Z, *et al*. Lysine-specific molecular tweezers are broad-spectrum inhibitors of assembly and toxicity of amyloid proteins. J. Am. Chem. Soc. 2011;133:16958–69.

33 Prabhudesai S, Sinha S, Attar A, *et al*. A novel "molecular tweezer" inhibitor of α-synuclein neurotoxicity in vitro and in vivo. Neurotherapeutics 2012;9(2):464–76.

34 Bier D, Rose R, Bravo-Rodriguez K, *et al*. Molecular tweezers modulate 14-3-3 protein-protein interactions. Nat. Chem. 2013;5(3):234–9.

35 Nimse SB, Kim T. Biological applications of functionalized calixarenes. Chem. Soc. Rev. 2013;42(1):366–86.

36 Dutt S, Wilch C, Schrader T. Artificial synthetic receptors as regulators of protein activity. Chem. Commun. 2011;47(19):5376–83.

37 Park HS, Lin Q, Hamilton AD. Protein surface recognition by synthetic receptors: A route to novel submicromolar inhibitors for α-chymotrypsin. J. Am. Chem. Soc. 1999;121(1):8–13.

38 Lin Q, Park HS, Hamuro Y, *et al*. Protein surface recognition by synthetic agents: Design and structural requirements of a family of artificial receptors that bind to cytochrome c. Biopolymers 1998;47:285–97.

39 Park HS, Lin Q, Hamilton AD. Modulation of protein-protein interactions by synthetic receptors: Design of molecules that disrupt serine protease-proteinaceous inhibitor interaction. Proc. Natl Acad. Sci. USA 2002;99(8):5105–9

40 Gordo S, Martos V, Santos E, *et al*. Stability and structural recovery of the tetramerization domain of p53-R337H mutant induced by a designed templating ligand. Proc. Natl Acad. Sci. USA 2008;105(43):16426–31.

41 McGovern RE, Fernandes H, Khan AR, *et al*. Protein camouflage in cytochrome c–calixarene complexes. Nat. Chem. 2012;4(7):527–33.

42 Wilson AJ. Inhibition of protein-protein interactions using designed molecules. Chem. Soc. Rev. 2009;38(12):3289–300.

43 Orner BP, Ernst JT, Hamilton AD. Toward proteomimetics: Terphenyl derivatives as structural and functional mimics of extended regions of an α-helix. J. Am. Chem. Soc. 2001;123(22):5382–3.

44 Yin H, Lee GI, Sedey KA, *et al*. Terphenyl-based Bak BH3 α-helical proteomimetics as low-molecular-weight antagonists of Bcl-xL. J. Am. Chem. Soc. 2005;127(29):10191–6.

45 Rodriguez JM, Nevola L, Ross NT, *et al*. Synthetic inhibitors of extended helix-protein interactions based on a biphenyl 4,4′-dicarboxamide scaffold. ChemBioChem 2009;10(5):829–33.

46 Saraogi I, Hebda JA, Becerril J, *et al*. Synthetic α-helix mimetics as agonists and antagonists of islet amyloid polypeptide aggregation. Angew. Chem. Int. Ed. 2010;49(4):736–9.

47 Plante JP, Burnley T, Malkova B, *et al*. Oligobenzamide proteomimetic inhibitors of the p53-hDM2 protein-protein interaction. Chem. Commun. 2009;44(34):5091–3.

48 Azzarito V, Long K, Murphy NS, Wilson AJ. Inhibition of α-helix-mediated protein-protein interactions using designed molecules. Nat. Chem. 2013 Mar;5(3):161–73.

49 Jiang Z, Le NDB, Gupta A, Rotello VM. Cell surface-based sensing with metallic nanoparticles. Chem. Soc. Rev. 2015;44(13):4264–74.

50 Gupta A, Landis RF, Rotello VM. Nanoparticle-based antimicrobials: Surface functionality is critical. F1000Res. 2016;5(0):1–10.

51 Rosi NL, Mirkin CA. Nanostructures in biodiagnostics. Chem. Rev. 2005;105(4):1547–62.

52 Peer D, Karp JM, Hong S, *et al.* Nanocarriers as an emerging platform for cancer therapy. Nat. Nanotechnol. 2007;2(12):751–60.

53 Moyano DF, Rotello VM. Nano meets biology: Structure and function at the nanoparticle interface. Langmuir 2011;27(17):10376–85.

54 De M, Miranda OR, Rana S, Rotello VM. Size and geometry dependent protein-nanoparticle self-assembly. Chem. Commun. 2009;(16):2157–9.

55 De M, You CC, Srivastava S, Rotello VM. Biomimetic interactions of proteins with functionalized nanoparticles: A thermodynamic study. J. Am. Chem. Soc. 2007;129(35):10747–53

56 Fischer NO, McIntosh CM, Simard JM, Rotello VM. Inhibition of chymotrypsin through surface binding using nanoparticle-based receptors. Proc. Natl Acad. Sci. USA 2002;99(8):5018–23.

57 Verma A, Simard JM, Rotello VM. Effect of ionic strength on the binding of alpha-chymotrypsin to nanoparticle receptors. Langmuir 2004;20(10):4178–81.

58 Fischer NO, McIntosh CM, Simard JM, Rotello VM. Inhibition of chymotrypsin through surface binding using nanoparticle-based receptors. Proc. Natl Acad. Sci. USA 2002;99(8):5018–23.

59 Hong R, Fischer NO, Verma A, *et al.* Control of protein structure and function through surface recognition by tailored nanoparticle scaffolds. J. Am. Chem. Soc. 2004;126(3):739–43.

60 You CC, De M, Han G, Rotello VM. Tunable inhibition and denaturation of α-chymotrypsin with amino acid-functionalized gold nanoparticles. J. Am. Chem. Soc. 2005;127(37):12873–81.

61 You CC, De M, Rotello VM. Contrasting effects of exterior and interior hydrophobic moieties in the complexation of amino acid functionalized gold clusters with α-chymotrypsin. Org. Lett. 2005;7(25):5685–8.

62 Hong R, Emrick T, Rotello VM. Monolayer-controlled substrate selectivity using noncovalent enzyme-nanoparticle conjugates. J. Am. Chem. Soc. 2004;126(42):13572–3.

63 You CC, Agasti SS, De M, *et al.* Modulation of the catalytic behavior of α-chymotrypsin at monolayer-protected nanoparticle surfaces. J. Am. Chem. Soc. 2006;128(45):14612–8.

64 De M, Rotello VM. Synthetic "chaperones": nanoparticle-mediated refolding of thermally denatured proteins. Chem. Commun. 2008;(30):3504–6.

65 Tenzer S, Docter D, Kuharev J, *et al.* Rapid formation of plasma protein corona critically affects nanoparticle pathophysiology. Nat. Nanotechnol. 2013;8(10):772–81.

66 Aggarwal P, Hall JB, McLeland CB, *et al*. Nanoparticle interaction with plasma proteins as it relates to particle biodistribution, biocompatibility and therapeutic efficacy. Adv. Drug Deliv. Rev. 2009;61(6):428–37.

67 Moyano DF, Saha K, Prakash G, *et al*. Fabrication of corona-free nanoparticles with tunable hydrophobicity. ACS Nano 2014;8(7):6748–55.

68 Saha K, Moyano DF, Rotello VM. Protein coronas suppress the hemolytic activity of hydrophilic and hydrophobic nanoparticles. Mater Horiz. 2014;2014(1):102–5.

69 Saha K, Rahimi M, Yazdani M, *et al*. Regulation of macrophage recognition through the interplay of nanoparticle surface functionality and protein corona. ACS Nano 2016;10(4):4421–30.

70 Gupta A, Moyano DF, Parnsubsakul A, *et al*. Ultrastable and biofunctionalizable gold nanoparticles. ACS Appl. Mater Interfaces 2016;8(22):14096–101.

71 You C-C, Miranda OR, Gider B, *et al*. Detection and identification of proteins using nanoparticle-fluorescent polymer "chemical nose" sensors. Nat. Nanotechnol. 2007;2(5):318–23.

72 De M, Rana S, Akpinar H, *et al*. Sensing of proteins in human serum using conjugates of nanoparticles and green fluorescent protein. Nat. Chem. 2009;1(6):461–5.

73 Miranda OR, Chen HT, You CC, *et al*. Enzyme-amplified array sensing of proteins in solution and in biofluids. J. Am. Chem. Soc. 2010;132(14):5285–9.

74 Yonamine Y, Hoshino Y, Shea KJ. ELISA-mimic screen for synthetic polymer nanoparticles with high affinity to target proteins. Biomacromolecules 2012;13(9):2952–7.

75 Peczuh MW, Hamilton AD. Peptide and protein recognition by designed molecules. Chem. Rev. 2000;100(7):2479–94

76 Fletcher S, Hamilton AD. Protein surface recognition and proteomimetics: Mimics of protein surface structure and function. Curr. Opin. Chem. Biol. 2005;9(6):632–8.

11

Multivalent Calixarenes for the Targeting of Biomacromolecules

Francesco Sansone and Alessandro Casnati

Department of Chemistry, Life Sciences and Environmental Sustainability, Università di Parma, Parco Area delle Scienze 17/a, 43124 Parma, Italy

11.1   Introduction

Multivalency [1] is an extremely powerful concept in supramolecular chemistry, nanotechnology and bioorganic chemistry since it allows not only to reinforce binding, but also to make it more specific [2–5]. In fact, the multiple presentation of several binding/ligating units (epitopes) properly arranged on a main scaffold might result in multivalent ligands possessing a much higher affinity (lower dissociation constant, K_D) than that of a single ligating unit. For an efficient binding, the design of the multivalent ligands is extremely important since the number, the distance between the ligating units and the flexibility of the linkers which connect them to the scaffold deeply influence the overall binding energy with the target receptor. This effect, also called the multivalent effect, is generally attributed to an entropic gain originated by a high local concentration and a high statistical (re)binding probability of the ligating groups that, after the first intermolecular binding event, result in close proximity to the receptor sites. Multivalency is also quite often used by nature to make recognition events more efficient and selective [1].

Calix[*n*]arenes (Figure 11.1, **1**: $n \geq 4$) [6–11] are the cyclic oligomers obtained by the condensation of phenol and formaldehyde under basic conditions. On account of their oligomeric structure and characteristics, in the last 20 years calixarenes have started to be intensively used as scaffolds for the construction of multivalent ligands [4,5,12] for biomacromolecules and, nowadays, they can be considered rather peculiar and important multivalent scaffolds.

Certainly one of the most interesting feature that makes these macrocyclic compounds so popular is their ease of preparation even on a kilo-scale [13], and the large number of consolidated synthetic procedures for their functionalization [14]. This gives direct access to multivalent scaffolds of proper valency and with a wide variety of functional groups to be used in the conjugation of the desired ligating units. Another important feature is the possibility to mould the shape of the smallest calix[4]arene in

Multivalency: Concepts, Research & Applications, First Edition. Edited by Jurriaan Huskens, Leonard J. Prins, Rainer Haag, and Bart Jan Ravoo.
© 2018 John Wiley & Sons Ltd. Published 2018 by John Wiley & Sons Ltd.

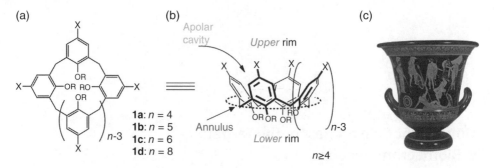

Figure 11.1 A planar (a) and a three-dimensional (b) representation of calix[*n*]arenes **1** (*n* ≥ 4) and (c) a Greek *calix* vase from which the name of these macrocycles derives.

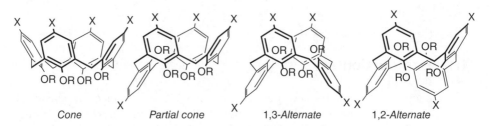

Figure 11.2 The four limiting conformations, when R = Me or Et, of calix[4]arenes **1a** become atropoisomers, when R groups are larger than ethyl.

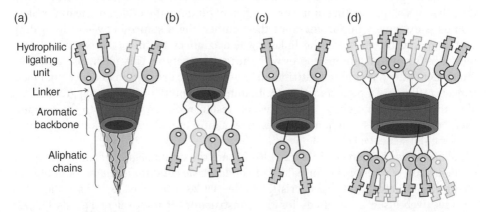

Figure 11.3 Schematic representation of multivalent ligands from calixarenes (truncated cones) exposing ligating units (key-shaped moieties). Tetravalent calix[4]arenes in cone structures and functionalized at the upper (a) or lower rim (b), a 1,3-alternate derivative (c), and a larger hexadecavalent calix[8]arene (d). Multivalent ligands (a) and (b) are facial amphiphiles with polar moieties at the upper rim (a) or at the lower rim (b), while derivatives (c) are bolaamphiphiles. Larger calix[*n*]arenes (*n* > 5) are conformationally mobile and might switch conformation and amphiphilicity depending on the surrondings.

one of the four structures (cone, partial cone, 1,3-alternate, 1,2-alternate; Figure 11.2) due to stereoselective syntheses that modify the lower rim of the macrocycle [10].

The multivalent calixarene ligands thus generated have different polarity, distance and stereochemical orientation of the ligating units (Figure 11.3).

(a)　　　　　(b)

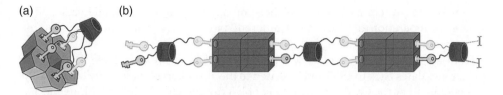

Figure 11.4 Different ways in which a multivalent calixarene ligand can interact with a multimeric protein: (a) a multivalent 1:1 complex; and (b) a cross-linked multivalent aggregate.

It has been demonstrated that especially this latter factor strongly influences the affinity for the target biomacromolecules. The cone calixarenes, having a facial disposition of ligating units (epitopes), are more suitable to bind to large surfaces or to multimeric proteins presenting binding domains on the same face of the macromolecule (Figure 11.4a). The 1,3-alternate derivatives, on the other hand, having a presentation of the epitopes on opposite sides of the multivalent ligand are suitable to span also receptor units far from each other, usually favouring large cross-linked multimeric aggregates (Figure 11.4b).

The lipophilic backbone also turns out to be a relevant feature of these macrocycles. This impairs the solubilization and compatibilization of calixarene ligands in water solution, but offers also different additional advantages. First, the amphiphilic character of calixarenes can be tuned at will. In fact, when calixarenes are properly functionalized with charged groups (carboxylic, sulfonate, phosphate or ammonium) or highly hydrophilic moieties (polyols, sugars and amino acids or small peptides) they present an amphiphilic character. The resulting amphiphiles can therefore be classified as facial amphiphiles when polar moieties are at one side (upper or lower) and the lipophilic backbone or aliphatic alkyl chains are at the other side of the molecule (Figure 11.3a and b) or as bolaamphiphiles when the polar moieties are fixed at both ends of the central aromatic calixarene block (Figure 11.3c,d).

These different situations might result in multivalent ligands endowed with peculiar self-assembly properties and able to originate micelles, solid lipid nanoparticles (SLNs) or liposomes. On the contrary, when the polar character predominates, that is when short alkyl chains (methyl to propyl) or even better polyether or charged groups are present on the side opposite to that of the ligating units or when the ligating units of facial amphiphiles are highly hydrophilic, the resulting multivalent ligands are usually present as monomers in water and even in highly saline aqueous solutions.

Another interesting feature related to the calixarene structure is that water soluble derivatives, having highly polar groups in close proximity to an apolar cavity or simply to the exterior hydrophobic surface of the macrocycle, also offer the opportunity to flank, during the interaction with biomacromolecules, groups able to establish electrostatic and/or hydrogen bonds and aromatic nuclei able to give rise to CH–π, π–π interactions or to exploit the hydrophobic effect [15]. This particular situation, often programmed in synthetic derivatives and referred to as 'hybrid character', somehow mimics the environment encountered in the binding sites of enzymes and proteins where the phenyl and indole nuclei of phenylalanine and tryptophan, forming highly lipophilic pockets, are in close contact with the hydrogen bond donor and acceptor groups of the protein backbone and those of the polar amino acid side chains. The synergistic operation of these quite different types of attracting interactions often strongly reinforces the binding.

From an historical point of view, the first reported example of biologically active calixarenes dates back to 1955 and is related to the antitubercolotic activity of a calix[8]-arene adorned at the lower rim with long poly(ethylene glycol) chains [16]. Its mechanism of action was only recently proposed but a specific target was not clearly identified [17]. An additional 40 years was required for the first example of a calixarene whose activity was designed for the recognition of a specific biological target (see below). Its antimicrobial activity was in fact connected to its specific ability to bind to peptidoglycan, one of the main constituents of the Gram-positive bacteria cell wall [18]. Following this report, a number of studies appeared in which it was demonstrated that calixarene ligands interacted with proteins and enzymes, oligonucleotides and nucleic acids or lipopolysaccharides (LPSs) with applications as inhibitors of the adhesion of viruses, bacteria and bacterial toxins on cell walls, as antitumour agents and as gene- or drug-delivery systems. This survey is not aimed at being comprehensive, quoting all the reported cases of biologically active compounds, already extensively reviewed in different recent reviews [5,19–23] but will describe, with selected relevant examples, the main characteristics that render calixarenes special scaffolds for the construction of multivalent ligands for biomacromolecules.

11.2 Binding to Proteins and Enzymes

Recognition of proteins or enzymes was successfully carried out using multivalent calixarenes exploiting three different modes of binding.

Hot spots in proteins are specific areas of a few hundreds of square angstroms (Figure 11.5a) which are responsible for protein–protein interactions (PPIs) that often trigger specific functions or are at the origin of pathological events. Calix[4]arene ligands fixed in the cone structure and functionalized at the upper rim with polar

(a)   (b)   (c)

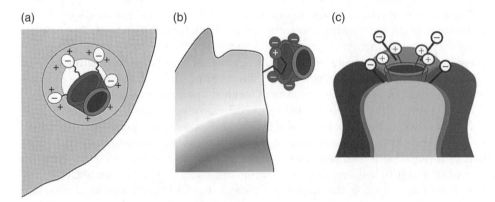

Figure 11.5 The different strategies used with multivalent calixarene ligands to bind proteins and/or enzymes: the (a) 'hot spot' and (b) 'single point' recognition; (c) the stoppering of lipophilic pockets.

moieties have been successfully used to recognize hot spots since they can offer a central apolar cavity to give rise to hydrophobic interactions with lipophilic areas on the protein surface and polar groups to bind complementary polar amino acid side chains or the peptide backbone surrounding the hydrophobic patches. Following this type of approach, Hamilton and coworkers proposed a series of calix[4]arene ligands able to disrupt PPIs [24]. The calixarene **2a** bearing four negatively charged GlyAspGlyAsp cyclopeptides at the upper rim was shown to cover the heme edge of cytochrome c (cyt c) [25]. The Asp units provide the negative charges necessary to interact with the ammonium side chain of the Lys residues present on the border of the lipophilic area containing the active site, while the 3-aminomethylbenzoyl units, together with the lipophilic calixarene backbone, establish hydrophobic interactions with the lipophilic patch of the protein. This results in the formation of a 1:1 complex between **2a** and the enzyme ($K_{ass} \sim 10^8 \, M^{-1}$) which prevents the approach of the reducing agent ascorbate to the Fe(III) heme edge of cyt c. Moreover, calixarene **2a**, covering an area of 450–500 $Å^2$, also prevents the interaction between cyt c and cyt c peroxidase [26]. Operating in a rather similar way, **2a** is also able to bind to the surface of α-chymotrypsin at nanomolar level, preventing the formation of the complex between the enzyme and some proteinaceous inhibitors [27]. More recently, a fluorescent sensor for cyt c was suggested [28], which is able to bind the protein close to the heme and forms a 1:1 complex of high affinity on account of electrostatic and hydrophobic forces.

Even more interestingly, the hot spot targeting can also be exploited to strongly inhibit angiogenesis and tumour progression by preventing the interaction between growth factors and their receptors. A topical example is that of the PDGF-PDGFR complex that is involved in the maintenance of blood vessels and is therefore of special relevance in oncogenesis. As in the cases of cyt c and α-chymotrypsin, in fact, the platelet derived growth factor (PDGF) exposes an hydrophobic area surrounded by a positively charged belt for the binding of the membrane bound receptor PDGFR [24,29]. Calixarene **2b**, characterized by the presence of a GlyAspGlyTyr cyclopeptide at the upper rim, binds to PDGF inhibiting the adhesion to the PDGFR with an $IC_{50} = 250 \, nM$.

2a: AA$_1$ = AA$_3$ = Gly; AA$_2$ = AA$_4$ = Asp
2b: AA$_1$ = AA$_3$ = Gly; AA$_2$ = Asp; AA$_4$ = Tyr

3

4a x Lys

5 x Lys-D-Ala-D-Ala

Remarkably, Hamilton and coworkers collected a series of data indicating that the growth of different human tumours implanted in nude mice was strongly inhibited by treatment with **2b**. A simplified version of **2b**, calixarene **3**, in which the pentapeptide loop is replaced by the monoacid monoester derivative of isophthalic acid, is able to inhibit both PDGF and vascular endothelial growth factor (VEGF) [30]. This result is extremely interesting since angiogenesis needs both growth factors for initiation and maintenance of blood vessels, respectively. Actually, the *in vivo* treatment of mice with calixarene **3** showed a marked suppression of human tumour growth and angiogenesis.

Calixarenes were also used to bind proteins by exploiting their ability to interact with amino acid residues whose side chains are exposed from the protein surface. This type of interaction is usually quite specific for and restricted to certain residues in the peptide sequence so that this mode of binding can be classified as 'single point' recognition (Figure 11.5b). Rather important in this case is the role exerted by the calixarene lipophilic cavity in aqueous environment with the inclusion of lipophilic and positively charged side chains of selected amino acids, due to the formation of hydrophobic or cation–π interactions, respectively. Polar and/or charged functional groups can be present at the upper rim in close proximity to the macrocyclic cavity, so that they can assist in binding through electrostatic interactions or hydrogen bonding. Evidence for the binding of free amino acids by the inclusion of their lateral residues in the cavity of hybrid water soluble calixarene ligands had already been obtained in solution for Leu, Ile, Phe and Trp [31] and in the solid state for lysine [32,33] and arginine [34]. Interestingly, the X-ray crystal structure of the **4a**-Lys complex showed the simultaneous operation of both electrostatic interactions, between the sulfonate groups and the ammonium head groups, and CH–π interactions, between the calixarene aromatic nuclei and the σ and ε methylene groups of Lys.

A prototype example of 'single point' recognition was first reported by us in 1996, when we prepared a series of calix[4]arenes in the cone structure bearing short peptide bridges at the upper rim (e.g. **5**) [18]. The macrobicyclic compound **5** which has a bridge formed by two L-alanines and a 1,3,5-diethylenetriamine unit showed the most interesting properties. It efficiently binds to *N*-acetyl-D-Ala-D-Ala [35], used as model of the terminal part of the short peptide chains of peptidoglycan, the main constituent of the cell wall of Gram-positive bacteria. It was proposed that the strong binding originates from the synergistic combination of a salt bridge, hydrogen bonds and hydrophobic interactions. While *N*-acetyl-D-Ala-D-Ala is interacting via electrostatic interactions and hydrogen bonds with the calixarene bridge, one of its methyl groups is included in the lipophilic cavity of the macrocycle originating CH–π interactions. This compound also possesses antimicrobial activity against Gram-positive bacteria very similar to that of vancomycin, a natural antibiotic whose activity is related to its ability to strongly bind to the L-Lys-D-Ala-D-Ala pendant of peptidoglycan, thus inhibiting the cross-linking of the bacteria cell wall constituents.

More recently, Crowley and coworkers investigated the potential of calixarene hybrid receptors in the recognition of the ε-amino groups of lysine in proteins. The tetrasulfonato calix[4]arene **4a**, which is fixed in the cone structure by the presence of an array of hydrogen bonds at the lower rim, was found to be able to bind to different lysine residues present in cyt c both in solution and in the solid state [36]. The X-ray crystal structure shows that sulfonato calixarenes bind three different Lys residues but always with nearly the same contacts. The Lys side chain is included in the calixarene apolar cavity with the ammonium group projected towards the negatively charged sulfonate units, while hydrophobic interactions of the -CH$_2$- groups with the calixarene aromatic nuclei further stabilize the complex. Interestingly, it could be evaluated that the calixarene binding to cyt c makes an area of 200–300 Å^2 inaccessible, suggesting the use of such ligands for protein camouflage. The same ligand **4a** was even more selective in the binding of lysozyme where only the C-terminal Arg128 was bound over the 11 Arg units present in the protein [37]. Remarkably, the 'single point' approach is particularly attractive for the recognition of proteins mutated in single positions. When the ε-amino group of lysines present in lysozyme undergoes methylation to dimethyl ammonium [-NH(Me)$_2$]$^+$, ligand **4a** shifts its selectivity from Arg to the dimethylated Lys [38]. A structural comparison between X-ray crystal structures indicates a marked similarity between the calixarene sulfonated cavity and the aromatic cages endowed with carboxylate groups that the chromodomains of different proteins employ to complex di- and trimethylated lysines. NMR and XRD data indicate that, contrary to what is observed in the case of **4a**-Lys-NH$_3$$^+$ complexes, complexation of dimethyl-lysine (Lys-NHMe$_2$$^+$) involves the inclusion of the trialkylammonium group in the macrocyclic cavity giving rise to cation–π interactions, while the methylene groups are essentially excluded from binding. Hof and coworkers also found that ligand **4a** presents increasing affinity as methylation of Lys increases in model peptides reproducing the histone tails [(39]. Since methylation of Lys ε-nitrogen is an important post-translational modification of proteins responsible for relevant effects on gene regulation and signalling pathways, the possibility to have small synthetic ligands able to selectively bind to such modified amino acid residues can constitute a rather important strategy for chemical biologists and medicinal chemists to modulate PPIs relevant in gene regulation and oncogenesis.

The third strategy used for calixarene protein binding can be referred to as the stoppering of lipophilic cavities. The aromatic backbone of, especially, calix[4]arenes fixed in the cone structure indeed offers an external lipophilic surface suitable to penetrate the apolar cavities/channels of proteins. This was the design followed by de Mendoza and coworkers who synthesized a series of tetraguanidinium cone calix[4]arenes **6** to inhibit the activity of voltage-dependent potassium channels [40], which are an important pharmacological target because of their involvement in several diseases. These derivatives were tested in the inhibition of the activity of the *Shaker* potassium channel. Compounds **6a** and **6b**, fixed in the cone structure by hydrogen bonds between the phenolic OH groups have a reversible inhibition activity at 50 μM. Even more active was the analogue **6c** which is fixed in a very rigid cone structure by the two short crown-3 ether bridges. In contrast, the two tetraalkylated derivatives **6d** and **6e**, being also in the cone structure but possessing a residual conformational flexibility and a marked amphiphilicity due to the presence of lipophilic groups at the lower rim, give rise to a significant decrease of the ionic current. Yet, this effect was irreversible due to a destructive denaturant action of the amphiphilic macrocycle. Importantly, a phenol moiety functionalized as in **6a** and used as monomeric model of the tetravalent calixarene, did not show any activity. This suggested an important role for the preorganized multivalent presentation of positive charges at the top of the calixarene stopper. All these data therefore suggest that the conical calixarenes should accommodate into the channel vestibule with the calix upper rim allowing the cationic guanidinium moieties to interact with the negatively charged lateral carboxylate groups of the Asp residues present around the vestibule of the channel (Figure 11.5c). This mechanism of action is also supported by molecular dynamics studies performed using the crystal structure of the potassium channel and a minimized calixarene.

Similar tetraguanidinium calix[4]arenes **7a** and **7b** were also satisfactorily used to restore the physiological activity of the homotetrameric p53 transcription factor [41]. As a consequence of an inherited mutation, Arg337 is replaced by a His in the p53 sequence, causing the weakening of the tetramerization process of the four peptide units. The substitution of Arg with His units results in fact in the presence of imidazole groups which, contrary to guanidine, are only partially protonated at physiological pH and cannot ensure a sufficiently strong interaction with the complementary carboxylates of the Asp units present on the opposite protein monomer. This mutation therefore determines a consistent weakening of the p53 activity against tumour cells, since its tetramerization triggers the expression of the DNA repair machinery or cell apoptosis. However, the presence of the carboxylate groups of Glu336 and Glu339 just above the hydrophobic pocket generated upon the tetramerization of p53 suggested to de Mendoza and coworkers to test cone shaped calixarenes **7** as promoter of the tetramerization process in mutated proteins. Data collected with different experimental techniques showed that calixarene **7a** stabilizes the tetrameric form of the mutated protein at 400 μM, while no effects were detected with the wild-type form. Molecular modelling studies also suggested that at both sides of the protein assembly a calixarene ligand is bound. Hydrophobic interactions arise between the lipophilic external calixarene surface and the hydrophobic pocket generated upon tetramerization by the lipophilic amino acid residues. Simultaneously, multivalent electrostatic interactions take place between the four guanidinium head groups and the four carboxylate anions of the Glu residues. Contrary to what was observed for ligands **6** and the potassium channel, in the tetramerization of p53 the more flexible compound **7b** is already active at 25 μM showing that, sometimes, a higher flexibility of the ligand, although detrimental in terms of entropy loss upon binding, can allow a more favourable induced fit binding to the protein due to an ideal optimization of the contact distances between interacting groups [42].

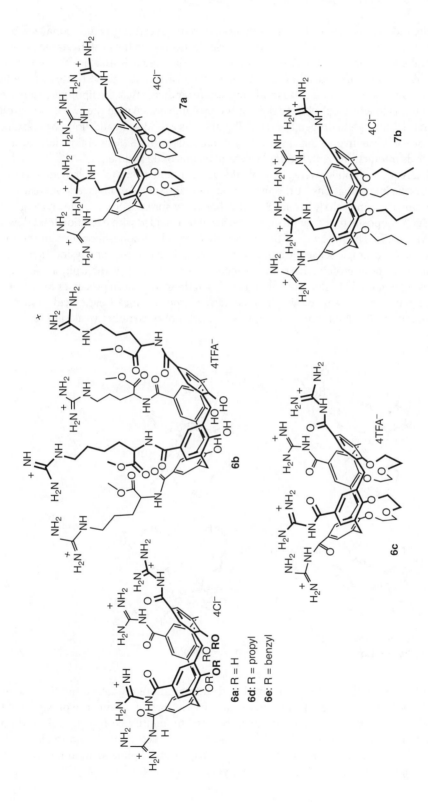

6a: R = H
6d: R = propyl
6e: R = benzyl

Quite recently, Neri and coworkers demonstrated that calixarene functionalized beads can be used for 'protein fishing' allowing the identification of the proteins targeted by the multivalent calixarenes [43]. A *p*-acetamidocalix[4]arene derivative, in fact, was shown to be able to identify a disulfide isomerase protein from HeLa cell lysates as the best partner. Interestingly, molecular modeling studies indicate that the lipophilic calixarene ligand binds to a hydrophobic pocket on the protein surface with the two amide groups interacting through hydrogen bonds to Trp and His residues. In addition, the calixarene exterior walls are involved in π-stacking interactions with a Phe lateral residue and Van der Waals interactions with the apolar lateral chains of Ile and Leu.

Calixarene amphiphiles were also embedded in Langmuir films to implement devices for protein sensing. Variations in the pressure/area diagrams of monolayers of stearic acid containing the cone calix[4]arenes **8**, **9** and **10a** allow, for instance, proteins having a surface charge opposite to that of the macrocycles to be revealed with detection limit down to 10^{-8} M [44,45]. The presence of the negatively charged tetraphosphonate **8** in the monolayer can for example sense the arginine-rich cyt c, chymotrypsin and lysine-rich histone H1 since all these proteins have an isoelectric point (pI) higher than physiological pH. Interestingly, it could be demonstrated that also in these interfacial processes multivalency plays a rather important role in the recognition process, since stearic acid monolayers containing anionic amphiphilic monophosphonates only give rather weak signals.

10a: R = Bu
10b: R = Dodecyl

11: R = Dodecyl

Similarly, monolayers functionalized with tetraammonium calixarene **9** are able to reveal acidic proteins (pI = 4) such as the acyl carrier protein, while the presence of the partially protonated tetraaniline **10a** can be used to identify both negatively and positively charged proteins. Remarkably, a 'naked eye' detection of the proteins could be achieved with monolayers of dimyristoylphosphatidylcholine containing the given calixarenes and doped with chromatic polydiacetylenes [46].

11.3 Recognition of Carbohydrate Binding Proteins (Lectins)

A special case of protein recognition is the recognition of lectins. Lectins are, by definition, carbohydrate binding proteins that do not show enzymatic or immunological activity. They are involved in different physiological and pathological processes that are generally characterized by important multivalent effects often referred to, in these cases, as glycoside cluster effects [47]. Carbohydrate–protein interactions are in fact usually rather weak (K_d in the millimolar range), so that nature exploits multivalency [1] to make them more efficient (K_d in the subnanomolar range) and selective. Lectins are therefore often multimeric, can associate in multimeric assemblies or are displayed as domains in a multivalent presentation on the surface of cells, viruses and bacteria. A plethora of polyglycosylated ligands based on different types of multivalent scaffold have been designed and synthesized to try to interfere with carbohydrate–protein interactions [48,49]. Calixarenes, as well, were intensively used as scaffolds to prepare multivalent glycosylated inhibitors of lectins. Glycosylated calixarenes, namely glycocalixarenes [5], are found to be important ligands because of their ease of functionalization and the possibility to modulate their valency. However, a very attractive and peculiar feature which distinguishes glycocalixarenes is that their selectivity and efficiency in protein binding is highly dependent on the stereochemical presentation of the saccharide units that can be easily tuned changing the conformational features of the macrocyclic scaffold. Stereoisomeric cone, 1,3-alternate and partial cone derivatives quite often show rather different efficiency in the binding of lectins of a given family, so that selectivity can be also finely modulated at will.

A nice example of the effects of the control of the stereochemical presentation of carbohydrates in the binding of lectins was obtained, for instance, in the inhibition of galectins. Galectins are lectins that selectively recognize galactosides on cell surfaces. Particularly relevant are human galectins since they are involved in the progression and migrations of tumours and metastases. The different subfamilies of human galectins present different activities and roles in the correlated pathologies. Moreover, human galectins gal-1, gal-3 and gal-4, which are examples of three different sub-families, show different structure and valency. Gal-1 and gal-4 display two carbohydrate binding sites (CBSs), facing opposite regions of the protein. Gal-3, on the other hand, is monomeric but may self-assemble in a pentameric, and consequently pentavalent, structure. To inhibit the adhesion of galectins to cells, we synthesized a small library of galactosyl- and lactosyl-calix[n]arenes (12–14) by using the click-chemistry reaction that yields thiourea moieties. While galactosyl-calixarenes are found to be weak inhibitors, lactosylated derivatives strongly inhibit the adhesion of galectins to surface immobilized asialofetuin also showing, in most of the cases, rather important multivalent effects [50]. Fluorescence assisted cell sorting tests indicate that, in terms of efficiency, the high-valent and conformationally mobile lactocalix[6]- (13d) and -[8] arenes (13f) are generally the most efficient inhibitors of adhesion of galectins to tumour cells. However, and even more interestingly, the 1,3-alternate calix[4]arene (14b) potently interferes with the adhesion of both gal-4 and gal-1 to human pancreatic carcinoma cells, probably because of a similar exposition geometry of the CBSs of the two proteins.

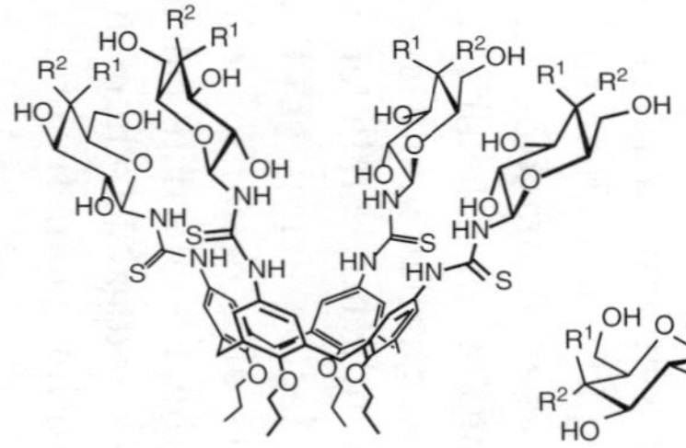

12a: R^1 = OH, R^2 = H
12b: R^1 = H, R^2 = O-β-D-Gal

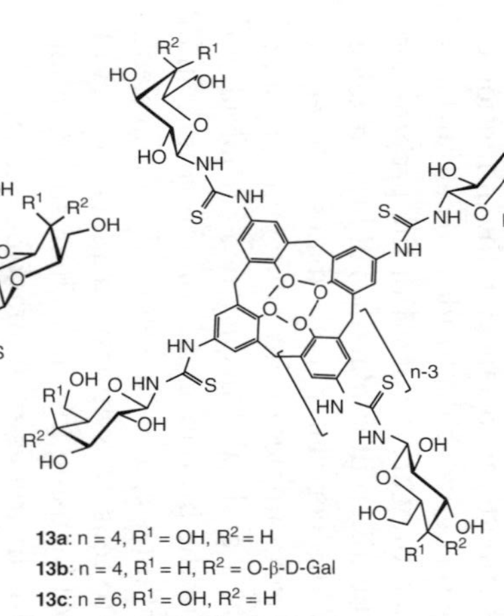

13a: n = 4, R^1 = OH, R^2 = H
13b: n = 4, R^1 = H, R^2 = O-β-D-Gal
13c: n = 6, R^1 = OH, R^2 = H
13d: n = 6, R^1 = H, R^2 = O-β-D-Gal
13e: n = 8, R^1 = OH, R^2 = H
13f: n = 8, R^1 = H, R^2 = O-β-D-Gal

16a:

16c:

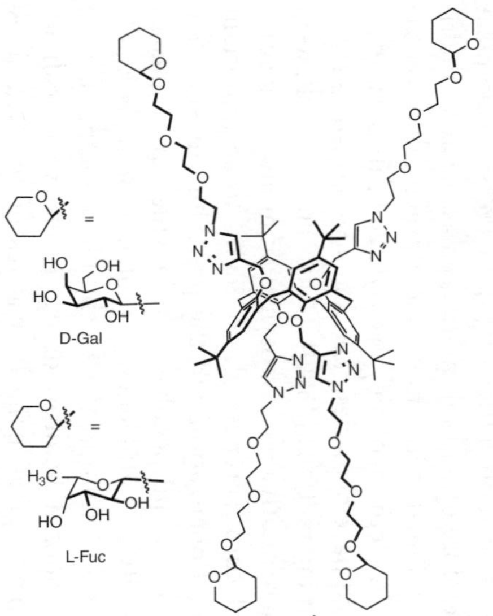

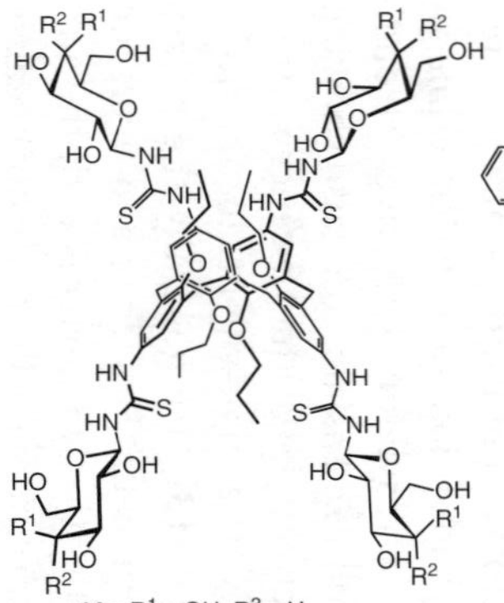

14a: R^1 = OH, R^2 = H
14b: R^1 = H, R^2 =O-β-D-Gal

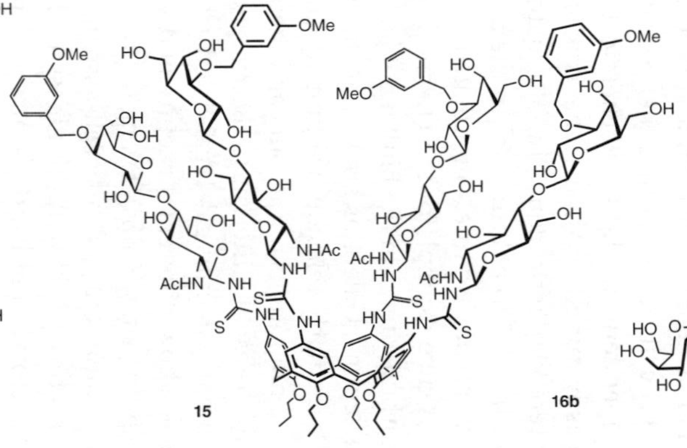

15

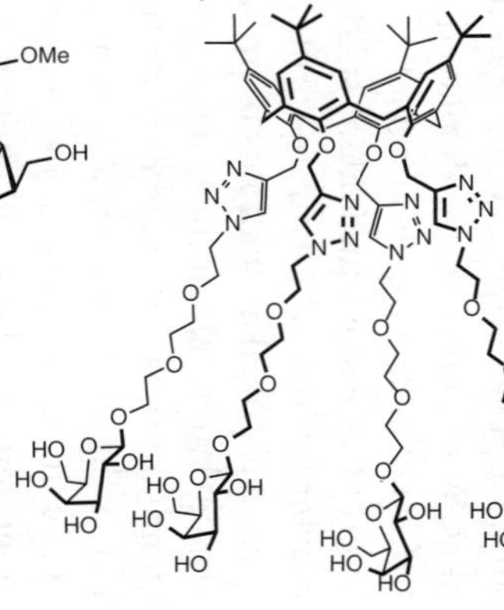

16b

This result appears of particular biomedical interest since both gal-1 and gal-4 are involved in the progression of the colon tumour. More interesting in terms of selectivity is that the cone isomer (**12b**), despite having the same valency as the 1,3-alternate derivative (**14b**), with the facial display of lactosyl units is a very poor inhibitor of gal-1 but it is the most active inhibitor of the gal-3 adhesion to human colon adenocarcinoma cells. In contrast, the 1,3-alternate (**14b**) is the worst inhibitor of gal-3 adhesion but, as previously mentioned, the best inhibitor for gal-1 [50]. This data indicates that inhibition of adhesion of galectins to cells is strictly dependent on the relative disposition of CBSs on the protein and of the lactosyl units on the multivalent ligand. When lactosyl units are replaced by the higher affinity LacNAc moieties bearing a benzyl group at the 3′-position, ligand **15a** in the cone structure shows an even higher selectivity for gal-3 over gal-1 as demonstrated by an amplified antiadhesion efficiency against gal-3 and a complete inactivity towards gal-1 [51]. Since gal-1 may have a potent pro-anoikis activity against some tumour cell lines, while gal-3 hinders its favourable action promoting the tumour progression, the possibility to block only the gal-3 not affecting the activity of gal-1 to induce a programmed tumour cell death might result in very attractive biomedical applications.

Another interesting example showing the importance of the stereochemical presentation of carbohydrate ligands on the calixarene scaffolds comes from the β-galactoside calix[4]arene derivatives **16a,b**, in the inhibition of the tetrameric galactose-binding lectin PA-IL from *Pseudomonas aeruginosa* (PA) [52]. This opportunistic bacterium is often associated with cystic fibrosis and is responsible for chronic respiratory diseases. The tetravalent derivative fixed in the 1,3-alternate structure **16a** shows for the PA-IL lectin a higher affinity ($K_D = 176$ nM) than its isomeric tetravalent analogue **16b** fixed in the cone structure ($K_D = 420$ nM). Also the stoichiometry (ligand to monomeric form of the lectin) of the two complexes is different being 1:4 for **16a** and 1:3 for **16b**. This indicates that, despite the rather long spacers between the galactose units and the calixarene scaffold, no more than three protein monomers can approach the same face of the macrocycle. Docking experiments and AFM images suggested that the two-facial disposition of galactose units in the 1,3-alternate derivative is more suitable to the arrangement of CBSs on the tetrameric lectin. Compound **16a**, in fact, can chelate two CBSs on one lectin tetramer and two other CBSs on another lectin tetramer thus resulting in a linear alternation of calixarene ligands and lectins (Fig. 11,4b). More recently, L-fucose units were inserted on the same 1,3-alternate calixarene scaffold to obtain a tetravalent inhibitor (**16c**) of PA-IIL, a different PA lectin which possesses selectivity for this saccharide [53]. Experiments *in vivo* on infected mice indicate that inhibition of both lectins by a mixture of the multivalent galactosylated **16a** and fucosylated **16c** 1,3-alternate calix[4]arenes seems a promising strategy to block PA infections and the bacterial biofilm proliferation.

An interesting case where an extremely high multivalent effect could be achieved with a calixarene derivative is that of cholera toxin (CT) inhibition. CT is a well-known AB$_5$ lectin with a CBS in each of its five B units. These units present a high affinity for a particular pentasaccharide **17** (GM1os) present on the surface of the cells. The CBSs are all located on the same side of the toroidal B$_5$ assembly (Figure 11.4a) so that CT can adhere to the cell surface also exploiting multivalency. In order to obtain a high affinity inhibitor of CT adhesion to cells we therefore designed and synthesized a pentavalent calix[5]arene (**18**) adorned at the upper rim with five arms terminating with a GM1os unit and long enough to allow the simultaneous interaction of all the five epitopes with

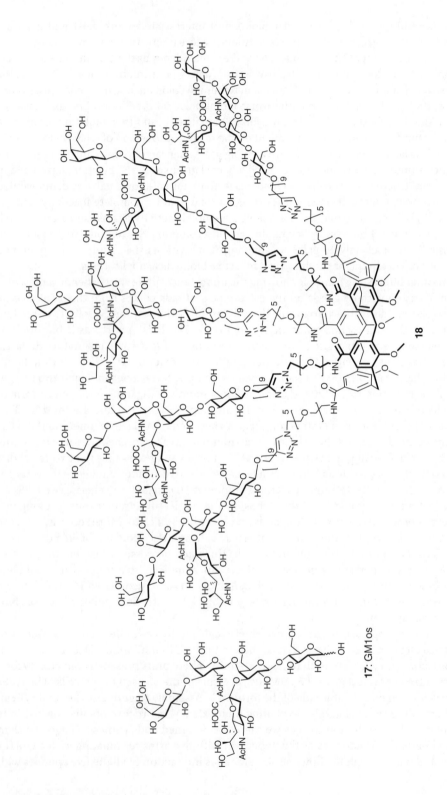

18

17: GM1os

the five CT CBSs. The compound **18**, prepared and studied in collaboration with Zuilhof and Wennekes, presents five GM1os units prepared through a chemoenzymatic synthesis and linked to the calixarene through a 31 atom chain. It presents an extremely high inhibition potency towards CT with an $IC_{50} = 450$ pM, as determined by ELISA tests. Not only does ligand **18** show the lowest IC_{50} value so far found for a pentavalent CT inhibitor but also a surprisingly high multivalent effect. Comparing the IC_{50} value of **18** with that of the monovalent GM1os (**17**), an astonishing gain in activity per single pentasaccharide installed on the macrocyclic core of 20 000-fold could be obtained [54].

All the previously reported characteristics of multivalent calixarene ligands to efficiently and selectively bind to proteins, together with the well know ability of these macrocycles to include small molecules, suggested the possibility of using calixarenes for targeted drug delivery [55]. Moreover, the amphiphilic structure [20] of water soluble calixarene ligands allows them to be embedded in different types of monolayers (self-assembled monolayers on metal surfaces, Langmuir films) or to self-organize them in aggregates such as micelles, SLNs [56] or vesicles and liposomes, potentially useful in bionanotechnology and nanomedicine [19,57].

Cone-shaped glycocalixarenes also represent well preorganized facial amphiphiles displaying a favourable attitude to form micelles or to be included in lipid mono- and bilayers. For example, we found that mannosylthioureidocalix[4]arene **19** blocked in the cone structure (Figure 11.6a) can be easily used in the noncovalent functionalization of dodecanthiol-stabilized gold nanoparticles (AuNPs) [58]. Simply by bringing into contact a chloroform layer of dodecanethiol-stabilized AuNPs and an aqueous solution

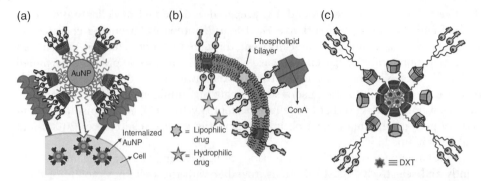

Figure 11.6 A clustered multivalent interaction (a) of the AuNPs functionalized with mannosylated calixarenes **19** targeting the human macrophage mannose receptors. (b) Section of a DOPC liposome functionalized with the bolaamphiphile **21** and exposing glucose units that can efficiently interact with concanavalin A (ConA): the functionalized liposomes can also be loaded with a lipophilic or hydrophilic cargo for targeted drug delivery. Self-assembled nanospheres (c) of calix[4] arene-β-cyclodextrin heterodimers **22** with a cargo of docetaxel (DXT) in the central core and noncovalently functionalized with mannosyl dendrons for targeted drug delivery. See colour section.

of the calix[4]arene **19**, the latter is adsorbed onto the monolayer at the surface of the AuNPs which become water dispersible and are transferred to the water layer. These mannosyl-coated nanoparticles present a rather low tendency to be coated by a protein corona ('stealth' properties), are rather stable and exhibit interesting targeting properties. Experiments with HeLa cells overexpressing a mannose receptor indicate that the uptake of mannosylcalixarene functionalized AuNPs is much higher than the uptake of AuNPs covered by an equimolar amount of monovalent mannosyl ligand **20**. Interestingly, these data also indicate that the AuNP uptake by cells exposing lectins on their external surface is strictly dependent on the disposition topology of the carbohydrate units on the nanoparticle and suggest that the high local sugar concentration ensured by the calixarene scaffold (clustered multivalency) might be extremely beneficial for an effective interaction with the cell membrane receptors. Structures other than cone calixarenes can also be used for incorporation into nanoobjects. The 1,3-alternate structure, for example, when properly modified at the upper rim with hydrophilic moieties becomes a bolaamphiphile [59]. The 1,3-alternate calix[4]arene bolaamphiphile **21**, having long spacers between the central multivalent core and the glucose units, can span the double layer of liposomes (Figure 11.6b) formed by 1,2-dioleoyl-*sn*-glycero-3-phosphocholine (DOPC). This represents also a facile and robust methodology to densely functionalize the liposome surface with sugar units. The obtained glucose functionalized liposomes efficiently interact with concanavalin A through rather specific multivalent interactions. This seems particularly interesting for the development of nanoobjects functionalized with glycocalixarenes, maybe decorated with more complex saccharide units, as drug-delivery systems that, when loaded with proper cargo, can selectively target cells exposing the complementary carbohydrate receptors.

Another possible use of calixarenes in drug delivery is related to their ability to self-assembly in micelles or SLNs and to act as central core of more complex multifunctional nanoobjects. In this way, calix[4]arenes were conjugated to a β-cyclodextrin derivative to form heterodimers **22a** and **22b** that, due to the self-assembly of 18–25 molecules (Figure 11.6c), give nanospheres of 20–25 nm in diameter [60], projecting the hydrophilic β-cyclodextrin moieties towards the external water solution. The highly lipophilic antimitotic docetaxel (DXT) can be loaded in these nanospheres and the drug release profile curves show a dual type of release. The part of DXT that is hosted in the β-cyclodextrin cavities is quickly released, while a slow kinetics of release is measured for DXT molecules included in the lipophilic nanosphere cores. The noncovalent functionalization of these nanospheres with trivalent mannosyl dendrons (Figure 11.6c), which insert the adamantyl moieties inside the β-cyclodextrin cavities, was used to efficiently direct the assemblies towards human macrophage mannose receptors.

11.4 Binding Polyphosphates, Oligonucleotides and Nucleic Acids

The well-known extremely important roles fulfilled by polyphosphates, oligonucleotides and particularly nucleic acids in living organisms make them important targets to be addressed. Being characterized by a high negative charge and a dense presentation of hydrogen bond acceptor groups along their structures, multivalent ligands adorned with positive charges and/or hydrogen bond donor head groups are predicted to be efficient binders of these molecules. Since the dawn of supramolecular chemistry, positively charged azamacrocycles or cyclophanes have been shown to efficiently bind to oligonucleotides and nucleic acids in aqueous solutions [61–63]. However, only several years later [64,65] did calixarenes start to be used as scaffolds for the synthesis of multivalent ligands bearing quaternary ammonium [62,66], N-methylimidazolium [67], primary ammonium and guanidinium groups [68–70] for nucleic acid binding. Our research group first explored the potential of the guanidinium ion as a binding moiety on calixarene scaffolds since, due to its known ability to interact with phosphate groups and its characteristic to be always protonated at physiological pH, it was expected to give rise to highly efficient multivalent ligands. The small library of multivalent p-guanidinocalixarenes **23**, characterized by different valency, conformational properties and length/lipophilicity of the lower rim alkyl chains were studied in detail [64,68].

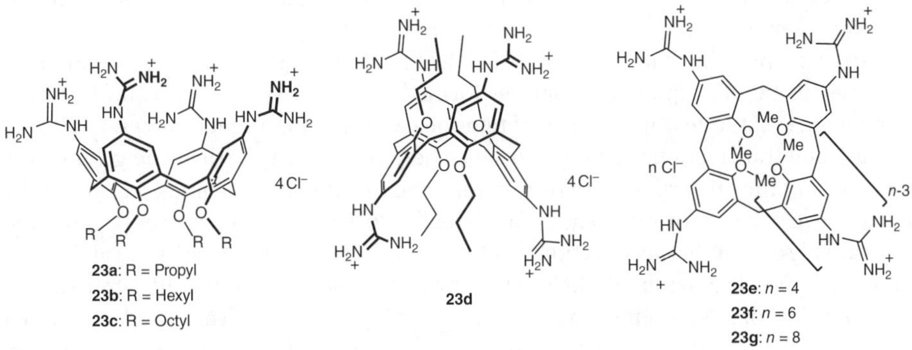

23a: R = Propyl
23b: R = Hexyl
23c: R = Octyl

23d

23e: n = 4
23f: n = 6
23g: n = 8

Interestingly, the data collected from a series of physico-chemical experiments (electrophoresis mobility shift assay, melting curves and ethidium bromide displacement assays and AFM) indicated a rather strong binding of these compounds with circular (plasmid) or linear DNA. In parallel studies, Schrader and coworkers compared the binding mode of a series of dimeric calix[4]arenes fixed in the cone structure and adorned with primary ammonium or guanidinium head groups [65,69]. By the combination of the data obtained by a different series of experimental techniques and modelling calculations, they concluded that short and destructured oligonucleotides interact with positively charged calixarenes mainly via hydrogen bonds and charge interactions to the negatively charged phosphate groups [71]. Also the guanidino derivatives were shown to target the phosphate anions of the B-DNA backbone while ligands bearing primary ammonium ions mainly bound to the nucleobases in the major groove. Remarkably, we first evidenced that *p*-guanidinocalixarenes **23** are able to transfer DNA into cells, thus paving the way for the use of macrocyclic amphiphiles as gene vectors [68]. In fact, together with some cyclodextrin-based gene-delivery systems which appeared nearly simultaneously [72] and despite several examples of cationic lipids and amphiphiles, to the best of our knowledge these calixarenes are the first examples of macrocyclic amphiphiles used as gene delivery systems. The design and optimization of gene vectors are rather critical processes since their success depends on the positive operation of a series of events: (i) an efficient interaction between the nucleic acid filament and the vector; (ii) a proper neutralization of charges; (iii) condensation of the nucleic acid and formation of carrier-nucleic acid aggregates of proper size to cross the cell membrane; (iv) activation of the endocytosis processes; (v) release of the nucleic acid from the endosomes into the cytoplasm; and (vi) its internalization into the cell nucleus. Since the carrier ability to compact the nucleic acid into small aggregates (40–80 nm) and to neutralize its charges is an important prerequisite, AFM can be successfully used to check the different carriers for DNA binding and condensation [68]. The perturbations observed in AFM images on the DNA folding upon incubation with variable amounts of macrocyclic amphiphiles (Figure 11.7), allow the calixarene ligands to be sorted into three different classes. Although inferred for *p*-guanidinocalixarenes **23** this classification can also be used for all the other calixarene amphiphiles studied up to now. To the first class of DNA-binders belong those amphiphilic calix[4]arenes – facial amphiphiles – fixed in the cone structure bearing ammonium groups at one rim and lipophilic chains/aromatic moieties at the other rim (e.g. **23a–c**). These ligands can strongly interact with the phosphodiester backbone of DNA through electrostatic interactions and condense the nucleic acid structure by operation of hydrophobic interactions among lipophilic tails (Figure 11.7a). This allows the formation of compact aggregates, named calixplexes (lipoplexes formed by calixarenes), which can be internalized into cells. Quite important to obtain the correct amphiphilic character for condensation and transfection is a proper balance between polarity of the head groups and lipophilicity of the tails, which can be programmed by inserting linear alkyl chains onto the macrocycle. Hexyl or octyl chains are usually sufficient to give compact and lipophilic calixplexes, while the insertion of longer alkyl chains often results in the formation of strongly self-assembled SLNs, which are often found to be highly cytotoxic. To the second class of DNA-binders belong those calix[4]arenes which are fixed by synthesis (e.g. **23d**) or adopt in aqueous solution (e.g. **23e**) the 1,3-alternate structure (Figure 11.7b). Such bolaamphiphiles usually bind a single DNA molecule but condensation is only driven by charge–charge interactions and thus is found to be not so efficient as in the previous case. For these reasons, it often requires a higher calixarene/DNA ratio and the use of a synergizer such as DOPE (dioleoyl-*sn*-phosphatidylethanolamine) which is known to help transfection by increasing the overall lipophilicity of the lipoplexes.

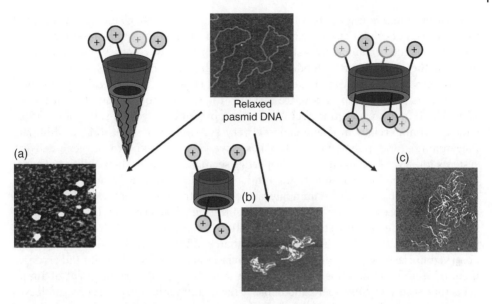

Figure 11.7 AFM images of the results of the different compaction modes of plasmid DNA (top image) with three different classes of DNA-binders in which amphiphilic calixarenes can be groups: (a) facial amphiphiles fixed in the cone structure; (b) bolaamphiphiles calix[4]arene in 1,3-alternate conformation; (c) globular amphiphiles having a conformationally mobile structure.

The third class of DNA-binders is formed by high-valency calix[6]- (e.g. **23f**) and calix[8] arenes (e.g. **23g**) which are characterized by a rather high conformational flexibility and a globular shape with positive charges on their surface. This morphology results in highly efficient ligands which are able to bind several DNA molecules, giving rise to gorgone-like aggregates (Figure 11.7c) which, however, are too large and usually not suitable for transfection.

24

25a: R = NHC(NH$_2$)NH$_2^+$
25b: R = CH$_2$NH$_3^+$

26: R = *n*-C$_8$H$_{17}$

Facial amphiphiles can be also obtained by the introduction of guanidinium groups at the lower rim of calix[4]arenes (e.g. **24**). Even if linear alkyl chains are absent, the lipophilic calixarene backbone gives sufficient lipophilicity to warrant a proper amphiphilicity for the gene vector to condense DNA and transfect cells [66,73]. Several other cationic binding groups were linked to calixarene. However, the best calixarene-based gene vectors prepared up to now are certainly the tetralysino- **25b** and tetraarginino- **25a** calix[4]arenes which transfect cells even more efficiently than some of the most effective commercially available gene-delivery systems such as Jet-PEI or LTX [70]. Polyarginino- and polylysino-amphiphiles have been well-known as gene-delivery systems for several years and their properties seem to be due to their ability to trigger the signals for endocytosis due to the interaction of the ammonium head groups with the negatively charged glycosaminoglycans of the extracellular matrix [74]. The facial calixarenes **25** present a very high local concentration (cluster) of arginines at the upper rim which, besides giving a strong binding with DNA, might similarly trigger the signals leading to activation of the endocytosis mechanism. Interestingly, contrary to what is observed in classical polyarginino gene-delivery systems where a head-to-tail arrangement of arginine moieties is present, in **25a** there is a parallel arrangement of amino acids that ensures a convergent and dense array of ammonium head groups and leaves the α-amino groups free to reinforce the binding with the nucleic acids and facilitate DNA release into the cytosol, due to the so-called proton sponge effect [75,76].

It was also possible to study the mechanism of DNA condensation by calixarene amphiphiles and it could be pointed out that this process depends on the type of amphiphile. For amphiphiles such as **26**, in which lipophilicity prevails over hydrophilicity, a hierarchical self-aggregation mechanism takes place (Figure 11.8c) [67]. Such calixarenes are already self-assembled in aqueous solution and these micelles bind to DNA. In other amphiphiles, such as **25a**, in which the hydrophilic character prevails over the lipophilic one, free amphiphiles or small oligomers are present in solution which progressively clot onto the DNA filaments (Figure 11.8a) [66,70]. Subsequently, hydrophobic interactions between the lipophilic groups of the carriers lead to DNA condensation. Such type of mechanism is also supported by a TEM image corresponding to the nanocomplexes obtained by mixing compound **25a** [60] with DNA (Figure 11.8b) showing a snake-like ultra-thin structure with dark regions of DNA filaments and with light stripes corresponding to the tail-to-tail calixarene double layers.

Self-assembly can be also used to prepare stable Langmuir monolayers simply composed of cone tetraaminocalixarenes **10b**, in the absence of any co-surfactant. The partially protonated aniline macrocycle of the layer interacts with the DNA present in the sub-phase originating a monolayer expansion and a phase transition that allow the detection of the nucleic acid [77]. On the other hand, when the monolayers are formed with cone p-guanidinododecyloxy-calix[4]arene **11** the isotherm changes depend on the duplex DNA sequence in the aqueous sub-phase. Larger increases of surface area are observed upon interaction with poly(AT), while poly(GC) or random DNA sequences give much smaller changes [78]. Similarly, SLNs of **11**, obtained by the solvent displacement method and characterized by diameters of roughly 190 nm and positive ζ-potentials, preferably interact with DNA consisting of $A_{30}^*T_{30}$ oligonucleotides compared with $G_{30}^*C_{30}$ oligonucleotides [79].

LPS endotoxins, which are produced by Gram-negative bacteria, are also an interesting target containing phosphate groups. LPS released by bacteria binds to a series of

(a)

(b)

(c)

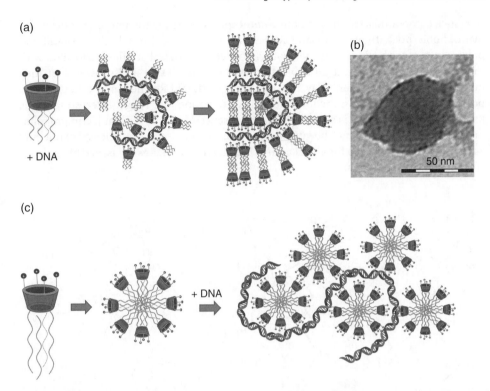

Figure 11.8 The two proposed mechanisms for DNA condensation with calixarenes **25a** and **26**. (a) Sequential binding and condensation: amphiphile **25a** binds to DNA as monomer or in small aggregates and then hydophobic interaction leads to condensation; (b) the snake-like ultra-thin TEM image of the **25a**-DNA aggregate; (c) hierarchical self-assembly, with micelles of amphiphile **26** interacting with a DNA filament. *Source*: Ref. [12]. Reproduced with permission of Royal Society of Chemistry. See colour section.

proteins such as TLR4, CD14 and MD2 triggering an important inflammatory response that might also end in sepsis and septic shock. With the idea to mimic the activity of antibacterial peptides, such as βpep peptides and dodecapeptide SC4, which possess an amphiphilic structure able to bind to LPS, Mayo and coworkers prepared a series of 23 amino- and guanidinocalixarenes (e.g. **27**, **28**) which are able to bind to Lipid A (**29**), the conserved region of all the different LPSs [80]. Among the most active inhibitors of the TLR4 activation, the cone compounds **27** and **28** have IC_{50} values in the nanomolar range. With a 25–100% of survival, compound **28c** also shows a fairly good efficacy in the protection of mice administered with a lethal dose of LPS. NMR studies also indicated that these compounds form 1:1 complexes with Lipid A due to strong electrostatic interaction of the ammonium groups with the phosphate anions. Although these data could suggest that the inhibition of the inflammatory process originates from the binding of the calixarene to Lipid A (**29**), the mechanism of protection *in vivo* should still be studied in detail. To this end, very recently we actually identified some guanidinocalixarenes as potential ligands not simply for LPS but for TLR4 or for co-receptors involved in the TLR4 signalling process, such as CD14 protein and TLR4.MD-2 dimer [81]. Preliminary molecular modelling studies evidenced that macrocycles **23a,b**, and also **28c** used as reference compounds, can bind to these coreceptors inserting the lipophilic tails in the hydrophobic pockets of CD14 and MD-2 and interacting through electrostatic and polar contacts with the rim of the protein pockets. Biological studies performed with these compounds ruled out an agonist activity towards TLR4 and showed that they inhibit in a dose-dependent way the activation of the TLR4 mediated inflammation process with IC_{50} values in the low micromolar to nanomolar range. Interestingly, this happens also when TLR4 is stimulated in the absence of LPS by a plant lectin phytohemagglutinin, suggesting that the guanidinocalixarenes do not act, or at least act not only as ligands for LPS but also for CD14 and/or TLR4.MD-2 dimer as envisaged by molecular modelling calculations.

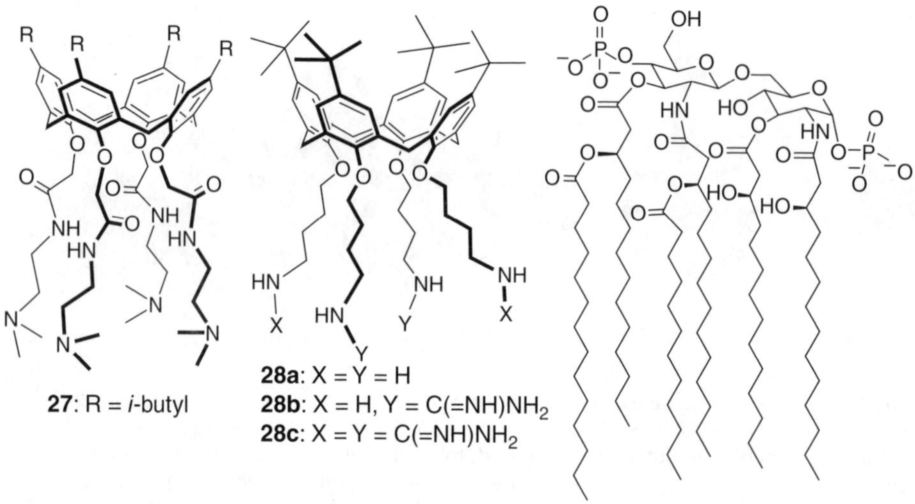

27: R = *i*-butyl

28a: X = Y = H
28b: X = H, Y = C(=NH)NH₂
28c: X = Y = C(=NH)NH₂

29: Lipid A

11.5 Conclusions

The possibility to easily introduce at the upper or lower rim of calixarene macrocycles a variable number of hydrophilic groups such as ammonium, sulfonate and phosphate or of more complex moieties such as amino acids, small peptides, mono- and oligosaccharides allows a wide variety of multivalent ligands to be obtained which proved to selectively and efficiently bind to biomacromolecules. The oligomeric nature of these macrocycles also allows to easily tune the valency of the ligands, their size and their structural properties. Particularly important is the stereochemical presentation of the ligating units which strongly influences the binding and selectivity observed for proteins. A relevant potential of these multivalent ligands, still not completely exploited, is related to their amphiphilic character and to the possibility to flank strong multivalent electrostatic and hydrogen-bonding interactions provided by the ligating groups present in their structure, with hydrophobic and Van der Waals interactions of the aromatic backbone and cavity with apolar moieties of proteins. The peculiar structures of facial- (cone calix[4]arenes) or bola-(1,3-alternate calix[4]arenes) amphiphiles strictly control many processes such as the condensation of nucleic acids and the self-assembly and incorporation into SLNs, AuNPs and liposomes. Most of the data collected up to now has shown the great potential of these multivalent ligands especially in targeted drug delivery, gene delivery and in the selective sensing of biomacromolecules with possible future applications in bionanotechnology and nanomedicine.

Acknowledgements

We thank the Italian Ministero dell'Istruzione, dell'Università e della Ricerca (MIUR, PRIN project 2010JMAZML), and the EU-COST Actions CM1102 'MultiGlycoNano' and CM1005 'Supramolecular Chemistry in Water'.

References

1 Mammen M, Choi S-K, Whitesides GM. Polyvalent interactions in biological systems: Implications for design and use of multivalent ligands and inhibitors. Angew. Chem. Int. Engl. Ed. 1998;37(20):2755–94.

2 Fasting C, Schalley CA, Weber M, *et al*. Multivalency as a chemical organization and action principle. Angew. Chem.Int. Ed. 2012;51(42):10472–98.

3 Mulder A, Huskens J, Reinhoudt DN. Multivalency in supramolecular chemistry and nanofabrication. Org. Biomol. Chem. 2004;2(23):3409–24.

4 Baldini L, Casnati A, Sansone F, Ungaro R. Calixarene-based multivalent ligands. Chem. Soc. Rev. 2007;36(2):254–66.

5 Sansone F, Casnati A. Multivalent glycocalixarenes for recognition of biological macromolecules: glycocalyx mimics capable of multitasking. Chem. Soc. Rev. 2013;42(11):4623–39.

6 Baldini L, Sansone F, Casnati A, Ungaro R. Calixarenes in molecular recognition. In: Steed JW, Gale PA, editors. Supramolecular Chemistry: from Molecules to Nanomaterials. Chichester: John Wiley & Sons, Ltd; 2012, pp. 863–94.

7 Shinkai S, Mori S, Tsubaki T, Sone T, Manabe O. New water-soluble host molecules derived from calix[6]arene. Tetrahedron Lett. 1984;25(46):5315–8.

8 Böhmer V. Calixarenes, Macrocycles with (almost) unlimited possibilities. Angew. Chem. Int. Engl. Ed. 1995;34(7):713–45.

9 Ungaro R, Arduini A, Casnati A, *et al*. New synthetic receptors based on calix[4]arenes for the selective recognition of ions and neutral molecules. Pure Appl. Chem. 1996;68(6):1213–8.

10 Casnati A. Calixarenes: From chemical curiosity to a rich source for molecular receptors. Gazz. Chim. Ital. 1997;127(11):637–49.

11 Sansone F, Baldini L, Casnati A, Ungaro R. Calixarenes: from biomimetic receptors to multivalent ligands for biomolecular recognition. New J. Chem. 2010;34(12):2715–28.

12 Giuliani M, Morbioli I, Sansone F, Casnati A. Moulding calixarenes for biomacromolecule targeting. Chem. Commun. 2015;51(75):14140–59.

13 Gutsche CD. Calixarenes: An Introduction. Cambridge: The Royal Society of Chemistry; 2008.

14 Casnati A. Calixarenes and cations: a time-lapse photography of the big-bang. Chem. Commun. 2013;49(61):6827–30.

15 Casnati A, Sansone F, Ungaro R. Peptido- and glycocalixarenes: Playing with hydrogen bonds around hydrophobic cavities. Acc. Chem. Res. 2003;36(4):246–54.

16 Cornforth JW, D'Arcy Hart P, *et al.* Antituberculous effects of certain surface-active polyoxyethylene ethers. Br. J. Pharmacol. Chemother. 1955;10(1):73–86.

17 Colston MJ, Hailes HC, Stavropoulos E, *et al.* Antimycobacterial calixarenes enhance innate defense mechanisms in murine macrophages and induce control of *Mycobacterium tuberculosis* infection in mice. Infect. Immun. 2004;72(11):6318–23.

18 Casnati A, Fabbi M, Pelizzi N, *et al.* Synthesis, antimicrobial activity and binding properties of calix[4]arene based vancomycin mimics. Bioorg. Med. Chem. Lett. 1996;6(22):2699–704.

19 Sansone F, Casnati A. Elsevier Reference Module in Chemistry, Molecular Sciences and Chemical Engineering. Waltham, MA: Elsevier; 2015.

20 Jie K, Zhou Y, Yao Y, Huang F. Macrocyclic amphiphiles. Chem. Soc. Rev. 2015;44(11):3568–87.

21 Zadmard R, Alavijeh NS. Protein surface recognition by calixarenes. RSC Adv. 2014;4(78):41529–42.

22 Nimse SB, Kim T. Biological applications of functionalized calixarenes. Chem. Soc. Rev. 2013;42(1):366–86.

23 Peters MS, Li M, Schrader T. Interactions of calix[n]arenes with nucleic acids. Nat. Prod. Commun. 2012;7(3), 409–17.

24 Sebti SM, Hamilton AD. Design of growth factor antagonists with antiangiogenic and antitumor properties. Oncogene 2000;19(56):6566–73.

25 Hamuro Y, Crego-Calama M, Park HS, Hamilton AD. A calixarene with four peptide loops: An antibody mimic for recognition of protein surfaces. Angew. Chem. Int. Engl. Ed. 1997;36(23):2680–3.

26 Wei Y, McLendon GL, Hamilton AD, *et al.* Disruption of protein-protein interactions: design of a synthetic receptor that blocks the binding of cytochrome c to cytochrome c peroxidase. Chem. Commun. 2001;1580–1.

27 Park HS, Lin Q, Hamilton AD. Protein surface recognition by synthetic receptors: A route to novel submicromolar inhibitors for alpha-chymotrypsin. J. Am. Chem. Soc. 1999;121(1):8–13.

28 Prata JV, Barata PD. Fostering protein-calixarene interactions: From molecular recognition to sensing. RSC Adv. 2016;6(2):1659–69.

29 Blaskovich MA, Lin Q, Delarue FL, *et al.* Design of GFB-111, a platelet-derived growth factor binding molecule with antiangiogenic and anticancer activity against human tumors in mice. Nat. Biotech. 2000;18(10):1065–70.

30 Sun JZ, Wang DA, Jain RK, *et al.* Inhibiting angiogenesis and tumorigenesis by a synthetic molecule that blocks binding of both VEGF and PDGF to their receptors. Oncogene 2005;24(29):4701–9.

31 Sansone F, Barboso S, Casnati A, *et al.* A new chiral rigid cone water soluble peptidocalix[4]arene and its inclusion complexes with alpha-amino acids and aromatic ammonium cations. Tetrahedron Lett. 1999;40(25):4741–4.

32 Douteau-Guevel N, Coleman AW, Morel JP, Morel-Desrosiers N. Complexation of the basic amino acids lysine and arginine by three sulfonatocalix[n]arenes (n = 4, 6 and 8) in water: microcalorimetric determination of the Gibbs energies, enthalpies and entropies of complexation. J. Chem. Soc., Perkin Trans. 2 1999;629–33.

33 Selkti M, Coleman AW, Nicolis I, *et al*. The first example of a substrate spanning the calix[4]arene bilayer: the solid state complex of p-sulfonatocalix[4]arene with L-lysine. Chem. Commun. 2000;161–2.

34 Lazar A, Da Silva E, Navaza A, *et al*. A new packing motif for para-sulfonatocalix[4] arene: the solid state structure of the para-sulfonatocalix[4]arene d-arginine complex. Chem. Commun. 2004;2162–3.

35 Frish L, Sansone F, Casnati A, *et al*. Complexation of a peptidocalix[4]arene, a vancomycin mimic, with alanine-containing guests by NMR diffusion measurements. J. Org. Chem. 2000;65(16):5026–30.

36 McGovern RE, Fernandes H, Khan AR, *et al*. Protein camouflage in cytochrome c-calixarene complexes. Nat. Chem. 2012;4(7):527–33.

37 McGovern RE, McCarthy AA, Crowley PB. Protein assembly mediated by sulfonatocalix[4]arene. Chem. Commun. 2014;50(72):10412–5.

38 McGovern RE, Snarr BD, Lyons JA, *et al*. Structural study of a small molecule receptor bound to dimethyllysine in lysozyme. Chem. Sci. 2015;6(1):442–9.

39 Daze KD, Pinter T, Beshara CS, *et al*. Supramolecular hosts that recognize methyllysines and disrupt the interaction between a modified histone tail and its epigenetic reader protein. Chem. Sci. 2012;3(9):2695–9.

40 Martos V, Bell SC, Santos E, *et al*. Calix[4]arene-based conical-shaped ligands for voltage-dependent potassium channels. Proc. Natl Acad. Sci. 2009;106(26):10482–6.

41 Gordo S, Martos V, Santos E, *et al*. Stability and structural recovery of the tetramerization domain of p53-R337H mutant induced by a designed templating ligand. Proc. Natl Acad. Sci. 2008;105(43):16426–31.

42 Gordo S, Martos V, Vilaseca M, *et al*. On the role of flexibility in protein-ligand interactions: The example of p53 tetramerization domain. Chem. Asian J. 2011;6(6): 1463–1469.

43 Tommasone S, Talotta C, Gaeta C, *et al*. Biomolecular fishing for calixarene partners by a chemoproteomic approach. Angew. Chem. Int. Ed. 2015;54(51):15405–9.

44 Zadmard R, Arendt M, Schrader T. Multipoint recognition of basic proteins at a membrane model. J. Am. Chem. Soc. 2004;126(25):7752–3.

45 Zadmard R, Schrader T. Nanomolar protein sensing with embedded receptor molecules. J. Am. Chem. Soc. 2005;127(3):904–15.

46 Kolusheva S, Zadmard R, Schrader T, Jelinek R. Color fingerprinting of proteins by calixarenes embedded in lipid/polydiacetylene vesicles. J. Am. Chem. Soc. 2006;128(41):13592–8.

47 Lee YC, Lee RT. Carbohydrate-protein interactions – Basis of glycobiology. Acc. Chem. Res. 1995;28(8):321–7.

48 Chabre YM, Roy R. Multivalent glycoconjugate syntheses and applications using aromatic scaffolds. Chem. Soc. Rev. 2013;42(11):4657–708.

49 Bernardi A, Jimenez-Barbero J, Casnati A, *et al*. Multivalent glycoconjugates as anti-pathogenic agents. Chem. Soc. Rev. 2013;42(11):4709–27.

50 André S, Sansone F, Kaltner H, *et al*. Calix[n]arene-based glycoclusters: bioactivity of thiourea-linked galactose/lactose moieties as inhibitors of binding of medically relevant lectins to a glycoprotein and cell-surface glycoconjugates and selectivity among human adhesion/growth-regulatory galectins. ChemBioChem 2008;9(10):1649–61.

51 André S, Grandjean C, Gautier FM, *et al*. Combining carbohydrate substitutions at bioinspired positions with multivalent presentation towards optimising lectin inhibitors: case study with calixarenes. Chem. Commun. 2011;47(21):6126–8.

52 Cecioni S, Lalor R, Blanchard B, *et al.* Achieving high affinity towards a bacterial lectin through multivalent topological isomers of calix[4]arene glycoconjugates. Chem. Eur. J. 2009;15(47):13232–40.

53 Boukerb AM, Rousset A, Galanos N, *et al.* Antiadhesive properties of glycoclusters against *Pseudomonas aeruginosa* lung infection. J. Med. Chem. 2014;57(24):10275–89.

54 Garcia-Hartjes J, Bernardi S, Weijers CAGM, *et al.* Picomolar inhibition of cholera toxin by a pentavalent ganglioside GM1os-calix[5]arene. Org. Biomol. Chem. 2013;11(26):4340–9.

55 Deshayes S, Gref R. Synthetic and bioinspired cage nanoparticles for drug delivery. Nanomedicine 2014;9(10):1545–64.

56 Shahgaldian P, Cesario M, Goreloff P, Coleman AW. Para-acyl calix[4]arenes: amphiphilic self-assembly from the molecular to the mesoscopic level. Chem. Commun. 2002;326–7.

57 Wang L, Li Ll, Fan YS, Wang H. Host-guest supramolecular nanosystems for cancer diagnostics and therapeutics. Adv. Mater. 2013;25(28):3888–98.

58 Avvakumova S, Fezzardi P, Pandolfi L, *et al.* Gold nanoparticles decorated by clustered multivalent cone-glycocalixarenes actively improve the targeting efficiency toward cancer cells. Chem. Commun. 2014;50(75):11029–32.

59 Aleandri S, Casnati A, Fantuzzi L, *et al.* Incorporation of a calixarene-based glucose functionalised bolaamphiphile into lipid bilayers for multivalent lectin recognition. Org. Biomol. Chem. 2013;11(29):4811–7.

60 Gallego-Yerga L, Lomazzi M, Franceschi V, *et al.* Cyclodextrin- and calixarene-based polycationic amphiphiles as gene delivery systems: a structure-activity relationship study. Org. Biomol. Chem. 2015;13:1708–23.

61 Dietrich B, Hosseini MW, Lehn JM, Sessions RB. Anion receptor molecules. Syntheis and anion-binding properties of polyammonium macrocycles. J. Am. Chem. Soc. 1981;103:1282–3.

62 Shi YH, Schneider H-J. Interactions between aminocalixarenes and nucleotides or nucleic acids. J. Chem. Soc., Perkin Trans. 2 1999;1797–803.

63 Ragunathan KG, Schneider H-J. Nucleotide complexes with azoniacyclophanes containing phenyl-, biphenyl- or bipyridyl- units. J. Chem. Soc., Perkin Trans. 2 1996;2597–600.

64 Dudic M, Colombo A, Sansone F, *et al.* A general synthesis of water soluble upper rim calix[n]arene guanidinium derivatives which bind to plasmid DNA. Tetrahedron 2004;60(50):11613–8.

65 Zadmard R, Schrader T. DNA recognition with large calixarene dimers. Angew. Chem. Int. Ed. 2006;45(17):2703–6.

66 Bagnacani V, Franceschi V, Fantuzzi L, *et al.* Lower rim guanidinocalix[4]arenes: Macrocyclic nonviral vectors for cell transfection. Bioconjugate Chem. 2012;23(5):993–1002.

67 Rodik RV, Klymchenko AS, Jain N, *et al.* Virus-sized DNA nanoparticles for gene delivery based on micelles of cationic calixarenes. Chem. Eur. J. 2011;17(20):5526–38.

68 Sansone F, Dudic M, Donofrio G, *et al.* DNA condensation and cell transfection properties of guanidinium calixarenes: Dependence on macrocycle lipophilicity, size, and conformation. J. Am. Chem. Soc. 2006;128(45):14528–36.

69 Hu W, Blecking C, Kralj M, *et al.* Dimeric calixarenes: A new family of major-groove binders. Chem. Eur. J. 2012;18(12):3589–97, S3589-1.

70 Bagnacani V, Franceschi V, Bassi M, *et al*. Arginine clustering on calix[4]arene macrocycles for improved cell penetration and DNA delivery. Nat. Commun. 2013;4:1721.

71 Zadmard R, Taghvaei-Ganjali S, Gorji B, Schrader T. Calixarene dimers as host molecules for biologically important di- and oligophosphates. Chem. Asian J. 2009;4(9):1458–64.

72 Ortiz Mellet C, Benito JM, Garcia Fernandez JM. Preorganized, macromolecular, gene-delivery systems. Chem. Eur. J. 2010;16(23):6728–42.

73 Bagnacani V, Sansone F, Donofrio G, *et al*. Macrocyclic nonviral vectors: High cell transfection efficiency and low toxicity in a lower rim guanidinium calix[4]arene. Org. Lett. 2008;10(18):3953–6.

74 Nakase I, Akita H, Kogure K, *et al*. Efficient intracellular delivery of nucleic acid pharmaceuticals using cell-penetrating peptides. Acc. Chem. Res. 2012;45(7):1132–9.

75 Akinc A, Thomas M, Klibanov AM, Langer RJ. Exploring polyethylenimine-mediated DNA transfection and the proton sponge hypothesis. J. Gene Med. 2005;7:657–63.

76 Behr J-P. Synthetic gene transfer vectors II: Back to the future. Acc. Chem. Res. 2012;45(7):980–4.

77 Shahgaldian P, Sciotti MA, Pieles U. Amino-substituted amphiphilic calixarenes: Self-assembly and interactions with DNA. Langmuir 2008;24(16):8522–6.

78 Rullaud V, Moridi N, Shahgaldian P. Sequence-specific DNA interactions with calixarene-based langmuir monolayers. Langmuir 2014;30(29):8675–9.

79 Rullaud V, Siragusa M, Cumbo A, *et al*. DNA surface coating of calixarene-based nanoparticles: a sequence-dependent binding mechanism. Chem. Commun. 2012;48(100):12186–8.

80 Chen XM, Dings RPM, Nesmelova I, *et al*. Topomimetics of amphipathic beta-sheet and helix-forming bactericidal peptides neutralize lipopolysaccharide endotoxins. J. Med. Chem. 2006;49(26):7754–65.

81 Sestito SE, Facchini FA, Morbioli I, *et al*. Amphiphilic guanidinocalixarenes inhibit lipopolysaccharide (LPS)- and lectin-stimulated Toll-like receptor 4 (TLR4) signaling. *J. Med. Chem.* 2017;60(12):4882–92.

12

Cucurbit[n]uril Assemblies for Biomolecular Applications

Emanuela Cavatorta[1,2], Luc Brunsveld[3], Jurriaan Huskens[2], and Pascal Jonkheijm[1,2]

[1] *Bioinspired Molecular Engineering Laboratory, MIRA Institute for Biomedical Technology and Technical Medicine, University of Twente, 7500 AE, Enschede, the Netherlands*
[2] *Molecular Nanofabrication Group, MESA+ Institute for Nanotechnology, University of Twente, 7500 AE, Enschede, the Netherlands*
[3] *Department of Biomedical Engineering, Laboratory of Chemical Biology and Institute of Complex Molecular Systems, Eindhoven University of Technology, Eindhoven, the Netherlands*

12.1   Introduction

Cucurbit[n]urils (CB[n]s) are a relatively young class of macrocycles that are able to form stable host–guest complexes with various guests. These guests range from small molecules such as drugs, dyes, and amino acids, to polymers, peptides and even protein domains, via their appended side-chains. The wide range of possible guests has stimulated an increasing number of fundamental molecular recognition studies that have found application in numerous fields, from catalysis to drug delivery and fabrication of (bio)materials [1–5].

The first synthetic report of cucurbituril originates from 1905 by Behrend and coworkers [5a], but it was only in 1981 that Mock's group revealed the molecular structure of this "condensation product", giving it the name of CB[6] [5b]. These compounds were named cucurbit[n]urils to highlight their symmetric shape resembling that of a pumpkin, botanically classified as a *Cucurbitacea*. In fact, the CB[n] structure is given by the repetition of n glycoluril units connected by 2n methylene bridges, so that the hydrophobic cavity is decorated by two symmetric carbonyl-fringed portals (Figure 12.1). Over the last 15 years, the conditions of the acid-catalyzed condensation of glycoluril and formaldehyde were varied leading to the successful purification of various homologs, including CB[5], CB[6], CB[7], CB[8], CB[10], and CB[14], which has a folded structure and is so far the largest member of this family (Table 12.1) [6–8]. In the last two decades several groups [8–14] succeeded in synthesizing functionalized CB[6] and CB[7] enabling further chemical modifications and often better solubility in either water or organic solvents. Various derivatives of CB[n] have been discovered up to now such as hemicucurbit[n]urils (one half of CB[n] cut along the molecular equator), inverted (containing one glycoluril unit directed toward the inside of the cavity) and

Multivalency: Concepts, Research & Applications, First Edition. Edited by Jurriaan Huskens,
Leonard J. Prins, Rainer Haag, and Bart Jan Ravoo.
© 2018 John Wiley & Sons Ltd. Published 2018 by John Wiley & Sons Ltd.

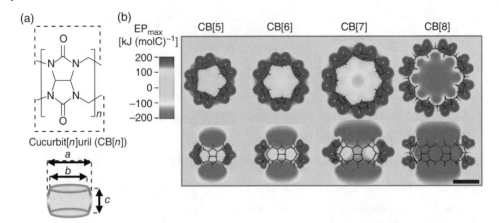

Figure 12.1 (a) Molecular and schematic structure of cucurbit[*n*]uril macrocycles. (b) Calculated electrostatic potential (EP) maps for CB[5], CB[6], CB[7], and CB[8] as cross-sectional and side views. Scale bar: approximately 4 Å. *Source*: Adapted from Ref. [16]. Reproduced with permission of American Chemical Society. See color section.

Table 12.1 Structural parameters of CB[*n*] as graphically shown in Figure 12.1a.

Host	Cavity diameter, a^a (Å)	Portal diameter, b^a (Å)	Height, c^b (Å)
CB[5]	4.4	2.4	9.1
CB[6]	5.8	3.9	9.1
CB[7]	7.3	5.4	9.1
CB[8]	8.8	6.9	9.1
CB[10]	11.7	10.0	9.1

[a] Data from Ref. [5].
[b] Data from Ref. [17].

nor-*sec*-CB[*n*] (missing one methylene bridge). For more information on the binding properties and applications of these CB[*n*] derivatives the reader is referred to other reviews [5,15].

The CB[*n*] family is a powerful supramolecular platform for developing innovative applications, mainly because of the following factors: (1) commercial availability in four different sizes; (2) high binding affinity and selectivity towards their guests; (3) accessibility of synthetic routes for several homologs and derivatives; (4) high stability; (5) solubility in both organic and aqueous solvents; (6) remarkably low or absence of toxicity *in vitro* and *in vivo*; and (7) temporal and spatial control of the molecular recognition process, such as by chemical, electrochemical and photochemical triggers [1,2].

In this chapter we first discuss the origins of the molecular recognition properties of the CB[*n*] family (Section 12.2) and methods to gain control over the binding affinity to CB[*n*] (Section 12.3). Afterwards, we describe the amino acid based recognition of CB[*n*]

(Section 12.4) and subsequently review the literature on how these properties can be successfully employed to dynamically probe the properties and function of peptides, proteins, and cells, which is important for bioanalytical and biomedical applications (Section 12.5).

12.2 Molecular Recognition Properties of CB[*n*]

12.2.1 Interactions with the Carbonyl Portals of CB[*n*]

In aqueous environment the molecular recognition properties of CB[*n*] originate from the unique combination of their structure and their cavity size. As mentioned before, the structure of the CB[*n*] family is highly symmetric, with a hydrophobic concave cavity laced on both sides by carbonyl rims (Figure 12.1). The electrostatic potential map visualizes the high density of electrons localized uniformly at the portals of CB[*n*] and explains the affinity of cucurbiturils for positively polarized species through cation–dipole and hydrogen bonding interactions (Figure 12.1) [16,17]. For example, optimal ion–dipole interactions and desolvated carbonyl portals were found in the crystal structure of the complex of CB[7] and the diamantane diammonium ion guest to form a complex of $K = 7.2 \times 10^{17}\,M^{-1}$, which exceeds the affinity of biotin and streptavidin by two orders of magnitude [18].

However, these ion–dipole and hydrophobic interactions cannot be the main driving force for the complex formation. Installing one or even two ammonium groups on a ferrocene molecule led to more stable interactions as expected (**1** and **2**, Figure 12.2), yet the overall enthalpy of binding with CB[7] was not increased when compared with a neutral ferrocene derivative (**3**, Figure 12.2) [19]. The differences in overall binding energy in these three guests were ascribed solely to the contribution of the solvation entropy. Desolvation of the guest upon complex formation is more favorable for the charged species. While on the one hand the cationic derivatives can interact more favorably with the carbonyl portals in the complex, on the other hand they also interact stronger with the aqueous environment prior to complex formation. The balance between favorable and unfavorable interactions is reversed for the complex between the neutral derivative **3** and CB[7], so that the overall enthalpies of the three complexes are similar [2].

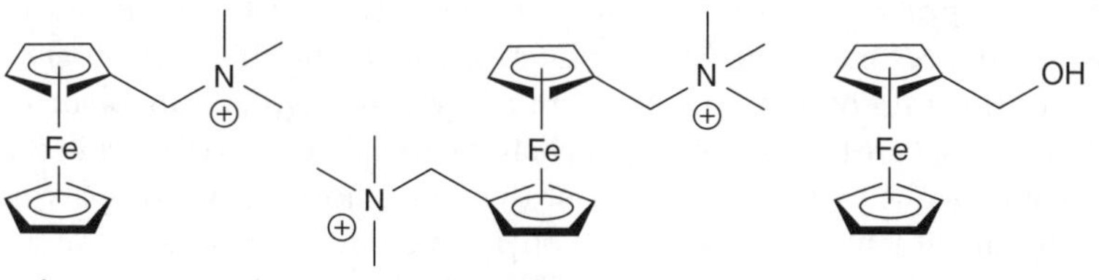

Figure 12.2 Chemical structures of neutral and cationic ferrocene guests for CB[7] with the following thermodynamic parameters [19] determined by calorimetry: $K(\mathbf{1}) = 3.2 \times 10^9\,M^{-1}$; $K(\mathbf{2}) = 4 \times 10^{12}\,M^{-1}$; $K(\mathbf{3}) = 3 \times 10^{15}\,M^{-1}$; $\Delta H^0(\mathbf{1}\text{–}\mathbf{3}) = 90\,kJ\,mol^{-1}$ and $T\Delta S^0(\mathbf{1}) = -36\,kJ\,mol^{-1}$; $T\Delta S^0(\mathbf{2}) = -18\,kJ\,mol^{-1}$; $T\Delta S^0(\mathbf{3}) = -2\,kJ\,mol^{-1}$.

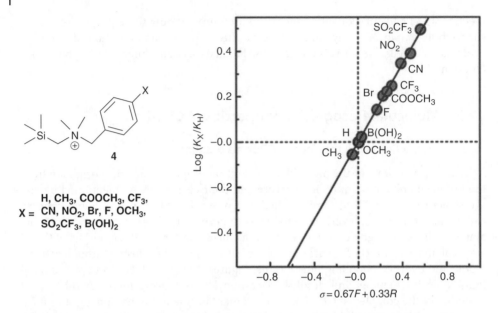

Figure 12.3 Binding affinities for CB[7] of silanes **4** (K_X) relative to silane **4** with X=H (K_H) as a function of a linear combination of Swain–Lupton field/inductive (*F*) and resonance (*R*) parameters that are derived from the Hammett parameters ($\sigma = 0.67F + 0.33R$). *Source*: http://pubs.rsc.org/-/content/articlehtml/2016/sc/c5sc04475h. Used under CC-BY 3.0 https://creativecommons.org/licenses/by/3.0/.

This concept was recently further studied in detail by Masson and coworkers. [20]. In this study the binding affinities of CB[7] were measured for a series of homologs of *N*-benzyl-trimethylsilylmethylammonium cationic guests in which only the para-substituent was varied systematically (Figure 12.3). The binding affinities of substituted silanes **4** relative to the unsubstituted silane **4** (X = H) towards CB[7] vary only weakly between 0.9 (in the case of **4**, X = CH$_3$) and 3.1 (in the case of **4**, X = SO$_2$CF$_3$) and correlate well with linearly combined Swain–Lupton field/induction and resonance parameters, which are derived from the Hammett parameters (Figure 12.3) [20]. Due to this concept and supported by calculations, Mason and coworkers found that the differences in binding affinities throughout the series are due to changes in the solvation of the ammonium unit (for the free guests) by water and changes in the interaction between the ammonium unit and the CB[7] rim. Although each of these factors is strongly dependent on the resonance and field/inductive effects that the substituents exert on the charge of the ammonium unit, they are of opposite sign and in balance, yielding binding affinities that are barely affected by changes in substituents (Figure 12.3). In this context, it is important that the solvation of the entire host–guest complex is hardly affected by the substituent effect as the carbonyl rim of CB[7] cancels out most of the positive charge on the ammonium group and CB[7] shields the ammonium group from any water solvation [20]. This means that the contribution of the entropic desolvation of the free guest can be used to adjust the binding affinities. The results also demonstrate that the binding properties of CB[*n*] in aqueous environment are only marginally affected by the coulombic interactions of the guest with the macrocycle portals, while the release of water molecules from the macrocycle upon guest binding is found to be driving the interactions (Section 12.2.2).

12.2.2 Release of High Energy Water Molecules from the CB[n] Cavity

In contrast to other macrocycles such as calixarenes and cyclodextrins, which have an open or cone (vase-like) shape and are polarizable, the concave, rigid and weakly polarizable CB[n] cavity provides further understanding of the exceptionally strong binding of CB[n], also towards neutral guests (Figure 12.1).

Due to the shape and structure of the CB[n] cavity, a limited number of water molecules occupy the cavity of CB[n] before complex formation, with weaker mutual interactions when compared with water molecules in the surrounding bulk (Figure 12.4). The average number of hydrogen bonds per water molecule ranges from 2.55 for CB[8], which resembles the value of bulk water, to 0.99 for CB[5] (Table 12.2). Also, the dispersion interactions between the weakly polarizable water molecules and the cavity of CB[n] are weaker with respect to the interactions between water molecules in bulk, and the water–water hydrogen bonds in the cavity are weaker when compared with bulk water due to the constrained space in the cavity for optimal hydrogen bond formation [17]. The energy needed to remove the first water molecule from the cavity, $E_{pot}(H_2O)$, was estimated by molecular dynamic simulations and varies from 15.1 kcal mol^{-1} for CB[5] up to 19.4 kcal mol^{-1} for CB[8] (Table 12.2 and Figure 12.4c) [17]. Consequently, the water molecules within the CB[n] cavity are high in energy and go to a lower energy level when being displaced from the cavity to the bulk water by encapsulation of the guest (Figure 12.4).

Biedermann and coworkers [17,21] demonstrated, using molecular dynamic simulations and isothermal titration calorimetry (ITC) experiments, that the release of these high-energy water molecules is *the* major driving force for the complex formation in CB[n], overcoming the direct host–guest interaction energy. The total energy needed for removing all internal water molecules ($\Delta E_{pot}(all)$ in Table 12.2) perfectly explains the molecular recognitions properties through the different sized CB[n] homolog series ($n = 5$–8). While in the case of the smallest cavities a higher energy gain is achieved for removing one water molecule ($E_{pot}(H_2O)$), the larger cavities contain a higher number of water molecules inside leading to the highest total gain in energy in the case of CB[7] (Table 12.2 and Figure 12.4c). These calculations are in agreement with the extremely high binding constants reported in the case of CB[7]. In addition, although the values of $E_{pot}(H_2O)$ for CB[8] are similar to the ones for bulk water, the removal of all internal water molecules from the CB[8] cavity is energetically favored with respect to the bulk and even with respect to the smaller homologs CB[5] and CB[6] ($\Delta E_{pot}(all)$, Table 12.2). According to simulations, the specific geometric constraints of the CB[8] cavity size create a void in the water network and favor the release of these high energy water molecules from the host–guest complexation [17,21].

12.2.3 Enthalpy-driven Hydrophobic Effect for CB[n]

The release of high energy water molecules is associated mostly with enthalpic contributions, as shown by systematic calorimetric studies of 1:1 complexes of CB[n] with neutral guests [17]. In the case of CB[7] and CB[8] the binding with guests is usually entropically unfavorable and enthalpy driven. There is a perfect correlation between the trends of solvent effects (the release of high energy water molecules) and in enthalpy contributions. In particular, according to the lock-and-key principle a tighter fit between the guest and the host provides the full release of water molecules from the cavity and

(a)

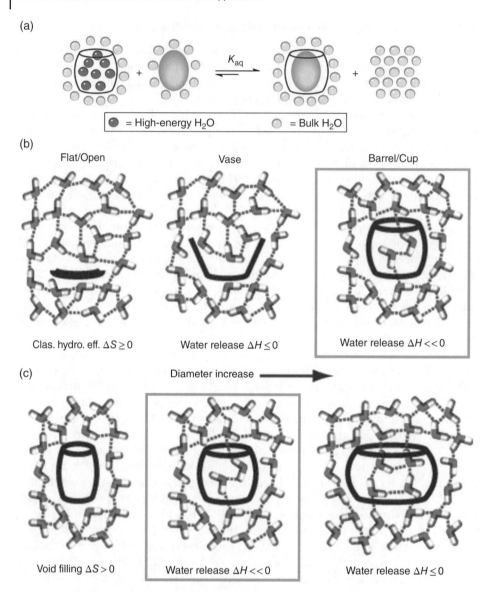

= High-energy H_2O = Bulk H_2O

(b)

Flat/Open   Vase   Barrel/Cup

Clas. hydro. eff. $\Delta S \geq 0$ Water release $\Delta H \leq 0$ Water release $\Delta H \lll 0$

(c)   Diameter increase ⟶

Void filling $\Delta S > 0$ Water release $\Delta H \lll 0$ Water release $\Delta H \leq 0$

Figure 12.4 Schematic representations of (a) the release of high energy water molecules upon complex formation and of the effect of the host shape (b) and size (c) on the water molecules network. *Source*: (a) Adapted from Ref. [17]. Reproduced with permission of American Chemical Society. (b and c) Ref. [3]. Reproduced with permission of John Wiley and Sons. See color section.

therefore a larger enthalpy gain. A tighter fit corresponds also to a decrease in the degrees of freedom of the guest and consequently to a larger unfavorable entropy contribution. Moreover, the entropic loss is in line with the model in which the water molecules are more mobile in the cavity due to a lower number of mutual hydrogen bonds formed with respect to the bulk. In the case of CB[6], the more restricted

Table 12.2 Calculated hydration properties of CB[*n*] with respect to bulk water.

	Cavity volume (Å³)	Number of cavity H₂O	Number of hydrogen bonds	$E_{pot}(H_2O)^a$ (kcal mol⁻¹)	$\Delta E_{pot}(all)^b$ (kcal mol⁻¹)
Bulk water	—	—	2.54	18.9	Reference
CB[5]	45	2.0	0.99	15.1	−9.9
CB[6]	118	3.3	1.31	15.4	−12.2
CB[7]	214	7.9	2.01	17.8	−24.5
β-CD	262ᶜ	4.4ᵈ	2.96ᵈ·ᵉ	—	—
CB[8]	356	13.1	2.55	19.4	−15.8
CB[8]·paraquat	155	5.5	1.85	—	−27.0

β-CD, β-cyclodextrin.
ᵃ Energy required to remove a single water molecule from the cavity.
ᵇ Energy required for removing all water molecules from the cavity and transfer them to the bulk water with respect to the bulk water itself taken as reference. Data taken from Refs [17,21].
ᶜ Data from Ref. [22].
ᵈ Data from Ref. [3].
ᵉ This value includes hydrogen bonds with the host.

orientation of the water molecules, solvents effects for CB[6] could generate either enthalpic or entropic contributions and therefore the overall binding can be enthalpy or entropy driven [17].

The enthalpy-driven hydrophobic effect for CB[7] seems to explain also its extremely high binding constants determined for some guests in comparison with the similarly sized β-cyclodextrin (β-CD; Figure 12.4b). For instance neutral and cationic ferrocene derivatives bind to CB[7] in the range of $10^9 - 10^{10}\,M^{-1}$ and $10^{12} - 10^{13}\,M^{-1}$, respectively, while they bind in the range of $10^3 - 10^4\,M^{-1}$ in the case of β-CD [22,23]. The CB[7] cavity, with its symmetric barrel structure, almost shields the encapsulated water molecules from interacting with the aqueous environment and forming hydrogen bonds with the bulk, thus creating high-energy water molecules and high binding enthalpies for the complex formation (Figure 12.4b and Table 12.2) [3]. In contrast, the β-CD macrocycle contains fewer water molecules while its cone shape allows more water molecules to participate in hydrogen bonding to the bulk water and the cavity, which means that the water molecules inside the cavity are stabilized and therefore a lower enthalpy gain occurs upon guest complexation and water release (Figure 12.4b and Table 12.2) [3].

12.2.4 Enthalpy-driven Hydrophobic Effect for CB[8] Heteroternary Complexes

The maximization of the release of energetically frustrated water molecules from the large cavity of CB[8] (Table 12.2) is achieved either by the complexation of one or two guests to form a 1:1 or 1:2 complex, respectively. When the two guests in the ternary complex are the same a so-called homoternary complex is formed, otherwise a

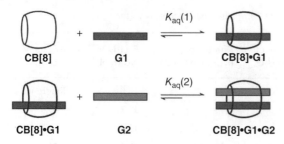

Figure 12.5 Stepwise formation of the heteroternary complex in CB[8] between an electron-poor guest G1 and an electron-rich guest G2 in aqueous solutions. *Source*: Adapted from Ref. [16]. Reproduced with permission of American Chemical Society.

heteroternary complex is formed (Figure 12.5). For example, CB[8] can bind to 1 eq of the electron-deficient paraquat and this 1:1 complex can then bind to 1 eq of the electron-rich 2,6-naphthol forming a heteroternary complex in aqueous solutions [9]. As in this example, the first guest is typically electron-poor and the second guest electron-rich, generating a charge-transfer (CT) donor–acceptor complex.

Biedermann and Scherman [16] demonstrated in a systematic study that the CT interaction contributes to the stabilization of the ternary complex and, in contrast to earlier reports [9,24], that the release of energetically frustrated water molecules is still energetically driving the complex formation (see below and Figure 12.6). From a UV-Vis spectroscopy study the authors concluded that polar solvents were able to stabilize the acceptor–donor pair in the absence of CB[8] and lower the charge repulsion (Figure 12.6a). This indicates that the donor and the acceptor in the CT excited state are in their (mono) cationic forms (Figure 12.6b) [16]. However, CB[8] can stabilize the CT complex much more strongly than any polar solvent used in the study, including water (Figure 12.6a). In the presence of CB[8] the absorption maximum of the CT band of the paraquat/2,6-naphthol CT complex red-shifted to 560 nm for the ternary complex, which is 94 nm more when compared with when the complex was dissolved in water in the absence of CB[8]. The authors attributed the findings to the large uniform negative electrostatic potential in the cavity of CB[8] (Figure 12.1b),which lowers the energy band gap of the CT complex by (1) raising the energy of the HOMO of the electron-rich donor and (2) stabilizing both cationic excited states of the donor and the acceptor (Figure 12.6b) [16].

The study of solvent effects for ternary complexes showed that after the complexation of the first guest, the residual water molecules in the cavity can form a much lower number of mutual hydrogen bonds and their energy is much higher with respect to the bulk and to the free CB[8] cavity (Figure 12.6c and Table 12.2) [21]. Even though the number of water molecules is reduced upon complexation of the first guest (from 13 H_2O for CB[8] to 5.5 H_2O for CB[8]·paraquat, Table 12.2), the overall energy for removing all residual encapsulated water molecules is twice as high when compared with free CB[8] (Table 12.2) [21]. After binding the first guest to CB[8] a smaller cavity volume remains and this leads to a distortion and a reduction of the number of hydrogen bonds formed per water molecule. In addition, due to the presence of the first guest only a part of the CB[8] portal is available for interactions between encapsulated and bulk water molecules [3,21].

The second guest should not only be electrostatically compatible with the first guest but also sterically able to displace all residual water molecules [21]. For example, while 1-naphthol is more electron-rich than the isomer 2-naphthol, the heteroternary

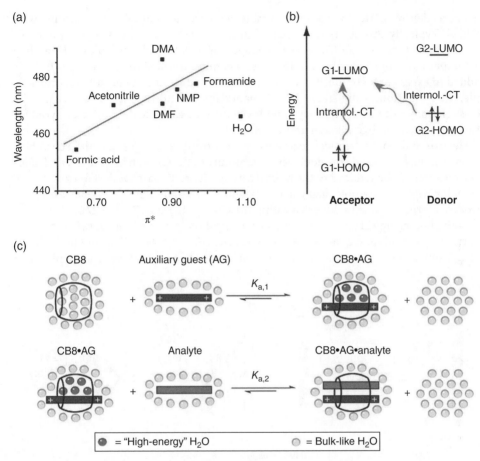

Figure 12.6 (a) Absorbance maximum of the charge-transfer band of the 1:1 pair paraquat/2, 6-naphthol in polar organic solvents as a function of their Taft π^* parameters. (b) Energy diagram for the charge-transfer interaction between G1 and G2. (c) Schematic representation of the solvent effects for ternary complex in CB[8] in aqueous environment. *Source*: (a and b) Adapted from Ref. [16]. Reproduced with permission of American Chemical Society. (c) Adapted from Ref. [21]. Reproduced with permission of American Chemical Society.

complex of CB[8]·paraquat·1-naphthol has a less favorable enthalpy contribution than that of 2-naphthol ($\Delta H^0 = -9.2$ and $-12.5\,\text{kcal}\,\text{mol}^{-1}$, respectively) [21]. Molecular dynamics simulations explained the differences in binding showing that the more electron-rich isomer is sterically unable to release all water molecules from the cavity [21]. This comparison again demonstrates that the molecular recognition properties of CB[8] are more strongly driven by solvent effects than by CT interactions.

12.3 Control Over the Binding Affinity with CB[*n*]

The diversity of the molecular structures of the guests for CB[*n*] makes their complexes ideal molecular switches for which the binding affinity can be changed depending on the chemical, electrochemical, or photochemical conditions applied. The simplest condition that can be exploited for lowering the affinity of the complex is pH, as deprotonation of a guest generally decreases the complexation affinity with CB[n]. Typically this strategy is applied in molecular machines such as switchable pseudorotaxanes [1]. Another way to change the affinity of the complex is by adding a competitively binding guest. In this case a competitor of similar binding affinity is added in excess or a competitor with a stronger binding affinity is added. For example, the formation of the stable 1:1 adamantylamine·CB[8] complex can trigger the dissociation of the heteroternary complex between a cationic derivative of pyrene, a functionalized methylviologen and CB[8] [25].

Electro- and photochemical reactions can control the charges and the steric hindrance of the guests and consequently their affinity to the CB[n] host. The methylviologen dication can be reduced to the radical cation (Figure 12.7a) that destabilizes the CT complex with, for example, azobenzene (Figure 12.7b and c) and – in contrast to the oxidized form – is able to dimerize within the CB[8] (Figure 12.7c) [26,27].

Both photochemical and electrochemical control was reported for a CB[8] heteroternary complex with paraquat as the first guest ($K_1 = 8.5 \times 10^5 \, \mathrm{M}^{-1}$) [27,28] and the light-responsive *trans*-azobenzene as the second guest ($K_2 = 1.4 \times 10^4 \, \mathrm{M}^{-1}$) as depicted in

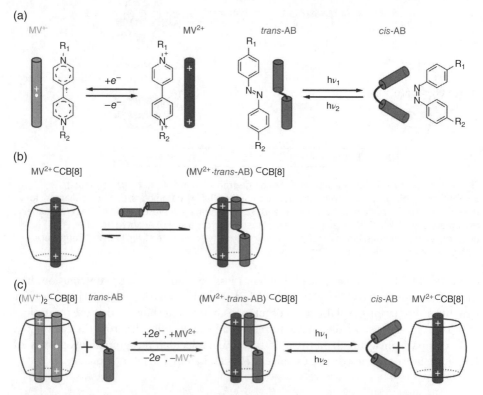

Figure 12.7 (a) Redox-controlled equilibrium between the dicationparaquat MV^{2+} and the radical cation MV$^{+·}$, and photoisomerization of azobenzene derivatives AB. (b) Stepwise formation of CB[8]-mediated heteroternary complex with MV^{2+} as the first guest and *trans*-azobenzene as the second guest. (c) Orthogonal reversible photochemical and electrochemical control over the equilibrium of formation of the heteroternary complex (MV^{2+}·*trans*-AB) ⊂ CB[8]. *Source*: Adapted from Ref. [27]. Reproduced with permission of Nature Publishing Group.

Figure 12.7 [29,30]. Upon UV light irradiation the photostationary state of the azoben-zene shifted towards the cis isomer (Figure 12.7a). The *cis*-azobenzene is too sterically hindered to participate in ternary complex formation, and its interactions with the cavity are too weak ($<10^3 \, M^{-1}$) to compete with the paraquat guest [4,27,29].

In the following section further molecular switches are presented in the context of amino acid and peptide recognition. The possibility to spatially and temporally control the assembly and disassembly of CB[n] complexes expands the toolbox for their applications.

12.4 CB[n] Recognition of Amino Acids, Peptides, and Proteins

Amino acid and peptide guests increase the overall biocompatibility and bioactivity and therefore lead to a broad field of application of CB[n] complexes under *in vitro* and *in vivo* conditions. Urbach and Inoue have made major contributions in finding new complexes between amino acids or peptides with CB[n] in an aqueous, pH neutral, physiological environment [31,32]. Among the 20 genetically encoded amino acids, CB[6] forms complexes with aliphatic cationic residues such as lysine (K) [32], while CB[7] and CB[8] bind with affinities higher than $10^3 \, M^{-1}$ selectively with the aromatic residues tryptophan (W), phenylalanine (F), and tyrosine (Y) [33]. Enthalpy is the driving force for binding aromatic amino acids with CB[7] (Table 12.3).

As derived from molecular dynamics simulations, the three aromatic amino acids are able to eliminate all water molecules from the CB[7] cavity [34]. Unfavorable free guest desolvation (for tyrosine) and restriction of motion (for tryptophan, see $-T\Delta S$ in Table 12.3) upon complexation are thought to be involved in the differences in binding affinity when compared with phenylalanine [34]. The presence of a positive charge,

Table 12.3 Thermodynamic binding constants for CB[7] complexes with amino acids, peptides, and proteins[a].

Guest	$K^{[b]}$ (M^{-1})	$\Delta G^{0[c]}$ $(kcal \, mol^{-1})$	$\Delta H^{0[b]}$ $(kcal \, mol^{-1})$	$-T\Delta S^{0[d]}$ $(kcal \, mol^{-1})$	Ref.
F	$1.8 \, (\pm 0.5) \times 10^5$	$-7.2 \, (\pm 0.9)$	$-7.3 \, (\pm 0.7)$	$0.14 \, (\pm 0.66)$	[34]
Y	$1.6 \, (\pm 0.3) \times 10^4$	$-5.7 \, (\pm 0.6)$	$-6.6 \, (\pm 0.4)$	$0.9 \, (\pm 0.4)$	[34]
W	$1.2 \, (\pm 0.1) \times 10^3$	$-4.2 \, (\pm 0.9)$	$-6.9 \, (\pm 0.1)$	$2.7 \, (\pm 0.1)$	[34]
FGG	$2.8 \, (\pm 0.1) \times 10^6$	$-8.9 \, (\pm 0.1)$	$-17.5 \, (\pm 0.1)$	$8.7 \, (\pm 0.1)$	[35]
GFG	$2.2 \, (\pm 0.1) \times 10^4$	$-6.0 \, (\pm 0.1)$	$-9.3 \, (\pm 0.1)$	$3.3 \, (\pm 0.1)$	[35]
GYG	$2.7 \, (\pm 0.1) \times 10^3$	$-4.7 \, (\pm 0.1)$	$-2.2 \, (\pm 0.1)$	$-2.5 \, (\pm 0.1)$	[35]
Human Insulin	$1.5 \, (\pm 0.4) \times 10^6$	$-8.5 \, (\pm 0.1)$	$-10.8 \, (\pm 0.5)$	$2.3 \, (\pm 0.4)$	[35]

[a] N-terminus is charged. Standard deviations are given in parentheses. For the peptides the residue in the cavity is underlined and the standard one letter code for amino acids is used.
[b] Values measured by ITC in phosphate buffer at pH 7.
[c] Gibbs free energy values calculated from K values.
[d] Entropic contributions to ΔG^0 calculated from K and ΔH^0 values.

which is the driving force for the molecular recognition in vacuum [34], has only an additional stabilizing role in solution and explains the selectivity of CB[7] toward N-terminal phenylalanine residues in the tripeptide FGG (G = glycine) with respect to a sequence with an internal, uncharged phenylalanine such as in GFG [35].

The N-terminal phenylalanine residue of human insulin was selectively recognized by CB[7] acting as an artificial receptor for the protein [35]. This concept was further explored in the development of a bioanalytical method for monitoring the protease activity of thermolysin on an unlabeled peptide [36]. This enzyme recognizes an internal phenylalanine residue while its lysis product has an N-terminal phenylalanine. The assay was based on the 100-fold difference in affinities of CB[7] for these positional isomers and their different ability of competing with the fluorescent dye acridine orange to bind to the macrocycle. The enzymatic reaction was monitored by fluorescence spectroscopy following the increase of the emission of the dye as it was displaced from the CB[7] cavity by the N-terminal phenylalanine lysis products [36].

In contrast to CB[7], CB[8] can either host homodimers of two aromatic amino acid residues to form a homoternary complex, or host heterodimers of paraquat and an aromatic amino acidic residue to form a heteroternary complex. The crystal structure of the homoternary complex $CB[8] \cdot (FGG)_2$ revealed the encapsulation of two aromatic residues into the cavity of the CB[8] in a face-to-face arrangement and additionally the coulombic interactions of the carbonyl fringed portals with the N-terminal phenylalanine residue and the amidic protons on the peptide backbone (Figure 12.8) [37]. As observed in the case of CB[7], the positive charge adjacent to the aromatic amino acid residue drives the selectivity of CB[8] towards the N-terminal phenylalanine and

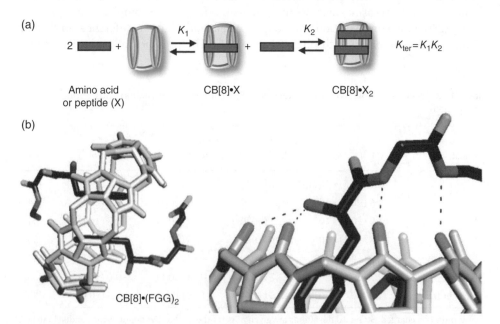

(a)

2 ☐ + Amino acid or peptide (X) $\xrightleftharpoons{K_1}$ CB[8]•X $+$ ☐ $\xrightleftharpoons{K_2}$ CB[8]•X$_2$ $K_{ter} = K_1 K_2$

(b)

CB[8]•(FGG)$_2$

Figure 12.8 (a) Formation of CB[8]-mediated homoternary complex with two molecules of peptide guest X. (b) Crystal structure of CB[8]·2(FGG), hydrogens and solvating water were removed; the dashed lines indicate key electrostatic interactions. *Source:* Adapted from Ref. [37]. Reproduced with permission of American Chemical Society.

Table 12.4 Equilibrium association constants for homoternary CB[8] complexes with amino acids and peptides[a].

Guest	K_1 (M^{-1})	K_2 (M^{-1})	K_{ter} (M^{-2})	Ref.
F[b]	—	—	$1.1\ (\pm0.2) \times 10^8$	[33]
Y[b]	—	—	$<10^3$	[33]
W[b]	—	—	$6.9\ (\pm1.3) \times 10^7$	[33]
FGG[b]	—	—	$1.5\ (\pm0.2) \times 10^{11}$	[37]
(FGGG)$_2$[c]	9.0×10^6	—	—	[42]
WGG	$1.3\ (\pm0.2) \times 10^5$	$2.8\ (\pm0.3) \times 10^4$	$3.6\ (\pm0.2) \times 10^9$	[37]
AEFRH	$7.6\ (\pm1.2) \times 10^5$	$4.9\ (\pm0.2) \times 10^4$	$3.7\ (\pm0.7) \times 10^{10}$	[38]
LVFIA	$1.3\ (\pm0.2) \times 10^5$	$6.4\ (\pm1.1) \times 10^4$	$7.7\ (\pm2.3) \times 10^9$	[38]
VIFAE	$3.7\ (\pm4.6) \times 10^7$	$4.2\ (\pm0.4) \times 10^5$	$1.6\ (\pm0.3) \times 10^{13}$	[38]

[a] N-terminus is charged. Standard deviations are given in parentheses. Values measured by ITC in phosphate buffer at neutral pH. For the peptides the residue in the cavity is underlined and the standard one letter code for amino acids is used. The peptides YGG as well as all scrambled peptide sequences GxG, GGx with x = F, Y, or W did not show affinities measurable by ITC [37].
[b] Values of K_1 and K_2 are not reported in the literature [37].
[c] Binding affinity measured for the 1:1 complex (FGGG)$_2$-penta(ethylene glycol) with CB[8].

tryptophan, $K_{ter} = 2.3 \times 10^{10}\,\text{M}^{-2}$ and $3.6 \times 10^9\,\text{M}^{-2}$ (Table 12.4), respectively, over the internal or C-terminal positions (no binding detectable by ITC) [37].

Only a few publications have reported the use of a homoternary complex with CB[8] involved in controlled biomolecular assemblies. The binding of 2 eq of FGG with CB[8] enabled the specific formation of discrete assemblies of peptides (Figure 12.8b and Table 12.4) [37,38] and protein dimers and tetramers (Figure 12.9) [39] and wires [40] in solution. The CB[8]-mediated dimerization of the monomeric caspase-9 tagged with a short N-terminal FGG motif provided a protein dimer with catalytic activity [41] Intrinsic affinities for dimerization of the caspases further aided in strengthening the homoternary complex formation, about 10-fold. A weak binding affinity between protein elements is sufficient to induce selective heterodimerization on the CB[8] scaffold. Split luciferase protein halves were efficiently assembled to functional enzymes via the recognition of CB[8] by way of their appended N-terminal phenylalanines [40]. A strategy for the supramolecular control over the formation of dimers of peptides and proteins was achieved by reversing the homodimerization of FGG-tagged fluorescent proteins with the bivalent guest (FGGG)$_2$ (Table 12.4) [42].

Heteroternary complexes offer easier applicability due to selective recognition by the host in a binding process divided into two stages. In the most reported case the first guest is paraquat, which is able to bind to CB[8] only in a 1:1 stoichiometry due to the electrostatic repulsion exerted by its double positive charge in its oxidized form. The second guest sequentially binds to the CB[8]·paraquat complex [28]. Examples of amino acid based heteroternary complexes are presented in Table 12.5. Heteroternary complex formation based on paraquat and methoxynaphthol has been used for selective induced protein heterodimerization [40].

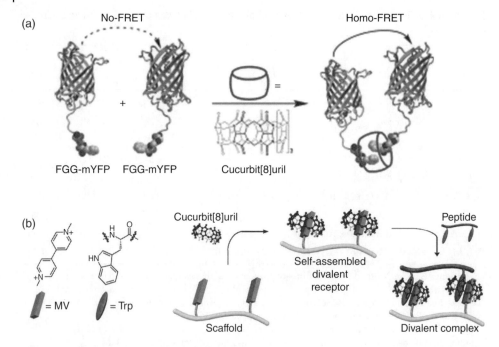

Figure 12.9 Schematic representation of a homoternary and a heteroternary CB[8]-mediated assembly. (a) Two monomeric yellow fluorescent proteins tagged with N-terminal FGG peptide motif dimerize in the presence of CB[8]. (b) A divalent scaffold displaying viologen groups forms with CB[8] a divalent receptor for a peptide with two tryptophan groups. *Source*: (a) Ref. [39]. Reproduced with permission of John Wiley and Sons. (b) Adapted from Ref. [43]. Reproduced with permission of American Chemical Society.

Table 12.5 Equilibrium association constants for heteroternary complexes of CB[8]·paraquat with amino acids and peptides[a].

Guest	K_2 (M^{-1})	K_{ter} [b] (M^{-2})
F	5.3 ($\pm$0.7) $\times$ 10^3	4.5 ($\pm$0.1) $\times$ 10^9
Y	2.2 ($\pm$0.1) $\times$ 10^3	1.9 ($\pm$0.1) $\times$ 10^9
W	4.3 ($\pm$0.3) $\times$ 10^4	3.6 ($\pm$0.1) $\times$ 10^{10}
<u>W</u>GG	1.3 ($\pm$0.3) $\times$ 10^5	1.1 ($\pm$0.2) $\times$ 10^{11}
G<u>W</u>G	2.1 ($\pm$0.1) $\times$ 10^4	1.8 ($\pm$0.1) $\times$ 10^{10}
GG<u>W</u>	3.1 ($\pm$0.4) $\times$ 10^3	2.6 ($\pm$0.1) $\times$ 10^9
GG<u>W</u>GG	2.5 ($\pm$0.2) $\times$ 10^4	2.1 ($\pm$0.1) $\times$ 10^{10}

[a] N-terminus is charged. Standard deviations are given in parentheses. For the peptides the residue in the cavity is underlined and the standard one letter code for amino acids is used. Values measured by ITC in phosphate buffer at pH 7 from Ref. [28].
[b] K_{ter} is calculated considering the equilibrium association constant for the complex CB[8]·paraquat is equal to $K_1 = 8.5$ ($\pm$0.3) $\times$ 10^5 M^{-1} [28].

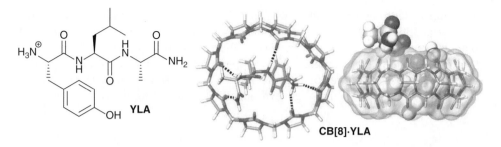

Figure 12.10 Peptide sequence of the peptide YLA and semi-empirical model of the 1:1 complex with CB[8]. *Source*: Adapted from Ref. [45]. Reproduced with permission of American Chemical Society.

Urbach and coworkers designed a polyvalent system where CB[8] mediated the "duplexing" of peptides containing one or more tryptophan residues with a series of synthetic analogs bearing one or more methylviologen side chains (Figure 12.9) [43]. Recently, a recombinant immunoglobulin with a tryptophan tag was selectively conjugated via CB[8] with methylviologen derivatives of a bioactive peptide [44].

Recently, also a 1:1 binding of CB[8] with the tripeptide YLA was found to feature a nanomolar binding affinity. Both the N-terminal tyrosine and the neighboring leucine residues are inside the cavity while the alanine interacts with the portals (Figure 12.10), boding for more selective protein recognition [45].

12.5 CB[*n*] for Bioanalytical and Biomedical Applications[1]

In this section selected examples from the literature are shown in which CB[*n*] is considered for bioanalytical and biomedical applications. While the whole CB[*n*] family has powerful self-assembly characteristics such as a large range of binding affinities and stimuli responsiveness to create platforms for the investigation of biological systems, the ability of the CB[8] homolog to simultaneously complex two guests is of particular interest and has initiated the development of many systems that exploit CB[8] to connect materials and surfaces ranging from polymers, hydrogels, and nanoparticles to planar surfaces.

12.5.1 CB[*n*]-mediated Assembly of Bioactive Polymers and Hydrogels

CB[8] is used as a supramolecular crosslinking unit to form biocompatible and dynamic hydrogels [46]. Self-assembled CB[8]-mediated hydrogels in water were formed by grafting Phe (F) or Trp (W) onto styrene copolymers. CB[*n*]s have also been used to supramolecularly graft bioactive ligands onto linear copolymers. Scherman and coworkers illustrated this concept using CB[8] to connect mannose-functionalized paraquat to pendant naphthol moieties on a methacrylate polymer through ternary complex formation [47]. A tetravalent lectin was used to crosslink the polymeric backbones by multivalent interactions with the mannose ligands [47]. Alternatively,

---

1 Part of this text is adapted from Ref. [29]. Reproduced with permission of Royal Society of Chemistry.

simply mixing CB[6]-grafted hyaluronic acid (CB[6]-HA) with spermidine-functionalized bioactive peptides resulted in the formation of supramolecular host–guest systems that could be used for *in vitro* and *in vivo* studies as shown by Kim and coworkers (Figure 12.11a) [48]. For example, when the spermidine-functionalized formyl peptide receptor-like 1 (FPRL1) specific peptide WKYMV was used in this system, elevated Ca^{2+} and extracellular signal-regulated kinase (pERK) phosphorylation levels in

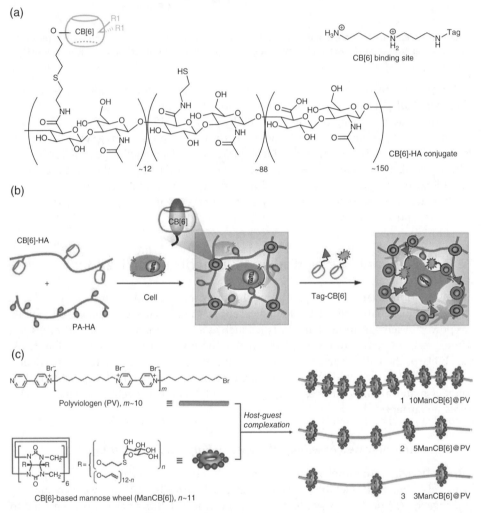

Figure 12.11 (a) The structures of the CB[6]-grafted hyaluronic acid (CB[6]-HA) and of the spermidine-tag where the tags can be an imaging probe as FITC and a peptide as WKYMV are reported. (b) Hydrogel formation of CB[6]-HA and polyamine-conjugated hyaluronic acid (PA-HA). Residual free guest sites are modularly modified with various tag-CB[6]s for cell probing. (c) Mannose-pseudopolyrotaxanes, composed of various densities of CB[6]-based mannose wheels threaded on polyviologen, showed the highest recognition activity towards *Escherichia coli* in the case of the lowest density of mannose-CB[6] relative to monomeric mannose. *Source*: (b) Adapted from Ref. [49]. Reproduced with permission of American Chemical Society. (c) Ref. [50]. Reproduced with permission of John Wiley and Sons.

FPRL1-expressing human breast adenocarcinoma cells were observed, in agreement with the therapeutic signal transduction of this specific peptide [48]. Another interesting hydrogel system was made by mixing CB[6]-HA with HA carrying 1,6-diaminohexane or spermine as pendant moieties (in their protonated forms) to make ultrastable host–guest complexes between the two HA-polymers (Figure 12.11b) [49]. This hydrogel was further modified modularly with, amongst others, a bioactive peptide-tagged CB[6], which was anchored to (residual) diaminohexane moieties in the hydrogel [49]. When an RGD-tagged CB[6] was incorporated into the hydrogel, human fibroblast cells were entrapped in the hydrogel and proliferated approximately 5-fold in 14 days. The cells showed a spread morphology, which matched characteristic cell adhesion and proliferation behavior in an RGD environment. In contrast, when hydrogels lacked the RGD-tagged CB[6], cell proliferation within the hydrogel network was relatively low and the cells retained a round morphology showing poor adhesion [49].

Another strategy to create supramolecular polymers decorated with bioactive ligands makes use of threading ligand-functionalized host molecules onto various polymers. For example Kim and coworkers threaded mannose-functionalized CB[6] onto the decamethylene segments of a linear chain of polyviologen [50]. The self-assembled mannose-pseudopolyrotaxanes not only induced bacterial aggregation effectively, but also exhibited high inhibitory activity against bacterial binding to host urinary epithelial cells (Figure 12.11c). The most potent inhibitor was the mannose-pseudorotaxane threaded with only three mannose-CB[6] with on average 33 mannose units. Its inhibitory potency was 300 times higher than compared with free mannose and approximately 1.6 times higher with respect to the compounds bearing 110 or 55 mannose units, indicating that the density of bioactive ligands along the rotaxane is of key importance for optimizing the interactions [50].

12.5.2 CB[n]-mediated Assembly of Bioactive Nanoparticles

CB[n] host–guest recognition can create supramolecular nanoparticles (SNPs). Different multivalent polymeric building blocks can be held together by specific noncovalent interactions allowing control of their size as well as their assembly and disassembly [51]. Four structural molecular elements were used to form the SNPs: CB[8], a methylviologen-grafted polymer, and mono- and multivalent azobenzene-functionalized molecules (Figure 12.12a) [51]. The higher the amount of the multivalent azobenzene component with respect to the monovalent one, the larger the size of the particles, yielding SNPs with sizes ranging from 55 to 110 nm in diameter. Electrochemically and photo-responsive SNPs were formed in a one-pot mixing process through heteroternary complexation [51]. A mesoporous silica nanoparticle core with a layer-by-layer coating of CB[7] alternated with polymeric layers with bis-amines was designed as a stimuli-responsive drug delivery system [52]. The recognition by the portals of CB[7] of two bis-amino functionalities on the polymer layers worked as a supramolecular glue and provided structural stability to the nanoparticle shell. The release of the cargo was achieved by controlling the dissociation of the complex, induced by adding amino-adamantane, or by acidifying the solution to endosomal pH in cancer cells. Cellular uptake and triggered cargo delivery was shown in several types of cancer cell lines and *in vivo* conditions using mice as a model system [52]. The results indicated that CB[n] nanoparticles provide a method for manipulating cellular behavior.

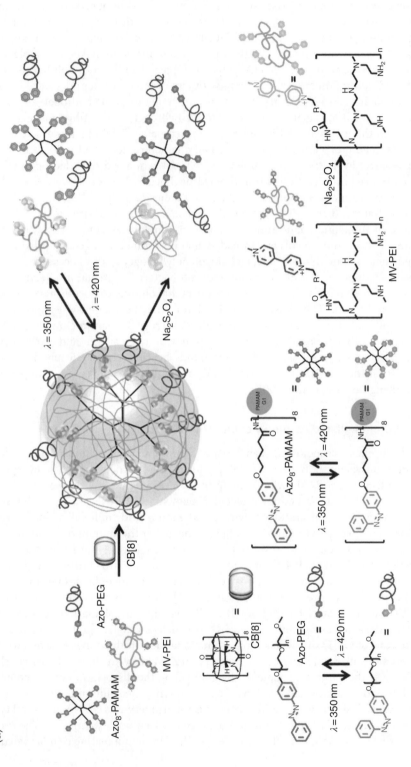

Figure 12.12 (a) Dual-stimuli responsive supramolecular nanoparticles self-assemble and disassemble by the formation and disruption of the ternary complex between CB[8], a polymer-grafted viologen (MV-PEI), and a combination of monovalent (Azo-PEG) and polyvalent azobenzene moieties (Azo$_8$-PAMAM).

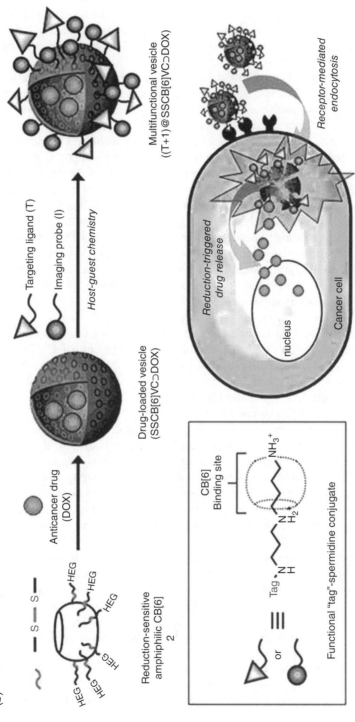

Figure 12.12 (Cont'd) (b) The surface of reduction-sensitive vesicles (SSCB[6]VC; HEG, hexaethylene glycol) formed by disulfide bridges between amphiphilic CB[6] derivatives is noncovalently modified for targeted drug delivery of doxorubicin (DOX). *Source:* (a) Ref. [51]. Reproduced with permission of John Wiley and Sons. (b) Ref. [56]. Reproduced with permission of John Wiley and Sons.

The use of the smaller homologs of CB[n] to display ligands on a nanoparticle surface relies on their chemical functionalization. In a contribution from Kim and coworkers, a covalently polymerized network of side chain-functionalized CB[6] was used as a host template to form nanocapsules [53,54]. Through host–guest interactions of two spermidine conjugates, a typical polyamine guest of CB[6], bearing galactose [53] or folate [54] as targeting ligands, was introduced to the spherical polymeric network of CB[n]s for receptor-mediated endocytosis.

The examples discussed above involve the dynamic display of ligands on the surface of nanoparticles by host–guest chemistry: such assembly allows the pre-organization of the ligands in an array of multiple binding sites for cell receptors, which leads to a more favorable multivalent interaction. For therapeutic applications additional characteristics are relevant, especially when considering *in vivo* applicability: (1) controlled cargo release; (2) persistence, clearance or accumulation in the body; and (3) enzyme degradability into non-toxic byproducts after the release of the load. Typically a decoration with hydrophilic polymers such as carboxybetaine analogs [55] improves the *in vitro* and *in vivo* stability of nanoparticles. Vesicles based on an amphiphilic CB[6] that was derivatized at its periphery with hexa(ethylene glycol) units containing disulfide bonds allowed for incorporation of spermine-modified cell targeting ligands and imaging probes through specific host–guest interactions between spermidine and the CB[6] host [56]. When these supramolecular vesicles were loaded with doxorubicin, both the internalization into cervical cancer (HeLa) cells by receptor-mediated endocytosis and the triggered release of entrapped drugs by the cytoplasmic reducing environment was demonstrated (Figure 12.12b) [56].

Instead of *displaying* ligands and stabilizing agents on the surface of nanoparticles, the high affinity and selectivity of the molecular recognition of CB[7] can *shield* cytotoxic moieties. A pyrene imidazolium-labeled hydrophilic peptide and a methyl-viologen-capped long alkyl chain were coupled by the CB[8] cavity to form a supramolecular peptide amphiphile (Figure 12.13a) [25]. In aqueous environment these amphiphiles can self-assemble further into nano-sized vesicles. After internalization in HeLa cells the vesicular structures triggered cell death by the addition of adamantylamine, which bound stronger to CB[8] and thus displaced both viologen and pyrene guests from the cavity. The disassembly of the complex and the particle was confirmed by the increase in fluorescence emission of the pyrene and concomitant cytotoxicity of the uncomplexed viologen in Hela cells. In contrast, the addition of electron-rich naphthol – which replaced only pyrene in the CB[8] cavity yielding a ternary complex with viologen – caused an increased pyrene emission but no cytotoxicity (Figure 12.13a) [25].

A similar concept was applied to mediate the therapeutic activity of gold nanoparticles inside living cells (Figure 12.13b) [57]. When capped by CB[7], diaminohexane-terminated gold nanoparticles showed a reduced toxicity by the shielding of the positive charges in the macrocycle cavity and were internalized in human breast cancer cells MCF-7. The administration of the stronger binding guest adamantylamine triggered the intracellular activation of the *in-situ* cytotoxicity of the polyamine moieties by dethreading of CB[7] from the nanoparticle surface (Figure 12.13b) [57].

Another field in which the strong, selective and reversible complexes with CB[7] can find application is bioanalytical chemistry. Plasma membrane proteins were selectively captured, extracted and recovered by a synthetic receptor–ligand pair between CB[7]

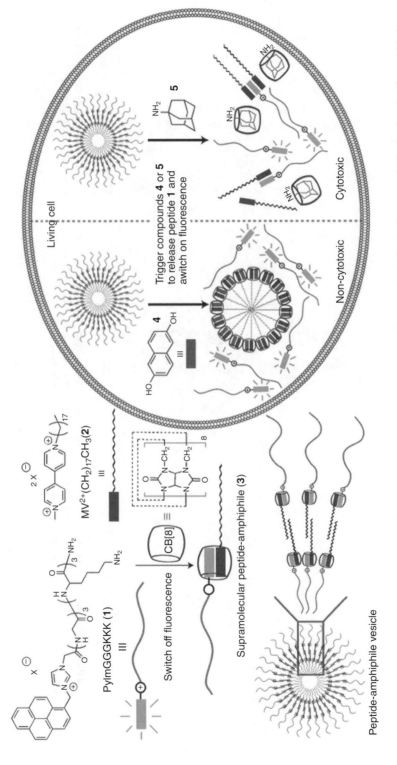

Figure 12.13 (a) Supramolecular peptide amphiphiles formed by CB[8] complexation of pyrene-labeled peptides and viologen lipid and self-assembled into vesicles with triggered fluorescence and cytotoxicity.

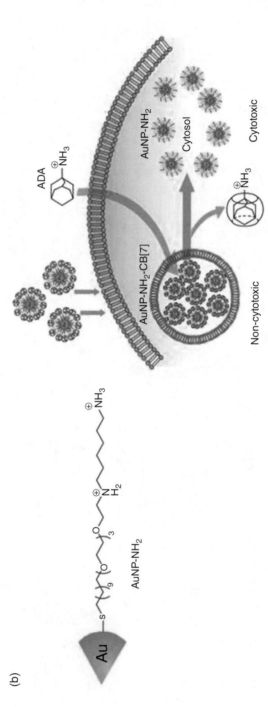

Figure 12.13 (Cont'd) (b) Structure of an diaminohexane-terminated gold nanoparticle (AuNP-NH₂) and the activation of the cytotoxicity of the nanoparticles capped with CB[7] (AuNP-NH₂-CB[7]) by reversing the complex by adamantylamine (ADA). *Source:* (a) Ref. [25]. Reproduced with permission of John Wiley and Sons. (b) Adapted from Ref. [57]. Reproduced with permission of Nature Publishing Group.

and 1-trimethylammoniummethylferrocene ($K \approx 10^{12} \, \mathrm{M^{-1}}$) [58]. Proteins of the plasma membrane of rat fibroblast Rat-1 cells were ferrocenylated on their amine residues by EDC/NHS chemistry, isolated from the cell lysate by CB[7]-immobilized sepharose beads and finally released, either by heat, or by treatment with a competitor (1,1-bis(trymethylammonium)methylferrocene, $K \approx 10^{15} \, \mathrm{M^{-1}}$) [58].

12.5.3 CB[n]-mediated Assembly on Bioactive Surfaces

Self-assembled CB[7] monolayers (CB[7] SAMs) with an approximate 40% surface coverage can be formed on gold surfaces simply by immersion of the substrate into a 0.1 mM aqueous solution of the macrocycle [59]. Binding studies with neutral adamantyl guests at a single molecule level by dynamic force spectroscopy showed that the binding affinity is similar to that in solution [60]. Stable and reversible immobilization of a fluorescent protein, which was selectively modified with a single ferrocene guest moiety, on CB[7] SAMs has been reported. This method allows the fabrication of stable protein monolayers with controlled protein orientation [61]. Similarly, synthetic integrin binding cyclic RGD peptides labeled with ferrocene were displayed on a CB[7] monolayer on gold [62]. Endothelial HUVEC cells specifically adhered to these surfaces under supramolecular control (Figure 12.14) [62].

The fabrication of CB[8] SAMs takes advantage of the possibility of including two guests in its large cavity. A monolayer of either the first [63] (Figure 12.15a), or the second guest [29] is assembled and subsequently the ternary complex can form on, for example, gold surfaces. Alternatively, an amino-terminated glass surface can be directly coated with CB[8] [64]. Supramolecular recognition of amino acid residues on native proteins, such as bovine hemoglobin and catalase, allowed the layer-by-layer assembly of CB[8]-protein stacks while maintaining the protein activity [64]. Another strategy to achieve stable CB[8]-modified surfaces was presented by Scherman and coworkers

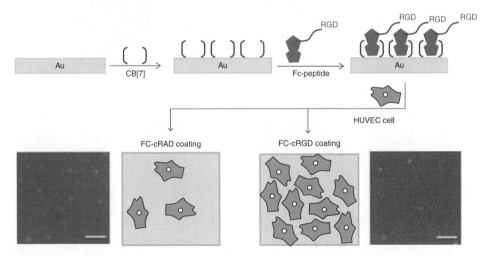

Figure 12.14 Supramolecularly controlled cell adhesion on a CB[7] monolayer on gold, pre-incubated with a ferrocene-modified cell adhesion ligand (Fc-cRGD) or the inactive compound Fc-cRAD. *Source:* http://pubs.rsc.org/is/content/articlehtml/2013/cc/c3cc37592g. Used under CC-BY 3.0 https://creativecommons.org/licenses/by/3.0/.

(a)

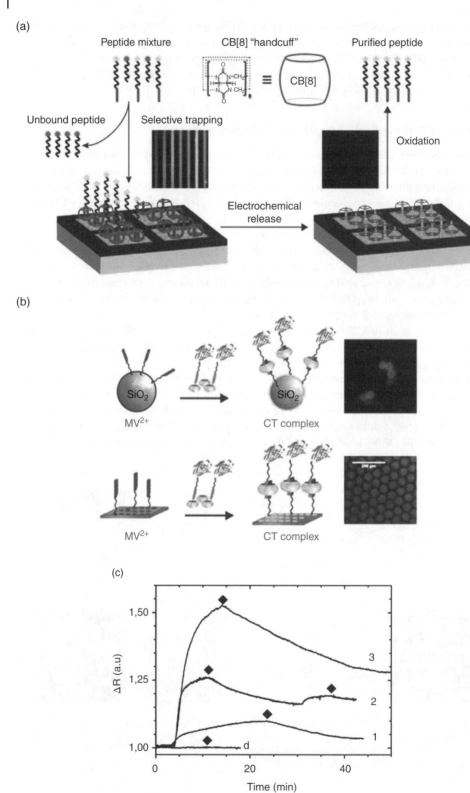

Peptide mixture   CB[8] "handcuff"   Purified peptide

CB[8]

Unbound peptide   Selective trapping

Oxidation

Electrochemical
release

(b)

SiO₂

MV²⁺   CT complex

MV²⁺   CT complex

(c)

ΔR (a.u)

1,50

1,25

1,00

3

2

1

d

0   20   40

Time (min)

preparing surface-mounted CB[8] rotaxanes yielding an interlocked architecture and preventing the dissociation of the first host–guest complex methylviologen·CB[8] [65].

Modifying a gold substrate in a stepwise manner, first with a thiolated methylviologen and subsequently with CB[8], allowed the selective supramolecular immobilization of peptides bearing an N-terminal tryptophan [63]. The electrochemical reduction of the methylviologen drove the release of the peptides [63]. A similar strategy was developed for creating methylviologen-modified gold and SiO_2 surfaces as well as SiO_2 nanoparticles (Figure 12.15b). In this work the surface assembly of the supramolecular ternary complex between methylviologen, CB[8] and 2-naphthol was investigated by surface plasma resonance. Three assembly routes were compared: (1) incubation of the gold sensor with a preassembled complex of methylviologen·CB[8], and then with a solution of naphthol; (2) stepwise incubation of a methylviologen monolayer with CB[8] followed by naphthol; and (3) incubation of a methylviologen monolayer with a mixed solution of naphthol and CB[8] (Figure 12.15c). Whereas the first route led to a higher surface coverage when compared with the second, the simultaneous incubation of CB[8] and the second guest naphthol by the third route gave the highest coverage values, by favorably making use of the high ternary association constant [66]. The latter assembly protocol allowed the oriented immobilization and micro-scale patterning of fluorescent proteins on glass surfaces through the grafting of an amino-methylviologen on a reactive *N,N*-carbonyldiimidazole monolayer [66].

In order to create a supramolecular system for the selective adhesion and electrochemically controlled release of cells, gold surfaces were coated with a cell-repellent monolayer, and a ternary complex was formed of thiolated methylviologen, CB[8] and tryptophan N-terminated peptide, bearing RGD ligands for integrin cell receptors (Figure 12.16a) [67]. A mixed monolayer of triethylene glycol and the labile 2-mercaptoethanol (99:1 v/v) was fabricated followed by the insertion of the preassembled ternary complex for dynamic display of the RGD ligands. Mouse myoblast C2C12 cells selectively adhered on these surfaces, and the electrochemical reduction of the surface-bound methylviologen led to the release of the WGGRGDS peptide from the ternary complex and subsequently to cell detachment (Figure 12.16a) [67].

The same strategy was extended to the selective dynamic immobilization of living *E. coli* bacteria [68]. The supramolecular functionality GGWGG was incorporated on the bacterial surface by genetically modifying a transmembrane protein to display a cysteine-stabilized mini-protein (knottin) containing the sequence in an accessible loop (Figure 12.16b). The bacteria selectively recognized the antifouling surfaces in the presence of methylviologen and CB[8] by heteroternary complexation. Moreover, the addition of CB[8] to a GGWGG-modified bacterial suspension caused aggregation by the formation of multiple supramolecular homoternary complexes between *E. coli* cells and CB[8] [68].

Figure 12.15 Fabrication of CB[8] SAMs using the first guest to covalently anchor the ternary complex to the surface. (a) Reversible trap-and-release of peptides bearing an N-terminal tryptophan was achieved onto a gold substrate modified in a stepwise manner with viologen and CB[8]. (b) Oriented positioning of proteins on silica particles and on glass or gold surfaces is mediated by CB[8] via incubation of the substrates with a solution of CB[8] and a naphthol guest moiety chemoselectively ligated to yellow fluorescent protein. (c) Surface plasma resonance sensograms of the three fabrication routes (see text for explanation). Diamonds indicate switching of the flow back to PBS buffer. *Source*: (a) Adapted from Ref. [63]. Reproduced with permission of American Chemical Society. (b and c) Adapted from Ref. [66]. Reproduced with permission of American Chemical Society.

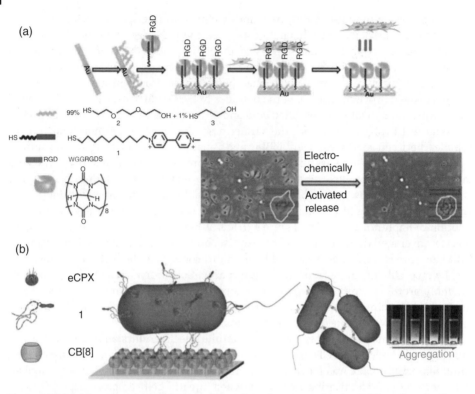

Figure 12.16 (a) Strategy for cell-repellent gold surfaces for supramolecular CB[8]-mediated cell adhesion. In the mixed monolayer of thiols **2** and **3** (99:1 v/v) the short, labile **3** is replaced by the preformed ternary complex of the thiolated methylviologen, CB[8] and WGGRGDS. The RGD ligand on the peptide promotes cell adhesion, and the electrochemical reduction of the viologen leads to the release of cells. (b) The peptide sequence GGWGG in one solvent-exposed loop of the mini-protein construct **1** is displayed on the outer membrane protein eCPX and binds CB[8] on a viologen·CB[8]-modified surface or promotes bacterial aggregation via multiple homoternary complexes. *Source*: (a) Ref. [67]. Reproduced with permission of John Wiley and Sons. (b) Adapted from Ref. [68]. Reproduced with permission of American Chemical Society.

When aiming for the supramolecular recognition of proteins, bacteria and cells at surfaces, the antifouling properties of the substrate need to be further optimized [69]. Mixed SAMs were prepared incubating gold-coated surfaces with the symmetric tetra(ethylene glycol)-terminated undecyl disulfide as the backfilling component and with the asymmetric maleimide and (triethylene glycol)-terminated undecyl disulfide as the reactive component [70]. A Michael addition was used to functionalize the maleimide groups at the surface covalently with a methylviologen thiol allowing for the specific supramolecular binding of a library of β-trypsin inhibitory knottins decorated with GGWGG guest tags for CB[8] at the N-termini, or at the C-termini, or at both ends [71]. As expected, the bivalent knottins showed a stronger binding affinity for CB[8] ($1.3 \times 10^6 \, M^{-1}$) on the surface when compared with either of the two monovalent knottin analogs under the same conditions, or the bivalent one when measured in solution in the presence of an excess of viologen to avoid homoternary complex formation ($1.6 \times 10^5 \, M^{-1}$). These knottins showed trypsin-inhibitory activity in solution and maintained it once supramolecularly immobilized on the surface [71].

Alternatively, supported lipid bilayers (SLBs) on glass can produce antifouling surfaces for CB[8]-mediated immobilization of proteins [72] and bacteria [30] (Figure 12.17). To achieve selective CB[8]-mediated ternary complex formation, the placement of viologen guest groups on the lipid bilayer was achieved by inserting a cholesterol-functionalized methylviologen into the bilayer as a 1:1 complex with CB[8], which enhances its incorporation into the bilayer (Figure 12.17a) [72]. In this case, a fluorescent protein was immobilized on the viologen·CB[8] SLB via a tryptophan residue positioned at the N-terminus of the protein in the presence of CB[8] as witnessed by quartz crystal microbalance (QCM) [72].

In another example, an SLB with maleimide functional groups was reacted with a thiolated methylviologen for CB[8]-mediated photoresponsive adhesion of bacteria [30]. As a second guest for CB[8], the light-sensitive switch azobenzene was conjugated with a mannose moiety that can be recognized by the tetravalent protein concanavalin A and by the FimH receptors of *E. coli*. The antifouling properties of the SLBs were investigated comparing ethylene glycol-based SAMs on gold, and gel and liquid-state SLBs for nonspecific bacterial adhesion. The liquid (Lα) phase is characteristic for unsaturated lipids (e.g., 1,2-dioleoyl-*sn*-glycero-3-phosphocholine, DOPC), with low melting temperature, generating SLBs with a disordered packing and a high lateral lipid mobility (on the order of $\mu m^2\ s^{-1}$). In contrast, saturated acyl chains (e.g., 1,2-dipalmitoyl-*sn*-glycero-3-phosphocholine, DPPC) provide a gel phase (Lβ) bilayer with a tight packing, increased melting temperature and reduced lateral mobility of at least one order of magnitude [30]. The zwitterionic and hydration properties of both SLBs resulted in better antifouling performance compared with the SAM. However, the liquid-state bilayer gave the best results in agreement with models that predict an out-of-plane mobility for liquid state lipids resulting in an undulatory motion of the bilayer [30]. On the liquid-state bilayer, bacterial cells were captured by the ternary complex formation between *trans*-azobenzene-anchored mannose groups and CB[8], and finally released upon UV photoirradiation and concomitant isomerization to the cis isomer (Figure 12.17b) [30].

A receptor-independent strategy for the CB[8]-mediated cell adhesion on a liquid-state SLB was prompted by (1) installing multiple naphthol ligands on the glycocalyx of Jurkat white blood cells via metabolic labeling and subsequent strain promoted azide–alkyne cycloaddition, and by (2) grafting methylviologen functionalities via Michael addition during the lipid hydration step that precedes the SLB fabrication on glass [73]. Only when CB[8] was present, cells dynamically assembled strongly on the SLB through the multivalent formation of supramolecular ternary complexes [73].

12.6 Conclusions and Outlook

CB[*n*]s provide a class of macrocycles that are able to form stable host–guest complexes with various guests. These guests range from small molecules such as drugs, dyes, and amino acids, to polymers, peptides, and even protein domains, via their appended side-chains. This wide range of potential guests has stimulated an increasing number of fundamental molecular recognition studies that led to various application in numerous fields, from catalysis to drug delivery and fabrication of (bio)material as a new line of development for biodevices. The systems are intrinsically dynamic and can rearrange,

(a)

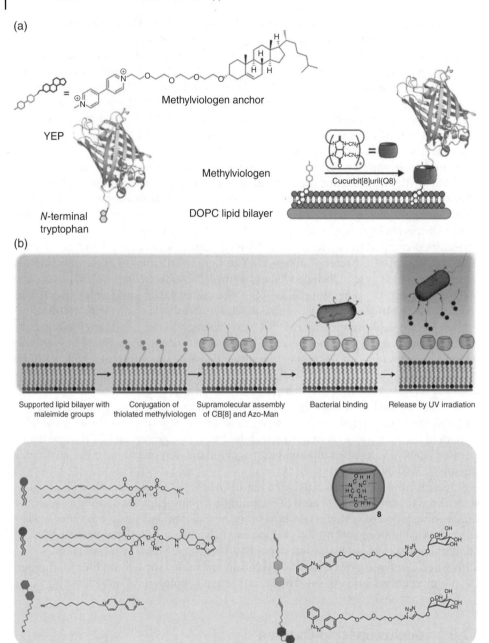

Methylviologen anchor

YEP

Methylviologen

Cucurbit[8]uril(Q8)

DOPC lipid bilayer

N-terminal
tryptophan

(b)

Supported lipid bilayer with   Conjugation of   Supramolecular assembly   Bacterial binding   Release by UV irradiation
maleimide groups   thiolated methylviologen   of CB[8] and Azo-Man

8

Figure 12.17 (a) SLBs for CB[8]-mediated immobilization of proteins via a methylviologen with a cholesterol anchor that inserts into the bilayer and an N-terminal tryptophan residue on yellow fluorescent protein (YFP). (b) Reversible immobilization on SLBs of bacteria via the naturally occurring receptors FimH for mannose which is displayed via CB[8] recognition of a methylviologen moiety conjugated onto the bilayer. *Source*: (a) Ref. [72]. Reproduced with permission of John Wiley and Sons. (b) Ref. [30]. Reproduced with permission of John Wiley and Sons.

while some of the interactions are reversible upon stimuli (pH, chemicals, electrochemistry, UV light). This opens the door for a greater control over the molecules and assemblies. We hope that by discussing the origins of the molecular recognition properties and binding affinities with a specific emphasis on amino acid based recognition of CB[*n*] that readers are intrigued as to how these properties can be successfully employed to dynamically probe the properties and function of peptides, proteins, and cells, which is important for the further development of bioanalytical and biomedical applications.

Acknowledgment

Funding to P.J. from the EU through Starting ERC grant Sumoman 259183 is gratefully acknowledged.

References

1 Lagona J, Mukhopadhyay P, Chakrabarti S, Isaacs L. The cucurbit[n]uril family. Angew. Chem. Int. Ed. 2005;44(31):4844–70.

2 Masson E, Ling X, Joseph R, *et al*. Cucurbituril chemistry: A tale of supramolecular success. RSC Adv. 2012;2(4):1213–47.

3 Biedermann F, Nau WM, Schneider HJ. The hydrophobic effect revisited - Studies with supramolecular complexes imply high-energy water as a noncovalent driving force. Angew. Chem. Int. Ed. 2014;53(42):11158–71.

4 Barrow SJ, Kasera S, Rowland MJ, *et al*. Cucurbituril-based molecular recognition. Chem Rev. 2015;115(22):12320–406.

5 (a)Behrend R, Meyer E, Rusche FI. Ueber condensationsprodukte aus glycoluril und formaldehyd. Liebigs Ann. Chem. 1905;339:1–37. (b)Assaf KI, Nau WM. Cucurbiturils: From synthesis to high-affinity binding and catalysis. Chem. Soc. Rev. 2015;44(2):394–418.

6 Kim J, Jung I-S, Kim S-Y, *et al*. New cucurbituril homologues: Syntheses, isolation, characterization, and X-ray crystal structures of cucurbit[n]uril (n = 5, 7, and 8). J. Am. Chem. Soc. 2000;122(3):540–1.

7 Day AI, Blanch RJ, Arnold AP, *et al*. A cucurbituril-based gyroscane: A new supramolecular form. Angew. Chem. Int. Ed. 2002;114(2):285–7.

8 Cheng X-J, Liang L-L, Chen K, *et al*. Twisted cucurbit[14]uril. Angew. Chem. Int. Ed. 2013;125(28):7393–6.

9 Kim H-J, Heo J, Jeon WS, *et al*. Selective inclusion of a hetero-guest pair in a molecular host: Formation of stable charge-transfer complexes in cucurbit[8]uril. Angew. Chem. Int. Ed. 2001;40(8):1526–9.

10 Jon SY, Selvapalam N, Oh DH, *et al*. Facile synthesis of cucurbit[n]uril derivatives via direct functionalization: Expanding utilization of cucurbit[n]uril. J. Am. Chem. Soc. 2003;125(34):10186–7.

11 Vinciguerra B, Cao L, Cannon JR, *et al*. Synthesis and self-assembly processes of monofunctionalized cucurbit[7]uril. J. Am. Chem. Soc. 2012;134(31):13133–40.

12 Zhao N, Lloyd GO, Scherman OA. Monofunctionalised cucurbit[6]uril synthesis using imidazolium host-guest complexation. Chem. Commun. 2012;48(25):3070–2.

13 Aav R, Shmatova E, Reile I, *et al.* New chiral cyclohexylhemicucurbit[6]uril. Org. Lett. 2013;15(14):3786–9.

14 Gilberg L, Khan MSA, Enderesova M, Sindelar V. Cucurbiturils substituted on the methylene bridge. Org. Lett. 2014;16(9):2446–9.

15 Isaacs L. The mechanism of cucurbituril formation. Isr. J Chem. 2011;51(5-6):578–91.

16 Biedermann F, Scherman OA. Cucurbit[8]uril mediated donor-acceptor ternary complexes: A model system for studying charge-transfer interactions. J. Phys. Chem. B. 2012;116(9):2842–9.

17 Biedermann F, Uzunova VD, Scherman OA, *et al.* Release of high-energy water as an essential driving force for the high-affinity binding of cucurbit[n]urils. J. Am. Chem. Soc. 2012;134(37):15318–23.

18 Cao L, Šekutor M, Zavalij PY, *et al.* Cucurbit[7]uril × guest pair with an attomolar dissociation constant. Angew. Chem. Int. Ed. 2014;53(4):988–93.

19 Rekharsky MV, Mori T, Yang C, *et al.* A synthetic host-guest system achieves avidin-biotin affinity by overcoming enthalpy - Entropy compensation. Proc. Natl Acad. Sci. USA 2007;104(52):20737–42.

20 Ling X, Saretz S, Xiao L, *et al.* Water: Vs. cucurbituril rim: A fierce competition for guest solvation. Chem. Sci. 2016;7(6):3569–73.

21 Biedermann F, Vendruscolo M, Scherman OA, *et al.* Cucurbit[8]uril and blue-box: High-energy water release overwhelms electrostatic interactions. J. Am. Chem. Soc. 2013;135(39):14879–88.

22 Szejtli J. Introduction and general overview of cyclodextrin chemistry. Chem. Rev. 1998;98(5):1743–53.

23 Jeon WS, Moon K, Park SH, et al. Complexation of ferrocene derivatives by the cucurbit[7]uril host: A comparative study of the cucurbituril and cyclodextrin host families. J. Am. Chem. Soc. 2005;127(37):12984–9.

24 Liu Y, Yu Y, Gao J, *et al.* Water-soluble supramolecular polymerization driven by multiple host-stabilized charge-transfer interactions. Angew. Chem. Int. Ed. 2010;122(37):6726–9.

25 Jiao D, Geng J, Loh XJ, *et al.* Supramolecular peptide amphiphile vesicles through host-guest complexation. Angew. Chem. Int. Ed. 2012;51(38):9633–7.

26 Jeon WS, Kim HJ, Lee C, Kim K. Control of the stoichiometry in host-guest complexation by redox chemistry of guests: Inclusion of methylviologen in cucurbit[8] uril. Chem. Commun. 2002;8(17):1828–9.

27 Tian F, Jiao D, Biedermann F, Scherman OA. Orthogonal switching of a single supramolecular complex. Nat. Commun. 2012;3.

28 Bush ME, Bouley ND, Urbach AR. Charge-mediated recognition of N-terminal tryptophan in aqueous solution by a synthetic host. J. Am. Chem. Soc. 2005;127(41):14511–7.

29 Brinkmann J, Cavatorta E, Sankaran S, Schmidt B, Van Weerd J, Jonkheijm P. About supramolecular systems for dynamically probing cells. Chem. Soc. Rev. 2014;43(13):4449–69.

30 Sankaran S, Van Weerd J, Voskuhl J, *et al.* Photoresponsive cucurbit[8]uril-mediated adhesion of bacteria on supported lipid bilayers. Small 2015;11(46):6187–96.

31 Urbach AR, Ramalingam V. Molecular recognition of amino acids, peptides, and proteins by cucurbit[n]uril receptors. Isr. J. Chem. 2011;51(5-6):664–78.

32 Rekharsky MV, Yamamura H, Ko YH, *et al.* Sequence recognition and self-sorting of a dipeptide by cucurbit[6]uril and cucurbit[7]uril. Chem. Commun. 2008(19):2236–8.

33 Rajgariah P, Urbach AR. Scope of amino acid recognition by cucurbit[8]uril. J. Incl. Phenom. Macrocycl. Chem. 2008;62(3-4):251–4.

34 Lee JW, Lee HHL, Ko YH, *et al.* Deciphering the specific high-affinity binding of cucurbit[7]uril to amino acids in water. J. Phys. Chem. B 2015;119(13):4628–36.

35 Chinai JM, Taylor AB, Ryno LM, *et al.* Molecular recognition of insulin by a synthetic receptor. J. Am. Chem. Soc. 2011;133(23):8810–3.

36 Ghale G, Ramalingam V, Urbach AR, Nau WM. Determining protease substrate selectivity and inhibition by label-free supramolecular tandem enzyme assays. J. Am. Chem. Soc. 2011;133(19):7528–35.

37 Heitmann LM, Taylor AB, Hart PJ, Urbach AR. Sequence-specific recognition and cooperative dimerization of N-terminal aromatic peptides in aqueous solution by a synthetic host. J. Am. Chem. Soc. 2006;128(38):12574–81.

38 Sonzini S, Ryan STJ, Scherman OA. Supramolecular dimerisation of middle-chain Phe pentapeptides via CB[8] host-guest homoternary complex formation. Chem. Commun. 2013;49(78):8779–81.

39 Nguyen HD, Dang DT, Van Dongen JLJ, Brunsveld L. Protein dimerization induced by supramolecular interactions with cucurbit[8]uril. Angew. Chem. Int. Ed. 2010;49(5):895–8.

40 Hou C, Li J, Zhao L, *et al.* Construction of protein nanowires through cucurbit[8] uril-based highly specific host-guest interactions: An approach to the assembly of functional proteins. Angew. Chem. Int. Ed. 2013;52(21):5590–3.

41 Dang DT, Nguyen HD, Merkx M, Brunsveld L. Supramolecular control of enzyme activity through cucurbit[8]uril-mediated dimerization. Angew. Chem. Int. Ed. 2013;52(10):2915–9.

42 Ramaekers M, Wijnands SPW, Van Dongen JLJ, *et al.* Cucurbit[8]uril templated supramolecular ring structure formation and protein assembly modulation. Chem. Commun. 2015;51(15):3147–50.

43 Reczek JJ, Kennedy AA, Halbert BT, Urbach AR. Multivalent recognition of peptides by modular self-assembled receptors. J. Am. Chem. Soc. 2009;131(6):2408–15.

44 Gubeli RJ, Sonzini S, Podmore A, *et al.* Selective, non-covalent conjugation of synthetic peptides with recombinant proteins mediated by host-guest chemistry. Chem. Commun. 2016;52(22):4235–8.

45 Smith LC, Leach DG, Blaylock BE, *et al.* Sequence-specific, nanomolar peptide binding via cucurbit[8]uril-induced folding and inclusion of neighboring side chains. J. Am. Chem. Soc. 2015;137(10):3663–9.

46 Rowland MJ, Appel EA, Coulston RJ, Scherman OA. Dynamically crosslinked materials via recognition of amino acids by cucurbit[8]uril. J. Mater. Chem. B 2013;1(23):2904–10.

47 Geng J, Biedermann F, Zayed JM, *et al.* Supramolecular glycopolymers in water: A reversible route toward multivalent carbohydrate-lectin conjugates using cucurbit[8]uril. Macromolecules 2011;44(11):4276–81.

48 Jung H, Park KM, Yang JA, *et al.* Theranostic systems assembled in situ on demand by host-guest chemistry. Biomaterials 2011;32(30):7687–94.

49 Park KM, Yang JA, Jung H, *et al.* In situ supramolecular assembly and modular modification of hyaluronic acid hydrogels for 3D cellular engineering. ACS Nano. 2012;6(4):2960–8.

50 Kim J, Ahn Y, Park KM, Lee DW, Kim K. Glyco-pseudopolyrotaxanes: Carbohydrate wheels threaded on a polymer string and their inhibition of bacterial adhesion. Chem. Eur. J. 2010;16(40):12168–73.

51 Stoffelen C, Voskuhl J, Jonkheijm P, Huskens J. Dual stimuli-responsive self-assembled supramolecular nanoparticles. Angew. Chem. Int. Ed. 2014;53(13):3400–4.

52 Li Q-L, Sun Y, Sun Y-L, *et al.* Mesoporous silica nanoparticles coated by layer-by-layer self-assembly using cucurbit[7]uril for in vitro and in vivo anticancer drug release. Chem. Mater. 2014;26(22):6418–31.

53 Kim E, Kim D, Jung H, Lee J, *et al.* Facile, template-free synthesis of stimuli-responsive polymer nanocapsules for targeted drug delivery. Angew. Chem. Int. Ed. 2010;49(26):4405–8.

54 Kim D, Kim E, Kim J, *et al.* Direct synthesis of polymer nanocapsules with a noncovalently tailorable surface. Angew. Chem. Int. Ed. 2007;46(19):3471–4.

55 Stoffelen C, Huskens J. Zwitterionic supramolecular nanoparticles: Self-assembly and responsive properties. Nanoscale 2015;7(17):7915–9.

56 Park KM, Lee DW, Sarkar B, *et al.* Reduction-sensitive, robust vesicles with a noncovalently modifiable surface as a multifunctional drug-delivery platform. Small 2010;6(13):1430–41.

57 Kim C, Agasti SS, Zhu Z, *et al.* Recognition-mediated activation of therapeutic gold nanoparticles inside living cells. Nat. Chem. 2010;2(11):962–6.

58 Lee DW, Park KM, Banerjee M, *et al.* Supramolecular fishing for plasma membrane proteins using an ultrastable synthetic host-guest binding pair. Nat. Chem. 2011;3(2):154–9.

59 An Q, Li G, Tao C, *et al.* A general and efficient method to form self-assembled cucurbit[n]uril monolayers on gold surfaces. Chem. Commun. 2008(17):1989–91.

60 Gomez-Casado A, Jonkheijm P, Huskens J. Recognition properties of cucurbit[7]uril self-assembled monolayers studied with force spectroscopy. Langmuir 2011;27(18):11508–13.

61 Young JF, Nguyen HD, Yang L, *et al.* Strong and reversible monovalent supramolecular protein immobilization. ChemBioChem 2010;11(2):180–3.

62 Neirynck P, Brinkmann J, An Q, *et al.* Supramolecular control of cell adhesion via ferrocene-cucurbit[7]uril host-guest binding on gold surfaces. Chem. Commun. 2013;49(35):3679–81.

63 Tian F, Cziferszky M, Jiao D, *et al.* Peptide separation through a CB[8]-mediated supramolecular trap-and-release process. Langmuir 2011;27(4):1387–90.

64 Yang H, An Q, Zhu W, *et al.* A new strategy for effective construction of protein stacks by using cucurbit[8]uril as a glue molecule. Chem. Commun. 2012;48(86):10633–5.

65 Hu C, Lan Y, Tian F, *et al.* Facile method for preparing surface-mounted cucurbit[8] uril-based rotaxanes. Langmuir 2014;30(36):10926–32.

66 González-Campo A, Brasch M, Uhlenheuer DA, *et al.* Supramolecularly oriented immobilization of proteins using cucurbit[8]uril. Langmuir 2012;28(47):16364–71.

67 An Q, Brinkmann J, Huskens J, *et al*. A supramolecular system for the electrochemically controlled release of cells. Angew. Chem. Int. Ed. 2012;51(49):12233–7.

68 Sankaran S, Kiren MC, Jonkheijm P. Incorporating bacteria as a living component in supramolecular self-assembled monolayers through dynamic nanoscale interactions. ACS Nano 2015;9(4):3579–86.

69 Sankaran S, Jaatinen L, Brinkmann J, *et al*. Cell adhesion on dynamic supramolecular surfaces probed by Fluid Force Microscopy-based single-cell force spectroscopy. ACS Nano 2017;11(4):3867–74.

70 Sobers CJ, Wood SE, Mrksich M. A gene expression-based comparison of cell adhesion to extracellular matrix and RGD-terminated monolayers. Biomaterials 2015;52(1):385–94.

71 Sankaran S, De Ruiter M, Cornelissen JJLM, Jonkheijm P. Supramolecular surface immobilization of knottin derivatives for dynamic display of high affinity binders. Bioconjugate Chem. 2015;26(9):1972–80.

72 Bosmans RPG, Hendriksen WE, Verheijden M, *et al*. Supramolecular protein immobilization on lipid bilayers. Chem. Eur. J. 2015;21:18466–73.

73 Cavatorta E, Verheijden ML, Van Roosmalen W, *et al*. Functionalizing the glycocalyx of living cells with supramolecular guest ligands for cucurbit[8]uril-mediated assembly. Chem. Commun. 2016;52(44):7146–9.

13

Multivalent Lectin–Glycan Interactions in the Immune System

João T. Monteiro and Bernd Lepenies

Immunology Unit & Research Center for Emerging Infections and Zoonoses (RIZ), University of Veterinary Medicine Hannover, Bünteweg 17, 30559 Hannover, Germany

13.1 Introduction

Accomplishing a more profound understanding of the structural and functional role of protein–glycan interactions in immunity is fundamental to gain insights into the pathogenesis of diseases such as allergies, cancer and autoimmune diseases [1,2]. Carbohydrates constitute the most abundant class of biomolecules on Earth and represent one of the most commonly added modifications to proteins, thereby fine-tuning protein functions [3,4]. It is estimated that more than half of the proteins in humans are glycosylated [5]. The role of glycans in nature is widespread, ranging from energy storage, to molecular recognition for intracellular trafficking, and host–pathogen interactions [6]. The biosynthesis of polysaccharides and glycoconjugates is not encoded directly in the genome, but is a result of the differential expression of glycoenzymes. Glycosyltransferases and glycosidases catalyze glycosylation reactions and the hydrolysis of glycosidic bonds, respectively, often leading to heterogeneous mixtures of glycoforms that possess distinct physiological activities [7]. Typically, glycans are covalently bound to proteins or lipids. Since the individual interactions between proteins and carbohydrates are usually weak, multiple simultaneous interactions are generally required to increase avidity in order to achieve biological effects [6]. The ability of glycans to arrange in multivalent displays is crucial for attaining strong, yet reversible interactions, that are involved in diverse biological processes like self-organization of matter, recognition between cells with subsequent signal transduction, and immune responses [8,9].

The immune system provides an ideal framework to describe multivalent glycan interactions with their binding partners, since they are often fundamental to trigger immune responses [10]. The immune system is a host defense system composed of cellular and humoral components that protect the host against a variety of agents, such as pathogens, foreign substances, and tissue damage [11]. The immune system can be subdivided into two subsystems: the innate immune system [12]; and the adaptive immune system [13].

Multivalency: Concepts, Research & Applications, First Edition. Edited by Jurriaan Huskens, Leonard J. Prins, Rainer Haag, and Bart Jan Ravoo.
© 2018 John Wiley & Sons Ltd. Published 2018 by John Wiley & Sons Ltd.

The innate immune system constitutes the first line of host defense against foreign agents and consists of a collection of physical barriers, for example the skin, and chemical barriers to infection, for example gastric acid, as well as different cell subsets capable of recognizing conserved structures on pathogens, termed pathogen-associated molecular patterns (PAMPs) [12,14,15]. The main cell subsets of innate immunity include macrophages, dendritic cells (DCs), monocytes, mast cells, granulocytes, and natural killer (NK) cells [16]. The primary role of innate immunity is to recognize and limit pathogen spread during the early stages of infection, for instance by complement activation (a cascade-like activation of plasma proteins that orchestrate phagocytosis, chemotaxis, opsonization, and pathogen lysis [17]), and activation of different families of pattern recognition receptors (PRRs) [18].

The adaptive immune system relies on clonal lymphocyte populations with randomly generated receptors [14] that are custom-tailored to produce highly specific and flexible immune responses. Two types of receptors can be found according to the lymphocyte subset: B cell receptors (BCRs) and T cell receptors (TCRs). B cells mature in the bone marrow and are antibody-producing cells, whereas T lymphocytes mature in the thymus and elicit cellular immune responses [13,14]. In the thymus, self-peptides presented by major histocompatibility complex (MHC) molecules to T cell precursors dictate the fate of the cells: thymocytes may survive if their TCRs show a sufficient affinity for MHC molecules (positive selection) [19], whereas apoptosis is induced if a strong binding occurs to self-peptide/MHC complexes, which indicates autoreactivity (negative selection) [20]. Numerous features of T cell biology, from development to differentiation to effector functions are influenced by the strength with which a T cell responds to MHC-bound peptides [21–23]. Positive thymocyte selection results in their differentiation into naïve $CD4^+$ or $CD8^+$ T cells, depending on the binding of the TCR to MHC-II or MHC-I molecules, respectively [24].

Since the immune system is responsible for surveying the host for PAMPs and self-antigens released by damaged cells (so-called danger associated molecular patterns, DAMPs) it requires a process to discriminate between self and non-self-antigens. To this end, the immune system possesses a set of PRRs that mediate PAMPs and DAMPs recognition [25,26]. Major PRR classes comprehend Toll-like receptors (TLRs) [27], retinoic acid-inducible gene (RIG)-I-like receptors (RLRs) [28], nucleotide-binding oligomerization domain (NOD)-like receptors (NLRs) [29], and DNA sensors [30]. Given that a large number of PAMPs and DAMPs are glycoconjugates, glycan-binding receptors, so-called lectins, have evolved as PRRs to bind to these glycoconjugates [31].

Lectins are carbohydrate-binding proteins that are specialized in the recognition and binding to sugar moieties displayed at cell surfaces and serve as interaction partners between cells and their environment [32,33]. Lectins have been categorized into different families, according to their conserved structure of sequence motifs for glycan binding and specificity, termed calnexin, M-type, L-type, C-type, P-type/mannose-6-phosphate receptors (MPRs), I-type/Siglecs (sialic-acid-binding immunoglobulin-like lectins), F-type (absent in mammals), R-type, chitinase-like lectins, galectins, and intelectins [34,35]. The intricate connection between glycan–lectin interactions and the subsequent triggering of immune responses will be discussed in the following sections with an emphasis on three classes of lectins: C-type lectin receptors (CLRs), galectins, and Siglecs. Moreover, antigen uptake and presentation to T cells will also be addressed to highlight how multivalent interactions may shape adaptive immunity.

13.2 Targeting Innate Immunity to Shape Adaptive Immunity

Antigen presenting cells (APCs) process exogenous antigens in endo-lysosomal compartments and present peptides by MHC-II to CD4$^+$ T cells, whereas endogenous proteins are cleaved in the cytosol and the resultant peptide fragments are finally displayed by MHC-I molecules to activate CD8$^+$ T cells [36] (Figure 13.1). Current vaccination strategies take advantage of the capacity of DCs to present exogenous antigens on MHC-I molecules in a process termed cross-presentation [37,38]. Thus, loading of DCs with tumor antigens can lead to activation of tumor specific CD8$^+$ T cells via MHC-I presentation. Antigen processing and presentation by APCs is essential to ensure T cell priming and complete T cell activation [39,40]. Priming of T cells by APCs also involves the expression of the co-stimulatory molecules CD80/CD86 and the production of cytokines by APCs [38,41,42]. Upon activation, CD4$^+$ T helper cells (Th) may differentiate into different subsets: (i) Th1 cells are responsible for macrophage activation and to increase their phagocytic activity via secretion of pro-inflammatory cytokines, namely IFN-γ; (ii) Th2 cells elicit humoral responses and enhance host defense against bacterial and parasitic infections mainly by IL-4, IL-5, and IL-6 production; and (iii) Th17 cells promote the recruitment and activation of neutrophils via IL-17 [39,43]. Antigen presentation of intravesicular pathogens, like mycobacteria, prompts the secretion of IL-12, thus triggering a Th1 response [44,45]. A humoral response is orchestrated by Th2 cells by induction of IL-4 production upon recognition of extracellular pathogens such as helminths by APCs [46]. A Th17 response can be elicited in the

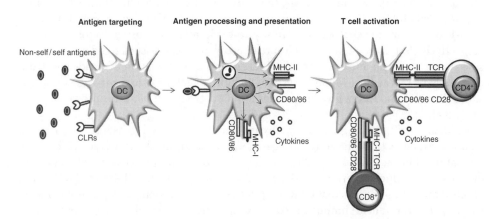

Figure 13.1 DC-dependent activation of T cells and modulation of effector functions in response to antigens. DCs express PRRs, like CLRs, that are able to recognize and bind to antigens present on the surface of pathogens. Next, internalization of the antigens, processing into peptides and loading on MHC molecules take place, for efficient priming of CD4$^+$ and CD8$^+$ T cells via MHC-II and MHC-I presentation, respectively. CLR engagement triggers DC maturation, resulting in upregulated expression of MHC and costimulatory molecules. Three signals are essential for T cell activation and differentiation by DCs: first, TCR recognition of a matching peptide/MHC complex; secondly, expression of the co-stimulatory molecules CD80/CD86 and their interaction with CD28 on T cells; and thirdly, production of cytokines by DCs. The combination of these three signals leads to T cell activation and differentiation into different T helper cell subsets.

presence of extracellular bacteria and fungi, which then leads to the recruitment of neutrophils to the site of infection [47]. Modulation of Th1, Th2, and Th17 responses is performed by regulatory T cells (Tregs), often by secretion of IL-10 [48]. In contrast, cytotoxic CD8$^+$ T cells, upon recognition of antigens presented by MHC-I molecules, promote the killing of virus-infected or malignant cells, mainly through the release of perforin and granzyme B from intracellular granules. The combination of the pore formation in cell membranes by perforin, followed by the insertion of granzyme B, triggers the activation of apoptotic signals via caspase activation inside target cells [49].

Since lectins in innate immunity play crucial roles in shaping adaptive immune responses, lectins represent promising targets for immune modulation. The following sections will focus on CLRs, galectins, and Siglecs.

13.3   C-type Lectin Receptors

C-type lectin receptors (CLRs) are carbohydrate-binding proteins that belong to one of the largest and most diverse lectin families found in animals, being composed of 17 groups that differ in their structural and functional properties [50–52]. CLRs in innate immunity are mainly expressed by APCs, namely DCs and macrophages, and recognize antigens on pathogens and self-antigens [53,54]. Their shared feature is the presence of a C-type lectin-like domain (CTLD), a key structural motif in CLRs that recognizes a variety of ligands, being mainly involved in Ca^{2+}-dependent carbohydrate binding through a compact module named a carbohydrate recognition domain (CRD) [52]. However, binding of CLRs to additional ligands, like cholesterol and uric acid crystals, also occurs [55,56]. Within the CRD, there are specific amino acid motifs that confer ligand specificity to CLRs. Two major motifs are the EPN (Glu-Pro-Asn) and the QPD (Gln-Pro-Asp) motifs that exhibit affinity towards mannose and galactose residues, respectively [52]. Detection of a widespread array of glycan-associated danger signals, such as β-1-3-glucan and mannan, by CLRs enables discrimination of a wide spectrum of pathogens by the host immune system, including bacteria, fungi, viruses, and parasites [57–59]. In addition, glycans present in blood group antigens are also recognized by CLRs [54].

Multivalent CLR engagement by glycans mediates a diverse set of downstream responses such as cell adhesion, pathogen uptake, and antigen presentation. Moreover, regulation of homeostasis, including the clearance of apoptotic cells, and cytokine and chemokine production are also triggered by CLR engagement [60,61]. Taking into account the manifold roles that CLRs play in the immune system, they represent attractive targets to modulate immune responses via activating or inhibiting signaling cascades [60,61]. Myeloid CLRs can be distributed in four main groups according to their intracellular signaling motifs (Figure 13.2) [60,62]: immunoreceptor tyrosine-based activating motif (ITAM)-bearing CLRs; hemITAM-bearing CLRs; immunoreceptor tyrosine-based inhibitory motif (ITIM)-bearing CLRs; and a group of CLRs lacking typical signaling motifs [60]. ITAMs consist of YxxL/I tandem repeats and most activating receptors associate with ITAM-bearing adaptor chains such as the FcRγ chain, including DC-associated C-type lectin (Dectin)-2, DC-immunoactivating receptor (DCAR), macrophage inducible C-type lectin (Mincle) and myeloid DAP12-associating lectin (MDL-1) [63,64]. In addition, a hemITAM motif (a single YxxL/I sequence) is

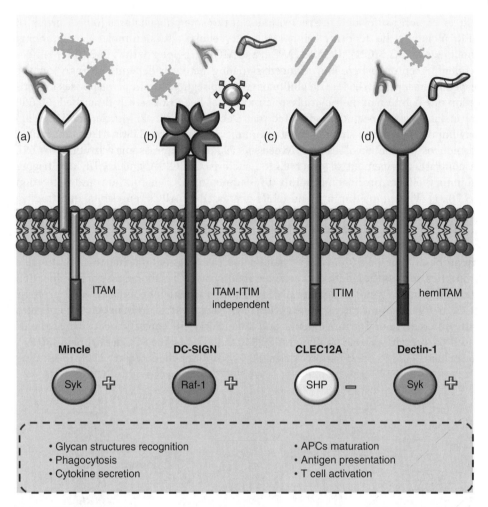

Figure 13.2 Myeloid CLR groups and respective signaling motifs. CLRs are capable of recognizing a vast array of antigens from either pathogens, such as bacteria, viruses, parasites, fungi, or self-antigens. Upon CLR engagement, a signaling cascade is triggered through binding of early adaptors and the recruitment of kinases or phosphatases. Myeloid CLRs can be grouped into four groups based on the cytoplasmic signaling motifs and early adaptors: (a) immunoreceptor tyrosine-based activating motif (ITAM)-coupled CLRs, like Mincle or Dectin-2, signal via spleen tyrosine kinase, Syk, by forming an association with ITAM-bearing adaptors, like DAP12 or FcRγ chain; (b) ITAM-immunoreceptor tyrosine-based inhibitory motif (ITIM) independent CLRs, such as DC-SIGN and mannose receptor, possess tyrosine-based motifs involved in endocytosis, however signal independently from Syk or phosphatases; (c) ITIM-coupled CLRs dampen immune responses by the recruitment of phosphatases SHP-1 and SHP-2 via the ITIM motif; and (d) hemITAM-coupled CLRs, such as Dectin-1 and SIGNR3, modulate immune responses by transducing the signal by a single tyrosine-based motif in their tail and Syk recruitment. See color section.

carried by CLRs such as Dectin-1, CLEC2, CLEC9A, and SIGNR3 [60]. Cellular activation by both groups of receptors is mediated by recruitment of the spleen tyrosine kinase Syk [65]. In contrast, ITIM-bearing (S/I/V/LxYxxI/V/L) CLRs such as the DC immunoreceptor (DCIR), CLEC12A, and CLEC12B, are responsible for inhibiting cellular responses through the recruitment of tyrosine phosphatases [60]. A group of CLRs including the dendritic cell-specific intercellular adhesion molecule-3-grabbing non integrin (DC-SIGN), SIGNR1, MCL, and MGL-1, belong to the ITAM/ITIM-independent receptors where the induced signaling pathway depends strongly on the respective receptor [60,64]. The ability of CLRs to regulate immune responses by activation or inhibition of their cytoplasmic motifs has been extensively described [60,66]. For instance, it has been demonstrated that Dectin-1, Dectin-2, Mincle, and DEC-205 play important roles in shaping innate immune responses by eliciting DC maturation, chemotaxis, production of reactive oxygen species, and inflammasome activation [67–69]. In contrast, engagement of other CLRs, such as CLEC12A and DCIR, can trigger inhibitory effects on host immunity by dampening DC maturation and activation [59,70,71]. Thus, targeting myeloid CLRs expressed by APCs represents an attractive approach for the cell-specific delivery of either drugs or vaccine antigens and to modulate immune responses [43]. The preferred strategy for the activation of DCs *in vivo* via CLRs targeting has been antibody-mediated antigen delivery [72–75]. However, glycan-based carrier systems have gained increasing attention due to their tunable properties, such as the length of the spacer, rigidity of the backbone, ease of modification of carbohydrate ligands and generally lower immunogenicity compared with proteins [9,43,76–79]. For multivalent presentation of glycans, a vast array of neoglycoconjugates with different spatial arrangements and valencies were constructed using scaffolds based on proteins [80,81], polymers [82,83], calixarenes [84,85], dendrimers [86,87], glycoclusters [88,89], and nanoparticles [90,91], among others (Figure 13.3).

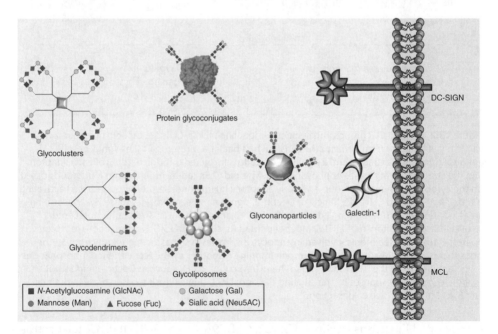

Figure 13.3 Schematic representation of the multivalent glycan carriers employed to target lectins. Glycan-carrier systems that enable multivalent presentation of glycans include glycoclusters, glycan-modified antigens (proteins or peptides), liposomes, dendrimers, and nanoparticles. See color section.

13.3.1 Multivalent Glycoconjugates Targeting DC-SIGN

DC-SIGN was first described as a transmembrane protein capable of binding to HIV glycoprotein gp120 and this interaction promoted HIV trans-infection of T cells [92–94]. DC-SIGN is mainly expressed by naïve DCs, macrophages and microglia, and it is constituted of an N-terminal cytoplasmic domain, a transmembrane region, a tandem repeat neck domain and a CRD [92,95,96]. Tetramerization of the receptor occurs due to the presence of an α-helix stalk in the neck domain, thus increasing the glycan binding avidity [97]. The crystal structure of DC-SIGN and the presence of an EPN motif indicate affinity towards mannose- and fucose-type oligosaccharides [98–101]. Ligand capture by DC-SIGN was inhibited by mannan and EDTA, denoting a mannose specificity and Ca^{2+} dependence of the glycan recognition by DC-SIGN [102]. DC-SIGN exhibits a broad specificity for pathogens, including viruses, such as HIV, Ebola, dengue virus, and hepatitis C virus, as well as different mycobacteria, fungi, and parasites [94,103–108]. Upon antigen uptake, internalized ligands are targeted to late endosomes and lysosomes, followed by subsequent presentation to T cells. In contrast, binding of HIV-derived gp120 to DC-SIGN results in uptake to a non-lysosomal intracellular compartment which enables the virus to remain competent to elicit T cell infection. Therefore, DC-SIGN can be exploited by pathogens for infection of different cell subsets [94,109,110]. Moreover, DC-SIGN engagement by mannose- or fucose-type ligands may promote pro-inflammatory responses or suppression of immune responses, respectively [111,112].

HIV infection is promoted by the interaction between DC-SIGN and glycoprotein gp120 in APCs present at the mucosal endothelium, followed by transfer of the virus to T cells where viral replication takes place [102]. Multivalent glycoconjugate systems have been designed to mimic the surface of the virus to compete with natural glycan ligands, with the aim of developing carbohydrate-based antiviral drugs against HIV [9,43].

In this regard, the multivalent display of N-linked high mannose glycans of the HIV gp120 represents an attractive approach. For instance, water-soluble gold nanoparticles displaying truncated high-mannose structures of the HIV undecasaccharide ($Man_9GlcNAc_2$) demonstrated a binding affinity to DC-SIGN ranging from the micro- to nanomolar range and impaired the trans-infection of human T cells by efficient inhibition of DC-SIGN interaction [113,114]. In other studies, a specific inhibition of the HIV infection of T lymphocytes by competitive binding of a tetravalent dendron display of four linear trimannosides to DC-SIGN was shown [115,116]. An additional example for DC-SIGN blockade was the use of a fullerene-based presentation of 36 mannose moieties displayed in a spherical manner for binding competition with pseudotyped Ebola virus particles [117]. Since DC-SIGN targeting approaches have gained remarkable emphasis in the past years, numerous approaches using different multivalent carrier systems have been employed. Besides anti-viral strategies, it was also demonstrated that human $CD8^+$ T cells could be efficiently primed using glycans displayed on antigen-loaded liposomes to specifically target and take advantage of the endocytic capacity of DC-SIGN [118].

13.3.2 Multivalent Glycoconjugates Targeting Other CLRs

The mannose receptor (MR) is a CLR predominantly expressed on the surface of DCs and macrophages that comprises eight CRDs, a cysteine rich domain, a fibronectin type II domain and a short cytoplasmic tail devoid of signaling motifs [119,120]. Similar to

DC-SIGN, MR recognizes mannose- and fucose-terminated glycans on the surface of pathogens, hence it is able to bind to bacteria, viruses, fungi, and also endogenous ligands [121]. In addition, a role for MR in antigen cross-presentation was described [41]. Dendrimers as vaccine carriers represent an attractive strategy to induce cellular immune responses [122]. For example, mannosylated dendrimers covalently linked to the model antigen ovalbumin (OVA) induced efficient DC maturation *in vitro* and led to robust antibody production and T cell responses upon immunization of mice. Reduced tumor growth in an OVA specific tumor model after prophylactic immunization of mice using these mannosylated dendrimers was observed, highlighting their utility as vaccine carriers to promote DC maturation and T cell activation [123].

A recent study reported that small structural changes in glycans, in this case a biantennary *N*-glycan and its xylosylated analog, can markedly influence CLR recognition, DC targeting, and subsequent T cell activation [124]. An additional approach to tackle the development of novel adjuvants for prophylactic or therapeutic cancer vaccination consists in the use of oligomannose-coated liposomes [76,118]. A comparison between mono-, di- and tetraantennary mannosyl lipid derivatives demonstrated a higher affinity and uptake by MR when multibranched glycoconjugates were displayed in the liposome [125].

Other CLRs selected for eliciting immune responses via multivalent ligand display were, for example, the macrophage galactose-type lectin (MGL) [126] and Dectin-1 [127]. In the former approach, a GalNAc modification strategy of antigens was employed to target DCs, while in the latter a complexation of antisense oligonucleotides with β-1,3-glucans enabled efficient incorporation into APCs expressing Dectin-1.

13.4 Galectins

Galectins are involved in numerous processes in the immune system and characterized by a common CRD composed of two extended antiparallel β-sheets that fold into a β-sandwich structure, which is responsible for glycan binding [128,129]. It is worth mentioning that within the CRD a highly conserved number of amino acids exists, accountable for glycan binding, termed carbohydrate-binding cassette [130]. Galectins are involved in a variety of functions, for example retaining pre-B cells in a developmental niche in the bone marrow stroma [131]; regulation of thymocyte selection via the strength of TCR signaling [132]; cell migration of DCs, neutrophils, and monocytes across the endothelium [133]; cytokine secretion and signaling [134]; priming or blockade of pathogen–host interactions [135]; and activation or inhibition of B and T cell apoptosis [136]. Based on their structure, there are three subfamilies of mammalian galectins [137–139] (Figure 13.4).

In the first subfamily (Gal-1, -2, -5, -7, -10, -11, -13, -14, -15), galectin-1 is the prototypical member that is present as a monomer and can associate into noncovalent homodimers. Dimerization takes place in an antiparallel manner and opposite to the CRD [137]. The second subfamily, termed chimera-type galectins, has a C-terminal CRD and a N-terminal domain containing 120–160 amino acids and is only constituted by galectin-3 in mammals [31]. Galectin-3 can either dimerize or multimerize (into pentamers) via interactions of the C-terminal CRD or the N-terminal domain, respectively [140,141]. The third subfamily (Gal-4, 6, 8, 9, and 12) encompasses tandem-repeat or bivalent galectins, which possess two distinct CRDs connected through a

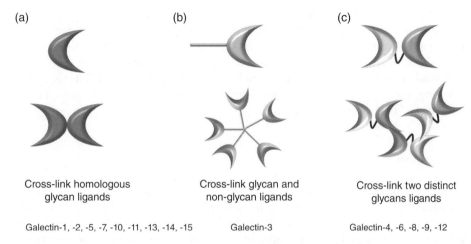

(a)

(b)

(c)

Cross-link homologous
glycan ligands

Cross-link glycan and
non-glycan ligands

Cross-link two distinct
glycans ligands

Galectin-1, -2, -5, -7, -10, -11, -13, -14, -15   Galectin-3   Galectin-4, -6, -8, -9, -12

Figure 13.4 Schematic representation of the three subfamilies of galectins. (a) Prototype galectins are able to form dimers (bottom), each monomer containing one carbohydrate recognition domain (top). (b) Chimera-type galectins can dimerize or multimerize (in pentamers, bottom) by interactions of the C-terminal CRD or N-terminal domain, respectively. (c) Tandem-repeat or bivalent galectins can bind to multiple glycan backbones in glycoproteins or glycolipids due to their ability to oligomerize via their N and C termini (bottom).

flexible peptide linker. The two CRDs in dimeric galectins may cross-link with adjacent dimeric galectins, potentiating their ligand recognition ability [142]. A key feature of galectins is their capacity to oligomerize, thus increasing binding avidity to multiple glycans on the backbone of glycoproteins and glycolipids [138]. Oligomerization and cross-linking of glycan ligands play a pivotal role in galectin-mediated cell–cell or cell–pathogen interactions [138,139]. The disaccharide lactosamine, Gal-β1,4GlcNAc, is the minimal glycan ligand for most mammalian galectins. Multiple binding of lactosamine sequences on complex *N*-glycans and O-glycans takes place in a biological context, driving either cell proliferation or cell cycle arrest depending on the number of glycans present and branches extension [138,139,143]. Moreover, galectins present affinity for β-galactosides involved in self- and non-self-antigen recognition [138].

Regarding self-antigen recognition, galectins play a prominent role in tumor progression by dampening DC maturation, impairing T cell functions and modulating NK cell activity, thus regulating tumor metastasis and tumor angiogenesis [144]. Galectin family members, such as galectin-1, galectin-3, and galectin-9, are positively associated with malignity of tumors, thus they represent attractive targets to disrupt galectin–glycan interactions to elicit antitumor immune responses [145,146].

Galectin-1 is secreted as a soluble protein and is involved in cell aggregation and tumor formation [144]. To target galectin-1, a lactose functionalized poly(amidoamine) (PAMAM) dendrimer was used as a multivalent framework in order to promote binding and further aggregation of galectin-1 [147]. The lactose glycodendrimers outcompeted galectin-1 binding to tumor cells and efficiently impaired galectin-1, thus preventing cellular aggregation and tumor formation [147]. In a different approach to target galectins, bivalent lactose-containing clusters were designed and structurally defined to enhance and promote optimal binding to galectins [148,149]. Two bivalent glycoclusters presented selectivity towards galectin-3 when an entire set of compounds was tested for binding to adhesion/growth-regulatory galectins from chicken [148,149].

Besides the antitumor strategies reported, galectins are also involved in viral evasion of the immune system, where galectin-1 facilitates HIV-1 attachment to host cell surface glycans [150]. Therefore, anti-viral strategies encompass the disruption of this interaction to inhibit viral invasion [151,152]. Moreover, galectin-1 was identified as a unique TCR modulator in thymocytes with the ability to recognize distinct TCR ligands, thus influencing positive or negative selection. This study highlights the role of galectin-1 in thymocyte fate determination [132].

13.5 Siglecs

The immunoglobulin superfamily (IgSF) comprises a family of receptors that exhibit a broad ligand specificity, as exemplified by antibodies and TCRs [153]. The immunoglobulin-type (I-type) lectins are a subset of the IgSF, in which the Siglecs represent the best characterized I-type lectins, being composed of an amino-terminal V-set immunoglobulin domain that binds to sialic acid and variable numbers of C2-set immunoglobulin domains [154–156]. Siglecs are type I membrane proteins that are categorized into two subsets according to their sequence homology and evolutionary conservation [154]. Siglecs remotely related ($\sim$ 25–30% sequence identity) have defined orthologs in all mammalian species and are represented by sialoadhesin (also known as Siglec-1 and CD169), Siglec-2 (CD22), myelin-associated glycoprotein (MAG; also known as Siglec-4), and Siglec-15 [153,154]. In addition, Siglecs that share $\sim$ 50–99% sequence identity, like the CD33-related Siglecs, also exist [157]. With the exception of resting T cells, Siglecs are expressed by most cell types in the human and murine immune system and they can be categorized into three different groups based on their features of the transmembrane and cytoplasmic tails [158]. Siglecs that lack inhibitory signaling cytosolic motifs and have neutral transmembrane domains constitute the first group, namely Siglec-1 and Siglec-4 [158]. Siglec-4 does not have an ITIM itself but has an ITIM-like tyrosine residue in its cytoplasmic tail. Both Siglecs are mainly involved in mediating cell adhesion events, whereas Siglec-1 also has a role in innate immunity and T cell activation [158]. The second group of Siglecs is the largest one and dampens immune responses via cytosolic ITIMs, with CD22 as the most prominent member that orchestrates both inhibitory and activating signals [159]. Ligand recognition in this family of Siglecs renders the cytosolic ITIM tyrosine and the ITIM-like tyrosine accessible to kinases that phosphorylate cytosolic ITIM tyrosine residues ultimately leading to the recruitment of tyrosine phosphatases, such as SHP-1 and SHP-2 [160]. The last category of Siglecs has a positively charged residue in the transmembrane anchor region and generally associates with a disulfide-linked homodimer of DAP12 (DNAX-associated protein of 12 kDa). These Siglec members contain a cytosolic ITAM that facilitates the recruitment of Syk family tyrosine kinases upon phosphorylation [155,158]. In humans, Siglecs belonging to this group are Siglec-14, -15, and -16 [158,161].

The main roles of Siglecs in innate immunity include: (i) internalization of sialic acid-expressing pathogens through phagocytosis; (ii) inhibitory signaling cascades to attenuate inflammatory responses to both sialic acid-expressing and non-sialylated pathogens; (iii) dampening of DAMP-mediated inflammation via inhibitory signaling; and (iv) exploiting sialic acid moieties on mucins and gangliosides to hinder NK cell activation [155,158,162].

Siglecs recognize glycans bearing a terminal sialic acid residue both on the same cell (in cis) or on other cells, glycoproteins, bacteria, parasites or viruses (in trans). Binding of Siglecs to different pathogens was reported, including HIV-1, *Campylobacter jejuni*, Group B *Streptococcus*, and *Neisseria meningitides* binding to Siglec-1, Siglec-7, Siglec-9, and Siglec-F, respectively [163–166]. Siglec–ligand interactions are of low affinity and due to competition with cis ligands, multivalent ligand presentation is necessary for trans interactions. Targeting of Siglecs has involved the use of multivalent polyacrylamide (PAA)-linked carbohydrate ligands [167]. Using viral capsids displaying a high affinity ligand for the Siglec CD22 in a spatially defined manner, it was possible to trigger an efficient targeting of B cells [168]. In conclusion, targeting Siglecs with multivalent constructs is a promising strategy to modulate immune responses. Substituents at position C5 and C9 of the sialic acid were shown to enhance or decrease Siglec binding affinity, therefore providing a basis to develop high-affinity ligands [169].

13.6 Conclusions

Glycan-based carrier systems, namely dendrimers, polymers, nanoparticles, and liposomes, have successfully been employed for targeting different families of lectins, such as CLRs, galectins, and Siglecs, to fine-tune functions of DCs or macrophages and to impact subsequent T cell activation. Glycan-based carrier systems have benefited from the advances made in the field of glycan array technology and novel synthetic strategies including chemoenzymatic methods and automated glycan synthesis, to enable access to more complex glycan structures for lectin targeting [170–172]. There are still open questions in the field, for instance: (i) the pros and cons of different glycan-based carrier systems; (ii) the nature of the carriers backbones (rigid or flexible); (iii) spacer length and display; and (iv) ligand density and its impact on targeting [43]. The basis for understanding the effects of multivalent glycan presentation towards lectin targeting *in vitro* and *in vivo* has been laid in recent years. The next steps should address in depth how multivalent lectin targeting can be turned into clinical applications.

Acknowledgment

We acknowledge funding from the European Union's Horizon 2020 research and innovation programme under the Marie Sklodowska-Curie grant agreement No. 642870 (ETN-Immunoshape).

References

1 Steinman, R. M. & Banchereau, J. Taking dendritic cells into medicine. *Nature* **449**, 419–426 (2007).
2 Seeberger, P. H. & Werz, D. B. Synthesis and medical applications of oligosaccharides. *Nature* **446**, 1046–1051 (2007).

3 Cummings, R. D. The repertoire of glycan determinants in the human glycome. *Mol. Biosyst.* **5**, 1087–1104 (2009).

4 Maverakis, E., Kim, K., Shimoda, M., *et al.* Glycans in the immune system and the altered glycan theory of autoimmunity: A critical review. *J. Autoimmun.* **57**, 1–13 (2015).

5 Apweiler, R., Hermjakob, H., & Sharon, N. On the frequency of protein glycosylation, as deduced from analysis of the SWISS-PROT database1. *Biochim. Biophys. Acta - Gen. Subj.* **1473**, 4–8 (1999).

6 Bernardi, A., Jiménez-Barbero, J., Casnati, A., *et al.* Multivalent glycoconjugates as anti-pathogenic agents. *Chem. Soc. Rev.* **42**, 4709–4727 (2013).

7 Krasnova, L. & Wong, C.-H. Understanding the chemistry and biology of glycosylation with glycan synthesis. *Annu. Rev. Biochem.* **85**, 599–630 (2016).

8 Haag, R. Multivalency as a chemical organization and action principle. *Beilstein J. Org. Chem.* **11**, 848–849 (2015).

9 Bhatia, S., Dimde, M., & Haag, R. Multivalent glycoconjugates as vaccines and potential drug candidates. *Med. Chem. Commun.* **5**, 862–878 (2014).

10 Rabinovich, G. A., van Kooyk, Y., & Cobb, B. A. Glycobiology of immune responses. *Ann. N. Y. Acad. Sci.* **1253**, 1–15 (2012).

11 Murphy, K., Travers, P., Walport, M., & Janeway, C. *Janeway's Immunobiology*. Garland Science (2012).

12 Mogensen, T. H. Pathogen recognition and inflammatory signaling in innate immune defenses. *Clin. Microbiol. Rev.* **22**, 240–273 (2009).

13 Bonilla, F. A. & Oettgen, H. C. Adaptive immunity. *J. Allergy Clin. Immunol.* **125**, S33–S40 (2010).

14 Chaplin, D. D. Overview of the immune response. *J. Allergy Clin. Immunol.* **125**, S3–S23 (2010).

15 Beutler, B. Innate immunity: An overview. *Mol. Immunol.* **40**, 845–859 (2004).

16 Janeway, C. A. & Medzhitov, R. Innate immune recognition. *Annu. Rev. Immunol.* **20**, 197–216 (2002).

17 Holers, V. M. Complement and its receptors: New insights into human disease. *Annu. Rev. Immunol.* **32**, 433–459 (2014).

18 Turvey, S. E. & Broide, D. H. Innate immunity. *J. Allergy Clin. Immunol.* **125**, S24–S32 (2010).

19 Klein, L., Hinterberger, M., Wirnsberger, G., & Kyewski, B. Antigen presentation in the thymus for positive selection and central tolerance induction. *Nat. Rev. Immunol.* **9**, 833–844 (2009).

20 Palmer, E. Negative selection – clearing out the bad apples from the T-cell repertoire. *Nat. Rev. Immunol.* **3**, 383–391 (2003).

21 Hogquist, K. A. & Jameson, S. C. The self-obsession of T cells: how TCR signaling thresholds affect fate 'decisions' and effector function. *Nat. Immunol.* **15**, 815–823 (2014).

22 Wucherpfennig, K. W., Gagnon, E., Call, M. J., *et al.* Structural biology of the T-cell receptor: insights into receptor assembly, ligand recognition, and initiation of signaling. *Cold Spring Harb. Perspect. Biol.* **2**, 1–14 (2010).

23 Hsieh, C.-S., Lee, H.-M., & Lio, C.-W. J. Selection of regulatory T cells in the thymus. *Nat. Rev. Immunol.* **12**, 157–167 (2012).

24 Germain, R. N. T-cell development and the CD4-CD8 lineage decision. *Nat. Rev. Immunol.* **2**, 309–322 (2002).

25 Lee, M. S. & Kim, Y.-J. Signaling pathways downstream of pattern-recognition receptors and their cross talk. *Annu. Rev. Biochem.* **76**, 447–480 (2007).

26 Takeuchi, O. & Akira, S. Pattern recognition receptors and inflammation. *Cell* **140**, 805–820 (2010).

27 Akira, S., Uematsu, S., & Takeuchi, O. Pathogen recognition and innate immunity. *Cell* **124**, 783–801 (2006).

28 Loo, Y.-M. & Gale Jr, M. Immune signaling by RIG-I-like receptors. *Immunity* **34**, 680–692 (2016).

29 Chen, G., Shaw, M. H., Kim, Y.-G., & Nuñez, G. NOD-like receptors: role in innate immunity and inflammatory disease. *Annu. Rev. Pathol.* **4**, 365–398 (2009).

30 Takaoka, A., Wang, Z., Choi, M. K., *et al.* DAI (DLM-1/ZBP1) is a cytosolic DNA sensor and an activator of innate immune response. *Nature* **448**, 501–505 (2007).

31 Gabius, H.-J. Animal lectins. *Eur. J. Biochem.* **243**, 543–576 (1997).

32 Dam, T. K. & Brewer, C. F. Lectins as pattern recognition molecules: The effects of epitope density in innate immunity. *Glycobiology* **20**, 270–279 (2009).

33 Smith, D. F. & Cummings, R. D. Application of microarrays for deciphering the structure and function of the human glycome. *Mol. Cell. Proteomics* **12**, 902–12 (2013).

34 Liu, Y., Liu, J., Pang, X., *et al.* The roles of direct recognition by animal lectins in antiviral immunity and viral pathogenesis. *Molecules* **20**, 2272–2295 (2015).

35 Gupta, G. S. Lectins: An overview. In: *Animal Lectins: Form, Function and Clinical Applications* (ed. Gupta, G. S.) 3–25. Springer (2012).

36 Steinman, R. M. & Hemmi, H. Dendritic cells: Translating innate to adaptive immunity. In: *From Innate Immunity to Immunological Memory* (eds Pulendran, B. & Ahmed, R.) 17–58. Springer (2006).

37 Joffre, O. P., Segura, E., Savina, A., & Amigorena, S. Cross-presentation by dendritic cells. *Nat. Rev. Immunol.* **12**, 557–569 (2012).

38 Blander, J. M. The comings and goings of MHC class I molecules herald a new dawn in cross-presentation. *Immunol. Rev.* **272**, 65–79 (2016).

39 Zhu, J. & Paul, W. E. CD4 T cells: fates, functions, and faults. *Blood* **112**, 1557–1569 (2008).

40 Reiser, J. & Banerjee, A. Effector, memory, and dysfunctional CD8(+) T cell fates in the antitumor immune response. *J. Immunol. Res.* **2016**, 8941260 (2016).

41 Burgdorf, S., Kautz, A., Böhnert, V., *et al.* Distinct pathways of antigen uptake and intracellular routing in CD4 and CD8 T cell activation. *Science* **316**, 612–616 (2007).

42 Annunziato, F., Romagnani, C., & Romagnani, S. The 3 major types of innate and adaptive cell-mediated effector immunity. *J. Allergy Clin. Immunol.* **135**, 626–635 (2015).

43 Lepenies, B., Lee, J., & Sonkaria, S. Targeting C-type lectin receptors with multivalent carbohydrate ligands. *Adv. Drug Deliv. Rev.* **65**, 1271–1281 (2013).

44 Hsieh, C. S., Macatonia, S. E., Tripp, C. S., *et al.* Development of TH1 CD4+ T cells through IL-12 produced by Listeria-induced macrophages. *Science* **260**, 547–549 (1993).

45 Macatonia, S. E., Hosken, N. A., Litton, M., *et al.* Dendritic cells produce IL-12 and direct the development of Th1 cells from naive CD4+ T cells. *J. Immunol.* **154**, 5071–5079 (1995).

46 Swain, S. L., Weinberg, A. D., English, M., & Huston, G. IL-4 directs the development of Th2-like helper effectors. *J. Immunol.* **145**, 3796–3806 (1990).

47 Curtis, M. M. & Way, S. S. Interleukin-17 in host defence against bacterial, mycobacterial and fungal pathogens. *Immunology* **126**, 177–185 (2009).

48 Jonuleit, H. & Schmitt, E. The regulatory T cell family: Distinct subsets and their interrelations. *J. Immunol.* **171**, 6323–6327 (2003).

49 Barry, M. & Bleackley, R. C. Cytotoxic T lymphocytes: all roads lead to death. *Nat. Rev. Immunol.* **2**, 401–409 (2002).

50 Drickamer, K. & Taylor, M. E. Recent insights into structures and functions of C-type lectins in the immune system. *Curr. Opin. Struct. Biol.* **34**, 26–34 (2015).

51 Drickamer, K. C-type lectin-like domains. *Curr. Opin. Struct. Biol.* **9**, 585–590 (1999).

52 Zelensky, A. N. & Gready, J. E. The C-type lectin-like domain superfamily. *FEBS J.* **272**, 6179–6217 (2005).

53 Dambuza, I. M. & Brown, G. D. C-type lectins in immunity: recent developments. *Curr. Opin. Immunol.* **32**, 21–27 (2015).

54 Robinson, M. J., Sancho, D., Slack, E. C., *et al.* Myeloid C-type lectins in innate immunity. *Nat. Immunol.* **7**, 1258–1265 (2006).

55 Neumann, K., Castiñeiras-Vilariño, M., Höckendorf, U., *et al.* Clec12a is an inhibitory receptor for uric acid crystals that regulates inflammation in response to cell death. *Immunity* **40**, 389–399 (2016).

56 Kiyotake, R., Oh-hora, M., Ishikawa, E., *et al.* Human Mincle binds to cholesterol crystals and triggers innate immune responses. *J. Biol. Chem.* **290**, 25322–25332 (2015).

57 Drummond, R. A. & Brown, G. D. Signalling C-type lectins in antimicrobial immunity. *PLoS Pathog.* **9**, e1003417 (2013).

58 Hardison, S. E. & Brown, G. D. C-type lectin receptors orchestrate anti-fungal immunity. *Nat. Immunol.* **13**, 817–822 (2012).

59 Yan, H., Ohno, N., & Tsuji, N. M. The role of C-type lectin receptors in immune homeostasis. *Int. Immunopharmacol.* **16**, 353–357 (2013).

60 Sancho, D. & Reis e Sousa, C. Signaling by myeloid C-type lectin receptors in immunity and homeostasis. *Annu. Rev. Immunol.* **30**, 491–529 (2012).

61 Geijtenbeek, T. B. H. & Gringhuis, S. I. Signalling through C-type lectin receptors: shaping immune responses. *Nat. Rev. Immunol.* **9**, 465–479 (2009).

62 Hoving, J. C., Wilson, G. J., & Brown, G. D. Signalling C-type lectin receptors, microbial recognition and immunity. *Cell. Microbiol.* **16**, 185–194 (2014).

63 Kerscher, B., Willment, J. A., & Brown, G. D. The Dectin-2 family of C-type lectin-like receptors: an update. *Int. Immunol.* **25**, 271–277 (2013).

64 Plato, A., Willment, J. A., & Brown, G. D. C-type lectin-like receptors of the Dectin-1 cluster: Ligands and signaling pathways. *Int. Rev. Immunol.* **32**, 134–156 (2013).

65 Kerrigan, A. M. & Brown, G. D. Syk-coupled C-type lectins in immunity. *Trends Immunol.* **32**, 151–156 (2011).

66 Osorio, F. & Reis e Sousa, C. Myeloid C-type lectin receptors in pathogen recognition and host defense. *Immunity* **34**, 651–664 (2011).

67 Zhu, L.-L., Zhao, X.-Q., Jiang, C., *et al.* C-type lectin receptors Dectin-3 and Dectin-2 form a heterodimeric pattern-recognition receptor for host defense against fungal infection. *Immunity* **39**, 324–334 (2013).

68 Wevers, B. A., Kaptein, T. M., Zijlstra-Willems, E. M., *et al.* Fungal engagement of the C-type lectin Mincle suppresses Dectin-1-induced antifungal immunity. *Cell Host Microbe* **15**, 494–505 (2014).

69 Ifrim, D. C., Joosten, L. A. B., Kullberg, B.-J., *et al. Candida albicans* primes TLR cytokine responses through a Dectin-1/Raf-1–mediated pathway. *J. Immunol.* **190**, 4129–4135 (2013).

70 Lambert, A. A., Barabé, F., Gilbert, C., & Tremblay, M. J. DCIR-mediated enhancement of HIV-1 infection requires the ITIM-associated signal transduction pathway. *Blood* **117**, 6589–6599 (2011).

71 Marshall, A. S. J., Willment, J. A., Pyż, E., *et al.* Human MICL (CLEC12A) is differentially glycosylated and is down-regulated following cellular activation. *Eur. J. Immunol.* **36**, 2159–2169 (2006).

72 Nuñez-Prado, N., Compte, M., Harwood, S., *et al.* The coming of age of engineered multivalent antibodies. *Drug Discov. Today* **20**, 588–594 (2015).

73 Tacken, P. J. & Figdor, C. G. Targeted antigen delivery and activation of dendritic cells in vivo: Steps towards cost effective vaccines. *Semin. Immunol.* **23**, 12–20 (2011).

74 Sancho, D., Mourão-Sá, D., Joffre, O.P., *et al.* Tumor therapy in mice via antigen targeting to a novel, DC-restricted C-type lectin. *J. Clin. Invest.* **118**, 2098–2110 (2008).

75 Klechevsky, E., Flamar, A.-L., Cao, Y., *et al.* Cross-priming CD8+ T cells by targeting antigens to human dendritic cells through DCIR. *Blood* **116**, 1685–1697 (2010).

76 Reddy, S. T., Swartz, M. A., & Hubbell, J. A. Targeting dendritic cells with biomaterials: developing the next generation of vaccines. *Trends Immunol.* **27**, 573–579 (2006).

77 Hubbell, J. A., Thomas, S. N., & Swartz, M. A. Materials engineering for immunomodulation. *Nature* **462**, 449–460 (2009).

78 Johannssen, T. & Lepenies, B. Identification and characterization of carbohydrate-based adjuvants. In: *Carbohydrate-Based Vaccines SE – 11* (ed. Lepenies, B.) **1331**, 173–187. Springer (2015).

79 Johannssen, T. & Lepenies, B. Glycan-based cell targeting to modulate immune responses. *Trends Biotechnol.* **35**, 334–346 (2017).

80 Payne, R. J. & Wong, C.-H. Advances in chemical ligation strategies for the synthesis of glycopeptides and glycoproteins. *Chem. Commun.* **46**, 21–43 (2010).

81 Rendle, P. M., Seger, A., Rodrigues, J., *et al.* Glycodendriproteins: A synthetic glycoprotein mimic enzyme with branched sugar-display potently inhibits bacterial aggregation. *J. Am. Chem. Soc.* **126**, 4750–4751 (2004).

82 Issei, O., Blanchard, B., Borsali, R., *et al.* Enhancement of plant and bacterial lectin binding affinities by three-dimensional organized cluster glycosides constructed on helical poly(phenylacetylene) backbones. *ChemBioChem* **11**, 2399–2408 (2010).

83 Ponader, D., Wojcik, F., Beceren-Braun, F., *et al.* Sequence-defined glycopolymer segments presenting mannose: Synthesis and lectin binding affinity. *Biomacromolecules* **13**, 1845–1852 (2012).

84 Andre, S., Grandjean, C., Gautier, F.-M., *et al.* Combining carbohydrate substitutions at bioinspired positions with multivalent presentation towards optimising lectin inhibitors: case study with calixarenes. *Chem. Commun.* **47**, 6126–6128 (2011).

85 Dondoni, A. & Marra, A. Calixarene and calixresorcarene glycosides: Their synthesis and biological applications. *Chem. Rev.* **110**, 4949–4977 (2010).

86 Heidecke, C. D. & Lindhorst, T. K. Iterative synthesis of spacered glycodendrons as oligomannoside mimetics and evaluation of their antiadhesive properties. *Chem. Eur. J.* **13**, 9056–9067 (2007).

87 Mintzer, M. A., Dane, E. L., O'Toole, G. A., & Grinstaff, M. W. Exploiting dendrimer multivalency to combat emerging and re-emerging infectious diseases. *Mol. Pharm.* **9**, 342–354 (2012).

88 Holler, M., Allenbach, N., Sonet, J., & Nierengarten, J.-F. The high yielding synthesis of pillar[5]arenes under Friedel-Crafts conditions explained by dynamic covalent bond formation. *Chem. Commun.* **48**, 2576–2578 (2012).

89 Vincent, S. P., Buffet, K., Nierengarten, I., *et al.* Biologically active heteroglycoclusters constructed on a pillar[5]arene-containing [2]rotaxane scaffold. *Chem. Eur. J.* **22**, 88–92 (2016).

90 Marradi, M., Martín-Lomas, M., & Penadés, S. Glyconanoparticles: Polyvalent tools to study carbohydrate-based interactions. In: *Advances in Carbohydrate Chemistry and Biochemistry* (ed. Horton, D.) **64**, 211–290. Academic Press (2010).

91 Wang, X., Matei, E., Deng, L., *et al.* Multivalent glyconanoparticles with enhanced affinity to the anti-viral lectin Cyanovirin-N. *Chem. Commun.* **47**, 8620–8622 (2011).

92 Curtis, B. M., Scharnowske, S., & Watson, A. J. Sequence and expression of a membrane-associated C-type lectin that exhibits CD4-independent binding of human immunodeficiency virus envelope glycoprotein gp120. *Proc. Natl Acad. Sci. USA* **89**, 8356–8360 (1992).

93 Geijtenbeek, T. B. H., Torensma, R., van Vliet, S. J., *et al.* Identification of DC-SIGN, a novel dendritic cell–specific ICAM-3 receptor that supports primary immune responses. *Cell* **100**, 575–585 (2000).

94 Geijtenbeek, T. B. H., Kwon, D. S., Torensma, R., *et al.* DC-SIGN, a dendritic cell–specific HIV-1-binding protein that enhances trans-infection of T cells. *Cell* **100**, 587–597 (2000).

95 Soilleux, E. J., Morris, L. S., Leslie, G., *et al.* Constitutive and induced expression of DC-SIGN on dendritic cell and macrophage subpopulations in situ and in vitro. *J. Leukoc. Biol.* **71**, 445–457 (2002).

96 García-Vallejo, J. J., Ilarregui, J. M., Kalay, H., *et al.* CNS myelin induces regulatory functions of DC-SIGN–expressing, antigen-presenting cells via cognate interaction with MOG. *J. Exp. Med.* **211**, 1465–1483 (2014).

97 Feinberg, H., Guo, Y., Mitchell, D. A., *et al.* Extended neck regions stabilize tetramers of the receptors DC-SIGN and DC-SIGNR. *J. Biol. Chem.* **280**, 1327–1335 (2005).

98 Guo, Y., Feinberg, H., Conroy, E., *et al.* Structural basis for distinct ligand-binding and targeting properties of the receptors DC-SIGN and DC-SIGNR. *Nat. Struct. Mol. Biol.* **11**, 591–598 (2004).

99 van Liempt, E., Bank, C. M. C., Mehta, P., *et al.* Specificity of DC-SIGN for mannose- and fucose-containing glycans. *FEBS Lett.* **580**, 6123–6131 (2006).

100 Feinberg, H., Mitchell, D. A., Drickamer, K., & Weis, W. I. Structural basis for selective recognition of oligosaccharides by DC-SIGN and DC-SIGNR. *Science* **294**, 2163–2166 (2001).

101 Adams, E. W., Ratner, D. M., Bokesch, H. R., *et al.* Oligosaccharide and glycoprotein microarrays as tools in HIV glycobiology: Glycan-dependent gp120/protein interactions. *Chem. Biol.* **11**, 875–881 (2004).

102 Engering, A., Geijtenbeek, T. B. H., van Vliet, S. J., *et al.* The dendritic cell-specific adhesion receptor DC-SIGN internalizes antigen for presentation to T cells. *J. Immunol.* **168**, 2118–2126 (2002).

103 Simmons, G., Reeves, J. D., Grogan, C. C., *et al.* DC-SIGN and DC-SIGNR bind Ebola glycoproteins and enhance infection of macrophages and endothelial cells. *Virology* **305**, 115–123 (2003).

104 Tassaneetrithep, B., Burgess, T. H., Granelli-Piperno, A., *et al.* DC-SIGN (CD209) mediates dengue virus infection of human dendritic cells. *J. Exp. Med.* **197**, 823–829 (2003).

105 Ludwig, I. S., Lekkerkerker, A. N., Depla, E., *et al.* Hepatitis C virus targets DC-SIGN and L-SIGN to escape lysosomal degradation. *J. Virol.* **78**, 8322–8332 (2004).

106 Tailleux, L., Schwartz, O., Herrmann, J.-L., *et al.* DC-SIGN is the major *Mycobacterium tuberculosis* receptor on human dendritic cells. *J. Exp. Med.* **197**, 121–127 (2003).

107 Cambi, A., Netea, M. G., Mora-Montes, H. M., *et al.* Dendritic cell interaction with *Candida albicans* critically depends on N-linked mannan. *J. Biol. Chem.* **283**, 20590–20599 (2008).

108 Meyer, S., van Liempt, E., Imberty, A., *et al.* DC-SIGN mediates binding of dendritic cells to authentic pseudo-LewisY glycolipids of *Schistosoma mansoni* Cercariae, the first parasite-specific ligand of DC-SIGN. *J. Biol. Chem.* **280**, 37349–37359 (2005).

109 Alvarez, C. P., Lasala, F., Carrillo, J., *et al.* C-type lectins DC-SIGN and L-SIGN mediate cellular entry by Ebola virus in cis and in trans. *J. Virol.* **76**, 6841–6844 (2002).

110 van Kooyk, Y. & Geijtenbeek, T. B. H. DC-SIGN: escape mechanism for pathogens. *Nat. Rev. Immunol.* **3**, 697–709 (2003).

111 Gringhuis, S. I., den Dunnen, J., Litjens, M., *et al.* Carbohydrate-specific signaling through the DC-SIGN signalosome tailors immunity to *Mycobacterium tuberculosis*, HIV-1 and *Helicobacter pylori. Nat. Immunol.* **10**, 1081–1088 (2009).

112 den Dunnen, J., Gringhuis, S. I., & Geijtenbeek, T. B. H. Innate signaling by the C-type lectin DC-SIGN dictates immune responses. *Cancer Immunol. Immunother.* **58**, 1149–1157 (2009).

113 Martínez-Ávila, O., Hijazi, K., Marradi, M., *et al.* Gold manno-glyconanoparticles: Multivalent systems to block HIV-1 gp120 binding to the lectin DC-SIGN. *Chem. Eur. J.* **15**, 9874–9888 (2009).

114 Arnáiz, B., Martínez-Ávila, O., Falcon-Perez, J. M., & Penadés, S. Cellular uptake of gold nanoparticles bearing HIV gp120 oligomannosides. *Bioconjug. Chem.* **23**, 814–825 (2012).

115 Berzi, A., Reina, J. J., Ottria, R., *et al.* A glycomimetic compound inhibits DC-SIGN-mediated HIV infection in cellular and cervical explant models. *AIDS* **26**, 127–137 (2012).

116 Sattin, S., Daghetti, A., Thépaut, M., *et al.* Inhibition of DC-SIGN-mediated HIV infection by a linear trimannoside mimic in a tetravalent presentation. *ACS Chem. Biol.* **5**, 301–312 (2010).

117 Luczkowiak, J., Muñoz, A., Sánchez-Navarro, M., *et al.* Glycofullerenes inhibit viral infection. *Biomacromolecules* **14**, 431–437 (2013).

118 Unger, W. W. J., van Beelen, A. J., Bruijns, S. C., *et al.* Glycan-modified liposomes boost CD4+ and CD8+ T-cell responses by targeting DC-SIGN on dendritic cells. *J. Control. Release* **160**, 88–95 (2012).

119 Taylor, M. E., Conary, J. T., Lennartz, M. R., *et al.* Primary structure of the mannose receptor contains multiple motifs resembling carbohydrate-recognition domains. *J. Biol. Chem.* **265**, 12156–12162 (1990).

120 McKenzie, E. J., Taylor, P. R., Stillion, R. J., *et al.* Mannose receptor expression and function define a new population of murine dendritic cells. *J. Immunol.* **178**, 4975–4983 (2007).

121 Vautier, S., MacCallum, D. M., & Brown, G. D. C-type lectin receptors and cytokines in fungal immunity. *Cytokine* **58**, 89–99 (2012).

122 Shiao, T. C. & Roy, R. Glycodendrimers as functional antigens and antitumor vaccines. *New J. Chem.* **36**, 324–339 (2012).

123 Sheng, K.-C., Kalkanidis, M., Pouniotis, D. S., *et al.* Delivery of antigen using a novel mannosylated dendrimer potentiates immunogenicity in vitro and in vivo. *Eur. J. Immunol.* **38**, 424–436 (2008).

124 Brzezicka, K., Vogel, U., Serna, S., *et al.* Influence of core β-1,2-xylosylation on glycoprotein recognition by murine C-type lectin receptors and its impact on dendritic cell targeting. *ACS Chem. Biol.* **11**, 2347–2356 (2016).

125 Espuelas, S., Thumann, C., Heurtault, B., *et al.* Influence of ligand valency on the targeting of immature human dendritic cells by mannosylated liposomes. *Bioconjug. Chem.* **19**, 2385–2393 (2008).

126 Freire, T., Zhang, X., Dériaud, E., *et al.* Glycosidic Tn-based vaccines targeting dermal dendritic cells favor germinal center B-cell development and potent antibody response in the absence of adjuvant. *Blood* **116**, 3526–3536 (2010).

127 Mochizuki, S. & Sakurai, K. Dectin-1 targeting delivery of TNF-α antisense ODNs complexed with β-1,3-glucan protects mice from LPS-induced hepatitis. *J. Control. Release* **151**, 155–161 (2011).

128 Di Lella, S., Sundblad, V., Cerliani, J. P., *et al.* When galectins recognize glycans: From biochemistry to physiology and back again. *Biochemistry* **50**, 7842–7857 (2011).

129 Vasta, G. R., Ahmed, H., Bianchet, M. A., *et al.* Diversity in recognition of glycans by F-type lectins and galectins: molecular, structural, and biophysical aspects. *Ann. N. Y. Acad. Sci.* **1253**, E14–E26 (2012).

130 Barondes, S. H., Castronovo, V., Cooper, D. N. W., *et al.* Galectins: A family of animal β-galactoside-binding lectins. *Cell* **76**, 597–598 (1994).

131 Bonzi, J., Bornet, O., Betzi, S., *et al.* Pre-B cell receptor binding to galectin-1 modifies galectin-1/carbohydrate affinity to modulate specific galectin-1/glycan lattice interactions. *Nat. Commun.* **6** (2015).

132 Liu, S. D., Whiting, C. C., Tomassian, T., *et al.* Endogenous galectin-1 enforces class I–restricted TCR functional fate decisions in thymocytes. *Blood* **112**, 120–130 (2008).

133 Cooper, D., Iqbal, A. J., Gittens, B. R., *et al.* The effect of galectins on leukocyte trafficking in inflammation: sweet or sour? *Ann. N. Y. Acad. Sci.* **1253**, 181–192 (2012).

134 Liu, F.-T. & Rabinovich, G. A. Galectins: regulators of acute and chronic inflammation. *Ann. N. Y. Acad. Sci.* **1183**, 158–182 (2010).

135 Baum, L. G., Garner, O. B., Schaefer, K., & Lee, B. Microbe–host interactions are positively and negatively regulated by galectin–glycan interactions. *Front. Immunol.* **5**, 284 (2014).

136 Hsu, D. K., Yang, R. & Liu, F. Galectins in apoptosis. In: *Methods in Enzymology* (eds Colowick, S. P. and Kaplan, N. O.) **417**, 256–273. Academic Press (2006).

137 Hirabayashi, J. & Kasai, K. The family of metazoan metal-independent β-galactoside-binding lectins: structure, function and molecular evolution. *Glycobiol.* **3**, 297–304 (1993).

138 Thiemann, S. & Baum, L. G. Galectins and immune responses—Just how do they do those things they do? *Annu. Rev. Immunol.* **34**, annurev-immunol-041015-055402 (2016).

139 Boscher, C., Dennis, J. W., & Nabi, I. R. Glycosylation, galectins and cellular signaling. *Curr. Opin. Cell Biol.* **23**, 383–392 (2011).

140 Halimi, H., Rigato, A., Byrne, D., *et al.* Glycan dependence of Galectin-3 self-association properties. *PLoS One* **9**, e111836 (2014).

141 Ahmad, N., Gabius, H.-J., André, S., *et al.* Galectin-3 precipitates as a pentamer with synthetic multivalent carbohydrates and forms heterogeneous cross-linked complexes. *J. Biol. Chem.* **279**, 10841–10847 (2004).

142 Brewer, C. F., Miceli, M. C., & Baum, L. G. Clusters, bundles, arrays and lattices: novel mechanisms for lectin–saccharide-mediated cellular interactions. *Curr. Opin. Struct. Biol.* **12**, 616–623 (2002).

143 Lau, K. S., Partridge, E. A., Grigorian, A., *et al.* Complex N-glycan number and degree of branching cooperate to regulate cell proliferation and differentiation. *Cell* **129**, 123–134 (2007).

144 Cousin, J. M. & Cloninger, M. J. The role of Galectin-1 in cancer progression, and synthetic multivalent systems for the study of Galectin-1. *Int. J. Mol. Sci.* **17**, 1566 (2016).

145 Hockl, P. F., Wolosiuk, A., Pérez-Sáez, J. M., *et al.* Glyco-nano-oncology: Novel therapeutic opportunities by combining small and sweet. *Pharmacol. Res.* **109**, 45–54 (2016).

146 Punt, S., Thijssen, V. L., Vrolijk, J., *et al.* Galectin-1, -3 and -9 expression and clinical significance in squamous cervical cancer. *PLoS One* **10**, e0129119 (2015).

147 Cousin, J. M. & Cloninger, M. J. Glycodendrimers: tools to explore multivalent galectin-1 interactions. *Beilstein J. Org. Chem.* **11**, 739–747 (2015).

148 Leyden, R., Velasco-Torrijos, T., André, S., *et al.* Synthesis of bivalent lactosides based on terephthalamide, N,N′-diglucosylterephthalamide, and glycophane scaffolds and assessment of their inhibitory capacity on medically relevant lectins. *J. Org. Chem.* **74**, 9010–9026 (2009).

149 André, S., Jarikote, D. V., Yan, D., *et al.* Synthesis of bivalent lactosides and their activity as sensors for differences between lectins in inter- and intrafamily comparisons. *Bioorg. Med. Chem. Lett.* **22**, 313–318 (2012).

150 Sato, S., Ouellet, M., St-Pierre, C., & Tremblay, M. J. Glycans, galectins, and HIV-1 infection. *Ann. N. Y. Acad. Sci.* **1253**, 133–148 (2012).

151 Ouellet, M., Mercier, S., Pelletier, I., *et al.* Galectin-1 Acts as a soluble host factor that promotes HIV-1 infectivity through stabilization of virus attachment to host cells. *J. Immunol.* **174**, 4120–4126 (2005).

152 Vasta, G. R. Roles of galectins in infection. *Nat. Rev. Microbiol.* **7**, 424–438 (2009).

153 Crocker, P. R., Paulson, J. C., & Varki, A. Siglecs and their roles in the immune system. *Nat. Rev. Immunol.* **7**, 255–266 (2007).

154 Varki, A. & Angata, T. Siglecs—the major subfamily of I-type lectins. *Glycobiol.* **16**, 1R–27R (2006).

155 Bochner, B. S. & Zimmermann, N. Role of siglecs and related glycan-binding proteins in immune responses and immunoregulation. *J. Allergy Clin. Immunol.* **135**, 598–608 (2015).

156 Angata, T. & Brinkman-Van der Linden, E. C. M. I-type lectins. *Biochim. Biophys. Acta - Gen. Subj.* **1572**, 294–316 (2002).

157 Angata, T., Margulies, E. H., Green, E. D., & Varki, A. Large-scale sequencing of the CD33-related Siglec gene cluster in five mammalian species reveals rapid evolution by multiple mechanisms. *Proc. Natl Acad. Sci. USA* **101**, 13251–13256 (2004).

158 Pillai, S., Netravali, I. A., Cariappa, A., & Mattoo, H. Siglecs and immune regulation. *Annu. Rev. Immunol.* **30**, 357–392 (2012).

159 Tsubata, T. Siglec-2 is a key molecule for immune response. In: *Experimental Glycoscience: Glycobiology* (eds Taniguchi, N. *et al.*) 167–170. Springer (2008).

160 Taylor, V. C., Buckley, C. D., Douglas, M., *et al.* The myeloid-specific sialic acid-binding receptor, CD33, associates with the protein-tyrosine phosphatases, SHP-1 and SHP-2. *J. Biol. Chem.* **274**, 11505–11512 (1999).

161 Angata, T., Tabuchi, Y., Nakamura, K., & Nakamura, M. Siglec-15: an immune system Siglec conserved throughout vertebrate evolution. *Glycobiol.* **17**, 838–846 (2007).

162 Paulson, J. C., Macauley, M. S., & Kawasaki, N. Siglecs as sensors of self in innate and adaptive immune responses. *Ann. N. Y. Acad. Sci.* **1253**, 37–48 (2012).

163 van der Kuyl, A. C., van den Burg, R., Zorgdrager, F., *et al.* Sialoadhesin (CD169) expression in CD14+ cells is upregulated early after HIV-1 infection and increases during disease progression. *PLoS One* **2**, e257 (2007).

164 Avril, T., Wagner, E. R., Willison, H. J., & Crocker, P. R. Sialic acid-binding immunoglobulin-like lectin 7 mediates selective recognition of sialylated glycans expressed on *Campylobacter jejuni* lipooligosaccharides. *Infect. Immun.* **74**, 4133–4141 (2006).

165 Carlin, A. F., Uchiyama, S., Chang, Y.-C., *et al.* Molecular mimicry of host sialylated glycans allows a bacterial pathogen to engage neutrophil Siglec-9 and dampen the innate immune response. *Blood* **113**, 3333–3336 (2009).

166 Jones, C., Virji, M., & Crocker, P. R. Recognition of sialylated meningococcal lipopolysaccharide by siglecs expressed on myeloid cells leads to enhanced bacterial uptake. *Mol. Microbiol.* **49**, 1213–1225 (2003).

167 Rapoport, E. M., Pazynina, G. V, Sablina, M. A., *et al.* Probing sialic acid binding Ig-like lectins (siglecs) with sulfated oligosaccharides. *Biochem.* **71**, 496–504 (2006).

168 Kaltgrad, E., O'Reilly, M. K., Liao, L., *et al.* On-virus construction of polyvalent glycan ligands for cell-surface receptors. *J. Am. Chem. Soc.* **130**, 4578–4579 (2008).

169 O'Reilly, M. K. & Paulson, J. C. Multivalent ligands for Siglecs. *Methods Enzymol.* **478**, 343–363 (2010).

170 Seeberger, P. H. Automated oligosaccharide synthesis. *Chem. Soc. Rev.* **37**, 19–28 (2008).

171 Miranda, O. R., Creran, B., & Rotello, V. M. Array-based sensing with nanoparticles: 'Chemical noses' for sensing biomolecules and cell surfaces. *Curr. Opin. Chem. Biol.* **14**, 728–736 (2010).

172 Arthur, C. M., Cummings, R. D., & Stowell, S. R. Using glycan microarrays to understand immunity. *Curr. Opin. Chem. Biol.* **18**, 55–61 (2014).

14

Blocking Disease Linked Lectins with Multivalent Carbohydrates

Marjon Stel and Roland J. Pieters

Department of Chemical Biology & Drug Discovery, Utrecht Institute for Pharmaceutical Sciences, Utrecht University, PO Box 80082, 3508 TB Utrecht, the Netherlands

14.1 Introduction

Lectins are carbohydrate-binding proteins involved in various biological phenomena of life. Since every cell surface is covered with oligosaccharide structures, a layer called the glycocalyx, lectins play an important role in signalling pathways and cell adhesion [1,2]. Carbohydrate–lectin binding often depends on multivalent interactions because the affinity of one single binding event between a lectin and a monovalent carbohydrate ligand is often quite low. However, several binding interactions together can reach high affinities [3,4]. For this reason, synthetic ligands targeting lectins benefit from multivalency, as we will see in this chapter.

Inhibiting binding of these lectins to their natural ligands can possibly prevent the diseases they are involved in. In this chapter we will discuss a selection of lectins and their inhibitors. For example, the lectins LecA and LecB from the bacterial pathogen *Pseudomonas aeruginosa*, the hemagglutinin protein of the influenza A virus and the cholera toxin (CT) from the bacterium *Vibrio cholerae* are all involved in infectious disease. Human galectins play a role in inflammation and numerous cancer mechanisms. In this chapter, we will cover a variety of ligands that bind these proteins, aiming to inhibit their natural function which will hopefully lead to new therapeutics. In addition, we will cover Concanavalin A as one of the most studied lectins and, finally, a selection of propeller lectins that can have up to ten carbohydrate binding domains arranged symmetrically resembling a propeller.

Binding of multivalent ligands to lectins can occur through a number of mechanisms, the most important of which are mentioned here (Figure 14.1) [5,6]. Chelation of a ligand is possible when the ligand can bridge the distance between two binding sites on the same protein or protein complex. Crystal structures of these proteins provide detailed information on the distance between binding sites, allowing for the design of sufficiently long linkers and scaffolds. An alternative mechanism is statistical rebinding, where one carbohydrate binding site is freed up by its carbohydrate ligand, but due to multiple ligands presented closely together, another ligand can bind immediately after,

Multivalency: Concepts, Research & Applications, First Edition. Edited by Jurriaan Huskens, Leonard J. Prins, Rainer Haag, and Bart Jan Ravoo.
© 2018 John Wiley & Sons Ltd. Published 2018 by John Wiley & Sons Ltd.

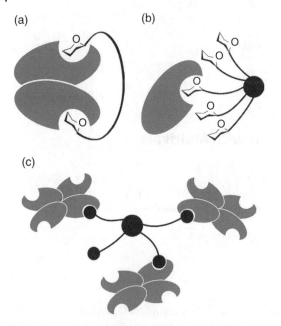

(a)

(b)

(c)

Figure 14.1 Three major binding mechanisms of lectin–carbohydrate systems: (a) chelation; (b) statistical rebinding; and (c) aggregation.

leading to low overall off rates. For this mechanism, only the ligand has to be multivalent, the protein does not, or may have multiple binding sites too far apart to be bridged. For these cases aggregation is also often observed, the ligand then bridges binding sites on different proteins, forming a network of proteins and ligands, leading to larger species that may precipitate out of solution. It is less clear what the possible contribution of the aggregation is to the inhibitory potency.

Detecting binding of ligands to these lectins can be done through widely used techniques such as isothermal calorimetry (ITC) providing thermodynamic parameters and stoichiometry, and surface plasmon resonance (SPR) providing insight into the association and dissociation rates. In the case of aggregation of ligands and lectins, sedimentation velocity analytical ultracentrifugation (SV-AUC) and dynamic light scattering (DLS) can provide information about the size of the particles formed. Atomic force microscopy (AFM) has been used for similar purposes, and is able to show the shape of the particles. NMR and X-ray crystallography provide atomic detail not only of the proteins themselves, allowing detailed design of ligands, but also of the specific interactions between protein and ligand. In addition, a variety of lectin specific assays have been developed [1,4]. As this chapter deals with carbohydrate–lectin binding, a short overview will be given here.

The haemagglutination inhibition assay (HIA) depends on the ability of lectins or viral particles that display lectins to agglutinate (aggregate) red blood cells. Adding a soluble ligand causes less precipitate to form, giving a minimum inhibitory concentration (MIC) or a concentration at the halfway point between fully aggregated and no aggregation at all (IC_{50}). The enzyme-linked lectin assay (ELLA) is a variant of the widely used ELISA (enzyme-linked immunosorbent assay). A natural carbohydrate ligand is immobilized on the surface of a microtiter plate, while the synthetic ligand and lectin are in solution. The lectin is labelled with a fluorescent tag or an enzyme such as horse-radish peroxidase (HRP). HRP catalyses the conversion of a prodye to a dye, thus providing colour to the well proportional to the amount of lectin that is absorbed on the surface. This assay gives an IC_{50} as well, based on the curve-fitting of absorbance and the concentration range of the synthetic ligand used.

The solid-phase inhibition assay, specific to galectins, depends on immobilized asialofetuin (ASF) or glycosylated bovine serum albumin (BSA). Asialofetuin is a glyco-protein containing three triantennary glycans terminated with LacNAc. The galectins that can recognize the terminal galactose, are in solution, where ligands will prevent binding to the glycoprotein on the surface [7].

Finally, instead of immobilizing the protein or natural ligand, the synthetic ligand can be immobilized. In the context of a microarray, this allows for testing a large number of potential ligands and even determining a fingerprint for the binding specificity of a protein. For example, glycodendrimers (which will be discussed later in this chapter for various lectins) with a valency of one to eight were outfitted with various carbohydrates such as glucose, galactose and mannose. Testing a variety of fluorescently labelled lectins against this panel provided insight into the specificity and the effect of multivalency [8]. Besides chemical immobilization, a DNA-directed microarray has been used, where ligands contain a stretch of DNA that binds to the complementary strand of DNA on the microarray surface. This allows for screening a large number of ligands of which initially only a very small amount has to be synthesized [9].

In determining the effect of multivalency, the fold of improvement of the multivalent ligand is generally compared as improvement *per sugar*, compared with either the monovalent monosaccharide, or an equivalent monovalent ligand that contains a similar linker to the multivalent ligand.

14.2 Haemagglutinin

Haemagglutinin (HA) and neuraminidase (NA) are the two major proteins on the surface of influenza A viruses (IAVs). With 18 types of HA and 11 types of NA, IAVs are referred to through these subtypes. Pandemic strains, such as H1N1 (the Spanish flu in 1918), H2N2 (Asian pandemic in 1957) and H3N2 (Hong Kong flu in 1986), have caused millions of deaths. The annual influenza is fortunately milder than these pandemic strains, but there is concern over more recently reported H5N1 and other H5 and H7 strains, that are very pathogenic and could emerge as a new pandemic [10,11].

HA is a trimeric membrane protein, with three carbohydrate binding sites, evenly spaced on one side of the protein complex (see Figure 14.2 for a protein monomer) [12–14]. It recognizes sialic acids (*N*-acetylneuraminic acid, Neu5Ac) and linked LacNAc sections, and specificity for the sialic acid linkage to the linked galactose determines the ability of the virus to infect certain organisms. Avian HA recognizes α2-3-linked sialic acids, while human HA recognizes predominantly α2-6-linked sialic acids.

Vaccines have to be updated yearly due to the antigenic shift: as the virus mutates over the year, the vaccines lose their potency [10,15,16]. Current therapies, such as zanamivir (Relenza) and oseltamivir (Tamiflu), aim to block NA activity, which cleaves terminal sialic acids to facilitate the release of viral particles [17,18]. Resistance against these antivirals has been reported for H5N1 and H1N1 strains, among others [19,20]. Amantadine targets an M2 ion channel of IAV, which maintains a favourable

(a)

(b)

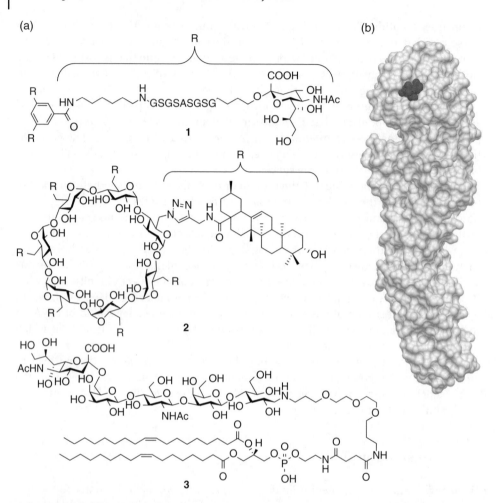

Figure 14.2 (a) Multivalent inhibitors against haemagglutinin. (b) A single H5 hemagglutinin with sialic acid (PDB: 1JSO). In the cell membrane, three haemagglutinin molecules associate and form a trimeric structure.

environment around the viral membrane, but many viral strains have become resistant to this antiviral drug [10]. No clinically approved therapies are available that target HA but blocking viral attachment might be advantageous to prevent infection in the first place and presents an opportunity to find new therapies.

The trivalent ligand **1** with a benzene core, alkyl-peptide linkers and sialic acids was presented to HA H5 from avian influenza [21]. SPR studies found a dissociation constant (K_d) of 450 nM for this ligand, which is a 4000-fold improvement compared with Neu5Ac-α2Me. A shorter peptide or alkyl linker led to K_ds in the millimolar range, and showed little improvement compared with a monovalent equivalent ligand containing the same peptide and alkyl linker. This indicates the importance of designing a ligand with sufficiently long spacers. Using molecular dynamics (MD) simulations, the ligand was fitted to H7, and found to bind with all three sialic acids in the binding sites, with some movement still allowed in the linker regions.

The heptavalent ligand **2**, consisting of a β-cyclodextrin scaffold with triterpene ligands attached through CuAAC reaction, was found to be active against influenza A in the early stage of the viral lifecycle [22]. Further experiments indicated an interaction between HA1 and the ligand, with a K_d of 2.08 µM as determined by SPR.

While most multivalent compounds discussed in this chapter are relatively small compounds, multivalency can also be achieved through supramolecular structures such as micelles and liposomes. Compound **3** containing an oligosaccharide terminated with an α2-6 linked sialic acid, chosen for its ability to bind to a number of influenza HAs, could assemble into liposomes and present this oligosaccharide [23]. The monovalent compound showed no inhibition of influenza A infection, while liposomes containing 7.5 mol% of **3** were able to inhibit infection at an effective concentration of 10 nM sialic acid. This inhibition was specific for α2-6 linked sialic acid bearing influenza A viruses. Liposomes and micelles containing an S-linked α2-6-sialolactoside phospholipid were able to interfere with H1N1 influenza virus entry into MDCK cells [24].

Sialic acid-containing polyacrylamide polymers bound influenza A virus particles in an agglutination assay of erythrocytes in concentrations below 15 nM [25]. Brush-polymers with α2-6-sialosides were also able to inhibit haemagglutination, where a longer spacer from scaffold to ligand was advantageous [26]. Having a larger amount of sialosides per polymer was beneficial. They were specific for influenza strains binding α2-6, while α2-3 influenza viruses in the haemagglutination assay were not inhibited.

Polyglycerol-based nanoparticles of 3–4 nm displaying approximately 20–35 sialic acids were found to have an IC_{50} in the micromolar range for a haemagglutination assay, while nanogels of 50 or 70 nm displaying 10 000–60 000 sialic acids had an IC_{50} in the low nanomolar range [27]. A study of sialoside-gold nanoparticles showed that the binding of a virus particle and nanoparticle depends on the size of this colloidal particle [28]. A small particle, smaller than the approximate 50 Å of a virion particle, benefits somewhat from multivalency. These small particles decorate the exterior of the virus. Large nanoparticles form clusters of virus and nanoparticles, thereby inhibiting cell adhesion and thus infection; in this study up to 70% inhibition of virus binding to erythrocytes.

14.3 LecA

LecA is one of the two lectins on the surface of *Pseudomonas aerigunosa* (PA) and is also known as PA-IL. LecA and LecB (or PA-IIL, discussed later in this chapter) are two important lectins in PA infection, a Gram-negative bacterium responsible for example for severe lung infections and forms a risk especially for cystic fibrosis patients [29,30]. LecA is a tetrameric, calcium-dependent lectin [31] (Figure 14.3) with a specificity for galactose and galactose derivatives, especially those attached to a hydrophobic aglycone [32–34]. In binding oligosaccharides on the cell surface of epithelial cells, it facilitates adhesion of PA to the patient's cell surface and the subsequent infection. Additionally, LecA is involved in the formation of PA biofilms [35], a protective polymeric matrix around the bacteria that makes it much more difficult for the immune system or therapeutics to reach the bacterial cells. Inhibition of both PA cell adhesion and biofilm formation can be achieved through LecA ligands, a promising approach to fighting these infections [36].

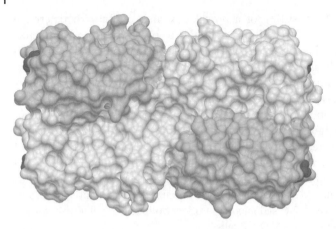

Figure 14.3 The structure of the tetrameric LecA with D-galactose in each binding site (PDB: 1OKO).

In order to facilitate a chelate binding mechanism of a multivalent ligand, the distance between the binding sites on the protein should be taken into account. The distance between the binding sites on LecA is approximately 26 Å [31], however, a flexible spacer such as poly(ethylene glycol) (PEG) needs to be about three times as long to bridge a given length [37]. By making the spacer much more rigid, a compound with a more defined length can be obtained. An initial exploration of these rigid spacers led to **4** (Figure 14.4a), where a series of glucose and triazole units were outfitted on both ends with galactose [38]. These were soluble in water, an important property for such ligands. The spacer length is crucial, especially for entirely rigid compounds, while a bit of flexibility by introducing some alkyl chains makes the ligand more forgiving for not being exactly the right length. The best binder **4** was found to have an IC_{50} of 220 nM, an improvement over the monovalent ligand of 545-fold. This assay used fluorescently labelled LecA in solution with a galactoside-functionalized chip surface.

Optimization of this divalent rigid ligand led to **5** which showed an improved IC_{50} of 2.7 nM and a K_d of 28 nM [39]. Molecular modelling showed the spacer had a length of 24 Å, a good match for the 26 Å between binding sites of LecA. An expected stoichiometry for all divalent compounds of 0.5 was found, indicating the chelate binding mechanism. Two types of variants of this compound were prepared. In one case, two triazoles were replaced by thiourea spacer units **6** [40], which had a K_d in the same range as **4**. Another series of modifications were at the 6 position of two of the spacer glucose moieties [41]. These substituents were positively and negatively charged, lipophilic and also aromatic as shown for compound **7** and were thought to interact with the protein surface. While found to be very good binders as well, they did not improve upon previous ones, suggesting that the groups were rotated towards the solvent, rather than the protein surface. A very different type of rigid ligand was **8** with alternating phenyl and alkyne moieties. The IC_{50} compared with the monovalent ligand improved 130-fold, as determined through an ELISA-like assay on a glycochip [42].

The binding mechanism of these divalent rigid ligands was confirmed by X-ray crystallography (Figure 14.4b) of **5** in which the entire molecule was visible, including the spacer. This is a clear indication that a more rigid spacer leads to a well-defined complex. The spacers in PEG linked multivalent ligands are rarely visible in X-ray

(a)

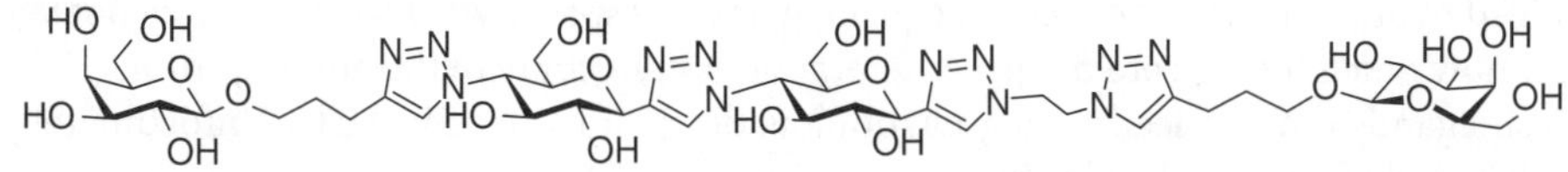

4

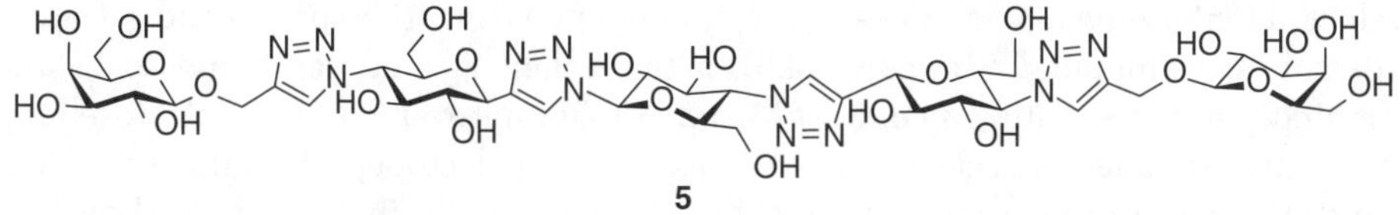

5

6

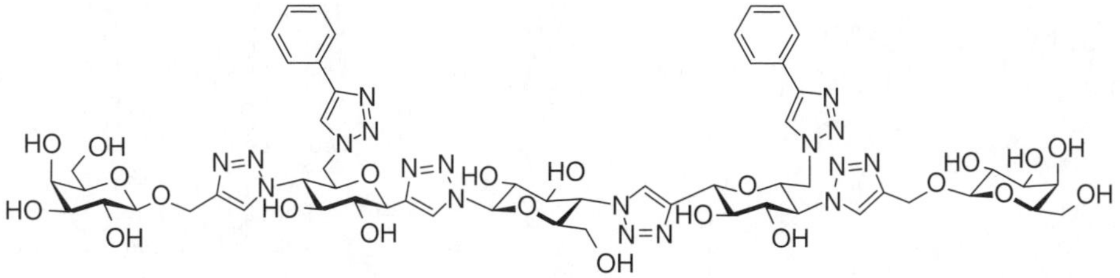

7

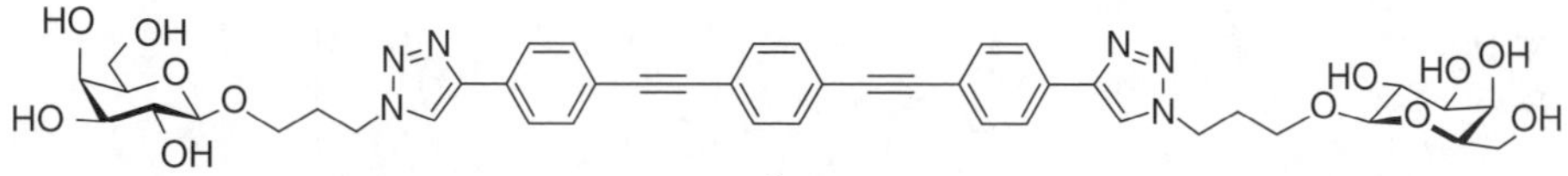

8

(b)

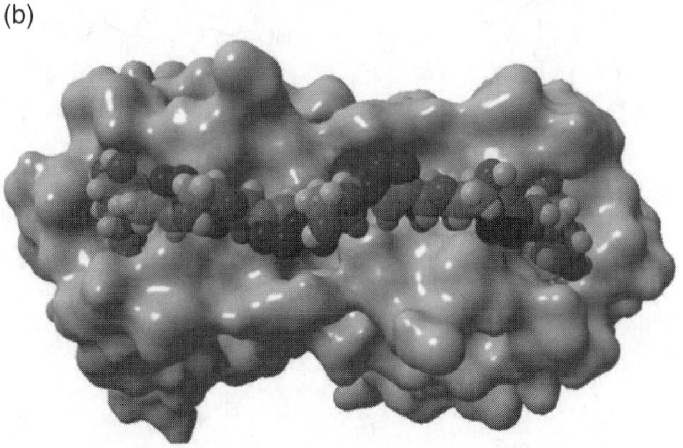

Figure 14.4 (a) Rigid divalent inhibitors of LecA. (b) X-ray structure of ligand **5** bound to LecA. See colour section.

crystallography. Interestingly, in the crystal structure a number of direct and water bridged hydrogen bonds between the protein and the spacer were observed. It is not yet clear how much these protein–spacer interactions contribute to the binding affinity [43]. These ligands did not lead to aggregation, as determined by SV-AUC. Unfortunately, these ligands do not inhibit biofilm formation.

Glycopeptide dendrimers that use lysine as a branching unit were synthesized and evaluated for binding to LecA [44–46]. They contained tripeptide spacers and galactose end-groups. Compound **9** (Figure 14.5), a tetravalent ligand, was found to be the best binder to LecA, with a K_d of 0.1 µM, a 219-fold improvement over galactose [44]. The multivalent effect was clear, as from a monovalent galactopeptide to the tetravalent ligand, the K_d improved with every step. The haemagglutination assay showed a 1000-fold improved inhibition for the tetravalent ligand, while the mono- and divalent ligands showed only a 40- to 50-fold improvement. More importantly in the context of infection, **9** can prevent formation of biofilms and disperse existing ones, but the dendrimers were not toxic by themselves.

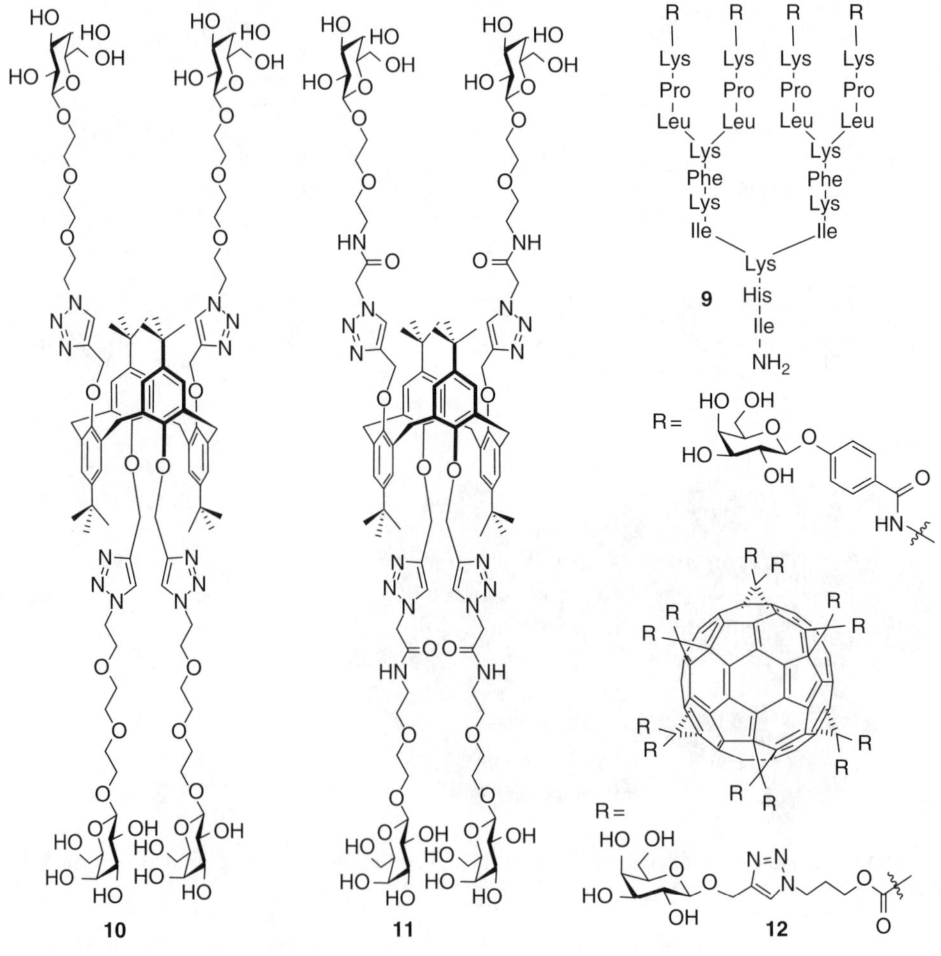

Figure 14.5 Inhibitors of LecA.

The X-ray structure of the monovalent galactose-tripeptide with LecA shows a T-shaped interaction of the phenyl with a CH of His50, explaining why LecA prefers aromatic galactosides, to the thioethyl galactoside linker dendrimers also synthesized [44]. An additional X-ray structure of the first generation peptide dendrimer found only the terminal tripeptide galactoside bound to LecA, while the rest of the dendrimer was not visible [43]. Optimization of the tripeptide linker, starting from **9** led to two galactose-tripeptides that have a slightly worse K_d and inhibition of biofilm formation, but improved biofilm dispersion, making them interesting from a therapeutic standpoint [45].

A study of calix[4]arene topology indicated that two ligand arms on either side of the calixarene scaffold was most advantageous **10** [47]. This 2:2 topology, with two carbohydrate ligands on one side of the calix[4]arene ring and two on the other side, had a K_d of 176 nM, a 852-fold improvement over the monovalent ligand. The 3:1 calix[4]arene showed a similar K_d of 200 nM, that is a very good ligand for LecA as well. Both had a stoichiometry of one ligand to one LecA monomer. This binding possibly occurs through a chelate binding mechanism, but some precipitation occurred during ITC experiments indicated aggregation as well. AFM studies using this 2:2 calix[4]arene revealed the formation of 90–500 nm filaments [48,49]. These filaments contain about 10–50 LecA tetramers linked by these glycoclusters. This supports the aggregative chelation mechanism of binding, likely in addition to chelation. Furthermore, it was found to inhibit cell adhesion between lung epithelial cells by around 70%, compared with 30% with the monovalent β-GalOMe, and additionally reduced biofilm formation of PA and reduced lung infection in an *in vivo* lung infection model [50].

In a somewhat more rigid linker between galactoside and calix[4]arene by introduction of amide bonds **11**, K_d again indicated this 2:2 topology as the most efficient ligand for LecA [51]. With a PEG-amide linker, a K_d of 90 nM was found.

Pillar[5]arene scaffolds displaying up to 10 galactosides reached nanomolar affinities [52,53]. Dodecavalent fullerenes improved binding, according to a HIA, by more than 12 000-fold, accounting for about 1000-fold improvement per sugar displayed. SPR and ELLA binding studies showed less dramatic improvements in binding [54].

Glycoclusters with a scaffold based on pyranosides allow for a variety of topologies and linkers to lead to generally tetravalent ligands. A broad approach, using mannose, galactose, and glucose as a central scaffold and a wide variety of phosphodiester linkers showed that having a mannose scaffold with sufficiently long linkers containing aromatic motifs was best for binding to LecA [55]. Following up with mannose as a scaffold, a variety of phosphodiester linkers containing thioethers, PEG, amides, and triazoles were tested for their influence on binding to LecA [56]. LecA prefers aromatic galactoside residues for binding, which was reflected in the performance of these linkers in a fluorescent microarray assay. The linker lacking this aromatic residue performed worse, and in general, solvated linkers were better than hydrophobic ones. However, the best binding was found with a previously synthesized linker containing a triazole moiety in addition to the phenyl, that is **13** (Figure 14.6). This was hypothesized to be due to the rigidity of the galactoside or the introduction of an interaction between Pro51 and the triazole in addition to the previously reported interaction of the aromatic moiety and His50. This interaction of an aromatic aglycone with Pro51 and His50 was confirmed when comparing a number of aromatic linkers. *O*-naphthyl and *O*-biphenyl showed the best affinity, compared with *O*- and *S*-benzyl and phenyl linkers [57]. Biofilm formation

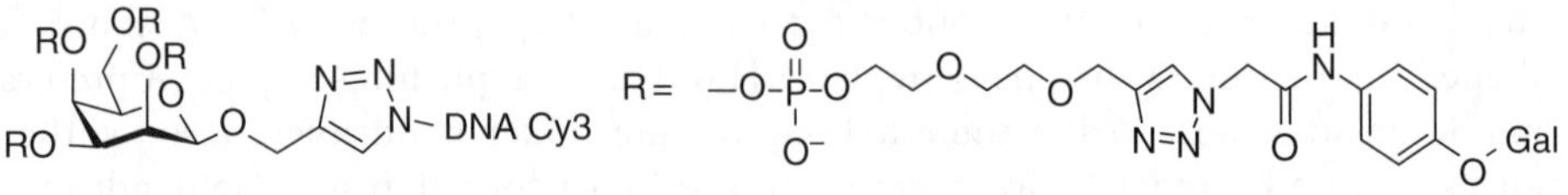

Figure 14.6 Inhibitors of LecA with mannose or cyclodextrin as a scaffold, and a divalent inhibitor found through the screening of a combinatorial library.

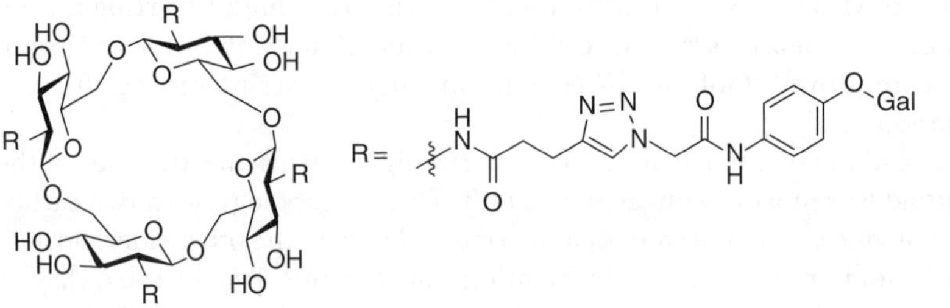

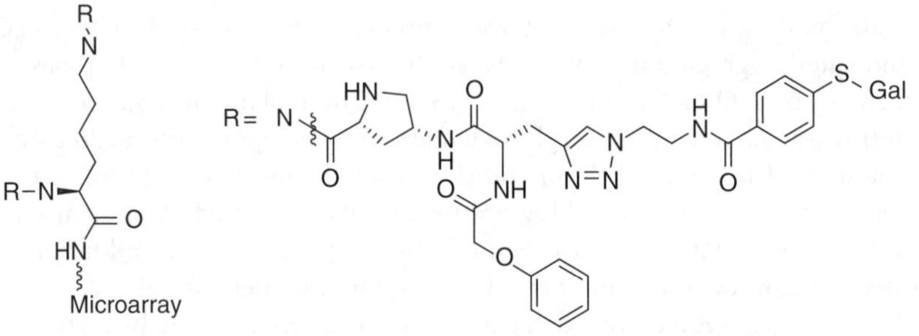

was inhibited with **13** up to 40%. A control experiment of a Δ*lecA* mutant of PA showed that biofilm formation cannot be entirely inhibited by this ligand, implying the role of other lectins in this process [9].

Divalent and tetravalent glycoclusters based on cyclic glycan **14** were found through ELLA to have IC$_{50}$s in the nanomolar range, with up to 1200-fold improvements compared with the monovalent compound [58]. Especially the rigid linker, containing phenyl and triazole moieties, contributed to a much better binding, findings that were reflected in HIA and ITC experiments.

Taking a combinatorial library approach, compound **15** was found to be a very good ligand to LecA with a K_d of 82 nM and remarkably showed a large reduction of the ability of PA to enter human epithelial lung cells [59]. Modelling suggests a chelate binding mechanism and interaction of the phenyl with His50.

14.4 LecB

LecB (or PA-IIL) is the other major soluble lectin produced by PA. Similar to LecA, this lectin is involved in cell adhesion and biofilm formation, and inhibition of both processes can be achieved through ligands targeting this fucose-specific lectin [60]. Like LecA, it

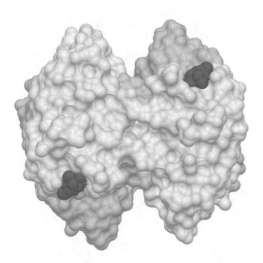

Figure 14.7 LecB binding α-L-fucose (PDB: 1OXC). Two binding sites are on the back of the protein in this view.

is a tetrameric protein with four carbohydrate binding sites that can bind both L-fucose and L-mannose [61–63] but in LecB these sites are more evenly spaced with approximately 40 Å between the sites [64–66], quite a distance to bridge with a multivalent ligand (Figure 14.7). An ELLA determined IC_{50} for α-FucOMe was found to be 430 nM [67], which is already a tightly binding ligand, so making even more tightly binding, multivalent ligands is not without challenge.

A number of tightly binding peptide dendrimers have been made with affinity in the nanomolar range (Figure 14.8). A *C*-fucosyl peptide dendrimer library led to the discovery of an octavalent **16** that has an IC_{50} of 25 nM as determined through ELLA, a 55-fold improvement per fucose. Lower valency showed much more modest improvements [68]. Molecular modelling showed that the peptide sequence did not interact extensively with LecB, implying that the multivalency was key to the improved potency. Dendrimer **17** has an affinity for LecB that is 46-fold higher than fucose and was shown capable of inhibiting biofilm formation. Since biofilm inhibition did not occur in a Δ*lecB* strain, this interaction between **17** and LecB seems crucial for biofilm inhibition [69]. Substituting the L-amino acids in **17** for D-amino acids altered the affinity for LecB, from 0.14 μM for **17** to a slightly higher 0.66 μM for **D-17** [70]. However, the D-variant was proteolytically stable while **17** was not and was still able to inhibit biofilm formation. A wide variety of tetravalent peptide dendrimers presenting both fucose and galactose, and Lewis[a] trisaccharide were synthesized [46]. In the crystal structure of **18** with LecB, two binding sites are bridged by ligands fused through a Cys-Cys dimerization, while for the other two binding sites only the terminal arm is found. No chelate binding mechanism is observed. These dendrimers were able to disperse biofilms and inhibit their formation, doing so in sub-inhibitory concentrations when used in conjunction with the antibiotic tobramycin. A K_d of 28 nM and ELLA determined IC_{50} of 0.6 nM was found for **19**, a hexadecavalent α-fucosylated glycocluster [67]. The improvement compared to αFucOMe was 65-fold per sugar. β-Fucosylated glycoclusters were notably worse ligands than the α-fucosides, however, multivalency benefits of these β-fucosylated glycoclusters were slightly larger, with a 86-fold improvement in IC_{50} for the best ligand.

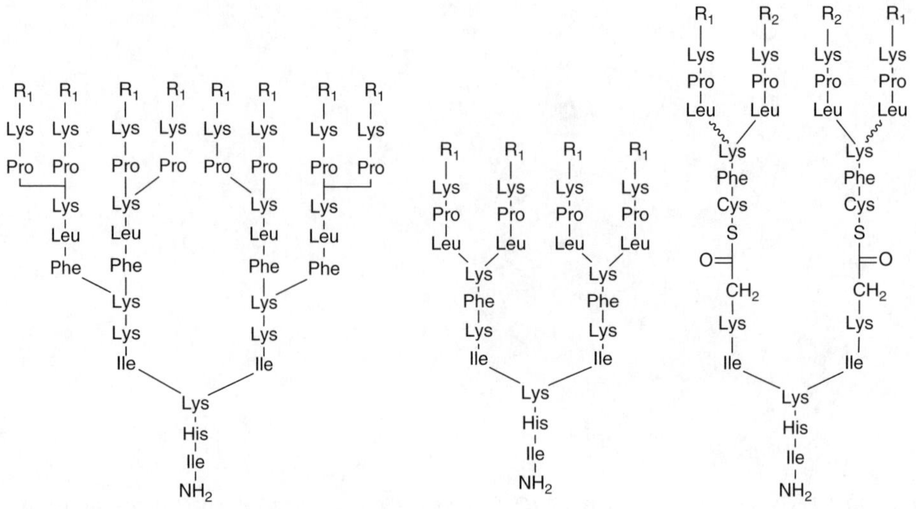

16 **17** **18**

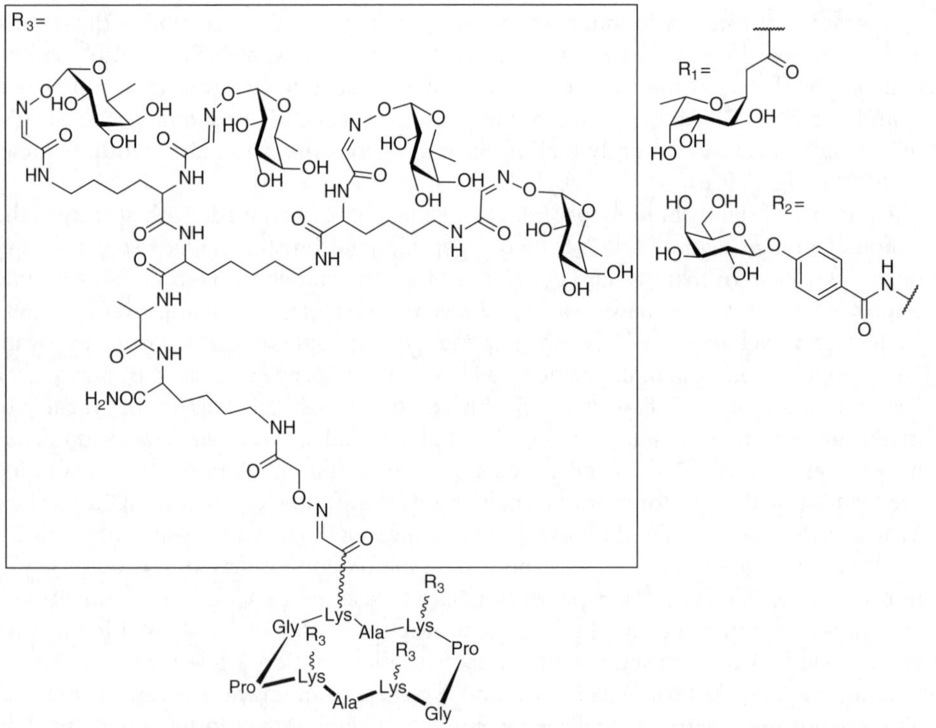

19

Figure 14.8 Multivalent peptide-based dendrimer ligands for LecB. The wavy lines indicate a bond with the side chain of the amino acid.

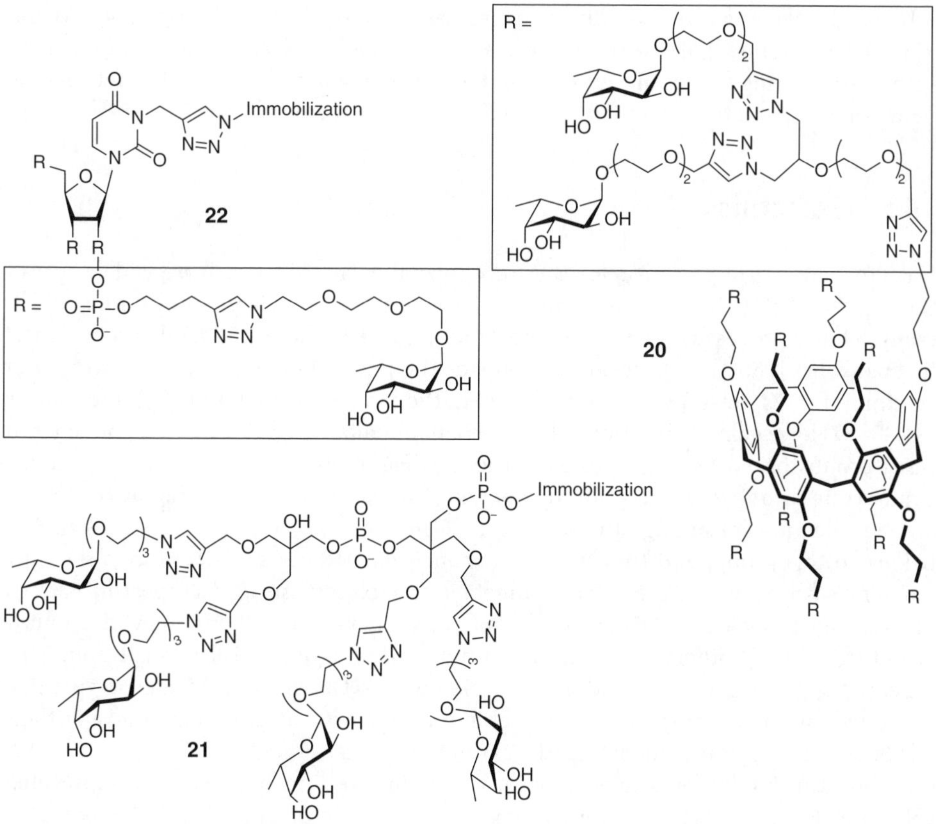

Figure 14.9 Multivalent ligands against LecB.

For the tetravalent 2:2 calix[4]arene **10** for LecB now presenting fucosides (Figure 14.5), a K_d of 304 nM was found with a stoichiometry of occupying three binding sites [50]. This compound cannot bridge the distances between LecB binding sites to facilitate a chelate binding mechanism, so a bind and jump mechanism was proposed [71], where LecB jumps from ligand to ligand, that is a rebinding mechanism. The calix[4]arene compound exhibited a 90% inhibition of PA adhesion to lung epithelial cells and reduced biofilm formation, but this happened in a LecB-independent manner as found out with the Δ*lecB* mutant. A similar structure is the pillar[5]arene **20** [52] (Figure 14.9). Aggregation of seven to eight LecB monomers to one decavalent ligand was found and ten LecB monomers for the icosavalent ligand. ELLA gave an IC_{50} of 6 nM, a 74-fold improved binding compared with the monovalent fucose; however, ITC experiments gave K_ds only up to a 3-fold improvement. C60 fullerenes were also effective, especially those presenting 24 instead of 12 fucosides, with a K_d of 23 nM [72]. Aggregation was found with a stoichiometry of three LecB lectins bound per fullerene, but only few of the fucoses present on the fullerene participate at any given time.

In varying the topology (antenna, linear or crown-like) of immobilized glycoclusters, high valency could be reached with **21** with a tetravalent antennary topology most preferred [73]. Since the glycoclusters were immobilized on a microarray, synthesis could be done on a picomole scale.

Using glycoclusters with various furanose-based cores, affinities between 56 nM and 95 nM were found [74], with ribose **22** and xylose-based scaffolds outperforming arabinose. Linking two mannose-centred ligands with a long linker performed slightly better on these microarrays than the single mannose-centred ligand [75].

14.5 Galectins

Galectins are a group of proteins that recognize β-galactosides and are widely spread throughout life [76,77]. Fifteen mammalian galectins are known and they fulfil a role in many cellular processes such as cell recognition [3,78], adhesion [79,80] and in signalling pathways [78,81], both inside and outside the cell. In humans, they are involved in the immune response [82,83] but especially their roles in infection [84] and cancer [81,85–87] have gained attention. The fifteen mammalian galectins are categorized into three groups based on their protein structure (Figure 14.10a). Prototypical galectins contain one carbohydrate recognition domain (CRD), a domain that is highly conserved among galectins consisting of two β-sheets. Galectins-1, 2, 5, 7, 10, 11, 13, 14 and 15 belong to this group, and they may form noncovalent dimers. Galectin-3 is the only chimeric galectin, where the CRD is connected to a tail consisting of a repeating pattern of amino acids. Galectin-3 can associate into pentamers when interacting with a multivalent ligand [88], otherwise it exists as a monomer (Figure 14.10b). Finally, tandem-repeat galectins (4, 6, 8, 9 and 12) contain two CRDs within a single polypeptide separated by an amino acid linker. Inhibitors against galectins can prevent their adhesion, and stop the signalling pathways in their tracks [76,89–92].

Monovalent inhibitors developed against galectins are usually based on a substituted galactose (Figure 14.11). Especially aromatic substituents on the C2 and C3 positions of galactose were found to have improved binding compared with free galactose or lactose. In galectins 1 and 3 (Gal-1 and Gal-3, respectively), this substituent has a favourable interaction with specific arginines. A large number of anomeric phenyl galactosides were synthesized to explore the role of the aryl–arginine interactions in Gal-1 and Gal-3 binding. Most had no or very little improved inhibitory properties compared with lactose, except **23**, which was around 160 times better than lactose [93]. Addition of a 4-phenoxyaryl group to the C3 of thiodilactoside **24** led to an inhibitor with almost 200 times improved binding

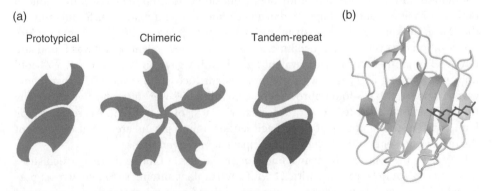

Figure 14.10 (a) Schematic representation of the three types of galectins. (b) A crystal structure image of galectin-3 in complex with LacNAc (PDB: 1KJL).

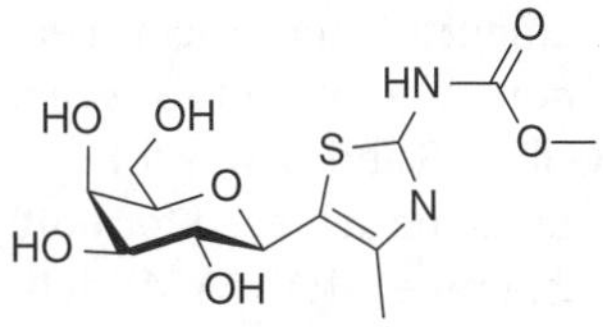

Figure 14.11 Monovalent and multivalent ligands for galectins.

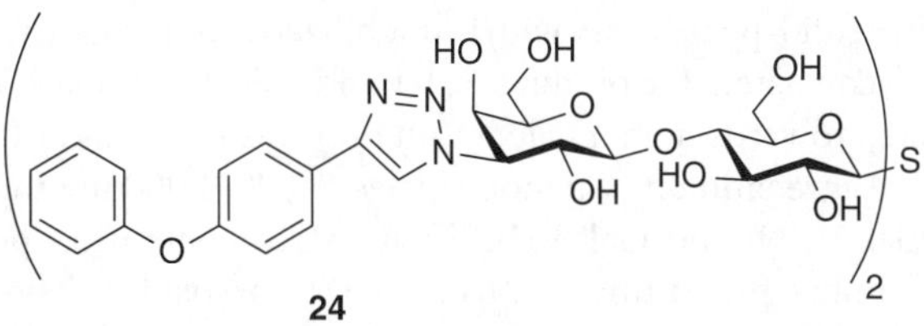

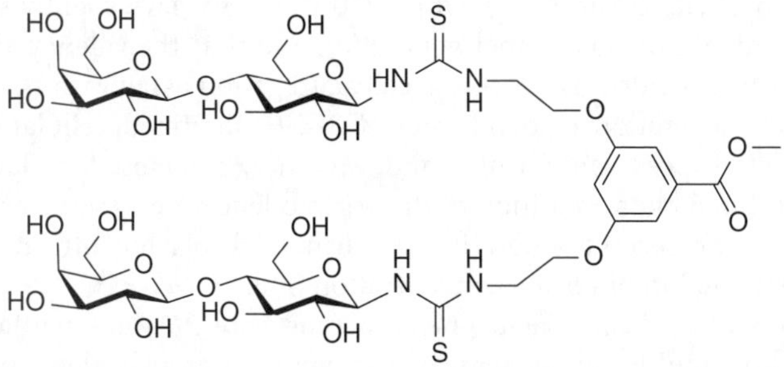

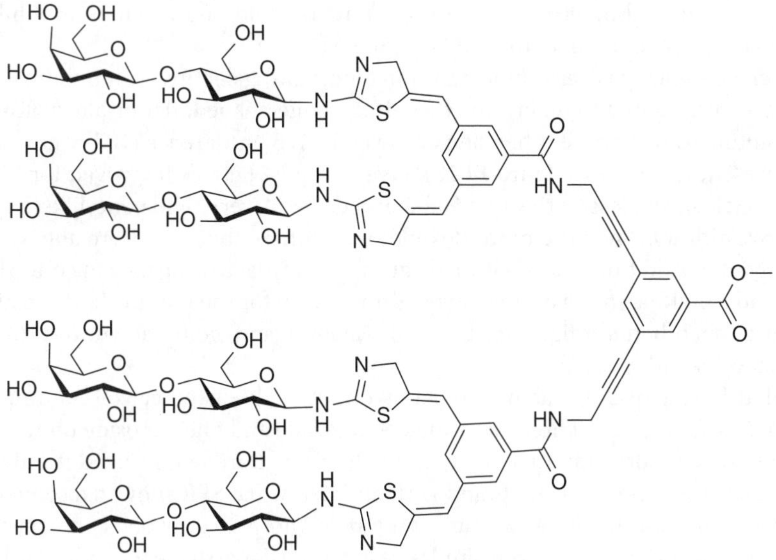

to Gal-3 compared with Gal-1, while the parent thiodigalactoside showed no such selectivity [94]. Glycopeptides containing a threonine-*O*-lactoside improved binding up to 66-fold compared with free lactose [95]. Glycomimetic peptides, such as the pentameric WYKYW [96] or a number of peptides based on a Tyr-Xaa-Tyr [97] motif were found to have two to three times better binding compared with free galactose.

Dendrimers based on 3,5-di-(2-aminoethoxy)benzoic acid **25** up to the third generation found an improvement in IC_{50} of up to 144-fold per lactose in binding to Gal-3 [98] in a solid-phase assay with immobilized glycoproteins. Improvements for these octavalent glycoclusters for binding Gal-1 and galectin 7 (Gal-7) were more modest, but for every step towards higher valency an improved IC_{50} and MIC for Gal-1, -3 and -7 was found.

Aggregation into nanoparticles of 500–2000 nm in diameter can be achieved for both Gal-1 [99] and Gal-3 [100] with sixth generation poly(amidoamine) (PAMAM) dendrimers presenting on average 130 lactosides. The ratio of galectin to dendrimeric ligand determines the size of the nanoparticles: a ratio of 9:1 Gal-1 or Gal-3 to ligand gave the largest particles (over 2 μm), while 220:1 and 3:1 gave slightly smaller ones. Additionally, the higher the dendrimer generation and thus the valency, the larger the particles. Using these dendrimers [99] in a cell aggregation assay with human prostate cancer cells, each generation of dendrimer (G2–G6) inhibited cellular aggregation. These cancer cells express both Gal-1 and Mucin-1, a ligand for Gal-1, and this interaction is inhibited by the addition of these glycodendrimers. More epitopes is not necessarily better: the second generation dendrimer, displaying fifteen galactosides, was most efficient and inhibited at a concentration of 0.4 mM.

Rigid dendrimers based on a rigid phenyl-alkyne core **26** were synthesized [101]. While for Gal-1 little multivalent effect was seen, with both the divalent and tetravalent dendrimers showing similar IC_{50}s in a solid-phase assay with asialofetuin, Gal-3 and Gal-5 did benefit from the more chemically rigid structure of the dendrimeric ligand. Especially towards Gal-3, tetravalency brought a 1071-fold improvement in inhibitory capacity per lactose unit, down to an IC_{50} of 70 nM.

Calixarene geometry is very important for binding to galectins. A number of calixarenes were investigated to determine this effect: cone-shaped, flat or alternating side chains pointing to one or the other side of the calix[4, 6 or 8]arenes [102] were synthesized to present lactose **27** (Figure 14.12). Overall, most ligands improved for binding galectin 4 (Gal-4), and less so towards Gal-1 and Gal-3. A very rigid cone-like shape did not improve affinity, while the more flexible ones (where the arms were able to move more freely) did. Addition of aromatic side groups to the lactose ligands in cone-shaped calix[4]- and calix[6]arenes gave improved affinity even for the monovalent substituted lactose, and much better affinity for the multivalent ligand, going down to 0.15–3 μM against Gal-3, -4 and -9 [103].

For Gal-1, having all four lactose moieties on one side of the calixarene improved affinity by 625-fold for **28**. Three lactosides on one side and one lactoside on the other, a 3:1 topology, saw some improvement, and a 2:2 calix[4]arene improved it only by 2-fold. However, these results were found with an HIA, while SPR showed a sequence of improvement that was the other way around [104]. Another study of a cone-shape and 1,3-alternate calix[4]arenes showed similar results: the cone-shape was a better binder than the 1,3-alternate [105].

A heptavalent cyclodextrin-based ligand **29** [106] successfully inhibited Gal-3, compared with Gal-1 and -7. This specificity was attributed to the ability of Gal-3 to form pentamers when in complex to a multivalent binder, while the Gal-1 and -7 binding sites are too far apart (~50 Å). Cyclodextrin galactosides that self-assembled like beads on an alkyl-phenyl string have a different architecture [107]. These fibres had a 30-fold improved binding to Gal-1 and were able to prevent cell adhesion in a T-cell agglutination assay.

LacdiNAc groups were attached to a thiourea linked alkyl spacer of varying length, to make a divalent ligand **30** [108] (Figure 14.13). In molecular modelling simulations the

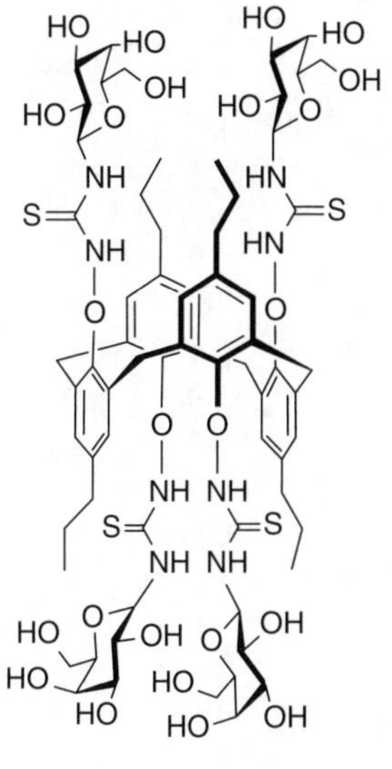

27

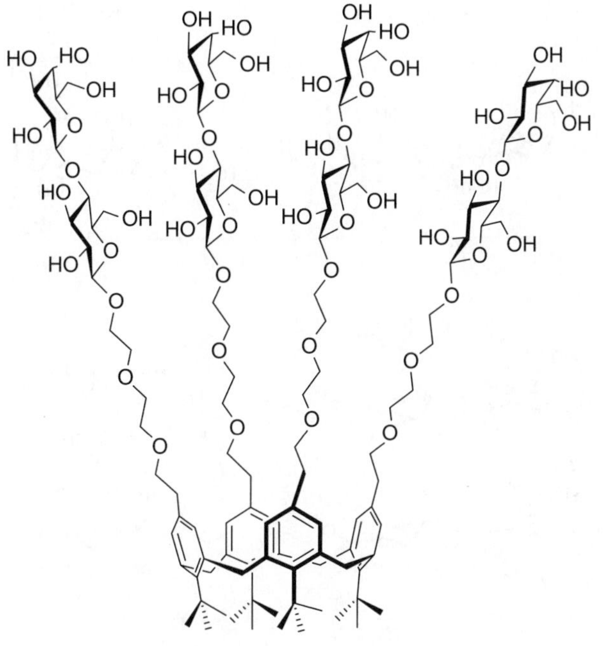

28

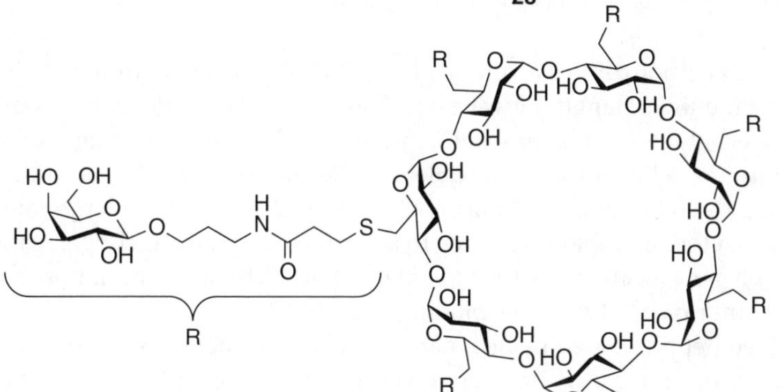

29

Figure 14.12 Multivalent ligands against galectins based on calixarene and cyclodextrin scaffolds.

optimal spacer length was slightly longer (eleven spacer units) than in ELLA assays (nine units). The ligand with this linker length was about three times as potent against Gal-3 with an IC_{50} of approximately 40–50 µM than a monovalent equivalent. Even ligands with a slightly shorter or longer spacer length (6–11 spacer lengths) were all about twice as potent as a monovalent control.

A variety of mono-, di-, tri- and tetravalent scaffolds with C-alkyl-lactosides were tested against Gal-3 [109]. Compound **31**, a divalent amino-derivative, showed the most improvement (21-fold, 10.8-fold per residue) in K_d (SPR, 15.8 µM) compared with lactose, and most multivalent ligands showed small improvements in affinity.

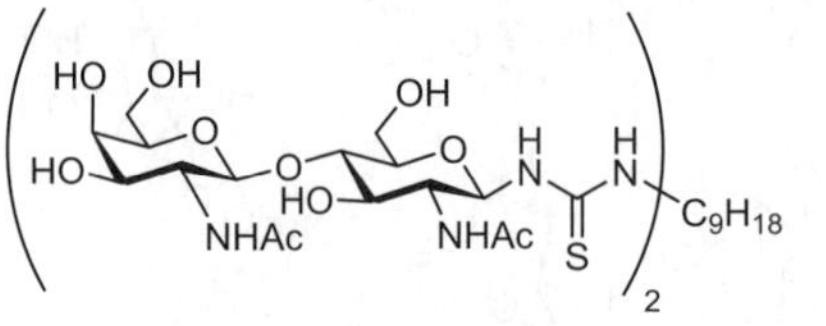

Figure 14.13 A variety of multivalent ligands against galectins.

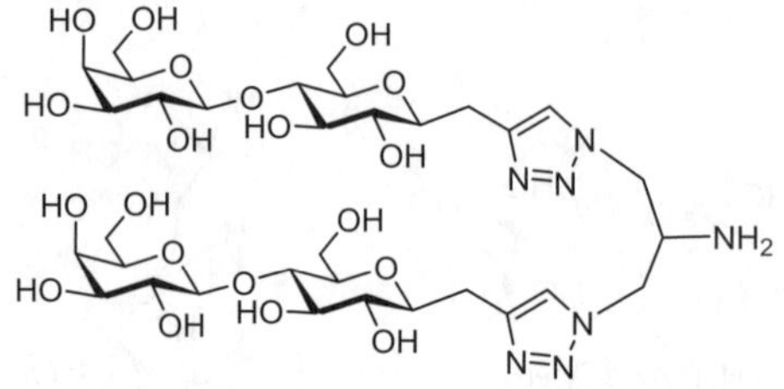

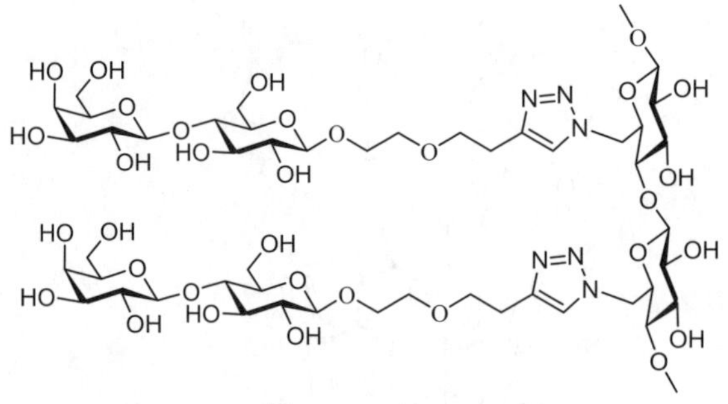

With a chain of carbohydrates as the scaffold **32** [110], the effect of multivalency up to tetravalency and linker length between scaffold and the lactoside epitope was investigated against Gal-1 and -3. Linker length, in this case, had little effect on binding, and in MD simulations the linkers are so flexible that the epitopes stay close to the scaffold. They argue that the rebinding mechanism would probably not play a large role, as the length between lactose epitopes is too long to make it efficient. Instead, aggregation was shown to be a bigger aspect, where the linker length needs to have a minimum length to prevent steric hindrance between the binding galectins.

Self-assembled peptide-based β-sheet nanofibres presenting LacNAc attached to the N-terminal asparagine had a K_d 10 times lower than monovalent lactose when assayed with Gal-1 [111]. They were also able to to inhibit tumour cell agglutination of Jurkat T cells. CdSeS/ZnS quantum dots with 108 LacNAc moieties presented to Gal-3 show a 184-fold improved binding compared with LacNAc, resulting in a K_d of 57 nM in an SPR assay [112]. Using human serum albumin tagged with azides as a scaffold, an alkyne-terminated N-glycan isolated from chicken was attached [113]. This glycoprotein containing approximately 30 glycans blocked binding of Gal-3 to cancer cells.

14.6 Concanavalin A

Concanavalin A (ConA) is a lectin from Jack bean and one of the first lectins discovered. It is a tetrameric lectin with four binding sites that recognizes α-mannosides, but can also bind α-glucosides [114] (Figure 14.14). The distance between the binding sites is very large compared with other lectins (ca. 70 Å) making a chelation binding

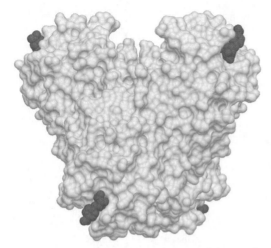

Figure 14.14 Structure of concanavalin A with 4-nitrophenyl-α-D-mannosides (PDB: 1VAM).

mechanism for a multivalent ligand difficult. A more recent sample of ConA ligands will be discussed here, as there is a long history of research on this lectin already [5].

An octavalent glycopeptide dendrimer **33** (Figure 14.15) with a branching core consisting of lysine residues and S-linked mannose residues was found to have an IC_{50} of 2.9 μM using an ELLA assay, a 57-fold improvement per sugar over the monovalent mannose [115]. The octavalent ligand was clearly the better ligand compared with the tetravalent variants, where IC_{50}s in the 50–750 μM range were found. A tyrosine residue close to mannose improved affinity. A different octavalent glycopeptide **34** had a slightly better IC_{50} of 2 μM [116].

C60 fullerenes with 12 or 24 mannosides were efficient ligands with up to 15 ConA molecules attached to each fullerene molecule according to ITC measurements [117]. Cyclodextrins (cyclic α1-4 oligoglucosides) have been used as multivalent scaffolds before, but novel cyclic α1-6 octaglycoside presenting mannosides were used also where tetravalency showed to be the most efficient at binding to ConA [118]. Another scaffold was the so-called aromatic asterisk **35** where an IC_{50} of 89 nM was found in a HIA for this hexavalent ligand [119]. This was a case of a very efficient rebinding mechanism.

Hyperbranched polyglycerol polymers were able to form aggregates with ConA, with a higher valency leading to larger aggregates [120]. An IC_{50} of 35 nM, as measured by a competitive SPR assay, was found for the polymer presenting 60 α-Man moieties which is a 357-fold improvement over α-ManOMe. This 60-valent polymer, as well as the 33-valent one, each bind 2.9 ConA tetramers, and the linker between the polymer proper and the carbohydrate ligand needed to be of sufficient length to allow for a high affinity.

A glycopolymer with mono- or diantennary side groups **36** was synthesized (Figure 14.16). Strong aggregation was found when using the tetrameric ConA at neutral pH but even at a lower pH where ConA is dimeric, aggregation was observed. The diantennary polymer, containing both α-Man and non-binding β-Glc on its antennae, bound as strong as the polymer containing only mannose, with a association constant (K_a) of 30 μM.

Artificial lipid rafts can be induced by addition of cholesterol to phospholipid vesicles. In this manner, the interaction of ConA with lipids presenting mannose could be

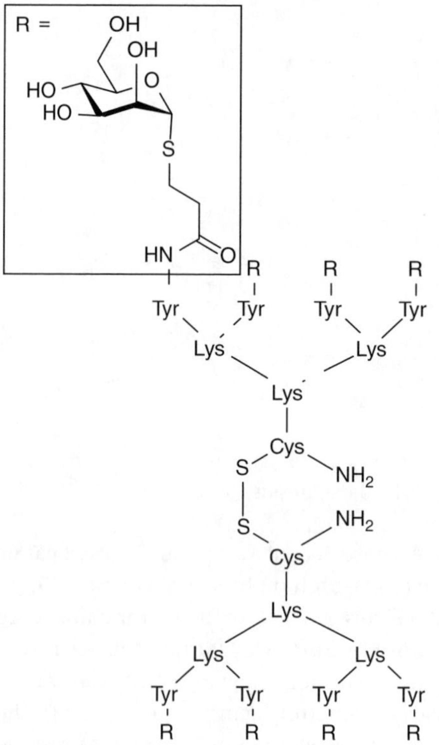

33

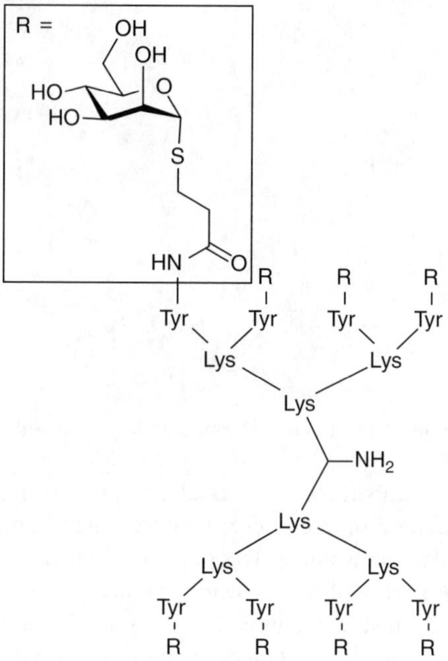

34

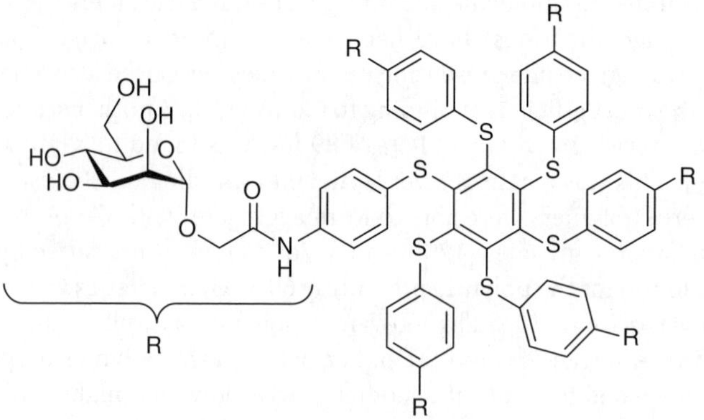

35

Figure 14.15 Multivalent ligands of concanavalin A.

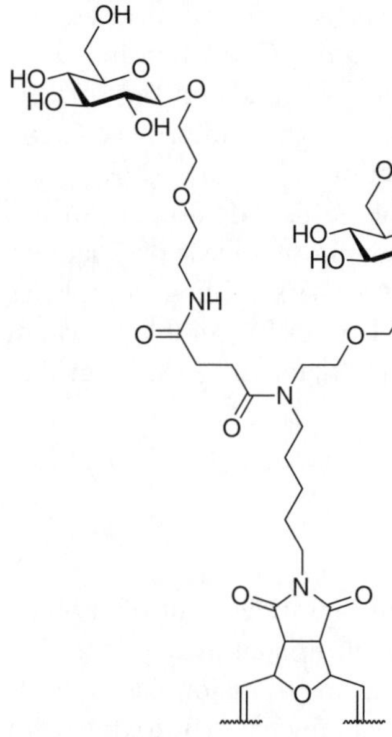

Figure 14.16 More multivalent ligands of concanavalin A.

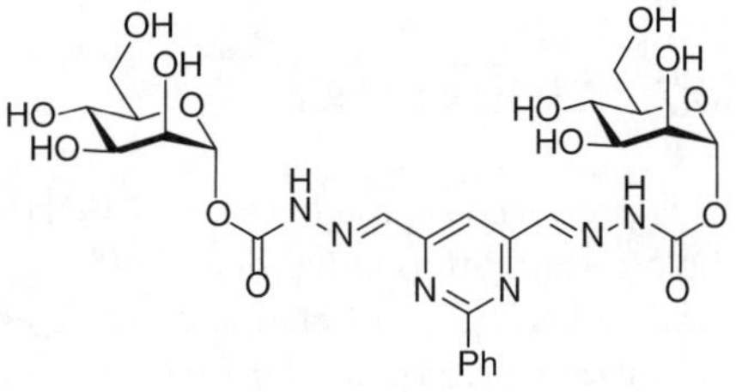

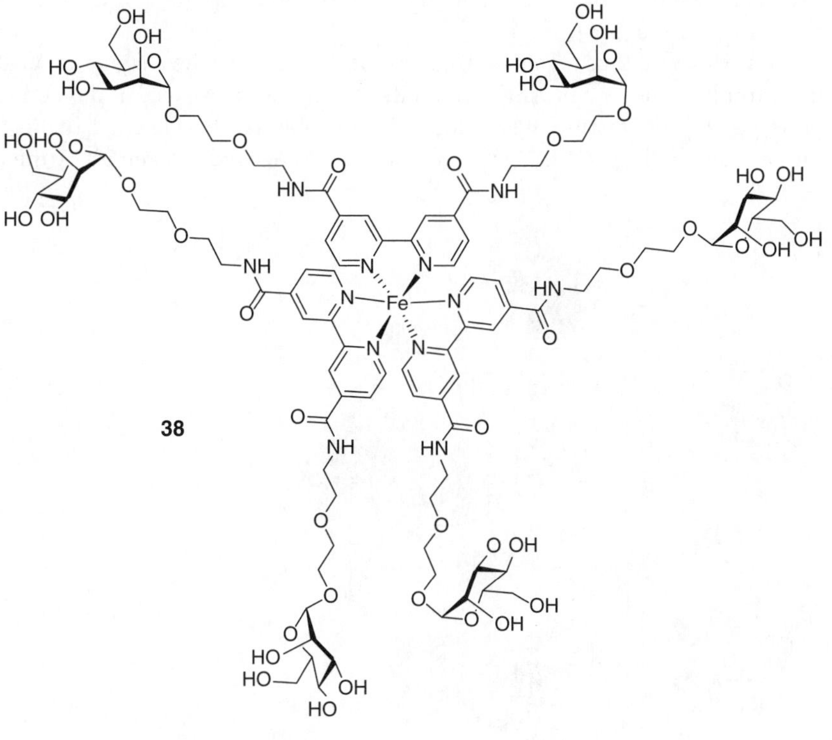

investigated with these groups dispersed over the vesicle surface or concentrated in a lipid raft [121]. Inducing lipid rafts did not significantly improve ConA binding compared with the dispersed ligands, possibly due to crowding of ligands on the surface. Only a higher concentration of dispersed mannosyl lipids slightly improved ConA binding, possibly due to some weak chelate binding interaction.

A dynamic combinatorial library of glycosylated bipyridine ligands **38** can form a complex with iron(II) [122]. Varying the ligands between mannose, fucose or galactose and exposing them to ConA, showed a bias for those self-assembling complexes containing mannosylated bipyridine ligands. An octavalent self-assembled system of ligand **37** containing zinc led to aggregation with ConA [123], forming a loose grid of ConA tetramers and the self-assembled complex.

14.7 Cholera Toxin

CT is produced by the bacterium *Vibrio cholera* and is the causative agent of cholera. Cholera is an often lethal disease that causes over 100 000 deaths annually across the world and is especially characterized by severe diarrhoea leading to major loss of fluids and nutrients [124]. Similar toxins exist and are collectively known as AB_5 toxins, with examples including Shiga toxin from *Shigella dysenteriae*, pertussis toxin from *Bordatella pertussis* and heat-labile enterotoxin from *Escherichia coli* [125]. All these toxins are taken up by the cell and are able to evade degradation through the normal cellular processes [126].

AB_5 toxins, and thus CT as well, consist of six subunits: the enzymatic A subunit, and five carbohydrate-recognizing B subunits that are evenly spaced and on one side of the protein [127]. The natural ligand for CT is the oligosaccharide GM1 that is present on the cell surface (Figure 14.17). CT-B binds to GM1os and is taken up by the cell, where

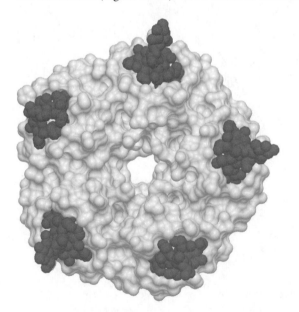

Figure 14.17 Structure of the pentameric cholera toxin B subunits in complex with GM1os (PDB: 3CHB).

the CT-A subunit induces a series of signals in the cell leading to secretion of fluids. Only one functional binding site is required for toxin activity [128].

Blocking the binding of CT-B to GM1os is an interesting approach for therapeutics to prevent the severe symptoms of cholera [129,130]. Two approaches towards multivalent ligands have been described: mismatch aggregation of non-pentavalent ligands and 1:1 binding of a pentavalent ligand to the pentavalent CT.

Dendrimers based on 3,5-di-(2-aminoethoxy)-benzoic acid, displaying lactose, galactose or GM1os, all show potent inhibition of CT [131–134] with an octavalent GM1os ligand showing the best IC_{50} of 50 pM of the variety of structures synthesized [134]. A study into the mechanism of the di- and tetravalent GM1os-presenting ligands investigated the aggregation of CT-B [135]. SV-AUC showed dimerization and higher aggregation upon addition of the divalent inhibitor.

To directly compare a tetra- versus a pentavalent dendrimer, an extra arm was attached to the dendrimer core to make a dendrimer-based pentavalent ligand **39** [136] (Figure 14.18). This arm was slightly shorter than the other linkers, to reach an approximate same distance from core to carbohydrate as the dendrimer proper. IC_{50}s of both the tetravalent and pentavalent compounds were in the same picomolar range. SV-AUC experiments indicate dimerization and larger aggregates for both compounds.

Taking the 1:1 approach, pentavalent and decavalent pentacyclen linked galactoside ligands were both shown to bind to CT with IC_{50}s in the low micromolar to nanomolar range [137,138] (Figure 14.19). Especially the decavalent ligand **40** with branched (divalent) ligand arms worked well, binding two CT pentamers at once, which was confirmed by X-ray studies. In the same vein, cyclic decapeptides that are slightly larger than the pentacyclen were found to have IC_{50}s in the same range as pentacyclen, with shorter linkers needed for optimal binding [139]. In this inhibition assay, GM1os was immobilized and both the inhibitor and CT were present in solution.

For the calix[5]arene presenting GM1os **41** an IC_{50} of 450 pM was found in the same type of assay as for the pentacyclen and cyclic decapeptide, which is a 100 000-fold improvement over the monovalent equivalent [140]. The importance of the terminal galactose for CT-B binding was confirmed, as a ligand lacking this sugar moiety had a much higher IC_{50} of 9 μM. Similarly, the corrannulene-based pentavalent ligand **42** had an IC_{50} in the low nanomolar range, depending on the optimal length of the linker between the scaffold and GM1os ligand [141].

In another ingenuous approach CT-B itself is used as a multivalent scaffold, although an inactive mutant, by modifying N-terminal threonine residues on each CT-B subunit and linking them with GM1os [142]. This pentavalent ligand has an IC_{50} of 104 pM in an ELLA assay and SV-AUC showed predominant association of CT-B and this protein-based ligand in a 1:1 heterodimer.

14.8 Propeller Lectins

Propeller lectins are lectins containing β-sheet carbohydrate binding domains arranged as blades of a propeller. They can contain 4–10 blades. A number of these lectins have been reported over the years, from organisms ranging from bacteria and fungi to humans.

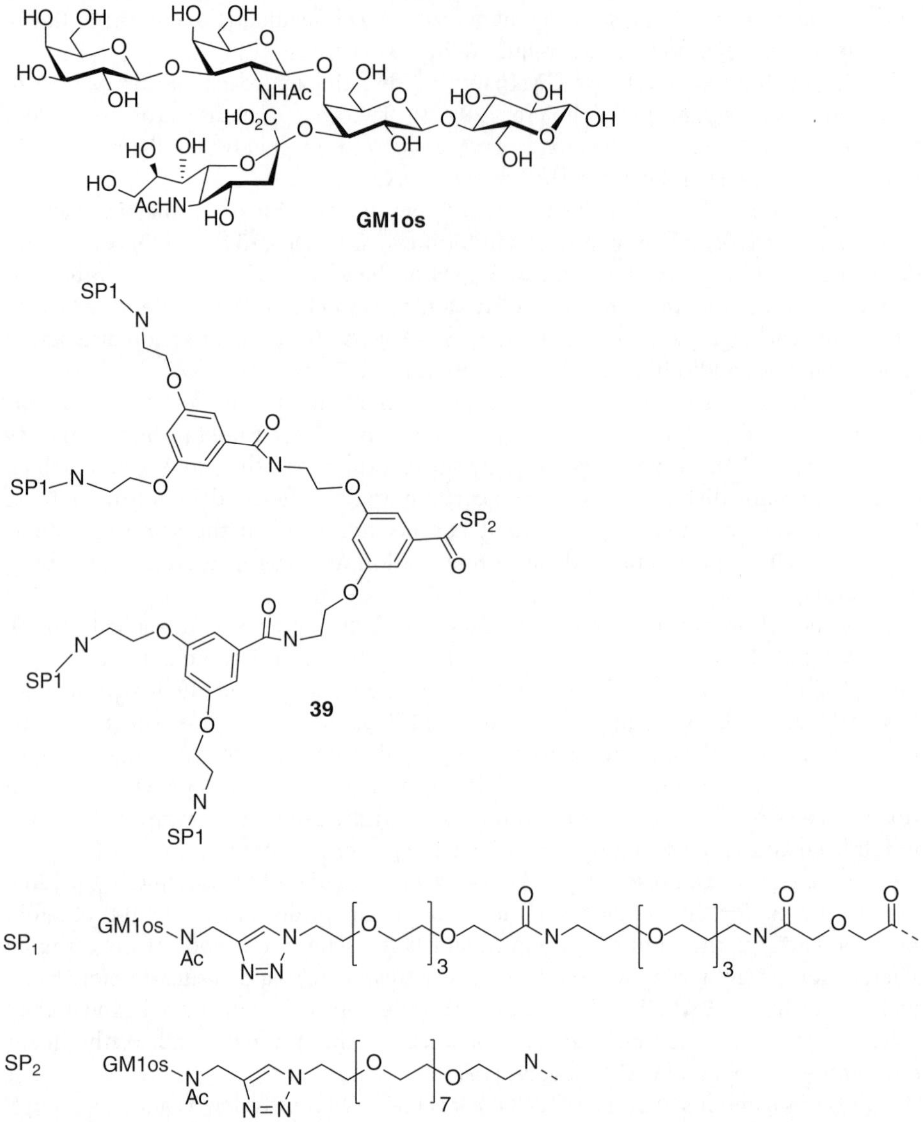

Figure 14.18 The structure of GM1os, the natural ligand of cholera toxin, and a pentavalent dendrimer inhibitor of cholera toxin.

Ralstonia solanacearum lectin (RSL) originates from a Gram-negative plant pathogen affecting many crops worldwide and binds L-fucose. RSL is a trimeric six-bladed lectin, with each monomer containing two β-sheet blades [143] (Figure 14.20a). RSL can be inhibited with a C60 fullerene presenting fucosides [72], as has been described earlier in this chapter for other lectins. While the monovalent fucoside already binds with a low K_d of approximately 300 nM, this can be improved through divalency **43** (K_d of 74 nM) or dodecavalency of a fullerene (K_d of 10 nM) (Figure 14.21). Due to steric hindrance many of the fucosides cannot reach the RSL binding sites, and as such the longer

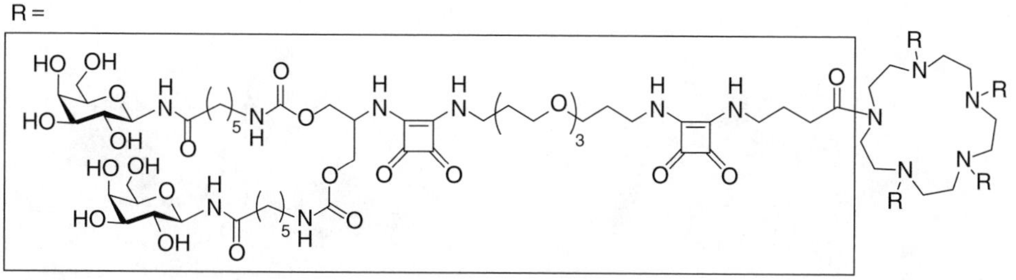

Figure 14.19 Multivalent inhibitors against cholera toxin, based on dendrimer, calixarene and pentacyclen scaffolds.

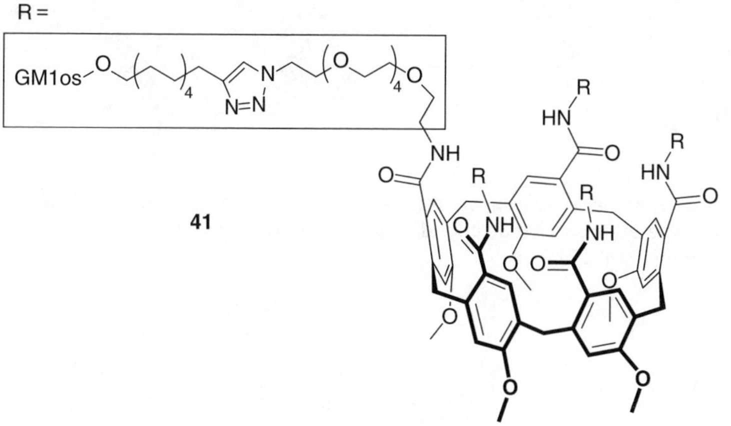

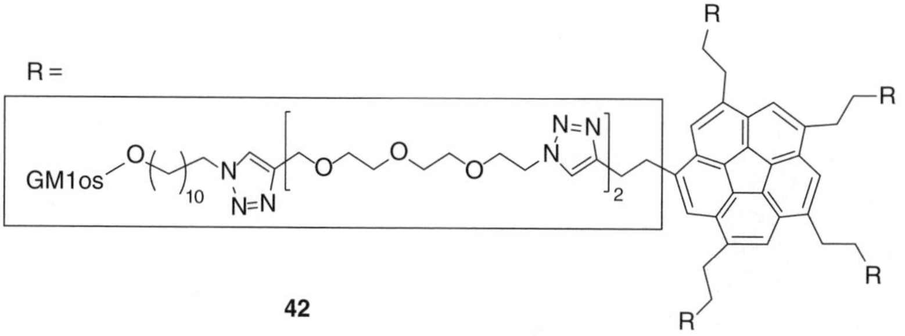

spacers have improved affinity compared with shorter spacers. In addition, the longer chains allow bridging between lectins and aggregation.

Burkholderia ambifaria lectin (BambL) is constructed in the same way as RSL as a trimer of tandem repeat β-sheets, resulting in a six-bladed lectin [144] (Figure 14.20b). It binds α- and β-L-fucose [145]. The Gram-negative bacterium has antifungal properties and protects plants from fungal infections, but can also pose a danger to immunocompromised patients, such as those with cystic fibrosis. They are capable of forming biofilms, as well, similar to PA [146]. Like other organisms discussed in this chapter, recognition by a lectin on the cell surface of a human cell is an important part of the infection. Several multivalent inhibitors have been developed against BambL.

(a)   (b)

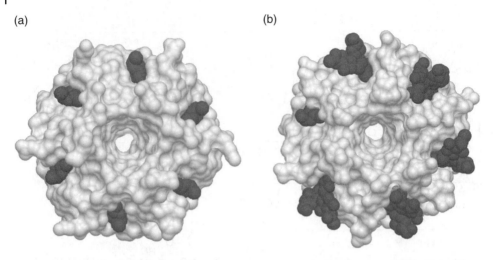

Figure 14.20 Structures of (a) RSL in complex with ʟ-fucose in all five binding sites (PDB: 2BT9) and (b) BambL with human H-type 2 tetrasaccharide bound in four of five binding sites, and α-ʟ-fucose bound to the fifth binding site (PDB: 3ZZV).

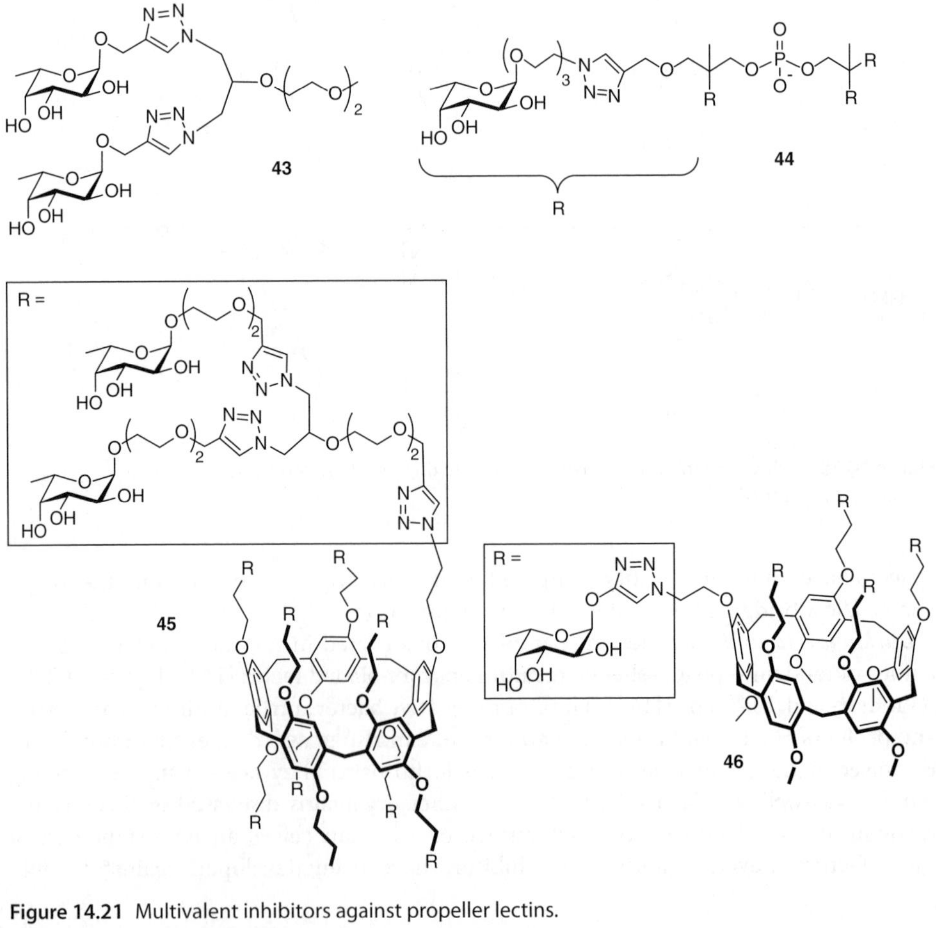

Figure 14.21 Multivalent inhibitors against propeller lectins.

The tetravalent ligand **44** was found to have an affinity of 43 nM, an improvement of 22-fold over the monovalent ligand [147]. A similar, but linear, decavalent ligand had a K_d of 13.8 nM [74]. Pillar[5]arenes, also tested for binding against LecA and LecB, were found to be very tight binders to BambL [52]. Long linker arms and icosavalency **45** led to an IC_{50} of 70 pM in an ELLA assay and a K_d of 27 nM. Interestingly, lowering the valency to pentavalent, with all fucosides presenting on the same side **46**, gave an even lower K_d of 21 nM (53).

The discussion in this chapter thus far has been on multivalency of a ligand to bind to a multivalent protein, in order to block its natural function. The multivalency of the protein itself is also of interest, however. Through genetic studies, neolectins can be created to study the effects of multivalency in carbohydrate–lectin binding. By inactivating the binding sites of RSL, 13 neolectins were created [148], with all different geometries of active and inactive binding sites, with a valency ranging from zero to six of the wild-type. While ITC experiments showed no change in affinity, SPR experiments showed a limited effect on avidity. Important was the ability to disrupt the membrane, a crucial step in infection, which was only possible when adjacent binding sites were capable of binding carbohydrates. Having alternating binding sites active made the lectin incapable of invagination of the membrane and delayed its uptake in lung epithelial cells dramatically [149].

14.9 Conclusion

Multivalency is crucial in carbohydrate recognition between cells by lectins. Taking advantage of a wide variety of scaffolds, such as dendrimers, peptides and calixarenes, these lectins can be targeted for inhibition by synthetic ligands. Through optimization and investigating the details in the interactions between lectin and ligand, tightly binding inhibitors can be designed that have the potential for therapeutic use against pathogens such as PA, *Vibrio cholerae* and influenza A viruses, or targeting cancer and inflammatory responses.

Acknowledgements

The Netherlands Organization for Scientific Research is gratefully acknowledged for a PhD scholarship to M.S.

References

1 Cecioni S, Imberty A, Vidal S. Glycomimetics versus multivalent glycoconjugates for the design of high affinity lectin ligands. Chem. Rev. 2015;115(1):525–61.
2 Gabius H-J, André S, Jiménez-Barbero J, *et al.* From lectin structure to functional glycomics: principles of the sugar code. Trends Biochem. Sci. 2011;36(6):298–313.
3 Lis H, Sharon N. Lectins: Carbohydrate-specific proteins that mediate cellular recognition. Chem. Rev. 1998;98(2):637–74.
4 Lundquist JJ, Toone EJ. The cluster glycoside effect. Chem. Rev. 2002;102(2):555–78.

5 Pieters RJ. Maximising multivalency effects in protein-carbohydrate interactions. Org. Biomol. Chem. 2009;7(10):2013–25.

6 Wittmann V, Pieters RJ. Bridging lectin binding sites by multivalent carbohydrates. Chem. Soc. Rev. 2013;42(10):4492–503.

7 Dam TK, Gabius HJ, André S, *et al*. Galectins bind to the multivalent glycoprotein asialofetuin with enhanced affinities and a gradient of decreasing binding constants. Biochemistry 2005;44(37):12564–71.

8 Parera Pera N, Branderhorst HM, Kooij R, *et al*. Rapid screening of lectins for multivalency effects with a glycodendrimer microarray. ChemBioChem 2010;11(13):1896–904.

9 Ligeour C, Vidal O, Dupin L, *et al*. Mannose-centered aromatic galactoclusters inhibit the biofilm formation of *Pseudomonas aeruginosa*. Org. Biomol. Chem. 2015;(31):1–3.

10 Salomon R, Webster RG. The influenza virus enigma. Cell 2009;136(3):402–10.

11 Böhm R, Haselhorst T, von Itzstein M. Influenza virus, overview: structures, infection mechanisms and antivirals. In: Taniguchi N, Endo T, Hart WG, *et al*., editors. Glycoscience: Biology and Medicine. Springer Japan; 2015;749–67.

12 Stevens J, Blixt O, Tumpey TM, *et al*. Structure and receptor specificity of the hemagglutinin from an H5N1 influenza virus. Science 2006;312(5772):404–10.

13 Wilson IA, Skehel JJ, Wiley DC. Structure of the haemagglutinin membrane glycoprotein of influenza virus at 3 Å resolution. Nature 1981;289(5796):366–73.

14 Sauter NK, Bednarski MD, Wurzburg BA, *et al*. Hemagglutinins from two influenza virus variants bind to sialic acid derivatives with millimolar dissociation constants: A 500-MHz proton nuclear magnetic resonance study. Biochemistry 1989;28(21):8388–96.

15 Krammer F, Palese P. Advances in the development of influenza virus vaccines. Nat. Rev. Drug Discov. 2015;14(3):167–82.

16 Schotsaert M, García-Sastre A. Influenza vaccines: a moving interdisciplinary field. Viruses 2014;6(10):3809–26.

17 Matrosovich MN, Matrosovich TY, Gray T, *et al*. Neuraminidase is important for the initiation of influenza virus infection in human airway epithelium. J. Virol. 2004;78(22):12665–7.

18 Ohuchi M, Asaoka N, Sakai T, Ohuchi R. Roles of neuraminidase in the initial stage of influenza virus infection. Microbes Infect. 2006;8(5):1287–93.

19 Van der Vries E, Stelma FF, Boucher CAB. Emergence of a multidrug-resistant pandemic influenza A (H1N1) virus. New Engl. J. Med. 2010;363(14):1381–2.

20 Le QM, Kiso M, Someya K, *et al*. Avian flu: isolation of drug-resistant H5N1 virus. Nature. 2005;437(7062):1108.

21 Waldmann M, Jirmann R, Hoelscher K, *et al*. A nanomolar multivalent ligand as entry inhibitor of the hemagglutinin of avian influenza. J. Am. Chem. Soc. 2014;136(2):783–8.

22 Xiao S, Si L, Tian Z, *et al*. Pentacyclic triterpenes grafted on CD cores to interfere with influenza virus entry: A dramatic multivalent effect. Biomaterials 2016;78:74–85.

23 Hendricks GL, Weirich KL, Viswanathan K, *et al*. Sialylneolacto-N-tetraose c (LSTc)-bearing liposomal decoys capture influenza A virus. J. Biol. Chem. 2013;288(12):8061–73.

24 Yeh H-W, Lin T-S, Wang H-W, *et al. S*-Linked sialyloligosaccharides bearing liposomes and micelles as influenza virus inhibitors. Org. Biomol. Chem. 2015;11518–28.

25 Sigal GB, Mammen M, Dahmann G, Whitesides GM. Polyacrylamides bearing pendant α-sialoside groups strongly inhibit agglutination of erythrocytes by influenza virus: the strong inhibition reflects enhanced binding through cooperative polyvalent interactions. J. Am. Chem. Soc. 1996;118(16):3789–800.

26 Tang S, Puryear WB, Seifried BM, *et al.* Antiviral agents from multivalent presentation of sialyl oligosaccharides on brush polymers. ACS Macro Lett. 2016;5(3):413–8.

27 Papp I, Sieben C, Sisson AL, *et al.* Inhibition of influenza virus activity by multivalent glycoarchitectures with matched sizes. ChemBioChem 2011;12(6):887–95.

28 Vonnemann J, Sieben C, Wolff C, *et al.* Virus inhibition induced by polyvalent nanoparticles of different sizes. Nanoscale 2014;6(4):2353–60.

29 Wagner VE, Iglewski BH. *P. aeruginosa* biofilms in CF infection. Clin. Rev. Allergy Immunol. 2008;35:124–34.

30 Chemani C, Imberty A, de Bentzmann S, *et al.* Role of LecA and LecB lectins in *Pseudomonas aeruginosa*-induced lung injury and effect of carbohydrate ligands. Infect Immun. 2009;77(5):2065–75.

31 Cioci G, Mitchell EP, Gautier C, *et al.* Structural basis of calcium and galactose recognition by the lectin PA-IL of *Pseudomonas aeruginosa*. FEBS Lett. 2003;555(2):297–301.

32 Garber N, Guempel U, Belz A, *et al.* On the specificity of the D-galactose-binding lectin (PA-I) of *Pseudomonas aeruginosa* and its strong binding to hydrophobic derivatives of D-galactose and thiogalactose. Biochim. Biophys. Acta – Gen. Subj. 1992;1116(3):331–3.

33 Chen C-P, Song S-C, Gilboa-Garber N, *et al.* Studies on the binding site of the galactose-specific agglutinin PA-IL from *Pseudomonas aeruginosa*. Glycobiology 1998;8(1):7–16.

34 Kadam RU, Garg D, Schwartz J, *et al.* CH-π "T-shape" interaction with histidine explains binding of aromatic galactosides to *Pseudomonas aeruginosa* lectin LecA. ACS Chem. Biol. 2013;8(9):1925–30.

35 Diggle SP, Stacey RE, Dodd C, *et al.* The galactophilic lectin, LecA, contributes to biofilm development in *Pseudomonas aeruginosa*. Environ. Microbiol. 2006;8(6):1095–104.

36 Grishin A V., Krivozubov MS, Karyagina AS, Gintsburg AL. *Pseudomonas aeruginosa* lectins as targets for novel antibacterials. Acta Naturae 2015;7(2):29–41.

37 Fan E, Zhang Z, Minke WE, *et al.* High-affinity pentavalent ligands of *Escherichia coli* heat-labile enterotoxin by modular structure-based design. J. Am. Chem. Soc. 2000;122(11):2663–4.

38 Pertici F, Pieters RJ. Potent divalent inhibitors with rigid glucose click spacers for *Pseudomonas aeruginosa* lectin LecA. Chem. Commun. 2012;48(33):4008.

39 Pertici F, De Mol NJ, Kemmink J, Pieters RJ. Optimizing divalent inhibitors of *Pseudomonas aeruginosa* lectin LecA by using a rigid spacer. Chem. Eur J. 2013;19(50):16923–7.

40 Pukin A V., Brouwer AJ, Koomen L, *et al.* Thiourea-based spacers in potent divalent inhibitors of *Pseudomonas aeruginosa* virulence lectin LecA. Org. Biomol. Chem. 2015;10923–8.

41 Fu O, Pukin A V., Quarlesvanufford HC, *et al.* Functionalization of a rigid divalent ligand for LecA, a bacterial adhesion lectin. ChemistryOpen 2015;4(4):463–70.

42 Pertici F, Varga N, Van Duijn A, *et al.* Efficient synthesis of phenylene-ethynylene rods and their use as rigid spacers in divalent inhibitors. Beilstein J. Org. Chem. 2013;9:215–22.

43 Visini R, Jin X, Bergmann M, *et al.* Structural insight into multivalent galactoside binding to *Pseudomonas aeruginosa* lectin LecA. ACS Chem. Biol. 2015;10(11):2455–62.

44 Kadam RU, Bergmann M, Hurley M, *et al.* A glycopeptide dendrimer inhibitor of the galactose-specific lectin LecA and of *Pseudomonas aeruginosa* biofilms. Angew. Chem. Int. Ed. 2011;50(45):10631–5.

45 Kadam RU, Bergmann M, Garg D, *et al.* Structure-based optimization of the terminal tripeptide in glycopeptide dendrimer inhibitors of *Pseudomonas aeruginosa* biofilms targeting LecA. Chem. Eur. J. 2013;19(50):17054–63.

46 Michaud G, Visini R, Bergmann M, *et al.* Overcoming antibiotic resistance in *Pseudomonas aeruginosa* biofilms using glycopeptide dendrimers. Chem. Sci. 2016;166–82.

47 Cecioni S, Lalor R, Blanchard B, *et al.* Achieving high affinity towards a bacterial lectin through multivalent topological isomers of calix[4]arene glycoconjugates. Chem. Eur. J. 2009;15(47):13232–40.

48 Sicard D, Cecioni S, Iazykov M, *et al.* AFM investigation of *Pseudomonas aeruginosa* lectin LecA (PA-IL) filaments induced by multivalent glycoclusters. Chem. Commun. 2011;47(33):9483–5.

49 Sicard D, Chevolot Y, Souteyrand E, *et al.* Molecular arrangement between multivalent glycocluster and *Pseudomonas aeruginosa* LecA (PA-IL) by atomic force microscopy: Influence of the glycocluster concentration. J. Mol. Recogn. 2013;26(12):694–9.

50 Boukerb AM, Rousset A, Galanos N, *et al.* Antiadhesive properties of glycoclusters against *Pseudomonas aeruginosa* lung infection. J. Med. Chem. 2014;57(24):10275–89.

51 Cecioni S, Praly JP, Matthews SE, *et al.* Rational design and synthesis of optimized glycoclusters for multivalent lectin-carbohydrate interactions: Influence of the linker arm. Chem. Eur. J. 2012;18(20):6250–63.

52 Buffet K, Nierengarten I, Galanos N, *et al.* Pillar[5]arene-Based Glycoclusters: Synthesis and Multivalent Binding to Pathogenic Bacterial Lectins. Chemistry. Wiley-VCH Verlag GmbH; 2016;22(9):2955–63.

53 Galanos N, Gillon E, Imberty A, *et al.* Pentavalent pillar[5]arene-based glycoclusters and their multivalent binding to pathogenic bacterial lectins. Org. Biomol. Chem. 2016;14(13):3476–81.

54 Cecioni S, Oerthel V, Iehl J, *et al.* Synthesis of dodecavalent fullerene-based glycoclusters and evaluation of their binding properties towards a bacterial lectin. Chem. Eur. J. 2011;17(11):3252–61.

55 Gerland B, Goudot A, Ligeour C, *et al.* Structure binding relationship of galactosylated glycoclusters toward *Pseudomonas aeruginosa* lectin LecA using a DNA-based carbohydrate microarray. Bioconjug. Chem. 2014;25(2):379–92.

56 Ligeour C, Dupin L, Marra A, *et al.* Synthesis of galactoclusters by metal-free thiol "click chemistry" and their binding affinities for *Pseudomonas aeruginosa* lectin LecA. Eur. J. Org. Chem. 2014;2014(34):7621–30.

57 Casoni F, Dupin L, Vergoten G, *et al.* The influence of the aromatic aglycon of galactoclusters on the binding of LecA: a case study with *O*-phenyl, *S*-phenyl, *O*-benzyl, *S*-benzyl, *O*-biphenyl and *O*-naphthyl aglycons. Org. Biomol. Chem.; 2014;12(45):9166–79.

58 Gening ML, Titov DV, Cecioni S, *et al.* Synthesis of multivalent carbohydrate-centered glycoclusters as nanomolar ligands of the bacterial lectin LecA from *Pseudomonas aeruginosa*. Chem. Eur. J. 2013;19(28):9272–85.

59 Novoa A, Eierhoff T, Topin J, *et al.* A LecA ligand identified from a galactoside-conjugate array inhibits host cell invasion by *Pseudomonas aeruginosa*. Angew. Chem. Int. Ed. 2014;53(34):8885–9.

60 Tielker D, Hacker S, Loris R, *et al. Pseudomonas aeruginosa* lectin LecB is located in the outer membrane and is involved in biofilm formation. Microbiology 2005;151(5):1313–23.

61 Wu AM, Gong YP, Li CC, Gilboa-Garber N. Duality of the carbohydrate-recognition system of *Pseudomonas aeruginosa*-II lectin (PA-IIL). FEBS Lett. 2010;584(11):2371–5.

62 Sabin C, Mitchell EP, Pokorná M, *et al.* Binding of different monosaccharides by lectin PA-IIL from *Pseudomonas aeruginosa*: Thermodynamics data correlated with X-ray structures. FEBS Lett. 2006;580(3):982–7.

63 Wu AM, Wu JH, Singh T, *et al.* Interactions of the fucose-specific *Pseudomonas aeruginosa* lectin, PA-IIL, with mammalian glycoconjugates bearing polyvalent Lewisa and ABH blood group glycotopes. Biochimie 2006;88(10):1479–92.

64 Imberty A, Wimmerová M, Mitchell EP, Gilboa-Garber N. Structures of the lectins from *Pseudomonas aeruginosa*: Insights into the molecular basis for host glycan recognition. Microbes Infect. 2004;6(2):221–8.

65 Marotte K, Sabin C, Préville C, *et al.* X-ray structures and thermodynamics of the interaction of PA-IIL from *Pseudomonas aeruginosa* with disaccharide derivatives. ChemMedChem 2007;2(9):1328–38.

66 Loris R, Tielker D, Jaeger K-E, Wyns L. Structural basis of carbohydrate recognition by the lectin LecB from *Pseudomonas aeruginosa*. J. Mol. Biol. 2003;331(4):861–70.

67 Berthet N, Thomas B, Bossu I, *et al.* High affinity glycodendrimers for the lectin LecB from *Pseudomonas aeruginosa*. Bioconjug. Chem. 2013;24(9):1598–611.

68 Kolomiets E, Swiderska MA, Kadam RU, *et al.* Glycopeptide dendrimers with high affinity for the fucose-binding lectin LecB from *Pseudomonas aeruginosa*. ChemMedChem 2009;4(4):562–9.

69 Johansson EM V, Crusz SA, Kolomiets E, *et al.* Inhibition and dispersion of *Pseudomonas aeruginosa* biofilms by glycopeptide dendrimers targeting the fucose-specific lectin LecB. Chem. Biol. 2008;15(12):1249–57.

70 Johansson EM V., Kadam RU, Rispoli G, *et al.* Inhibition of *Pseudomonas aeruginosa* biofilms with a glycopeptide dendrimer containing D-amino acids. MedChemComm 2011;2(5):418.

71 Dam TK, Gerken TA, Brewer CF. Thermodynamics of multivalent carbohydrate–lectin cross-linking interactions: Importance of entropy in the bind and jump mechanism. Biochemistry 2009;48(18):3822–7.

72 Buffet K, Gillon E, Holler M, *et al.* Fucofullerenes as tight ligands of RSL and LecB, two bacterial lectins. Org. Biomol. Chem. 2015;13(23):6482–92.

73 Gerland B, Goudot A, Pourceau G, *et al.* Synthesis of a library of fucosylated glycoclusters and determination of their binding toward *Pseudomonas aeruginosa* lectin B (PA-IIL) using a DNA-based carbohydrate microarray. Bioconjug. Chem. 2012;23(8):1534–47.

74 Ligeour C, Dupin L, Angeli A, *et al*. Importance of topology for glycocluster binding to *Pseudomonas aeruginosa* and *Burkholderia ambifaria* bacterial lectins. Org. Biomol. Chem.; 2015;11244–54.

75 Gerland B, Goudot A, Pourceau G, *et al*. Synthesis of homo- and heterofunctionalized glycoclusters and binding to *Pseudomonas aeruginosa* lectins PA-IL and PA-IIL. J. Org. Chem. 2012;77(17):7620–6.

76 Pieters RJ. Inhibition and detection of galectins. ChemBioChem 2006;7(5):721–8.

77 Leffler H, Carlsson S, Hedlund M, *et al*. Introduction to galectins. Glycoconj. J. 2004;19(7-9):433–40.

78 Di Lella S, Sundblad V, Cerliani JP, *et al*. When galectins recognize glycans: from biochemistry to physiology and back again. Biochemistry 2011;50(37):7842–57.

79 Hughes RC. Galectins as modulators of cell adhesion. Biochimie 2001;83(7):667–76.

80 Pieters RJ. Intervention with bacterial adhesion by multivalent carbohydrates. Med. Res. Rev. 2007;27(6):796–816.

81 Takenaka Y, Fukumori T, Raz A. Galectin-3 and metastasis. Glycoconj. J. 2004;19(7–9):543–9.

82 Rabinovich GA, Baum LG, Tinari N, *et al*. Galectins and their ligands: amplifiers, silencers or tuners of the inflammatory response? Trends Immunol. 2002;23(6):313–20.

83 Rubinstein N, Ilarregui JM, Toscano MA, Rabinovich GA. The role of galectins in the initiation, amplification and resolution of the inflammatory response. Tissue Antigens 2004;64(1):1–12.

84 Vasta GR. Roles of galectins in infection. Nat. Rev. Microbiol. 2009;7(6):424–38.

85 Califice S, Castronovo V, Van Den Brûle F. Galectin-3 and cancer. Int. J. Oncol. 2004;25:983–92.

86 Liu F-T, Rabinovich GA. Galectins as modulators of tumour progression. Nat. Rev. Cance. 2005;5(1):29–41.

87 Nangia-Makker P, Honjo Y, Sarvis R, *et al*. Galectin-3 induces endothelial cell morphogenesis and angiogenesis. Am. J. Pathol. 2000;156(3):899–909.

88 Ahmad N, Gabius HJ, André S, *et al*. Galectin-3 precipitates as a pentamer with synthetic multivalent carbohydrates and forms heterogeneous cross-linked complexes. J. Biol. Chem. 2004;279(12):10841–7.

89 Ingrassia L, Camby I, Lefranc F, *et al*. Anti-Galectin compounds as potential anti-cancer drugs. Curr. Med. Chem. 2006;13(29):3513–27.

90 Mackinnon A, Chen WS, Leffler H, *et al*. Carbohydrates as drugs. In: Seeberger PH, Rademacher C, editors. Topics in Medicinal Chemistry. Springer International Publishing; 2014;12:95–122.

91 Blanchard H, Bum-Erdene K, Bohari MH, Yu X. Galectin-1 inhibitors and their potential therapeutic applications: a patent review. Expert Opin. Ther. Pat. 2016;26(5):537–54.

92 Blanchard H, Yu X, Collins PM, Bum-Erdene K. Galectin-3 inhibitors: a patent review (2008-present). Expert Opin. Ther. Pat. 2014;24(10):1053–65.

93 Giguère D, Bonin MA, Cloutier P, *et al*. Synthesis of stable and selective inhibitors of human galectins-1 and -3. Bioorg. Med. Chem. 2008;16(16):7811–23.

94 Van Hattum H, Branderhorst HM, Moret EE, *et al*. Tuning the preference of thiodigalactoside- and lactosamine-based ligands to galectin-3 over galectin-1. J. Med. Chem. 2013;56(3):1350–4.

95 Maljaars CEP, André S, Halkes KM, *et al.* Assessing the inhibitory potency of galectin ligands identified from combinatorial (glyco)peptide libraries using surface plasmon resonance spectroscopy. Anal. Biochem. 2008;378(2):190–6.

96 André S, Arnusch CJ, Kuwabara I, *et al.* Identification of peptide ligands for malignancy- and growth-regulating galectins using random phage-display and designed combinatorial peptide libraries. Bioorg. Med. Chem. 2005;13(2):563–73.

97 Arnusch CJ, André S, Valentini P, *et al.* Interference of the galactose-dependent binding of lectins by novel pentapeptide ligands. Bioorg. Med. Chem. Lett. 2004;14(6):1437–40.

98 André S, Pieters RJ, Vrasidas I, *et al.* Wedgelike glycodendrimers as inhibitors of binding of mammalian galectins to glycoproteins, lactose maxiclusters, and cell surface glycoconjugates. ChemBioChem 2001;2:822–30.

99 Cousin JM, Cloninger MJ. Glycodendrimers: tools to explore multivalent galectin-1 interactions. Beilstein J. Org. Chem. 2015;11:739–47.

100 Goodman CK, Wolfenden ML, Nangia-Makker P, *et al.* Multivalent scaffolds induce galectin-3 aggregation into nanoparticles. Beilstein J. Org. Chem. 2014;10:1570–7.

101 Vrasidas I, André S, Valentini P, *et al.* Rigidified multivalent lactose molecules and their interactions with mammalian galectins: a route to selective inhibitors. Org. Biomol. Chem. 2003;1(5):803–10.

102 André S, Sansone F, Kaltner H, *et al.* Calix[n]arene-based glycoclusters: Bioactivity of thiourea-linked galactose/lactose moieties as inhibitors of binding of medically relevant lectins to a glycoprotein and cell-surface glycoconjugates and selectivity among human adhesion/growth-regulatory galectins. ChemBioChem 2008;9(10):1649–61.

103 André S, Grandjean C, Gautier F-M, *et al.* Combining carbohydrate substitutions at bioinspired positions with multivalent presentation towards optimising lectin inhibitors: case study with calixarenes. Chem. Commun. 2011;47:6126–8.

104 Cecioni S, Faure S, Darbost U, *et al.* Selectivity among two lectins: Probing the effect of topology, multivalency and flexibility of "clicked" multivalent glycoclusters. Chem. Eur. J. 2011;17(7):2146–59.

105 Bernardi S, Fezzardi P, Rispoli G, *et al.* Clicked and long spaced galactosyl- and lactosylcalix[4]arenes: New multivalent galectin-3 ligands. Beilstein J. Org. Chem. 2014;10:1672–80.

106 André S, Kaltner H, Furuike T, *et al.* Persubstituted cyclodextrin-based glycoclusters as inhibitors of protein-carbohydrate recognition using purified plant and mammalian lectins and wild-type and lectin-gene-transfected tumor cells as targets. Bioconjug. Chem. 2004;15(1):87–98.

107 Belitsky JM, Nelson A, Hernandez JD, *et al.* Multivalent interactions between lectins and supramolecular complexes: Galectin-1 and self-assembled pseudopolyrotaxanes. Chem. Biol. 2007;14(10):1140–51.

108 Šimonová A, Kupper CE, Böcker S, *et al.* Chemo-enzymatic synthesis of LacdiNAc dimers of varying length as novel galectin ligands. J. Mol. Catal. B Enzym. 2014;101:47–55.

109 Yao W, Xia M-J, Meng X-B, *et al.* Adaptable synthesis of *C*-lactosyl glycoclusters and their binding properties with galectin-3. Org. Biomol. Chem. 2014;12(41):8180–95.

110 Gouin SG, Fernández JMG, Vanquelef E, *et al.* Multimeric lactoside "click clusters" as tools to investigate the effect of linker length in specific interactions with peanut lectin, galectin-1, and -3. ChemBioChem 2010;11(10):1430–42.

111 Restuccia A, Tian YF, Collier JH, Hudalla GA. Self-assembled glycopeptide nanofibers as modulators of Galectin-1 bioactivity. Cell Mol. Bioeng. 2015;8(3):471–87.

112 Yang Y, Xue XC, Jin XF, *et al*. Synthesis of multivalent N-acetyl lactosamine modified quantum dots for the study of carbohydrate and galectin-3 interactions. Tetrahedron 2012;68(35):7148–54.

113 Wang H, Huang W, Orwenyo J, *et al*. Design and synthesis of glycoprotein-based multivalent glyco-ligands for influenza hemagglutinin and human galectin-3. Bioorg. Med. Chem. 2013;21(7):2037–44.

114 Mandal DK, Bhattacharyya L, Koenig SH, *et al*. Studies of the binding specificity of concanavalin A. Nature of the extended binding site for asparagine-linked carbohydrates. Biochemistry 1994;33(5):1157–62.

115 Euzen R, Reymond J-L. Glycopeptide dendrimers: tuning carbohydrate-lectin interactions with amino acids. Mol. Biosyst. 2011;7(2):411–21.

116 Euzen R, Reymond J-L. Synthesis of glycopeptide dendrimers, dimerization and affinity for Concanavalin A. Bioorg. Med. Chem. 2011;19(9):2879–87.

117 Sánchez-Navarro M, Muñoz A, Illescas BM, *et al*. [60]Fullerene as multivalent scaffold: Efficient molecular recognition of globular glycofullerenes by concanavalin A. Chem. Eur. J. 2011;17(3):766–9.

118 Yang LY, Haraguchi T, Inazawa T, *et al*. Synthesis of a novel class of glycocluster with a cyclic α-(1→6)-octaglucoside as a scaffold and their binding abilities to concanavalin A. Carbohydr. Res. 2010;345(15):2124–32.

119 Sleiman M, Varrot A, Raimundo J-M, *et al*. Glycosylated asterisks are among the most potent low valency inducers of Concanavalin A aggregation. Chem. Commun. 2008;(48):6507–9.

120 Papp I, Dernedde J, Enders S, *et al*. Multivalent presentation of mannose on hyperbranched polyglycerol and their interaction with concanavalin A lectin. ChemBioChem 2011;12(7):1075–83.

121 Noble GT, Flitsch SL, Liem KP, Webb SJ. Assessing the cluster glycoside effect during the binding of concanavalin A to mannosylated artificial lipid rafts. Org. Biomol. Chem. 2009;7(24):5245–54.

122 Reeh P, De Mendoza J. Dynamic multivalency for carbohydrate-protein recognition through dynamic combinatorial libraries based on FeII-bipyridine complexes. Chem. Eur. J. 2013;19(17):5259–62.

123 Chmielewski MJ, Buhler E, Candau J, Lehn JM. Multivalency by self-assembly: binding of concanavalin A to metallosupramolecular architectures decorated with multiple carbohydrate groups. Chemistry 2014;20(23):6960–77.

124 Harris JB, LaRocque RC, Qadri F, *et al*. Cholera. Lancet 2012;379(9835):2466–76.

125 Beddoe T, Paton AW, Le Nours J, *et al*. Structure, biological functions and applications of the AB5 toxins. Trends Biochem. Sci. 2010;35(7):411–8.

126 Mukhopadhyay S, Linstedt AD. Retrograde trafficking of AB5 toxins: Mechanisms to therapeutics. J. Mol. Med. 2013;91(10):1131–41.

127 Merritt EA, Kuhn P, Sarfaty S, *et al*. The 1.25 Å resolution refinement of the cholera toxin B-pentamer: evidence of peptide backbone strain at the receptor-binding site. J. Mol. Biol. 1998;282(5):1043–59.

128 Jobling MG, Yang Z, Kam WR, *et al*. A single native ganglioside GM1-binding site is sufficient for cholera toxin to bind to cells and complete the intoxication pathway. MBio 2012;3(6):1–9.

129 Branson TR, Turnbull WB. Bacterial toxin inhibitors based on multivalent scaffolds. Chem. Soc. Rev. 2013;42(11):4613–22.

130 Zuilhof H. Fighting cholera one-on-one: The development and efficacy of multivalent cholera-toxin-binding molecules. Acc. Chem. Res. 2016;49(2):274–85.

131 Vrasidas I, de Mol NJ, Liskamp RMJ, Pieters RJ. Synthesis of lactose dendrimers and multivalency effects in binding to the Cholera Toxin B subunit. Eur. J. Org. Chem. 2001;2001(24):4685–92.

132 Arosio D, Vrasidas I, Valentini P, *et al*. Synthesis and cholera toxin binding properties of multivalent GM1 mimics. Org. Biomol. Chem. 2004;2(14):2113–24.

133 Branderhorst HM, Ruijtenbeek R, Liskamp RMJ, Pieters RJ. Multivalent carbohydrate recognition on a glycodendrimer-functionalized flow-through chip. ChemBioChem 2008;9(11):1836–44.

134 Pukin A V., Branderhorst HM, Sisu C, *et al*. Strong inhibition of cholera toxin by multivalent GM1 derivatives. ChemBioChem 2007;8(13):1500–3.

135 Sisu C, Baron AJ, Branderhorst HM, *et al*. The influence of ligand valency on aggregation mechanisms for inhibiting bacterial toxins. ChemBioChem 2009;10(2):329–37.

136 Fu O, Pukin A V., Vanufford HCQ, *et al*. Tetra- versus pentavalent inhibitors of Cholera Toxin. ChemistryOpen 2015;4(4):471–7.

137 Merritt EA, Zhang Z, Pickens JC, *et al*. Characterization and crystal structure of a high-affinity pentavalent receptor-binding inhibitor for cholera toxin and *E. coli* heat-labile enterotoxin. J. Am. Chem. Soc. 2002;124(30):8818–24.

138 Zhang Z, Merritt EA, Ahn M, *et al*. Solution and crystallographic studies of branched multivalent ligands that inhibit the receptor-binding of cholera toxin. J. Am. Chem. Soc. 2002;124(44):12991–8.

139 Zhang Z, Liu J. Verlinde CL, *et al*. Large cyclic peptides as cores of multivalent ligands : Application to inhibitors of receptor binding by cholera toxin. J. Org. Chem. 2004;(20):7737–40.

140 Garcia-Hartjes J, Bernardi S, Weijers CAGM, *et al*. Picomolar inhibition of cholera toxin by a pentavalent ganglioside GM1os-calix[5]arene. Org. Biomol. Chem. 2013;11(26):4340–9.

141 Mattarella M, Garcia-Hartjes J, Wennekes T, *et al*. Nanomolar cholera toxin inhibitors based on symmetrical pentavalent ganglioside GM1os-*sym*-corannulenes. Org. Biomol. Chem. 2013;11(26):4333–9.

142 Branson TR, McAllister TE, Garcia-Hartjes J, *et al*. A protein-based pentavalent inhibitor of the cholera toxin B-subunit. Angew. Chem. Int. Ed. 2014;53(32):8323–7.

143 Kostlánová N, Mitchell EP, Lortat-Jacob H, *et al*. The fucose-binding Lectin from *Ralstonia solanacearum*: A new type of β-propeller architecture formed by oligomerization and interacting with fucoside, fucosyllactose, and plant xyloglucan. J. Biol. Chem. 2005;280(30):27839–49.

144 Audfray A, Claudinon J, Abounit S, *et al*. Fucose-binding lectin from opportunistic pathogen *Burkholderia ambifaria* binds to both plant and human oligosaccharidic epitopes. J. Biol. Chem. 2012;287(6):4335–47.

145 Antonik PM, Volkov AN, Broder UN, *et al*. Anomer-specific recognition and dynamics in a fucose-binding lectin. Biochemistry 2016;55(8):1195–203.

146 Valvano MA, Keith KE, Cardona ST. Survival and persistence of opportunistic *Burkholderia* species in host cells. Curr. Opin. Microbiol. 2005;8(1):99–105.

147 Ligeour C, Audfray A, Gillon E, *et al*. Synthesis of branched-phosphodiester and mannose-centered fucosylated glycoclusters and their binding studies with *Burkholderia ambifaria* lectin (BambL). RSC Adv. 2013;3(42):19515.

148 Arnaud J, Tröndle K, Claudinon J, *et al*. Membrane deformation by neolectins with engineered glycolipid binding sites. Angew. Chem. Int. Ed. 2014;53(35):9267–70.

149 Arnaud J, Claudinon J, Tröndle K, *et al*. Reduction of lectin valency drastically changes glycolipid dynamics in membranes but not surface avidity. ACS Chem. Biol. 2013;8(9):1918–24.

Index

Multivalency: Concepts, Research & Applications, First Edition. Edited by Jurriaan Huskens,
Leonard J. Prins, Rainer Haag, and Bart Jan Ravoo.
© 2018 John Wiley & Sons Ltd. Published 2018 by John Wiley & Sons Ltd.